THE BASIC PRACTICE OF STATISTICS

THIRD EDITION

David S. Moore

Purdue University

W. H. Freeman and Company
New York

Executive Editor: Craig Bleyer
Marketing Manager: Jeffrey Rucker
Senior Acquisitions Editor: Patrick Farace
Development Editors: Danielle Swearengin; Amanda McCorquodale
Project Editors: Mary Louise Byrd; Wendy Druck, TechBooks
Media Editor: Brian Donnellan
Cover and Text Designer: Diana Blume
Cover Illustration: Mark Chickinelli
Illustration Coordinator: Bill Page
Text Illustrations: TechBooks, Mark Chickinelli
Photo Editor: Nicole Villamora; Nigel Assam
Production Coordinator: Susan Wein
Composition: TechBooks
Manufacturing: Quebecor World

TI-83^{TM} screen shots are used with permission of the publisher: © 1996, Texas Instruments Incorporated. TI-83^{TM} Graphic Calculator is a registered trademark of Texas Instruments Incorporated. Minitab is a registered trademark of Minitab, Inc. Microsoft © and Windows © are registered trademarks of the Microsoft Corporation in the United States and other countries. Excel screen shots are reprinted with permission from the Microsoft Corporation. S-PLUS is a registered trademark of the Insightful Corporation.

Library of Congress Cataloging-in-Publication Data

Moore, David S.
    The basic practice of statistics / David S. Moore.—3d ed.
      p. cm.
    Includes index.
    ISBN 0-7167-9623-6 (Hardcover)
    ISBN 0-7167-0223-1 (Paperback)
    1. Statistics.   I. Title.
  QA276.12.M648   2003
  519.5—dc21   2003043894

Printed in the United States of America

First printing 2003

W. H. Freeman and Company
41 Madison Avenue
New York, NY 10010
Houndmills, Basingstoke RG21 6XS, England
www.whfreeman.com

About the Cover: Who Is Number #1?— The famous "Pepsi Challenge" in the 1980s dramatically changed the public perception of the "Cola Wars" and set off a star-studded advertising battle between Pepsi and Coke that is as intense as ever. So what do data tell us about the most popular soft drinks today? (See p. 157.) For that matter, statistically speaking, was the original Pepsi/Coke challenge a fair test? (See p. 210.) © The Coca-Cola Company. All Rights Reserved.

Brief Contents

* Starred material is optional

Contents

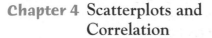

**Part III: Inference about
Variables 408**

Part V: Optional Companion Chapters (on CD and printed separately)

TO THE INSTRUCTOR:
About This Book

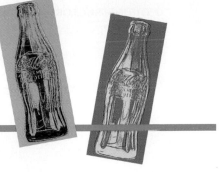

The Basic Practice of Statistics (BPS) is an introduction to statistics for students in two-year and four-year colleges and universities that emphasizes working with data and statistical ideas. It is designed to be accessible to students with limited quantitative background—just "algebra" in the sense of being able to read and use simple equations. The book is usable with almost any level of technology for calculating and graphing—from a simple "two-variables statistics" calculator through a graphing calculator or spreadsheet program through full statistical software. BPS was the pioneer in presenting a modern approach to statistics in a genuinely elementary text. In the following I describe for instructors the nature and features of the book and the considerable changes in this third edition.

Guiding principles

BPS is based on three principles: balanced content, experience with data, and the importance of ideas.

Balanced content. Once upon a time, basic statistics courses taught probability and inference almost exclusively, often preceded by just a week of histograms, means, and medians. Such unbalanced content does not match the actual practice of statistics, where data analysis and design of data production join with probability-based inference to produce a coherent science of data. There are also good pedagogical reasons for beginning with data analysis (Chapters 1 to 6), then moving to data production (Chapters 7 and 8), and then to probability (Chapters 9 to 12) and inference (Chapters 13 to 24). In studying data analysis, students learn useful skills immediately and get over some of their fear of statistics. Data analysis is a necessary preliminary to inference in practice, because inference requires clean data. Designed data production is the surest foundation for inference, and the deliberate use of chance in random sampling and randomized comparative experiments motivates the study of probability in a course that emphasizes data-oriented statistics. BPS gives a full presentation of basic probability and inference (16 of the 24 chapters) but places it in the context of statistics as a whole.

Experience with data. The study of statistics is supposed to help students work with data in their varied academic disciplines and in their unpredictable later employment. Students learn to work with data by working with data. BPS is full of data from many fields of study and from everyday life. Data are more than mere numbers—they are numbers with a context that should play a role in making sense of the numbers and in stating conclusions. Examples and exercises in BPS, though intended for beginners, use real data and give enough background to allow students to consider the meaning of their calculations. Even the simplest examples carry a message, as when a pie chart of municipal waste by material (page 7) and a bar graph of the percent recycled (page 8) suggest that plastics are a problem. I often ask for conclusions that are more

1. **Emphasize the elements of statistical thinking:**

   (a) the need for data;
   (b) the importance of data production;
   (c) the omnipresence of variability;
   (d) the measuring and modeling of variability.

2. **Incorporate more data and concepts, fewer recipes and derivations. Wherever possible, automate computations and graphics.** An introductory course should:

   (a) rely heavily on *real* (not merely realistic) data;
   (b) emphasize *statistical* concepts, e.g., causation vs. association, experimental vs. observational, and longitudinal vs. cross-sectional studies;
   (c) rely on computers rather than computational recipes;
   (d) treat formal derivations as secondary in importance.

3. **Foster active learning,** through the following alternatives to lecturing:

   (a) group problem solving and discussion;
   (b) laboratory exercises;
   (c) demonstrations based on class-generated data;
   (d) written and oral presentations;
   (e) projects, either group or individual.

Figure 1 Recommendations of the ASA/MAA Joint Curriculum Committee.

than a number (or "reject H_0"). Some exercises require judgment in addition to right-or-wrong calculations and conclusions. Statistics, more than mathematics, depends on judgment for effective use. I think it is proper to begin to develop students' judgment about statistical studies.

The importance of ideas. A first course in statistics introduces many skills, from making a stemplot and calculating a correlation to choosing and carrying out a significance test. In practice (even if not always in the course), calculations and graphs are automated. And anyone who makes serious use of statistics is likely to need some specific procedures not taught in her college stat course. *BPS* therefore tries to make clear the larger patterns and big ideas of statistics, not in the abstract, but in the context of learning specific skills and working with specific data. Many of the big ideas are then summarized in graphical outlines: for example, the process of data analysis (page 152), the design of a randomized comparative experiment (page 203), the idea of a sampling distribution (page 254), and of a confidence interval (page 390). Formulas without guiding principles do students little good once the final exam is past, so it is worth the time to slow down a bit and explain the ideas.

These three principles are widely accepted by statisticians concerned about teaching. In fact, statisticians have reached a broad consensus that first courses should reflect how statistics is actually used. As Richard Scheaffer says in discussing a survey paper of mine, "With regard to the content of an introductory statistics course, statisticians are in closer agreement today than at any previous time in my career."[1]* Figure 1 is an outline of the consensus as summarized by the Joint Curriculum Committee of the American Statistical Association and the Mathematical Association of America.[2] I was a member of the ASA/MAA committee, and I agree with their conclusions. Fostering

* All notes are collected in the Notes and Data Sources section at the end of the book.

active learning is the business of the teacher (though an emphasis on working with data helps). *BPS* is guided by the first two recommendations.

Accessibility

The intent of *BPS* is to be modern *and* accessible. Helping students learn better was the main motivation for the new format of this third edition. The subject matter is presented in **24 shorter chapters.** This helps students see the major content blocks clearly and offers them more stopping points to pull together their new knowledge. Each of the first three parts of the book ends with a **review chapter** that includes a point-by-point outline of skills learned and many review exercises. (Instructors can choose to cover any or none of the chapters in Parts IV and V, so each of these chapters includes a skills outline.)

I find the **three levels of exercises** a helpful teaching tool. The short "Apply Your Knowledge" exercise sections that follow every major new idea allow a quick check of basic mastery. Each chapter ends with many more exercises on the full content of the chapter. The review chapters for Parts I to III expand the selection of exercises available for each chapter. More important, they present exercises without the "I just studied that" context, thus asking for another level of learning. I think it is important to make use of the review exercises. Look at Exercises 1 to 5 of the Part III Review (page 514) to see the advantage of the part reviews. Many instructors will find that the review chapters appear at the right points for pre-examination review.

BPS has always featured straightforward explanations of statistical topics. This is one reason for the success of previous editions. I have rewritten as needed, striving for **greater clarity.** The rethinking extends to small points: I now capitalize "Normal" distributions to remind students that these are not "normal" in the usual sense.

APPLY YOUR KNOWLEDGE

Technology

Automating calculations increases students' ability to complete problems, reduces their frustration, and helps them concentrate on ideas and problem recognition rather than mechanics. *All students should have at least a "two-variables statistics" calculator* with functions for correlation and the least-squares regression line as well as for the mean and standard deviation. Because students have calculators, the text doesn't discuss out-of-date "computing formulas" for the sample standard deviation or the least-squares regression line.

Many instructors will take advantage of more elaborate technology. And many students who don't use technology in their college statistics course will find themselves using (for example) Excel on the job. *BPS* does not assume or require use of software. It does try to accommodate software use and to convince students that they are gaining knowledge that will enable them to read and use output from almost any source. There are regular "Using Technology" sections throughout the text. Each of these displays and comments on output from the same three technologies, representing graphing calculators (the Texas Instruments TI-83), spreadsheets (Microsoft Excel), and statistical software (Minitab). The output always concerns one of the main teaching examples, so that students can compare text and output.

The three chapters of Part IV present more elaborate statistical methods. It would be quite unpleasant to implement these with a basic calculator. Even here, use of software is not quite required because many exercises present output to be used and interpreted.

Using technology

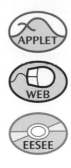

A quite different use of technology appears in the interactive **applets** created to my specifications and available online and on CD. These are designed primarily to help in learning statistics rather than in doing statistics. Animation and interaction make these applets the most effective way I know to convey ideas such as influence in regression and the concept of a confidence interval. I suggest using them as classroom demonstrations even if you do not ask students to work with them. "**Media exercises**" at the end of chapters suggest student work with the applets and also with the **EESEE library of case studies** available online and on the text CD.

What's new?

The first two editions of *BPS* have been very successful. I have nonetheless made a major revision, while keeping much the same spirit, content, and sequence. The **new format** of shorter chapters with part review chapters has already been mentioned. **Exercises** is another emphasis: this third edition has almost 50% more exercises than the second edition, and half of these exercises are new. (Most of the deleted exercises remain available online and on the CD, increasing yet more the choice available to teachers.)

There is also some **new content,** much of it requested by users. Readers will find brief introductions to personal probability (Chapter 9), $\bar{x}$ control charts (Chapter 10), tree diagrams (Chapter 11), the chi-square test for fit (Chapter 20), and a new table of critical points for the correlation r to allow testing correlation without technology (Chapter 21). For those who seek a compact modern discussion of statistical process control, a new Optional Companion Chapter 24 on this topic is available, joining a revision of the existing Optional Companion Chapter 23 on nonparametric tests.

One item of new content deserves separate discussion. Recent computational and theoretical work has demonstrated convincingly that the standard confidence intervals for proportions cannot be trusted. It is hard to abandon old friends, but I think that a look at the graphs in Section 2 of the paper by Brown, Cai, and DasGupta in the May 2001 issue of *Statistical Science* is both distressing and persuasive.[3] The standard intervals often have a true confidence level much less than what was requested, and requiring large samples encounters a maze of "lucky" and "unlucky" sample sizes until very large samples are reached. Fortunately, Alan Agresti and his students have found a simple cure: just add two successes and two failures to your data. These **plus four intervals** are presented in Chapters 18 and 19. Although they are marked optional, I think that up-to-date instruction requires that we teach them.

Why did you do that?

There is no single best way to organize a presentation of statistics to beginners. That said, my choices reflect thinking about both content and pedagogy. Here are comments on several "frequently asked questions" about the order and selection of material in *BPS*.

Why does the distinction between population and sample not appear in Part I? This is a sign that there is more to statistics than inference. In fact, statistical inference is appropriate only in rather special circumstances. The chapters in Part I present tools and tactics for describing data—any data. These tools and tactics do not depend on the idea of inference from sample to population. Many data sets in these chapters (for example, the several sets of data about the 50 states) do not lend themselves to inference because they represent an entire population. John Tukey of Bell Labs and Princeton, the philosopher of modern data analysis, insists that the population/sample distinction be avoided when it is not relevant. He uses the word "batch" for data sets in general. I see no need for a special word, but I think Tukey is right.

Why not begin with data production? It is certainly reasonable to do so—the natural flow of a planned study is from design to data analysis to inference. I choose to place the design of data production (Chapters 7 and 8) after data analysis to emphasize that data-analytic techniques apply to any data. One of the primary purposes of statistical designs for producing data is to make inference possible, so the discussion in Chapters 7 and 8 opens Part II and leads on to probability and the basics of inference.

Why do Normal distributions appear in Part I? Density curves such as the Normal curves are just another tool to describe the distribution of a quantitative variable, along with stemplots, histograms, and boxplots. It is becoming common for software to offer to make density curves from data just as it offers histograms. I prefer not to suggest that this material is essentially tied to probability, as the traditional order does. I also want students to think about the flow from graphs to numerical summaries to mathematical models, and density curves are the common mathematical model for the overall pattern of a distribution. Finally, I would like to break up the indigestible lump of probability that troubles students so much. Meeting Normal distributions early does this and strengthens the "probability distributions are like data distributions" way of approaching probability.

Why not delay correlation and regression until late in the course, as is traditional? *BPS* begins by offering experience working with data and gives a conceptual structure for this nonmathematical but very important part of statistics. Students profit from more experience with data and from seeing the conceptual structure worked out in relations among variables as well as in describing single-variable data. Moreover, correlation and regression as descriptive tools (Chapters 4 and 5) have wider scope than an emphasis on inference (Chapter 21) allows. The very important discussion of lurking variables, for example, fits poorly with inference. I consider Chapters 4 and 5 essential and Chapter 21 optional.

What about probability? Much of the usual formal probability appears in the *optional* Chapters 11 and 12. Chapters 9 and 10 present in a less formal way the ideas of probability and sampling distributions that are needed to understand inference. These two chapters follow a straight line from the idea of probability as long-term regularity, through concrete ways of assigning probabilities, to the central idea of the sampling distribution of a statistic. The law of large numbers and the central limit theorem appear in the context of discussing the sampling distribution of a sample mean. What is left to Chapters 11 and 12 is mostly "general probability rules," including conditional probability, and the binomial distributions.

I suggest that you omit Chapters 11 and 12 unless you are constrained by external forces. Experienced teachers recognize that students find probability difficult. Research on learning confirms our experience. Even students who can do formally posed probability problems often have a very fragile conceptual grasp of probability ideas. Attempting to present a substantial introduction to probability in a data-oriented statistics course for students who are not mathematically trained is in my opinion unwise. Formal probability does not help these students master the ideas of inference (at least not as much as we teachers imagine), and it depletes reserves of mental energy that might better be applied to essentially statistical ideas.

Why didn't you cover Topic X? Introductory texts ought not to be encyclopedic. Including each reader's favorite special topic results in a text that is formidable in size and intimidating to students. I chose topics on two grounds: they are the most commonly used in practice, and they are suitable vehicles for learning broader statistical ideas. There are studies of usage in many fields of application. For example, Emerson and Colditz[4] report that just descriptive statistics, t procedures, and two-way tables would give "full access" to 73% of the articles in the *New England Journal of Medicine*. That suggests a reasonable semester course from *BPS*: Chapters 1 to 10 and 13 to 20.

I am grateful to the many colleagues from two-year and four-year colleges and universities who commented on successive drafts of the manuscript:

Michael Allen
Glendale Community College

Diana J. Asmus
Greenville Technical College

Brigitte Baldi
University of California at Irvine

Joseph Cavanaugh
East Stroudsburg University

Smiley Cheng
University of Manitoba

Elizabeth Clarkson
Wichita State University

James C. Curl
Modesto Junior College

Joe DeMaio
Kennesaw State University

John Dye
California State University, Northridge

Mark D. Ecker
University of Northern Iowa

Michael E. Eraas
Iowa State University

T. Henry Jablonski, Jr.
East Tennessee State University

Rita Kottmeyer
Lindenwood University

Gary Kulis
Mohawk Valley Community College

Samantha C. Montgomery
Iowa State University

Hari Mukerjee
Wichita State University

Ronald F. Patterson
Georgia State University

Aileen Solomon
Trident Technical College

Mike Turegun
Oklahoma City Community College

James Wright
Bucknell University

I am particularly grateful to Patrick Farace, Diana Blume, Mary Louise Byrd, Danielle Swearengin, Brian Donnellan, Wendy Druck, Pam Bruton, and the other editorial and design professionals who have contributed greatly to the attractiveness of this book.

Finally, I am indebted to many statistics teachers with whom I have discussed the teaching of our subject over many years; to people from diverse fields with whom I have worked to understand data; and especially to students whose compliments and complaints have changed and improved my teaching. Working with teachers, colleagues in other disciplines, and students constantly reminds me of the importance of hands-on experience with data and of statistical thinking in an era when computer routines quickly handle statistical details.

David S. Moore

Media and Supplements for Students

A full range of supplements and media is available to help students get the most out of **BPS**:

Printed Supplements

Study Guide (0-7167-5886-5) prepared by William I. Notz and Michael A. Fligner of The Ohio State University. The study guide helps students learn and review the basic concepts of the textbook in a printed format. It explains crucial concepts in each section of the text and provides solutions to key text problems and step-through models of important statistical techniques.

Statistical Software Manuals will guide students in the use of particular statistical software with *BPS*. The chapters of each manual correlate to those of *BPS* and include exercises specific to each chapter's concepts. These manuals are:

- **Excel Manual** (0-7167-5891-1) by Fred M. Hoppe, McMaster University. Providing exercises and applications for each chapter, this manual demonstrates how Excel's ability to organize data into spreadsheets allows easy analysis and graphic exploration. Each chapter focuses on the manner in which Excel displays and analyzes data, the basic themes of *BPS*.

- **JMP Manual** (0-7167-5893-8) by Thomas F. Devlin, Montclair State University. This manual guides students through JMP INTRO, a task-oriented statistical software.

- **Minitab Manual** (0-7167-5887-3) by Betsy S. Greenberg, University of Texas. Written specifically for students who are using Minitab with *BPS*, this manual offers careful illustrations and exercises that allow the student to unlock the power of Minitab for statistical analysis. Appendices include lists of Minitab functions, commands, and macros for easy reference.

- **S-PLUS Manual** (0-7167-5885-7) by Greg Snow, Brigham Young University, and Laura Chihara, Carleton College. This manual explains how S-PLUS can be used to perform statistical techniques described in *BPS*.

- **SPSS Manual** (0-7167-5884-9) by Paul L. Stephenson, Neal T. Rogness, and Patricia A. B. Stephenson, Grand Valley State University. Written specifically for students who are using SPSS with *BPS*, this manual demonstrates the software's ability to perform a wide variety of statistical techniques ranging from descriptive statistics to complex multivariate procedures.

- **TI-83 Graphing Calculator Manual** (0-7167-5883-0) by David K. Neal, Western Kentucky University. Offering detailed instructions on the use of the TI-83 graphing calculator with *BPS*, this manual references text exercises to demonstrate solutions with TI-83 functions. Figures illustrate the calculator's output. A detailed list of TI-83 statistical functions is included.

- **Activities and Projects for the Freeman Statistics Series** (0-7167-9809-3) by Ran Millard, Shaunee Mission South High School and John C. Turner, U.S. Naval Academy. This supplement demonstrates the extensiveness of statistics with activities that show the discipline in a hands-on environment and articles that put statistics in the context of a wide range of fields.

Chapters 23 and 24 written by David S. Moore contain optional instruction on nonparametric tests and statistical process control. These chapters are available both as a printed supplement (0-7167-0232-0) and on the CD.

Telecourse Study Guide (0-7167-5892-X) by Yenphi Dang of Joliet Junior College.

Media

The Student Web site at **www.whfreeman.com/bps3e** seamlessly integrates topics from the text. On the Web site students can find:

- **Interactive Statistical Applets** that allow students to manipulate data and see the corresponding results graphically. End-of-chapter media exercises from the text require students to work with these applets in order to derive their answers.

- **Data Sets** in ASCII (plain text), Minitab, TI, SPSS, S-Plus, JMP, and Excel formats.

- **Interactive Exercises and Self-Quizzes** to help students prepare for tests.

- **Student version of the Electronic Encyclopedia of Statistical Examples and Exercises (EESEE)**, a rich repository of case studies that apply the concepts of the text in various real-world venues such as the mass media, sports, natural sciences, social sciences, and medicine. EESEE was developed by a consortium at Ohio State University dedicated to statistical education.

- **All Tables** from the text in .pdf format for quick, easy reference.

- **Additional Exercises** for every chapter written by David Moore, giving students more opportunities to make sure they understand key concepts. Solutions to odd-number additional exercises are on the book's Web site.

Interactive Student CD-ROM developed by Sumanas Multimedia Development Services, featuring EESEE (Electronic Encyclopedia of Statistical Examples and Exercises) case studies, is packaged with each copy of the textbook. The CD contains:

- **Additional Exercises** for every chapter written by David Moore, giving students more opportunities to make sure they understand key concepts. Solutions to odd-number additional exercises are on the book's Web site.

- **Interactive Statistical Applets** that all students to manipulate data and see the corresponding results graphically. End-of-chapter media exercises from the text require students to work with these applets in order to derive their answers.

- **"Q&A" Interactive Chapter Self-Quizzes** for each chapter constructed by instructors with many years of experience to anticipate the errors that students typically make.

- **Data Sets** in ASCII (plain text), Minitab, TI, SPSS, JMP, S-Plus, and Excel formats.

- **Chapters 23 and 24,** on nonparametric tests and statistical process control, available on CD or printed separately, give you the option of including this material in the course.

Statistical Software Packages

Student versions of JMP, Minitab, SPLUS, and SPSS can be packaged with *BPS*.

Media Support for Instructors

A full range of supplements and media support is available to help instructors teach from *BPS*:

Instructor's CD

Instructor's Resource CD-ROM (0-7167-5890-3) *contains all the student CD material* plus:

- **Instructor's version of the Encyclopedia of Statistical Examples and Exercises** (EESEE), which provides solutions to student exercises.
- **All *BPS* Figures** in an exportable presentation format, JPEG for Windows.
- **All Tables** from the text in .pdf format.
- **Instructor's Guide with Solutions** to exercises in .pdf format.
- **Presentation Manager Pro,** which creates presentations for all the figures and selected tables within the text. Extra material can be imported from locally saved files and the Web.
- **Power Point Slides** of all figures and tables from the text that can be used directly or customized.
- **Test Bank** in MS Word format.

Assessment Tools

Instructor's Guide with Solutions, prepared by Darryl K. Nester of Bluffton College. Includes solutions to all exercises, teaching suggestions, and chapter comments.

Test Bank (0-7167-5889-X) prepared by Michael A. Fligner and William I. Notz of Ohio State University is a printed supplement that contains all test questions and answers from the computerized Test Bank. The easy-to-use CD version lets you add, edit, and resequence questions to suit your needs. Windows and Mac versions are available on a single disk (0-7167-5882-2).

Online Testing powered by Diploma from the Brownstone Research Group offers instructors the ability to easily create and administer secure exams over a network and over the Internet, with questions that incorporate multimedia and interactive exercises. The program lets you restrict tests to specific computers or blocks of time and includes an impressive suite of grade book and result-analysis features.

Online Quizzing, powered by Question Mark and accessed via the *BPS* Web site uses Question Mark's Perception to enable instructors to easily and securely quiz students online using prewritten, multiple-choice questions for each text chapter, separate from those appearing in the Test Bank. Students receive instant feedback and can take the quizzes multiple times. Instructors can view results by quiz, student, or question or can get weekly results via email.

iSolve Homework Service powered by Brownstone EDU. About 500 Exercises from the Book are available online in W. H Freeman's iSolve homework service. This service, by randomizing values in the Exercise Statements, offers each student a different version of every exercise. Student performance can be collected in a gradebook that goes with system. For more information visit the book's Website at http://www.whfreeman.com/bps3e

BPS Web Site

The instructor's Web site at www.whfreeman.com/bps3e *contains all features available to students* plus:

- **Instructor version of the Electronic Encyclopedia of Statistical Examples and Exercises (EESEE),** with solutions to the exercises in the student version.
- **Instructor's Guide with Solutions** available in an Adobe .pdf electronic format.
- **Power Point Slides** that can be used as is or customized. Each slide offers outlined text to match the textbook's table of contents along with a wide range of figures and charts culled from the book and other sources.

Course Management

Online Course Materials (WebCT, Blackboard) can be provided as a service for adopters. We offer electronic content of *BPS*, including the complete Test Bank and all Website materials in either WebCT or Blackboard format.

Statistical Software Packages

Student versions of JMP, Minitab, SPSS, and S-Plus can be bundled with *BPS* for those instructors who wish to use a statistical software package in the course.

Applications

The Basic Practice of Statistics presents a wide variety of applications from diverse disciplines. The list below indicates the number of exercises and examples which relate to different fields.

Examples:

Agriculture: 6
Biological and environmental sciences: 17
Business and economics: 49
Education: 19
History and public policy: 4
People and places: 28
Physical sciences: 8
Psychology and behavioral sciences: 23
Public health and medicine: 32
Sports: 7

Exercises:

Agriculture: 19
Biological and environmental sciences: 160
Business and economics: 182
Education: 83
History and public policy: 18
Paleontology: 5
People and places: 168
Physical sciences: 44
Psychology and behavioral sciences: 83
Public health and medicine: 153
Sports: 13

For a complete breakdown list of examples and exercises by chapter and number, please see the *Instructor's Guide* or the web site: **www.whfreeman.com/bps3e**.

TO THE STUDENTS:
Statistical Thinking

(Stockbyte/PictureQuest)

Statistics is about data. Data are numbers, but they are not "just numbers." *Data are numbers with a context*. The number 10.5, for example, carries no information by itself. But if we hear that a friend's new baby weighed 10.5 pounds at birth, we congratulate her on the healthy size of the child. The context engages our background knowledge and allows us to make judgments. We know that a baby weighing 10.5 pounds is quite large, and that a human baby is unlikely to weigh 10.5 ounces or 10.5 kilograms. The context makes the number informative.

Statistics uses data to gain insight and to draw conclusions. Our tools are graphs and calculations, but the tools are guided by ways of thinking that amount to educated common sense. Let's begin our study of statistics with an informal look at some principles of statistical thinking.

Data Beat Anecdotes

An anecdote is a striking story that sticks in our minds exactly because it is striking. Anecdotes humanize an issue, but they can be misleading.

Does living near power lines cause leukemia in children? The National Cancer Institute spent 5 years and $5 million gathering data on the question. The researchers compared 638 children who had leukemia and 620 who did not. They went into the homes and actually measured the magnetic fields in the children's bedrooms, in other rooms, and at the front door. They recorded facts about power lines near the family home and also near the mother's residence when she was pregnant. Result: no connection between leukemia and exposure to magnetic fields of the kind produced by power lines. The editorial that accompanied the study report in the *New England Journal of Medicine* thundered, "It is time to stop wasting our research resources" on the question.[1]

Now compare the effectiveness of a television news report of a 5-year, $5 million investigation against a televised interview with an articulate mother whose child has leukemia and who happens to live near a power line. In the public mind, the anecdote wins every time. A statistically literate person, however, knows that data are more reliable than anecdotes because they systematically describe an overall picture rather than focus on a few incidents.

Beware the Lurking Variable

(Leslie McFarland/Picturesque/
PictureQuest)

The Kalamazoo (Michigan) Symphony once advertised a "Mozart for Minors" program with this statement: "Question: Which students scored 51 points higher in verbal skills and 39 points higher in math? Answer: Students who had experience in music."[2]

Who would dispute that early experience with music is good for you? The skeptical statistician, that's who. Children who take music lessons and attend concerts tend to have prosperous and well-educated parents. These same children are also likely to attend good schools, get good health care, and be encouraged to study hard. No wonder they score well on tests. The children's family background is a *lurking variable* when we talk about the relationship between music and test scores. It is lurking behind the scenes, unmentioned in the symphony's publicity. Yet family background, more than anything else we can measure, influences children's academic performance.

Perhaps the Kalamazoo Youth Soccer League should advertise that students who play soccer score higher on tests. After all, children who play soccer, like those who have experience in music, tend to have educated and prosperous parents. The message is worth repeating: beware the lurking variable, because almost all relationships between two variables are influenced by other variables lurking in the background.

(Paul A. Souders/CORBIS)

Where the data come from is important

The advice columnist Ann Landers once asked her readers, "If you had it to do over again, would you have children?" A few weeks later, her column was headlined "70% OF PARENTS SAY KIDS NOT WORTH IT." Indeed, 70% of the nearly 10,000 parents who wrote in said they would not have children if they could make the choice again. Do you believe that 70% of all parents regret having children?

You shouldn't. The people who took the trouble to write Ann Landers are not representative of all parents. Their letters showed that many of them were angry at their children. All we know from these data is that there are some unhappy parents out there. A statistically designed poll, unlike Ann Landers's appeal, targets specific people chosen in a way that gives all parents the same chance to be asked. Such a poll showed that 91% of parents *would* have children again. It matters a lot whether data come from a haphazard poll or a statistically designed survey: if you are careless about how you get your data, you may announce 70% "No" when the truth is close to 90% "Yes."

Here's another question: Should women take hormones such as estrogen after menopause, when natural production of these hormones ends? In 1992, several major medical organizations said "Yes." In particular, women who took hormones seemed to reduce their risk of a heart attack by 35% to 50%. The risks of taking hormones appeared small compared with the benefits.

The evidence in favor of hormone replacement came from a number of studies that simply compared women who were taking hormones with others who were not. Beware the lurking variable: women who choose to take hormones are richer and better educated and see doctors more often than women who do not. These women do many things to maintain their health. It isn't surprising that they have fewer heart attacks.

To get convincing data on the link between hormone replacement and heart attacks, do an *experiment*. Experiments don't let women decide what to do. They assign women to either hormone replacement or to dummy pills that look and taste the same as the hormone pills. The assignment is done by a coin toss, so that all kinds of women are equally likely to get either treatment. By 2002, several experiments with women of different ages agreed that hormone replacement does *not* reduce the risk of heart attacks. The National Institutes of Health, after reviewing the evidence, concluded that the first studies were wrong. Taking hormones after menopause quickly fell out of favor.[3] It matters a lot whether data come from observation, even careful observation, or experiments that compare several treatments. Only experiments can completely defeat the lurking variable and give convincing evidence that an alleged cause really does account for an observed effect.

VARIATION IS EVERYWHERE

The company's sales reps file into their monthly meeting. The sales manager rises. "Congratulations! Our sales were up 2% last month, so we're all drinking champagne this morning. You remember that when sales were down 1% last month I fired half of our reps." This picture is only slightly exaggerated. Many managers overreact to small short-term variations in key figures. Here is Arthur Nielsen, head of the country's largest market research firm, describing his experience:

> *Too many business people assign equal validity to all numbers printed on paper. They accept numbers as representing Truth and find it difficult to work with the concept of probability. They do not see a number as a kind of shorthand for a range that describes our actual knowledge of the underlying condition.*[4]

Business data such as sales and prices vary from month to month for reasons ranging from the weather to a customer's financial difficulties to the inevitable errors in gathering the data. The manager's challenge is to say when there is a real pattern behind the variation. Statistical tools can help. Often, simply gathering and plotting data bring understanding. Figure 1 plots the average price of a gallon of regular unleaded gasoline each month from January 1988 to February 2003.[5] There certainly is

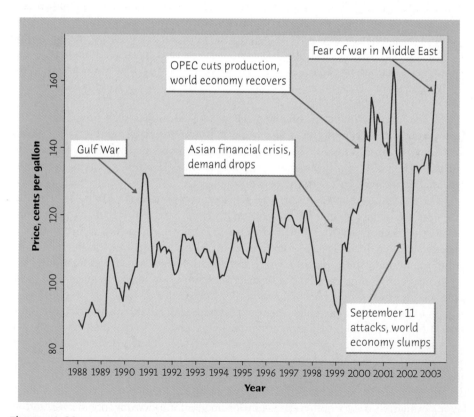

Figure 1 Variation is everywhere: the average price at the pump of regular unleaded gasoline, 1988 to early 2003. World events bring rapid changes on top of the regular up-in-summer, down-in-winter variation.

variation! But a close look shows a pattern: gas prices normally go up during the summer driving season each year, then down as demand drops in the fall. Against this regular pattern we see the effects of international events: prices rose because of the 1990 Gulf War, droped because of the 1998 financial crisis in Asia, and rose sharply from 1999 to 2003 due to turmoil in the Middle East and a general strike in Venezuela. The data carry an important message: because the United States imports much of its oil, we can't control the price we pay for gasoline.

Variation is everywhere. Individuals vary; repeated measurements on the same individual vary; almost everything varies over time. One reason we need to know some statistics is that statistics helps us deal with variation.

(AP/Wide World Photos)

CONCLUSIONS ARE NOT CERTAIN

Most women who reach middle age have regular mammograms to detect breast cancer. Do mammograms really reduce the risk of dying of breast cancer? To defeat the lurking variable, doctors rely on experiments (called "clinical trials" in medicine) that compare different ways of screening for breast cancer. The conclusion from 13 such trials is that mammograms reduce the risk of death in women aged 50 to 64 years by 26%.[6]

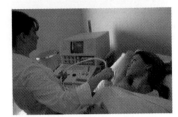

(Bob Llewellyn/Image State-Pictor/
PictureQuest)

On the average, then, women who have regular mammograms are less likely to die of breast cancer. But because variation is everywhere, the results are different for different women. Some women who have yearly mammograms die of breast cancer, and some who never have mammograms live to 100 and die when they crash their motorcycles. Statistical conclusions are "on the average" statements only. Well then, can we be sure that mammograms reduce risk on the average? No. We can be pretty confident, but we can't be sure. *Because variation is everywhere, conclusions are uncertain.*

Statistics gives us a language for talking about uncertainty that is used and understood by statistically literate people everywhere. In the case of mammograms, the doctors use that language to tell us that "mammography reduces the risk of dying of breast cancer by 26 percent (95 percent confidence interval, 17 to 34 percent)." That 26% is, in Arthur Nielsen's words, "a shorthand for a range that describes our actual knowledge of the underlying condition." The range is 17% to 34%, and we are 95% confident that the truth lies in that range. We will soon learn to understand this language. We can't escape variation and uncertainty. Learning statistics enables us to live more comfortably with these realities.

Statistical Thinking and You

What Lies Ahead in This Book

The purpose of this book is to give you a working knowledge of the ideas and tools of practical statistics. We will approach practical statistics in an order that reflects our short introduction to statistical thinking.

1. **Data analysis** concerns methods and strategies for exploring, organizing, and describing data using graphs and numerical summaries. Only organized data can illuminate reality. Only thoughtful exploration of data can defeat the lurking variable. Part I of this book (Chapters 1 to 6) discusses data analysis.

2. **Data production** provides methods for producing data that can give clear answers to specific questions. Where the data come from really is important—basic concepts about how to select samples and design experiments are the most influential ideas in statistics. These concepts are the subject of Chapters 7 and 8.

3. **Statistical inference** moves beyond the data in hand to draw conclusions about some wider universe, taking into account that variation is everywhere and that conclusions are uncertain. To describe variation and uncertainty, inference uses the language of probability, which is introduced in Chapters 9 and 10. Because we are concerned with practice rather than theory, we can function with a limited knowledge of probability. Chapters 11 and 12 offer more probability for those who want it. Chapters 13 to 15 discuss the reasoning of statistical inference. These chapters are the key to the rest of the book. Chapters 16 to 19 present inference as used in practice in the most common settings. Chapters 20 to 22, and the separate Optional Companion Chapters 23 and 24, concern more advanced or specialized kinds of inference.

Because data are numbers with a context, doing statistics means more than manipulating numbers. *The Basic Practice of Statistics* is full of data, and each set of data has some brief background to help you understand what the data say. Many exercises ask you to express briefly some understanding gained from the data. In practice, you would know much more about the background of the data you work with and about the questions you hope the data will answer. No textbook can be fully realistic. But it is important to form the habit of asking, "What do the data tell me?" rather than just concentrating on making graphs and doing calculations. This book tries to encourage good habits.

Nonetheless, statistics involves lots of calculating and graphing. The text presents the techniques you need, but you should use a calculator or software to automate calculations and graphs as much as possible. Many kinds of software, from spreadsheets to large specialized programs for advanced users, can do statistical calculations and graphs. The technology available to learners varies a great deal from place to place—but the big ideas of statistics don't depend on any particular level of access to computing. We encourage use of software or a graphing calculator, but this book does not require software and is not tied to any specific software. Even if you make little use of technology, you should look at the "Using Technology" sections throughout the book. You will see at once that this book prepares you to read and use the output from almost any technology used for statistical calculations. The ideas really are more important than the details of how to do the calculations.

This book does require that you have a calculator with some built-in statistical functions. Specifically, you need a calculator that will find means and standard deviations and calculate correlations and regression lines. Look for a calculator that claims to do "two-variables statistics" or mentions "regression."

Calculators and computers can follow recipes for graphs and calculations both more quickly and more accurately than humans can. Because graphing and calculating are automated in statistical practice, the most important assets you can gain from the study of statistics are an understanding of the big ideas and the beginnings of good judgment in working with data. Ideas and judgment can't (at least yet) be automated. They guide you in telling the computer what to do and in interpreting its output. This book tries to explain the most important ideas of statistics, not just teach methods. Some examples of big ideas that you will meet (one from each of the three areas of statistics) are "always plot your data," "randomized comparative experiments," and "statistical significance."

You learn statistics by doing statistical problems. This book offers three levels of problems, arranged to help you learn. Short "Apply Your Knowledge" problem sets appear after each major idea. These are straightforward exercises that help you solidify the main points as you read. Pause for a few of these exercises before going on. The Chapter Exercises help you combine all the ideas of a chapter. Finally, each of the three part review chapters looks back over a major block of learning, with many review exercises. At each step you are given less advance knowledge of exactly what statistical ideas and skills the problems will require, so each type of exercise requires more understanding. Each part review chapter (and the individual chapters in Part IV) includes a point-by-point list of specific things you should now be able to do. Go through that list, and be sure you can say "I can do that" to each item. Then try some of the review exercises. The book ends with a review titled "Statistical Thinking Revisited," which you should read and think about no matter where in the book your course ends.

The basic principle of learning is persistence. The main ideas of statistics, like the main ideas of any important subject, took a long time to discover and take some time to master. The gain will be worth the pain.

Exploring Data

The first step in understanding data is to hear what the data say, to "let the statistics speak for themselves." But numbers speak clearly only when we help them speak by organizing, displaying, summarizing, and asking questions. That's *data analysis*. The six chapters in Part I present the ideas and tools of statistical data analysis. They equip you with skills that are immediately useful whenever you deal with numbers.

These chapters reflect the strong emphasis on exploring data that characterizes modern statistics. Although careful exploration of data is essential if we are to trust the results of inference, data analysis isn't just preparation for inference. To think about inference, we carefully distinguish between the data we actually have and the larger universe we want conclusions about. The Bureau of Labor Statistics, for example, has data about employment in the 55,000 households contacted by its Current Population Survey. The bureau wants to draw conclusions about employment in all 110 million U.S. households. That's a complex problem. From the viewpoint of data analysis, things are simpler. We want to explore and understand only the data in hand. The distinctions that inference requires don't concern us in Chapters 1 to 6. What does concern us is a systematic strategy for examining data and the tools that we use to carry out that strategy.

Part of that strategy is to first look at one thing at a time and then at relationships. In Chapters 1, 2, and 3 you will study **variables and their distributions.** Chapters 4, 5, and 6 concern **relationships among variables.**

"Tonight, we're going to let the statistics speak for themselves."

EXPLORING DATA: VARIABLES AND DISTRIBUTIONS

EXPLORING DATA REVIEW

Picturing Distributions with Graphs

Statistics is the science of data. The volume of data available to us is overwhelming. Each March, for example, the Census Bureau collects economic and employment data from more than 200,000 people. From the bureau's Web site you can choose to examine more than 300 items of data for each person (and more for households): child care assistance, child care support, hours worked, usual weekly earnings, and much more. The first step in dealing with such a flood of data is to organize our thinking about data.

Individuals and variables

Any set of data contains information about some group of *individuals*. The information is organized in *variables*.

INDIVIDUALS AND VARIABLES

Individuals are the objects described by a set of data. Individuals may be people, but they may also be animals or things.

A **variable** is any characteristic of an individual. A variable can take different values for different individuals.

Are data artistic?

David Galenson, an economist at the University of Chicago, uses data and statistical analysis to study innovation among painters from the nineteenth century to the present. Economics journals publish his work. Art history journals send it back unread. "Fundamentally antagonistic to the way humanists do their work," said the chair of art history at Chicago. If you are a student of the humanities, reading this statistics text may help you start a new wave in your field.

A college's student data base, for example, includes data about every currently enrolled student. The students are the individuals described by the data set. For each individual, the data contain the values of variables such as date of birth, gender (female or male), choice of major, and grade point average. In practice, any set of data is accompanied by background information that helps us understand the data. When you plan a statistical study or explore data from someone else's work, ask yourself the following questions:

1. **Who?** What **individuals** do the data describe? **How many** individuals appear in the data?

2. **What?** How many **variables** do the data contain? What are the **exact definitions** of these variables? In what **units of measurement** is each variable recorded? Weights, for example, might be recorded in pounds, in thousands of pounds, or in kilograms.

3. **Why?** **What purpose** do the data have? Do we hope to answer some specific questions? Do we want to draw conclusions about individuals other than the ones we actually have data for? Are the variables suitable for the intended purpose?

Some variables, like gender and college major, simply place individuals into categories. Others, like height and grade point average, take numerical values for which we can do arithmetic. It makes sense to give an average income for a company's employees, but it does not make sense to give an "average" gender. We can, however, count the numbers of female and male employees and do arithmetic with these counts.

CATEGORICAL AND QUANTITATIVE VARIABLES

A **categorical variable** places an individual into one of several groups or categories.

A **quantitative variable** takes numerical values for which arithmetic operations such as adding and averaging make sense.

The **distribution** of a variable tells us what values it takes and how often it takes these values.

EXAMPLE 1.1   A professor's data set

Here is part of the data set in which a professor records information about student performance in a course:

	A	B	C	D	E	F	G	H	I
1	Name	School	Major	HW total	Midterm	Final Exam	Total	Grade	
2	Arroyo, Juan	EDU	EdPsych	95	80	88	263	A	
3	Arthur, Brenda	LA	Psych	32	61	54	147	D	
4	Bai, Jingyi	SCI	Biol	74	68	70	212	B	
5	Boggs, Amanda	SCI	Math	86	75	94	255	A	

Microsoft Excel - grades.xls
File Edit View Insert Format Tools Data Window Help
Fall semester grades   Sheet2   Sheet3

The *individuals* described are the students. Each row records data on one individual. Each column contains the values of one *variable* for all the individuals. In addition to the student's name, there are 7 variables. School and major are categorical variables. Scores on homework, the midterm, and the final exam and the total score are quantitative. Grade is recorded as a category (A, B, and so on), but each grade also corresponds to a quantitative score (A = 4, B = 3, and so on) that is used to calculate student grade point averages.

Most data tables follow this format—each row is an individual, and each column is a variable. This data set appears in a **spreadsheet** program that has rows and columns ready for your use. Spreadsheets are commonly used to enter and transmit data and to do simple calculations such as adding homework, midterm, and final scores to get total points.

spreadsheet

APPLY YOUR KNOWLEDGE

1.1 **Fuel economy.** Here is a small part of a data set that describes the fuel economy (in miles per gallon) of 2002 model motor vehicles:

Make and model	Vehicle type	Transmission type	Number of cylinders	City MPG	Highway MPG
⋮					
Acura NSX	Two-seater	Automatic	6	17	24
Audi A4	Compact	Manual	4	22	31
Buick Century	Midsize	Automatic	6	20	29
Dodge Ram 1500	Standard pickup truck	Automatic	8	15	20
⋮					

(a) What are the individuals in this data set?

(b) For each individual, what variables are given? Which of these variables are categorical and which are quantitative?

1.2 **A medical study.** Data from a medical study contain values of many variables for each of the people who were the subjects of the study. Which of the following variables are categorical and which are quantitative?

(a) Gender (female or male)

(b) Age (years)

(c) Race (Asian, black, white, or other)

(d) Smoker (yes or no)

(e) Systolic blood pressure (millimeters of mercury)

(f) Level of calcium in the blood (micrograms per milliliter)

Categorical variables: pie charts and bar graphs

exploratory data analysis

Statistical tools and ideas help us examine data in order to describe their main features. This examination is called **exploratory data analysis.** Like an explorer crossing unknown lands, we want first to simply describe what we see. Here are two basic strategies that help us organize our exploration of a set of data:

- Begin by examining each variable by itself. Then move on to study the relationships among the variables.

- Begin with a graph or graphs. Then add numerical summaries of specific aspects of the data.

We will follow these principles in organizing our learning. Chapters 1 to 3 present methods for describing a single variable. We study relationships among several variables in Chapters 4 to 6. In each case, we begin with graphical displays, then add numerical summaries for more complete description.

The proper choice of graph depends on the nature of the variable. The values of a categorical variable are labels for the categories, such as "male" and "female." The distribution of a categorical variable lists the categories and gives either the **count** or the **percent** of individuals who fall in each category.

EXAMPLE 1.2   Garbage

The formal name for garbage is "municipal solid waste." Here is a breakdown of the materials that made up American municipal solid waste in 2000.[1]

Material	Weight (million tons)	Percent of total
Food scraps	25.9	11.2%
Glass	12.8	5.5%
Metals	18.0	7.8%
Paper, paperboard	86.7	37.4%
Plastics	24.7	10.7%
Rubber, leather, textiles	15.8	6.8%
Wood	12.7	5.5%
Yard trimmings	27.7	11.9%
Other	7.5	3.2%
Total	231.9	100.0

roundoff error

It's a good idea to check data for consistency. The weights of the nine materials add to 231.8 million tons, not exactly equal to the total of 231.9 million tons given in the table. What happened? **Roundoff error:** Each entry is rounded to the nearest tenth, and the total is rounded separately. The exact values would add exactly, but the rounded values don't quite.

pie chart

The **pie chart** in Figure 1.1 shows us each material as a part of the whole. For example, the "plastics" slice makes up 10.7% of the pie because 10.7% of municipal solid waste consists of plastics. The graph shows more clearly than the numbers the predominance of paper and the importance of food scraps,

Figure 1.1 Pie chart of materials in municipal solid waste, by weight.

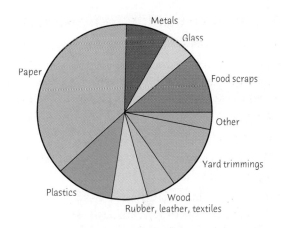

plastics, and yard trimmings in our garbage. Pie charts are awkward to make by hand, but software will do the job for you.

We could also make a **bar graph** that represents each material's weight by the height of a bar. To make a pie chart, you must include all the categories that make up a whole. Bar graphs are more flexible. Figure 1.2(a) is a bar graph of the percent of each material that was recycled or composted in 2000. These percents are not part of a whole because each refers to a different material. We could replace the pie chart in Figure 1.1 by a bar graph, but we can't make a pie chart to replace Figure 1.2(a). We can often improve a bar graph by changing the order of the groups we are comparing. Figure 1.2(b) displays the recycling data with the materials in order of percent recycled or composted. Figures 1.1 and 1.2 together suggest that we might pay more attention to recycling plastics.

Bar graphs and pie charts help an audience grasp the distribution quickly. They are, however, of limited use for data analysis because it is easy to understand data on a single categorical variable without a graph. We will move on to quantitative variables, where graphs are essential tools.

bar graph

APPLY YOUR KNOWLEDGE

1.3 **The color of your car.** Here is a breakdown of the most popular colors for vehicles made in North America during the 2001 model year:[2]

Color	Percent	Color	Percent
Silver	21.0%	Medium red	6.9%
White	15.6%	Brown	5.6%
Black	11.2%	Gold	4.5%
Blue	9.9%	Bright red	4.3%
Green	7.6%	Grey	2.0%

(a) What percent of vehicles are some other color?

(b) Make a bar graph of the color data. Would it be correct to make a pie chart if you added an "Other" category?

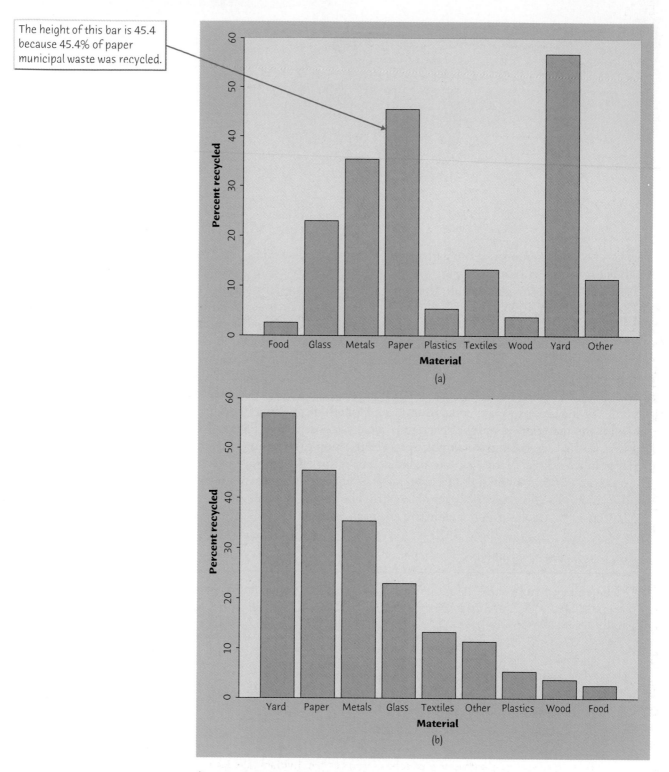

The height of this bar is 45.4 because 45.4% of paper municipal waste was recycled.

Figure 1.2 Bar graphs comparing the percents of each material in municipal solid waste that were recycled or composted.

1.4 Never on Sunday? Births are not, as you might think, evenly distributed across the days of the week. Here are the average numbers of babies born on each day of the week in 1999:[3]

Day	Births
Sunday	7,731
Monday	11,018
Tuesday	12,424
Wednesday	12,183
Thursday	11,893
Friday	12,012
Saturday	8,654

Present these data in a well-labeled bar graph. Would it also be correct to make a pie chart? Suggest some possible reasons why there are fewer births on weekends.

Quantitative variables: histograms

Quantitative variables often take many values. A graph of the distribution is clearer if nearby values are grouped together. The most common graph of the distribution of one quantitative variable is a **histogram.**

histogram

EXAMPLE 1.3   Making a histogram

One of the most striking findings of the 2000 census was the growth of the Hispanic population of the United States. Table 1.1 presents the percent of residents in each of the 50 states who identified themselves in the 2000 census as "Spanish/Hispanic/Latino."[4] The *individuals* in this data set are the 50 states. The *variable* is the percent of Hispanics in a state's population. To make a histogram of the distribution of this variable, proceed as follows:

Step 1. Choose the classes. Divide the range of the data into classes of equal width. The data in Table 1.1 range from 0.7 to 42.1, so we decide to choose these classes:

$$0.0 \leq \text{percent Hispanic} < 5.0$$
$$5.0 \leq \text{percent Hispanic} < 10.0$$
$$\vdots$$
$$40.0 \leq \text{percent Hispanic} < 45.0$$

Be sure to specify the classes precisely so that each individual falls into exactly one class. A state with 4.9% Hispanic residents would fall into the first class, but a state with 5.0% falls into the second.

TABLE 1.1 Percent of population of Hispanic origin, by state (2000)

State	Percent	State	Percent	State	Percent
Alabama	1.5	Louisiana	2.4	Ohio	1.9
Alaska	4.1	Maine	0.7	Oklahoma	5.2
Arizona	25.3	Maryland	4.3	Oregon	8.0
Arkansas	2.8	Massachusetts	6.8	Pennsylvania	3.2
California	32.4	Michigan	3.3	Rhode Island	8.7
Colorado	17.1	Minnesota	2.9	South Carolina	2.4
Connecticut	9.4	Mississippi	1.3	South Dakota	1.4
Delaware	4.8	Missouri	2.1	Tennessee	2.0
Florida	16.8	Montana	2.0	Texas	32.0
Georgia	5.3	Nebraska	5.5	Utah	9.0
Hawaii	7.2	Nevada	19.7	Vermont	0.9
Idaho	7.9	New Hampshire	1.7	Virginia	4.7
Illinois	10.7	New Jersey	13.3	Washington	7.2
Indiana	3.5	New Mexico	42.1	West Virginia	0.7
Iowa	2.8	New York	15.1	Wisconsin	3.6
Kansas	7.0	North Carolina	4.7	Wyoming	6.4
Kentucky	1.5	North Dakota	1.2		

Step 2. Count the individuals in each class. Here are the counts:

Class	Count	Class	Count	Class	Count
0.0 to 4.9	27	15.0 to 19.9	4	30.0 to 34.9	2
5.0 to 9.9	13	20.0 to 24.9	0	35.0 to 39.9	0
10.0 to 14.9	2	25.0 to 29.9	1	40.0 to 44.9	1

Step 3. Draw the histogram. Mark the scale for the variable whose distribution you are displaying on the horizontal axis. That's the percent of a state's population who are Hispanic. The scale runs from 0 to 45 because that is the span of the classes we chose. The vertical axis contains the scale of counts. Each bar represents a class. The base of the bar covers the class, and the bar height is the class count. There is no horizontal space between the bars unless a class is empty, so that its bar has height zero. Figure 1.3 is our histogram.

The bars of a histogram should cover the entire range of values of a variable. When the possible values of a variable have gaps between them, extend the bases of the bars to meet halfway between two adjacent possible values. For example, in a histogram of the ages in years of university faculty, the bars representing 25 to 29 years and 30 to 34 years would meet at 29.5.

Our eyes respond to the *area* of the bars in a histogram.[5] Because the classes are all the same width, area is determined by height and all classes are fairly represented. There is no one right choice of the classes in a histogram. Too

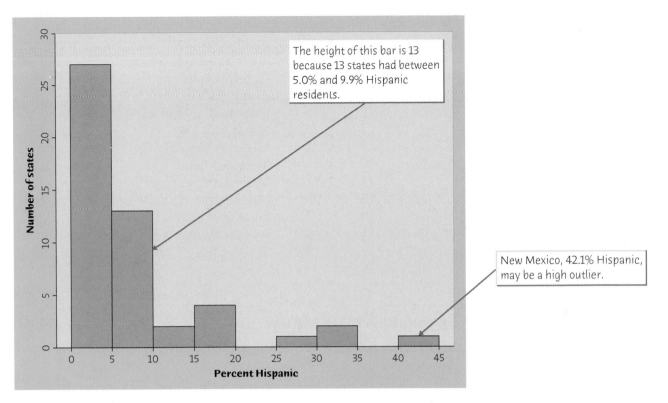

The height of this bar is 13 because 13 states had between 5.0% and 9.9% Hispanic residents.

New Mexico, 42.1% Hispanic, may be a high outlier.

Figure 1.3 Histogram of the distribution of the percent of Hispanics among the residents of the 50 states. This distribution is skewed to the right.

few classes will give a "skyscraper" graph, with all values in a few classes with tall bars. Too many will produce a "pancake" graph, with most classes having one or no observations. Neither choice will give a good picture of the shape of the distribution. You must use your judgment in choosing classes to display the shape. Statistics software will choose the classes for you. The software's choice is usually a good one, but you can change it if you want.

APPLY YOUR KNOWLEDGE

1.5 **Sports car fuel economy.** Interested in a sports car? The Environmental Protection Agency lists most such vehicles in its "two-seater" category. Table 1.2 gives the city and highway mileages (miles per gallon) for the 22 two-seaters listed for the 2002 model year.[6] Make a histogram of the highway mileages of these cars using classes with width 5 miles per gallon.

Interpreting histograms

Making a statistical graph is not an end in itself. The purpose of the graph is to help us understand the data. After you make a graph, always ask, "What do I see?" Once you have displayed a distribution, you can see its important features as follows.

TABLE 1.2 Gas mileage (miles per gallon) for 2002 model two-seater cars

Model	City	Highway	Model	City	Highway
Acura NSX	17	24	Honda Insight	57	56
Audi TT Quattro	20	28	Honda S2000	20	26
Audi TT Roadster	22	31	Lamborghini Murcielago	9	13
BMW M Coupe	17	25	Mazda Miata	22	28
BMW Z3 Coupe	19	27	Mercedes-Benz SL500	16	23
BMW Z3 Roadster	20	27	Mercedes-Benz SL600	13	19
BMW Z8	13	21	Mercedes-Benz SLK230	23	30
Chevrolet Corvette	18	25	Mercedes-Benz SLK320	20	26
Chrysler Prowler	18	23	Porsche 911 GT2	15	22
Ferrari 360 Modena	11	16	Porsche Boxster	19	27
Ford Thunderbird	17	23	Toyota MR2	25	30

EXAMINING A DISTRIBUTION

In any graph of data, look for the **overall pattern** and for striking **deviations** from that pattern.

You can describe the overall pattern of a histogram by its **shape, center,** and **spread.**

An important kind of deviation is an **outlier,** an individual value that falls outside the overall pattern.

We will learn how to describe center and spread numerically in Chapter 2. For now, we can describe the center of a distribution by its *midpoint*, the value with roughly half the observations taking smaller values and half taking larger values. We can describe the spread of a distribution by giving the *smallest and largest values.*

EXAMPLE 1.4   Describing a distribution

Look again at the histogram in Figure 1.3. **Shape:** The distribution has a *single peak*, which represents states that are less than 5% Hispanic. The distribution is *skewed to the right*. Most states have no more than 10% Hispanics, but some states have much higher percentages, so that the graph trails off to the right. **Center:** Table 1.1 shows that about half the states have less than 4.7% Hispanics among their residents and half have more. So the midpoint of the distribution is close to 4.7%. **Spread:** The spread is from about 0% to 42%, but only four states fall above 20%.

Outliers: Arizona, California, New Mexico, and Texas stand out. Whether these are outliers or just part of the long right tail of the distribution is a matter of judgment. There is no rule for calling an observation an outlier. Once you have spotted possible outliers, look for an explanation. Some outliers are due to mistakes, such as typing 4.2 as 42. Other outliers point to the special nature of some observations. These four states are heavily Hispanic by history and location.

When you describe a distribution, concentrate on the main features. Look for major peaks, not for minor ups and downs in the bars of the histogram. Look for clear outliers, not just for the smallest and largest observations. Look for rough *symmetry* or clear *skewness*.

SYMMETRIC AND SKEWED DISTRIBUTIONS

A distribution is **symmetric** if the right and left sides of the histogram are approximately mirror images of each other.

A distribution is **skewed to the right** if the right side of the histogram (containing the half of the observations with larger values) extends much farther out than the left side. It is **skewed to the left** if the left side of the histogram extends much farther out than the right side.

Here are more examples of describing the overall pattern of a histogram.

EXAMPLE 1.5   Iowa Test scores

Figure 1.4 displays the scores of all 947 seventh-grade students in the public schools of Gary, Indiana, on the vocabulary part of the Iowa Test of Basic Skills. The

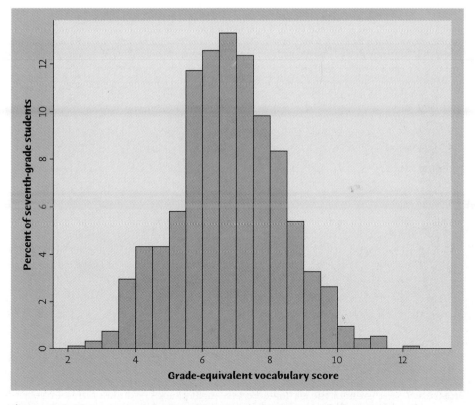

Figure 1.4 Histogram of the Iowa Test vocabulary scores of all seventh-grade students in Gary, Indiana. This distribution is single-peaked and symmetric.

distribution is *single-peaked* and *symmetric*. In mathematics, the two sides of symmetric patterns are exact mirror images. Real data are almost never exactly symmetric. We are content to describe Figure 1.4 as symmetric. The center (half above, half below) is close to 7. This is seventh-grade reading level. The scores range from 2.0 (second-grade level) to 12.1 (twelfth-grade level).

Notice that the vertical scale in Figure 1.4 is not the *count* of students but the *percent* of Gary students in each histogram class. A histogram of percents rather than counts is convenient when we want to compare several distributions. To compare Gary with Los Angeles, a much bigger city, we would use percents so that both histograms have the same vertical scale.

EXAMPLE 1.6   College costs

clusters

Jeanna plans to attend college in her home state of Massachusetts. In the College Board's *Annual Survey of Colleges*, she finds data on estimated college costs for the 2002–2003 academic year. Figure 1.5 displays the costs for all 56 four-year colleges in Massachusetts (omitting art schools and other special colleges). As is often the case, we can't call this irregular distribution either symmetric or skewed. The big feature of the overall pattern is two separate **clusters** of colleges, 11 costing less than $16,000 and the remaining 45 costing more than $20,000. Clusters suggest that two types of individuals are mixed in the data set. In fact, the histogram distinguishes the 11 state colleges in Massachusetts from the 45 private colleges, which charge much more.

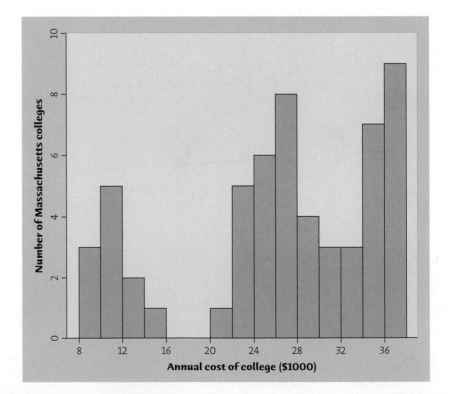

Figure 1.5 Histogram of the estimated costs (in thousands of dollars) for four-year colleges in Massachusetts. The two clusters distinguish public from private institutions.

The overall shape of a distribution is important information about a variable. Some types of data regularly produce distributions that are symmetric or skewed. For example, the sizes of living things of the same species (like lengths of crickets) tend to be symmetric. Data on incomes (whether of individuals, companies, or nations) are usually strongly skewed to the right. There are many moderate incomes, some large incomes, and a few very large incomes. Many distributions have irregular shapes that are neither symmetric nor skewed. Some data show other patterns, such as the clusters in Figure 1.5. Use your eyes and describe what you see.

APPLY YOUR KNOWLEDGE

1.6 **Sports car fuel economy.** Table 1.2 (page 12) gives data on the fuel economy of 2002 model sports cars. Your histogram (Exercise 1.5) shows an extreme high outlier. This is the Honda Insight, a hybrid gas-electric car that is quite different from the others listed. Make a new histogram of highway mileage, leaving out the Insight. Classes that are about 2 miles per gallon wide work well.

   (a) Describe the main features (shape, center, spread, outliers) of the distribution of highway mileage.

   (b) The government imposes a "gas guzzler" tax on cars with low gas mileage. Which of these cars do you think may be subject to the gas guzzler tax?

1.7 **College costs.** Describe the center (midpoint) and spread (smallest to largest) of the distribution of Massachusetts college costs in Figure 1.5. An overall description works poorly because of the clusters. A better description gives the center and spread of each cluster (public and private colleges) separately. Do this.

Quantitative variables: stemplots

Histograms are not the only graphical display of distributions. For small data sets, a *stemplot* is quicker to make and presents more detailed information.

STEMPLOT

To make a **stemplot:**

1. Separate each observation into a **stem,** consisting of all but the final (rightmost) digit, and a **leaf,** the final digit. Stems may have as many digits as needed, but each leaf contains only a single digit.

2. Write the stems in a vertical column with the smallest at the top, and draw a vertical line at the right of this column.

3. Write each leaf in the row to the right of its stem, in increasing order out from the stem.

Figure 1.6 Stemplot of the percents of Hispanic residents in the states. Each stem is a percent and leaves are tenths of a percent.

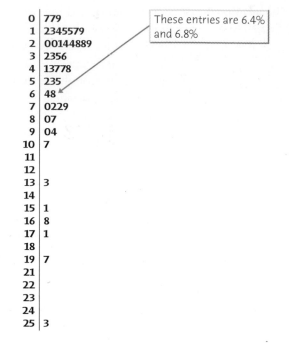

0	779
1	2345579
2	00144889
3	2356
4	13778
5	235
6	48
7	0229
8	07
9	04
10	7
11	
12	
13	3
14	
15	1
16	8
17	1
18	
19	7
21	
22	
23	
24	
25	3

These entries are 6.4% and 6.8%

The vital few?

Skewed distributions can show us where to concentrate our efforts. Ten percent of the cars on the road account for half of all carbon dioxide emissions. A histogram of CO_2 emissions would show many cars with small or moderate values and a few with very high values. Cleaning up or replacing these cars would reduce pollution at a cost much lower than that of programs aimed at all cars. Statisticians who work at improving quality in industry make a principle of this: distinguish "the vital few" from "the trivial many."

EXAMPLE 1.7   Making a stemplot

For the percents of Hispanic residents in Table 1.1, take the whole-number part of the percent as the stem and the final digit (tenths) as the leaf. The Massachusetts entry, 6.8%, has stem 6 and leaf 8. Wyoming, at 6.4%, places leaf 4 on the same stem. These are the only observations on this stem. We then arrange the leaves in order, as 48, so that 6 | 48 is one row in the stemplot. Figure 1.6 is the complete stemplot for the data in Table 1.1. To save space, we left out California, Texas, and New Mexico, which have stems 32 and 42.

A stemplot looks like a histogram turned on end. Compare the stemplot in Figure 1.6 with the histogram of the same data in Figure 1.3. Both show a single-peaked distribution that is strongly right-skewed and has some observations that we would probably call high outliers (three of these are left out of Figure 1.6). You can choose the classes in a histogram. The classes (the stems) of a stemplot are given to you. Figure 1.6 has more stems than there are classes in Figure 1.3. So histograms are more flexible. But the stemplot, unlike the histogram, preserves the actual value of each observation. Stemplots work well for small sets of data. Use a histogram to display larger data sets, like the 947 Iowa Test scores in Figure 1.4.

EXAMPLE 1.8   Pulling wood apart

Student engineers learn that although handbooks give the strength of a material as a single number, in fact the strength varies from piece to piece. A vital lesson in all fields of study is that "variation is everywhere." Here are data from a typical student

Figure 1.7 Stemplot of breaking strength of pieces of wood, rounded to the nearest hundred pounds. Stems are thousands of pounds and leaves are hundreds of pounds.

thousands   *hundreds*

```
23 | 0
24 | 0
25 |
26 | 5
27 |
28 | 7
29 |
30 | 259
31 | 399
32 | 033677
33 | 0236
```

laboratory exercise: the load in pounds needed to pull apart pieces of Douglas fir 4 inches long and 1.5 inches square.

33,190	31,860	32,590	26,520	33,280
32,320	33,020	32,030	30,460	32,700
23,040	30,930	32,720	33,650	32,340
24,050	30,170	31,300	28,730	31,920

We want to make a stemplot to display the distribution of breaking strength. To avoid many stems with only one leaf each, first **round** the data to the nearest hundred pounds. The rounded data are

rounding

332	319	326	265	333	323	330	320	305	327
230	309	327	336	323	240	302	313	287	319

Now it is easy to make a stemplot with the first two digits (thousands of pounds) as stems and the third digit (hundreds of pounds) as leaves. Figure 1.7 is the stemplot. The distribution is skewed to the left, with midpoint around 320 (32,000 pounds) and spread from 230 to 336.

You can also **split stems** to double the number of stems when all the leaves would otherwise fall on just a few stems. Each stem then appears twice. Leaves 0 to 4 go on the upper stem, and leaves 5 to 9 go on the lower stem. If you split the stems in the stemplot of Figure 1.7, for example, the 32 and 33 stems become

splitting stems

```
32 | 033
32 | 677
33 | 023
33 | 6
```

Rounding and splitting stems are matters for judgment, like choosing the classes in a histogram. The wood strength data require rounding but don't require splitting stems.

APPLY YOUR KNOWLEDGE

1.8 Students' attitudes. The Survey of Study Habits and Attitudes (SSHA) is a psychological test that evaluates college students' motivation, study habits, and attitudes toward school. A private college gives the SSHA to 18 of its incoming first-year women students. Their scores are

$$154 \quad 109 \quad 137 \quad 115 \quad 152 \quad 140 \quad 154 \quad 178 \quad 101$$
$$103 \quad 126 \quad 126 \quad 137 \quad 165 \quad 165 \quad 129 \quad 200 \quad 148$$

Make a stemplot of these data. The overall shape of the distribution is irregular, as often happens when only a few observations are available. Are there any outliers? About where is the center of the distribution (the score with half the scores above it and half below)? What is the spread of the scores (ignoring any outliers)?

1.9   **Alternative stemplots.** Return to the Hispanics data in Table 1.1 and Figure 1.6. Round each state's percent Hispanic to the nearest whole percent. Make a stemplot using tens of percents as stems and percents as leaves. All of the leaves fall on just five stems, 0, 1, 2, 3, and 4. Make another stemplot using split stems to increase the number of classes. With Figure 1.6, you now have three stemplots of the Hispanics data. Which do you prefer? Why?

Time plots

Many variables are measured at intervals over time. We might, for example, measure the height of a growing child or the price of a stock at the end of each month. In these examples, our main interest is change over time. To display change over time, make a *time plot*.

TIME PLOT

A **time plot** of a variable plots each observation against the time at which it was measured. Always put time on the horizontal scale of your plot and the variable you are measuring on the vertical scale. Connecting the data points by lines helps emphasize any change over time.

EXAMPLE 1.9   More on the cost of college

How have college tuition and fees changed over time? Table 1.3 gives the average tuition and fees paid by college students at four-year colleges, both public and private, from the 1971–1972 academic year to the 2001–2002 academic year. To compare dollar amounts across time, we must adjust for the changing buying power of the dollar. Table 1.3 gives tuition in *real dollars*, dollars that have constant buying power.[7] Average tuition in real dollars goes up only when the actual tuition rises by more than the overall cost of living. Figure 1.8 is a time plot of both public and private tuition.

TABLE 1.3 Average college tuition and fees, 1971–1972 to 2001–2002, in real dollars

Year	Private colleges	Public colleges	Year	Private colleges	Public colleges	Year	Private colleges	Public colleges
1971	7,851	1,622	1982	8,389	1,865	1992	13,012	2,907
1972	7,870	1,688	1983	8,882	2,002	1993	13,362	3,077
1973	7,572	1,667	1984	9,324	2,061	1994	13,830	3,192
1974	7,255	1,481	1985	9,984	2,150	1995	14,035	3,229
1975	7,272	1,386	1986	10,502	2,051	1996	14,514	3,323
1976	7,664	1,866	1987	10,799	2,275	1997	15,128	3,414
1977	7,652	1,856	1988	11,723	2,311	1998	15,881	3,506
1978	7,665	1,783	1989	12,110	2,371	1999	16,289	3,529
1979	7,374	1,687	1990	12,380	2,529	2000	16,456	3,535
1980	7,411	1,647	1991	12,601	2,706	2001	17,123	3,754
1981	7,758	1,714						

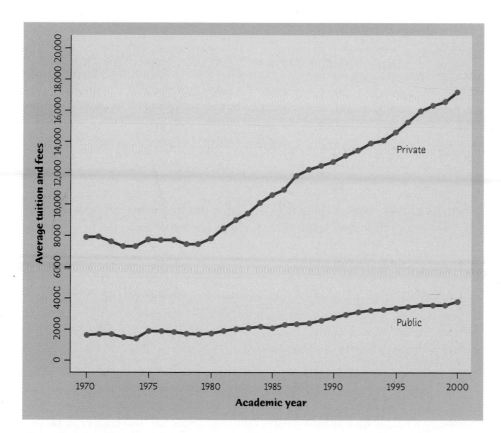

Figure 1.8 Time plot of the average tuition paid by students at public and private colleges for academic years 1970–1971 to 2001–2002.

trend

When you examine a time plot, look once again for an overall pattern and for strong deviations from the pattern. One common overall pattern is a **trend,** a long-term upward or downward movement over time. Figure 1.8 shows an upward trend in real college tuition costs, with no striking deviations such as short-term drops. It also shows that, beginning around 1980, private colleges raised tuition faster than public institutions, increasing the gap in costs between the two types of colleges.

time series
cross-sectional

Figures 1.5 and 1.8 both give information about college costs. The data for the time plot in Figure 1.8 are **time series data** that show the change in average tuition over time. The data for the histogram in Figure 1.5 are **cross-sectional data** that show the variation in costs (in one state) at a fixed time.

APPLY YOUR KNOWLEDGE

(Lester Lefkowitz/Corbis)

1.10 Vanishing landfills. The bar graphs in Figure 1.2 give cross-sectional data on municipal solid waste in 2000. Garbage that is not recycled is buried in landfills. Here are time series data that emphasize the need for recycling: the number of landfills operating in the United States in the years 1988 to 2000.[8]

Year	Landfills	Year	Landfills	Year	Landfills
1988	7924	1993	4482	1997	2514
1989	7379	1994	3558	1998	2314
1990	6326	1995	3197	1999	2216
1991	5812	1996	3091	2000	1967
1992	5386				

Make a time plot of these data. Describe the trend that your plot shows. Why does the trend emphasize the need for recycling?

Chapter 1 SUMMARY

A data set contains information on a number of **individuals.** Individuals may be people, animals, or things. For each individual, the data give values for one or more **variables.** A variable describes some characteristic of an individual, such as a person's height, gender, or salary.

Some variables are **categorical** and others are **quantitative.** A categorical variable places each individual into a category, like male or female. A quantitative variable has numerical values that measure some characteristic of each individual, like height in centimeters or salary in dollars per year.

Exploratory data analysis uses graphs and numerical summaries to describe the variables in a data set and the relations among them.

The **distribution** of a variable describes what values the variable takes and how often it takes these values.

To describe a distribution, begin with a graph. **Bar graphs** and **pie charts** describe the distribution of a categorical variable. **Histograms** and **stemplots** graph the distribution of a quantitative variable.

When examining any graph, look for an **overall pattern** and for notable **deviations** from the pattern.

Shape, center, and **spread** describe the overall pattern of a distribution. Some distributions have simple shapes, such as **symmetric** or **skewed.** Not all distributions have a simple overall shape, especially when there are few observations.

Outliers are observations that lie outside the overall pattern of a distribution. Always look for outliers and try to explain them.

When observations on a variable are taken over time, make a **time plot** that graphs time horizontally and the values of the variable vertically. A time plot can reveal **trends** or other changes over time.

Chapter 1 EXERCISES

1.11 Car colors in Japan. Exercise 1.3 (page 7) gives data on the most popular colors for motor vehicles made in North America. Here are similar data for 2001 model year vehicles made in Japan:[9]

Color	Percent
Gray	43%
White	35%
Black	8%
Blue	7%
Red	4%
Green	2%

What percent of Japanese vehicles have other colors? Make a graph of these data. What are the most important differences between choice of vehicle color in Japan and North America?

1.12 Deaths among young people. The number of deaths among persons aged 15 to 24 years in the United States in 2000 due to the leading causes of death for this age group were: accidents, 13,616; homicide, 4796; suicide, 3877; cancer, 1668; heart disease, 931; congenital defects, 425.[10]
(a) Make a bar graph to display these data.
(b) What additional information do you need to make a pie chart?

1.13 Athletes' salaries. Here is a small part of a data set that describes Major League Baseball players as of opening day of the 2002 season:

Player	Team	Position	Age	Salary
:				
Sosa, Jorge	Devil Rays	Pitcher	24	200,000
Sosa, Sammy	Cubs	Outfield	33	15,000,000
Speier, Justin	Rockies	Pitcher	28	310,000
Spivey, Junior	Diamondbacks	Infield	27	215,000
:				

(a) What individuals does this data set describe?

(b) In addition to the player's name, how many variables does the data set contain? Which of these variables are categorical and which are quantitative?

(c) Based on the data in the table, what do you think are the units of measurement for each of the quantitative variables?

1.14 Mutual funds. Here is information on several Vanguard Group mutual funds:

Fund	Number of stocks held	Largest holding	Annual return (10 years)
500 Index Fund	508	General Electric	10.01%
Equity Income Fund	167	ExxonMobil	11.96%
Health Care Fund	128	Pharmacia	20.27%
International Value Fund	84	Mazda Motor	5.04%
Precious Metals Fund	26	Barrick Gold	2.50%

In addition to the fund name, how many variables are recorded for each fund? Which variables are categorical and which are quantitative?

1.15 Reading a pie chart. Figure 1.9 is a pie chart prepared by the Census Bureau to show the origin of the 35.3 million Hispanics in the United States, according to the 2000 census.[11] About what percent of Hispanics are Mexican? Puerto Rican? You see that it is hard to read numbers from a pie chart. Bar graphs are much easier to use.

1.16 Do adolescent girls eat fruit? We all know that fruit is good for us. Many of us don't eat enough. Figure 1.10 is a histogram of the number of servings of fruit per day claimed by 74 seventeen-year-old girls in a study in Pennsylvania.[12] Describe the shape, center, and spread of this distribution. What percent of these girls ate fewer than two servings per day?

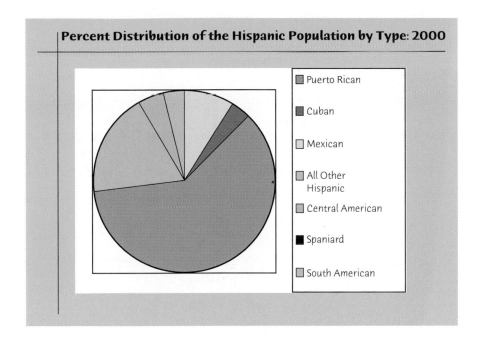

Figure 1.9 Pie chart of the origins of Hispanic residents of the United States, for Exercise 1.15. (Data from U.S. Census Bureau.)

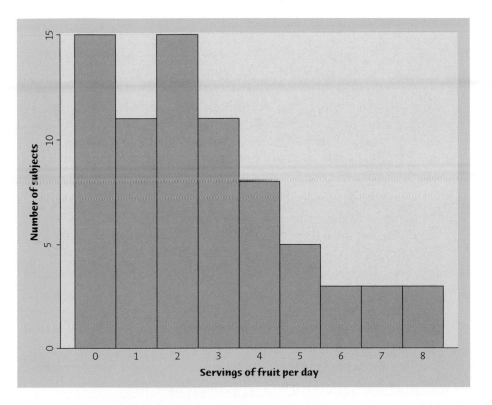

Figure 1.10 The distribution of fruit consumption in a sample of 74 seventeen-year-old girls, for Exercise 1.16.

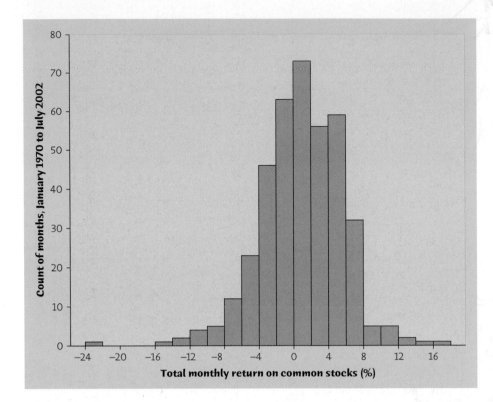

Figure 1.11 The distribution of monthly percent returns on U.S. common stocks from January 1970 to July 2002, for Exercise 1.17.

1.17 **Returns on common stocks.** The return on a stock is the change in its market price plus any dividend payments made. Total return is usually expressed as a percent of the beginning price. Figure 1.11 is a histogram of the distribution of the monthly returns for all stocks listed on U.S. markets from January 1970 to July 2002 (391 months).[13] The low outlier is the market crash of October 1987, when stocks lost more than 22% of their value in one month.

(a) Describe the overall shape of the distribution of monthly returns.

(b) What is the approximate center of this distribution? (For now, take the center to be the value with roughly half the months having lower returns and half having higher returns.)

(c) Approximately what were the smallest and largest monthly returns, leaving out the outlier? (This describes the spread of the distribution.)

(d) A return less than zero means that stocks lost value in that month. About what percent of all months had returns less than zero?

1.18 **Weight of newborns.** Here is the distribution of the weight at birth for all babies born in the United States in 1999:[14]

Weight	Count	Weight	Count
Less than 500 grams	5,912	3,000 to 3,499 grams	1,470,019
500 to 999 grams	22,815	3,500 to 3,999 grams	1,137,401
1,000 to 1,499 grams	28,750	4,000 to 4,499 grams	332,863
1,500 to 1,999 grams	59,531	4,500 to 4,999 grams	53,751
2,000 to 2,499 grams	184,175	5,000 to 5,499 grams	6,069
2,500 to 2,999 grams	653,327		

(a) For comparison with other years and with other countries, we prefer a histogram of the *percents* in each weight class rather than the counts. Explain why.

(b) Make a histogram of the distribution, using percents on the vertical scale.

(c) A "low-birth-weight" baby is one weighing less than 2,500 grams. Low birth weight is tied to many health problems. What percent of all births were low birth weight babies?

1.19 Marijuana and traffic accidents. Researchers in New Zealand interviewed 907 drivers at age 21. They had data on traffic accidents and they asked their subjects about marijuana use. Here are data on the numbers of accidents caused by these drivers at age 19, broken down by marijuana use at the same age:[15]

	Marijuana use per year			
	Never	1–10 times	11–50 times	51+ times
Drivers	452	229	70	156
Accidents caused	59	36	15	50

(a) Explain carefully why a useful graph must compare *rates* (accidents per driver) rather than counts of accidents in the four marijuana use classes.

(b) Make a graph that displays the accident rate for each class. What do you conclude? (You can't conclude that marijuana use *causes* accidents, because risk-takers are more likely both to drive aggressively and to use marijuana.)

1.20 Lions feeding. Feeding at a carcass leads to competition among lions. Ecologists collected data on feeding contests in Serengeti National Park, Tanzania.[16] In each contest, a lion feeding at a carcass is challenged by another lion seeking to take its place. Who wins these contests tells us something about lion society. The following table presents data on

396 contests between an adult lion (female or male) and an opponent of a different class:

Opponent	Adult male		Adult female	
	Contests	Male wins	Contests	Female wins
Adult female	136	113		
Subadult	8	8	53	20
Yearling	13	12	8	5
Cub	25	22	153	115

Make separate graphs for males and females that compare their success against different classes of opponent. Explain why you decided to make the type of graph you chose. Then describe the most important differences between the behavior of female and male lions in feeding contests.

1.21 Name that variable. Figure 1.12 displays four histograms without axis markings. They are the distributions of these four variables:[17]

1. The gender of the students in a large college course, recorded as 0 for male and 1 for female.

2. The heights of the students in the same class.

3. The handedness of students in the class, recorded as 0 for right-handed and 1 for left-handed.

4. The lengths of words used in Shakespeare's plays.

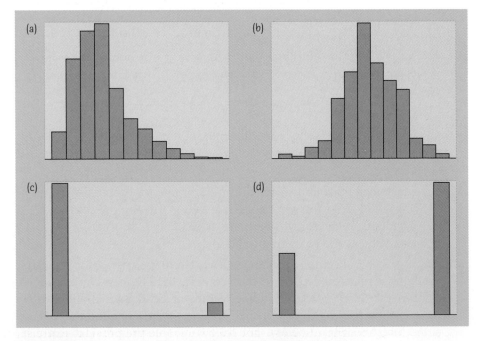

Figure 1.12 Histograms of four distributions, for Exercise 1.21.

TABLE 1.4 Percent of state residents living in poverty, 2000

State	Percent	State	Percent	State	Percent
Alabama	14.6	Maryland	7.3	Pennsylvania	9.9
Connecticut	7.6	Massachusetts	10.2	Rhode Island	10.0
Delaware	9.8	Michigan	10.2	South Carolina	11.9
Florida	12.1	Mississippi	15.5	Tennessee	13.3
Georgia	12.6	New Hampshire	7.4	Vermont	10.1
Illinois	10.5	New Jersey	8.1	Virginia	8.1
Indiana	8.2	New York	14.7	West Virginia	15.8
Kentucky	12.5	North Carolina	13.2	Wisconsin	8.8
Maine	9.8	Ohio	11.1		

Identify which distribution each histogram describes. Explain the reasoning behind your choices.

1.22 Dates on coins. Sketch a histogram for a distribution that is skewed to the left. Suppose that you and your friends emptied your pockets of coins and recorded the year marked on each coin. The distribution of dates would be skewed to the left. Explain why.

1.23 Poverty in the states. Table 1.4 gives the percents of people living below the poverty line in the 26 states east of the Mississippi River.[18] Make a stemplot of these data. Is the distribution roughly symmetric, skewed to the right, or skewed to the left? Which states (if any) are outliers?

1.24 Split the stems. Make another stemplot of the poverty data in Table 1.4, splitting the stems to double the number of classes. Do you prefer this stemplot or that from the previous exercise? Why?

1.25 Babe Ruth's home runs. Here are the numbers of home runs that Babe Ruth hit in his 15 years with the New York Yankees, 1920 to 1934:

54 59 35 41 46 25 47 60 54 46 49 46 41 34 22

Make a stemplot for these data. Is the distribution roughly symmetric, clearly skewed, or neither? About how many home runs did Ruth hit in a typical year? Is his famous 60 home runs in 1927 an outlier?

1.26 Back-to-back stemplot. A leading contemporary home run hitter is Mark McGwire, who retired after the 2001 season. Here are McGwire's home run counts for 1987 to 2001.

49 32 33 39 22 42 9 9 39 52 58 70 65 32 29

A *back-to-back stemplot* helps us compare two distributions. Write the stems as usual, but with a vertical line both to their left and to their right. On the right, put leaves for Babe Ruth (see the previous exercise). On the left, put leaves for McGwire. Arrange the leaves on each stem in

(National Baseball Hall of Fame and Museum, Inc., Cooperstown, N.Y.)

TABLE 1.5 Women's winning times (minutes) in the Boston Marathon

Year	Time	Year	Time	Year	Time
1972	190	1982	150	1992	144
1973	186	1983	143	1993	145
1974	167	1984	149	1994	142
1975	162	1985	154	1995	145
1976	167	1986	145	1996	147
1977	168	1987	146	1997	146
1978	165	1988	145	1998	143
1979	155	1989	144	1999	143
1980	154	1990	145	2000	146
1981	147	1991	144	2001	144
				2002	141

increasing order out from the stem. Now write a brief comparison of Ruth and McGwire as home run hitters. McGwire was injured in 1993 and there was a baseball strike in 1994. How do these events appear in the data?

1.27 **The Boston Marathon.** Women were allowed to enter the Boston Marathon in 1972. The times (in minutes, rounded to the nearest minute) for the winning woman from 1972 to 2002 appear in Table 1.5. In 2002, Margaret Okayo of Kenya set a new women's record for the race of 2 hours, 20 minutes, and 43 seconds.

(a) Make a time plot of the winning times.

(b) Give a brief description of the pattern of Boston Marathon winning times over these years. Has the rate of improvement in times slowed in recent years?

1.28 **Watch those scales!** The impression that a time plot gives depends on the scales you use on the two axes. If you stretch the vertical axis and compress the time axis, change appears to be more rapid. Compressing the vertical axis and stretching the time axis make change appear slower. Make two more time plots of the data on public college tuition in Table 1.3, one that makes tuition appear to increase very rapidly and one that shows only a gentle increase. The moral of this exercise is: pay close attention to the scales when you look at a time plot.

1.29 **Where are the doctors?** Table 1.6 gives the number of medical doctors per 100,000 people in each state.[19]

(a) Why is the number of doctors per 100,000 people a better measure of the availability of health care than a simple count of the number of doctors in a state?

(b) Make a graph that displays the distribution of doctors per 100,000 people. Write a brief description of the distribution. Are there any outliers? If so, can you explain them?

TABLE 1.6 Medical doctors per 100,000 people, by state (1999)

State	Doctors	State	Doctors	State	Doctors
Alabama	200	Louisiana	251	Ohio	237
Alaska	170	Maine	232	Oklahoma	167
Arizona	203	Maryland	379	Oregon	227
Arkansas	192	Massachusetts	422	Pennsylvania	293
California	248	Michigan	226	Rhode Island	339
Colorado	244	Minnesota	254	South Carolina	213
Connecticut	361	Mississippi	164	South Dakota	188
Delaware	238	Missouri	232	Tennessee	248
Florida	243	Montana	191	Texas	205
Georgia	211	Nebraska	221	Utah	202
Hawaii	269	Nevada	177	Vermont	313
Idaho	155	New Hampshire	234	Virginia	243
Illinois	263	New Jersey	301	Washington	237
Indiana	198	New Mexico	214	West Virginia	219
Iowa	175	New York	395	Wisconsin	232
Kansas	204	North Carolina	237	Wyoming	172
Kentucky	212	North Dakota	224	D.C.	758

1.30 Orange prices. Figure 1.13 is a time plot of the average price of fresh oranges each month during the decade from 1992 to 2002.[20] The prices are "index numbers" given as percents of the average price during 1982 to 1984.

(a) The most notable pattern in this time plot is **seasonal variation,** regular up-and-down movements that occur at about the same time each year. Why should we expect the price of fresh oranges to show seasonal variation?

seasonal variation

(b) Is there a longer-term trend visible under the seasonal variation? If so, describe it.

Chapter 1 MEDIA EXERCISES

The Internet is now the first place to look for data. Many of the data sets in this chapter were found online. Exercises 1.31 to 1.33 illustrate some places to find data on the Web.

1.31 No place like home? The Census Bureau Web site, www.census.gov, is the mother lode of data about America and Americans. On the home page, you can select a state and then a county within that state. Select your home county and then the county in which your school is located. For each county, record the population, the percent population change in the decade 1990 to 2000, and the percents of Asian, black, and white residents. Also calculate the number of people age 25 and over with

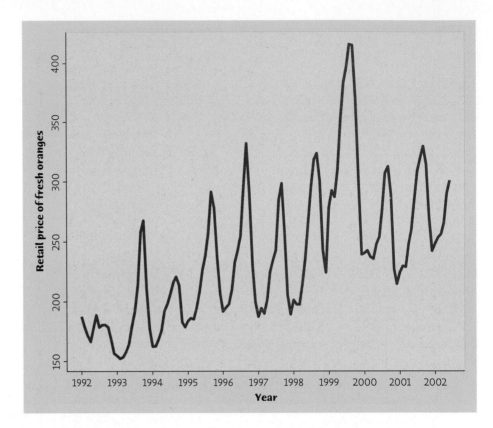

Figure 1.13 Time plot of the monthly average price of fresh oranges, for Exercise 1.30.

college degrees as a percent of the county's population. What are the main differences between your home county and your school's county?

1.32 Canada and the U.S.A. What are the very latest estimates of the populations of Canada and the United States? Find them at Statistics Canada, www.statcan.ca , and the U.S. Census Bureau, www.census.gov . What are the latest unemployment rates in the two countries? Canada, like most nations, has a single statistical agency, so the Statistics Canada site is the place to look. In the United States, statistical agencies are attached to many government departments. For unemployment data, go to the Bureau of Labor Statistics, www.bls.gov .

1.33 Current issues. The University of Chicago's National Opinion Research Center carries out polls of public opinion on many issues, as well as other studies. At www.norc.org you will find recent reports and press releases. For example, late in 2002 the site featured reports on opinions about owning and regulating guns, reactions to the September 11, 2001, attack on the World Trade Center, and rating hospitals. Choose a topic that interests you among those on the home page and browse until you find data that you think shed light on the topic. (For example, 76.9% of those polled in 2001 wanted mandatory registration of handguns.) Report the data and say why you think they are important.

1.34 How histograms behave. The data set menu that accompanies the *One-Variable Statistical Calculator* applet includes the data on Hispanics in the states from Table 1.1. Choose these data, then click on the "Histogram" tab to see a histogram.

(a) How many classes does the applet choose to use? (You can click on the graph outside the bars to get a count of classes.)

(b) Click on the graph and drag to the left. What is the smallest number of classes you can get? What are the lower and upper bounds of each class? (Click on the bar to find out.) Make a rough sketch of this histogram.

(c) Click and drag to the right. What is the greatest number of classes you can get? How many observations does the largest class have?

(d) You see that the choice of classes changes the appearance of a histogram. Drag back and forth until you get the histogram you think best displays the distribution. How many classes did you use?

1.35 Choices in a stemplot. The data set menu that accompanies the *One-Variable Statistical Calculator* applet includes the data on Hispanics in the states from Table 1.1. Choose these data, then click on the "Stemplot" tab to see a stemplot.

(a) The stemplot looks quite different from that in Figure 1.6. Make a copy of this stemplot, and explain carefully the reason for the difference.

(b) Figure 1.6 has 26 stems and would have 43 stems if we extended it to include New Mexico. The applet's plot has many fewer. Check the "Split stems" box to increase the number of stems used by the applet. Make a copy of this stemplot as well. You now have three stemplots for these data. Which do you prefer, and why?

1.36 Acorns. How big are acorns? It depends on the species of oak tree that produces them. The EESEE story "Acorn Size and Oak Tree Range" contains data on the average size (in cubic centimeters) of acorns from 39 species of oaks. Make a stemplot of the acorn size data. Describe the distribution carefully (shape, center, spread, outliers).

1.37 Eruptions of Old Faithful. The EESEE story "Is Old Faithful Faithful?" contains data on eruptions of the famous Old Faithful geyser in Yellowstone National Park. The variable named "Duration" records how long 299 of these eruptions lasted, in minutes. Use your software to make a histogram of the durations. The shape of the distribution is distinctive and interesting. Describe the shape, center, and spread of the distribution.

(Tony Arruza/Corbis)

Describing Distributions with Numbers

How much do people with a bachelor's degree (but no higher degree) earn? Here are the incomes of 15 such people, chosen at random by the Census Bureau in March 2002 and asked how much they earned in 2001.[1] Most people reported their incomes to the nearest thousand dollars, so we have rounded their responses to thousands of dollars.

    110   25   50   50   55   30   35   30   4   32   50   30   32   74   60

Figure 2.1 is a stemplot of these amounts. The distribution is irregular in shape, as is common when we have only a few observations. There is one high outlier, a person who made $110,000. Our goal in this chapter is to describe with numbers the center and spread of this and other distributions.

Measuring center: the mean

The most common measure of center is the ordinary arithmetic average, or *mean*.

Figure 2.1 Stemplot of the earnings (in thousands of dollars) of 15 people chosen at random from all people with a bachelor's degree but no higher degree.

```
 0 | 4
 1 |
 2 | 5
 3 | 000125
 4 |
 5 | 0005
 6 | 0
 7 | 4
 8 |
 9 |
10 |
11 | 0
```

THE MEAN $\bar{x}$

To find the **mean** of a set of observations, add their values and divide by the number of observations. If the n observations are $x_1, x_2, \ldots, x_n$, their mean is

$$\bar{x} = \frac{x_1 + x_2 + \cdots + x_n}{n}$$

or in more compact notation,

$$\bar{x} = \frac{1}{n} \sum x_i$$

The Σ (capital Greek sigma) in the formula for the mean is short for "add them all up." The subscripts on the observations x_i are just a way of keeping the n observations distinct. They do not necessarily indicate order or any other special facts about the data. The bar over the x indicates the mean of all the x-values. Pronounce the mean $\bar{x}$ as "x-bar." This notation is very common. When writers who are discussing data use $\bar{x}$ or $\bar{y}$, they are talking about a mean.

EXAMPLE 2.1   Earnings of college graduates

The mean earnings for our 15 college graduates are

$$\bar{x} = \frac{r_1 + r_2 + \cdots + r_n}{n}$$

$$= \frac{110 + 25 + \cdots + 60}{15}$$

$$= \frac{666}{15} = 44.4, \text{ or } \$44,400$$

In practice, you can key the data into your calculator and hit the mean key. You don't have to actually add and divide. But you should know that this is what the calculator is doing.

If we leave out the one high income, $110,000, the mean for the remaining 14 people is $39,700. The lone outlier raises the mean income of the group by $4700.

resistant measure

Example 2.1 illustrates an important fact about the mean as a measure of center: it is sensitive to the influence of a few extreme observations. These may be outliers, but a skewed distribution that has no outliers will also pull the mean toward its long tail. Because the mean cannot resist the influence of extreme observations, we say that it is not a **resistant** measure of center.

APPLY YOUR KNOWLEDGE

2.1   **Sports car gas mileage.** Table 1.2 (page 12) gives the gas mileages for the 22 two-seater cars listed in the government's fuel economy guide.

(a) Find the mean highway gas mileage from the formula for the mean. Then enter the data into your calculator and use the calculator's $\overline{x}$ button to obtain the mean. Verify that you get the same result.

(b) The Honda Insight is an outlier that doesn't belong with the other cars. Use your calculator to find the mean of the 21 cars that remain if we leave out the Insight. How does the outlier change the mean?

Measuring center: the median

In Chapter 1, we used the midpoint of a distribution as an informal measure of center. The *median* is the formal version of the midpoint, with a specific rule for calculation.

THE MEDIAN M

The **median** M is the midpoint of a distribution, the number such that half the observations are smaller and the other half are larger. To find the median of a distribution:

1.   Arrange all observations in order of size, from smallest to largest.

2.   If the number of observations n is odd, the median M is the center observation in the ordered list. Find the location of the median by counting $(n + 1)/2$ observations up from the bottom of the list.

3.   If the number of observations n is even, the median M is the mean of the two center observations in the ordered list. The location of the median is again $(n + 1)/2$ from the bottom of the list.

Note that the formula $(n + 1)/2$ does *not* give the median, just the location of the median in the ordered list. Medians require little arithmetic, so they are easy to find by hand for small sets of data. Arranging even a moderate number of observations in order is very tedious, however, so that finding the median by hand for larger sets of data is unpleasant. Even simple calculators have an $\overline{x}$

button, but you will need to use software or a graphing calculator to automate finding the median.

EXAMPLE 2.2   Finding the median: odd n

What are the median earnings for our 15 college graduates? Here are the data arranged in order:

4   25   30   30   30   31   32   **35**   50   50   50   55   60   74   110

The count of observations $n = 15$ is odd. The bold **35** is the center observation in the ordered list, with 7 observations to its left and 7 to its right. This is the median, $M = 35$.

Because $n = 15$, our rule for the location of the median gives

$$\text{location of M} = \frac{n+1}{2} = \frac{16}{2} = 8$$

That is, the median is the 8th observation in the ordered list. It is faster to use this rule than to locate the center by eye.

EXAMPLE 2.3   Finding the median: even n

How much does the high outlier affect the median? Drop the 110 from the ordered list and find the median of the remaining 14 incomes. The data are

4   25   30   30   30   31   **32** | **35**   50   50   50   55   60   74

There is no center observation, but there is a center pair. These are the bold **32** and **35** in the list, which have 6 observations to their left in the list and 6 to their right. The median is midway between these two observations:

$$M = \frac{32 + 35}{2} = 33.5$$

With $n = 14$, the rule for locating the median in the list gives

$$\text{location of M} = \frac{n+1}{2} = \frac{15}{2} = 7.5$$

The location 7.5 means "halfway between the 7th and 8th observations in the ordered list." That agrees with what we found by eye.

Comparing the mean and the median

Examples 2.1 to 2.3 illustrate an important difference between the mean and the median. The single high income pulls the mean up by $4700. It moves the median by only $1500. The median, unlike the mean, is *resistant*. If the high earner's income rose from 110 to 1100 (that is, from $110,000 to $1,100,000) the median would not change at all. The 1100 just counts as one observation above the center, no matter how far above the center it lies. The mean uses the actual value of each observation and so will chase a single large observation upward. The *Mean and Median* applet is an excellent way to compare the resistance of M and $\bar{x}$.

Poor New York?

Is New York a rich state? New York's mean income per person ranks eighth among the states, not far behind its rich neighbors Connecticut and New Jersey, which rank first and second. But while Connecticut and New Jersey rank third and fifth in median household income, New York stands 27th, below the national average. What's going on? Just another example of mean versus median. New York has many very highly paid people, who pull up its mean income per person. But it also has a higher proportion of poor households than do New Jersey and Connecticut, and this brings the median down. New York is not a rich state—it's a state with extremes of wealth and poverty.

APPLET

> **COMPARING THE MEAN AND THE MEDIAN**
>
> The mean and median of a symmetric distribution are close together. If the distribution is exactly symmetric, the mean and median are exactly the same. In a skewed distribution, the mean is farther out in the long tail than is the median.

Distributions of incomes are usually skewed to the right—there are many modest incomes and a few very high incomes. For example, the Census Bureau survey in March 2002 interviewed 16,018 people aged 25 to 65 who were in the labor force full-time in 2001 and who were college graduates but had only a bachelor's degree. We used 15 of these 16,018 incomes to introduce the mean and median. The median income for the entire group was $45,769. The mean of the same 16,018 incomes was much higher, $59,852. Reports about incomes and other strongly skewed distributions usually give the median ("midpoint") rather than the mean ("arithmetic average"). However, a county that is about to impose a tax of 1% on the incomes of its residents cares about the mean income, not the median. The tax revenue will be 1% of total income, and the total is the mean times the number of residents. The mean and median measure center in different ways, and both are useful.

APPLY YOUR KNOWLEDGE

2.2   Sports car gas mileage. What is the median highway mileage for the 22 two-seater cars listed in Table 1.2 (page 12)? What is the median of the 21 cars that remain if we remove the Honda Insight? Compare the effect of the Insight on mean mileage (Exercise 2.1) and on the median mileage. What general fact about the mean and median does this comparison illustrate?

2.3   House prices. The mean and median selling price of existing single-family homes sold in June 2002 were $163,900 and $210,900. Which of these numbers is the mean and which is the median? Explain how you know.

2.4   Barry Bonds. The major league baseball single-season home run record is held by Barry Bonds of the San Francisco Giants, who hit 73 in 2001. Here are Bonds's home run totals from 1986 (his first year) to 2002:

16   25   24   19   33   25   34   46   37   33   42   40   37   34   49   73   46

Bonds's record year is a high outlier. How do his career mean and median number of home runs change when we drop the record 73? What general fact about the mean and median does your result illustrate?

Measuring spread: the quartiles

The mean and median provide two different measures of the center of a distribution. But a measure of center alone can be misleading. The Census Bureau reports that in 2001 the median income of American households was $42,228. Half of all households had incomes below $42,228, and half had higher incomes. The mean income of these same households was $58,208. The mean is higher than the median because the distribution of incomes is skewed to the right. But the median and mean don't tell the whole story. The bottom 20% of households had incomes less than $17,970 and households in the top 5% took in more than $150,499.[2] We are interested in the *spread* or *variability* of incomes as well as their center. The simplest useful numerical description of a distribution consists of both a measure of center and a measure of spread.

One way to measure spread is to give the smallest and largest observations. For example, the incomes of our 15 college graduates range from $4000 to $110,000. These single observations show the full spread of the data, but they may be outliers. We can improve our description of spread by also looking at the spread of the middle half of the data. The *quartiles* mark out the middle half. Count up the ordered list of observations, starting from the smallest. The *first quartile* lies one-quarter of the way up the list. The *third quartile* lies three-quarters of the way up the list. In other words, the first quartile is larger than 25% of the observations, and the third quartile is larger than 75% of the observations. The second quartile is the median, which is larger than 50% of the observations. That is the idea of quartiles. We need a rule to make the idea exact. The rule for calculating the quartiles uses the rule for the median.

THE QUARTILES Q_1 AND Q_3

To calculate the **quartiles:**

1. Arrange the observations in increasing order and locate the median M in the ordered list of observations.

2. The **first quartile Q_1** is the median of the observations whose position in the ordered list is to the left of the location of the overall median.

3. The **third quartile Q_3** is the median of the observations whose position in the ordered list is to the right of the location of the overall median.

Here are examples that show how the rules for the quartiles work for both odd and even numbers of observations.

EXAMPLE 2.4   Finding the quartiles: odd *n*

Our sample of 15 incomes of college graduates, arranged in increasing order, is

4   25   30   (30)   30   31   32   (**35**)   50   50   50   (55)   60   74   110

There is an odd number of observations, so the median is the middle one, the bold 35 in the list. The first quartile is the median of the 7 observations to the left of the median. This is the 4th of these 7 observations, so $Q_1 = 30$. If you want, you can use the recipe for the location of the median with $n = 7$:

$$\text{location of } Q_1 = \frac{n+1}{2} = \frac{7+1}{2} = 4$$

The third quartile is the median of the 7 observations to the right of the median, $Q_3 = 55$. The overall median is left out of the calculation of the quartiles when there is an odd number of observations.

Notice that the quartiles are resistant. For example, Q_3 would have the same value if the outlier were 1100 rather than 110.

EXAMPLE 2.5   Finding the quartiles: even n

Here, from the same government survey, are the earnings in 2001 of 16 randomly chosen people who have high school diplomas but no college. For convenience we have arranged the incomes in increasing order.

5   6   12   19   20   21   22   24 | 25   31   32   40   43   43   47   67

There is an even number of observations, so the median lies midway between the middle pair, the 8th and 9th in the list. Its value is M = 24.5. We have marked the location of the median by |. The first quartile is the median of the first 8 observations, because these are the observations to the left of the location of the median. Check that $Q_1 = 19.5$ and $Q_3 = 41.5$. When the number of observations is even, all the observations enter into the calculation of the quartiles.

Be careful when, as in these examples, several observations take the same numerical value. Write down all of the observations and apply the rules just as if they all had distinct values. Some software packages use a slightly different rule to find the quartiles, so computer results may be a bit different from your own work. Don't worry about this. The differences will always be too small to be important.

The five-number summary and boxplots

The smallest and largest observations tell us little about the distribution as a whole, but they give information about the tails of the distribution that is missing if we know only Q_1, M, and Q_3. To get a quick summary of both center and spread, combine all five numbers.

THE FIVE-NUMBER SUMMARY

The **five-number summary** of a distribution consists of the smallest observation, the first quartile, the median, the third quartile, and the largest observation, written in order from smallest to largest. In symbols, the five-number summary is

Minimum Q_1 M Q_3 Maximum

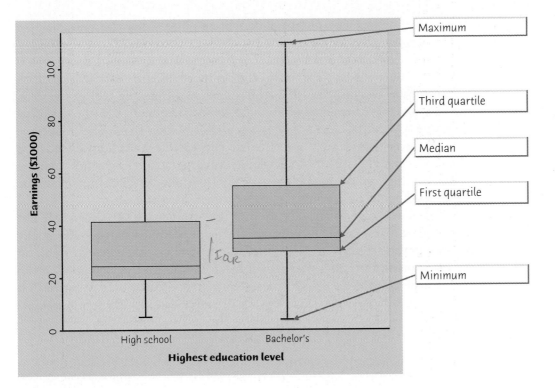

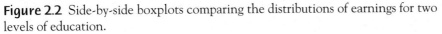

Figure 2.2 Side-by-side boxplots comparing the distributions of earnings for two levels of education.

These five numbers offer a reasonably complete description of center and spread. The five-number summaries from Examples 2.4 and 2.5 are

$$4 \quad 30 \quad 35 \quad 55 \quad 110$$

for people with a bachelor's degree, and

$$5 \quad 19.5 \quad 24.5 \quad 41.5 \quad 67$$

for people with only a high school diploma. The five-number summary of a distribution leads to a new graph, the *boxplot*. Figure 2.2 shows boxplots for the comparison of earnings.

BOXPLOT

A **boxplot** is a graph of the five-number summary.

- A central box spans the quartiles Q_1 and Q_3.
- A line in the box marks the median M.
- Lines extend from the box out to the smallest and largest observations.

Because boxplots show less detail than histograms or stemplots, they are best used for side-by-side comparison of more than one distribution, as in Figure 2.2. Be sure to include a numerical scale in the graph. When you look at a boxplot, first locate the median, which marks the center of the distribution. Then look at the spread. The quartiles show the spread of the middle half of the data, and the extremes (the smallest and largest observations) show the spread of the entire data set. We see from Figure 2.2 that holders of a bachelor's degree as a group earn considerably more than people with no education beyond high school. For example, the first quartile for college graduates is higher than the median for high school grads. The spread of the middle half of incomes (the box in the boxplot) is roughly the same for both groups.

A boxplot also gives an indication of the symmetry or skewness of a distribution. In a symmetric distribution, the first and third quartiles are equally distant from the median. In most distributions that are skewed to the right, on the other hand, the third quartile will be farther above the median than the first quartile is below it. That is the case for both distributions in Figure 2.2. The extremes behave the same way, but remember that they are just single observations and may say little about the distribution as a whole.

APPLY YOUR KNOWLEDGE

2.5 Pulling wood apart. Example 1.8 (page 16) gives the breaking strengths of 20 pieces of Douglas fir.

  (a) Give the five-number summary of the distribution of breaking strengths. (The stemplot, Figure 1.7, helps, because it arranges the data in order, but you should use the unrounded values in numerical work.)

  (b) The stemplot shows that the distribution is skewed to the left. Does the five-number summary show the skew? Remember that only a graph gives a clear picture of the shape of a distribution.

2.6 Midsize car gas mileage. Table 2.1 gives the city and highway gas mileage for 36 midsize cars from the 2002 model year.[3] There is one low outlier, the 12-cylinder Rolls-Royce. We wonder if midsize sedans get better mileage than sports cars.

  (a) Find the five-number summaries for both city and highway mileage for the midsize cars in Table 2.1 and for the two-seater cars in Table 1.2. (Leave out the Honda Insight.)

  (b) Make four side-by-side boxplots to display the summaries. Write a brief description of city versus highway mileage and two-seaters versus midsize cars.

2.7 How old are presidents? How old are presidents at their inauguration? Was Bill Clinton, at age 46, unusually young? Table 2.2 gives the data, the ages of all U.S. presidents when they took office.

  (a) Make a stemplot of the distribution of ages. From the shape of the distribution, do you expect the median to be much less than the

TABLE 2.1 Gas mileage (miles per gallon) for 2002 model midsize cars

Model	City	Highway	Model	City	Highway
Acura 3.2TL	19	29	Mazda 626	22	28
Audi A6	19	25	Mercedes-Benz E320	20	28
Buick Century	20	29	Mercedes-Benz E430	17	24
Buick Regal	18	27	Mercury Sable	20	28
Cadillac Seville	18	27	Mitsubishi Diamante	18	25
Chevrolet Malibu	20	29	Mitsubishi Galant	21	28
Chevrolet Monte Carlo	21	32	Nissan Altima	23	29
Chrysler Sebring	21	30	Nissan Maxima	20	26
Daewoo Leganza	20	28	Oldsmobile Aurora	18	27
Honda Accord	23	30	Oldsmobile Intrigue	20	30
Hyundai Sonata	22	28	Pontiac Grand Prix	18	28
Infiniti I35	20	26	Rolls-Royce Silver Seraph	12	16
Infiniti Q45	17	25	Saab 9-3	12	29
Jaguar S-Type	18	25	Saab 9-5	20	30
Kia Optima	20	27	Saturn L100/200	24	33
Lexus ES300	21	29	Saturn L300	21	29
Lexus LS430	18	25	Toyota Camry	23	32
Lincoln LS	17	24	Volkswagen Passat	22	31

TABLE 2.2 Ages of the presidents at inauguration

President	Age	President	Age	President	Age
Washington	57	Lincoln	52	Hoover	54
J. Adams	61	A. Johnson	56	F. D. Roosevelt	51
Jefferson	57	Grant	46	Truman	60
Madison	57	Hayes	54	Eisenhower	61
Monroe	58	Garfield	49	Kennedy	43
J. Q. Adams	57	Arthur	51	L. B. Johnson	55
Jackson	61	Cleveland	47	Nixon	56
Van Buren	54	B. Harrison	55	Ford	61
W. H. Harrison	68	Cleveland	55	Carter	52
Tyler	51	McKinley	54	Reagan	69
Polk	49	T. Roosevelt	42	G. H. W. Bush	64
Taylor	64	Taft	51	Clinton	46
Fillmore	50	Wilson	56	G. W. Bush	54
Pierce	48	Harding	55		
Buchanan	65	Coolidge	51		

mean, about the same as the mean, or much greater than the mean?

(b) Find the mean and the five-number summary. Verify your expectation about the median.

(c) What is the range of the middle half of the ages of new presidents? Was Bill Clinton in the youngest 25%?

Measuring spread: the standard deviation

The five-number summary is not the most common numerical description of a distribution. That distinction belongs to the combination of the mean to measure center and the *standard deviation* to measure spread. The standard deviation measures spread by looking at how far the observations are from their mean.

THE STANDARD DEVIATION s

The **variance s^2** of a set of observations is an average of the squares of the deviations of the observations from their mean. In symbols, the variance of n observations $x_1, x_2, \ldots, x_n$ is

$$s^2 = \frac{(x_1 - \bar{x})^2 + (x_2 - \bar{x})^2 + \cdots + (x_n - \bar{x})^2}{n - 1}$$

or, more compactly,

$$s^2 = \frac{1}{n - 1} \sum (x_i - \bar{x})^2$$

The **standard deviation s** is the square root of the variance s^2:

$$s = \sqrt{\frac{1}{n - 1} \sum (x_i - \bar{x})^2}$$

In practice, use software or your calculator to obtain the standard deviation from keyed-in data. Doing an example step-by-step will help you understand how the variance and standard deviation work, however.

EXAMPLE 2.6   *Calculating the standard deviation*

A person's metabolic rate is the rate at which the body consumes energy. Metabolic rate is important in studies of weight gain, dieting, and exercise. Here are the metabolic rates of 7 men who took part in a study of dieting. (The units are calories per 24 hours. These are the same calories used to describe the energy content of foods.)

1792	1666	1362	1614	1460	1867	1439

The researchers reported $\bar{x}$ and s for these men.
First find the mean:

$$\bar{x} = \frac{1792 + 1666 + 1362 + 1614 + 1460 + 1867 + 1439}{7}$$

$$= \frac{11,200}{7} = 1600 \text{ calories}$$

Figure 2.3 displays the data as points above the number line, with their mean marked by an asterisk (*). The arrows mark two of the deviations from the mean. These deviations show how spread out the data are about their mean. They are the starting point for calculating the variance and the standard deviation.

Observations x_i	Deviations $x_i - \bar{x}$	Squared deviations $(x_i - \bar{x})^2$
1792	$1792 - 1600 = 192$	$192^2 = 36,864$
1666	$1666 - 1600 = 66$	$66^2 = 4,356$
1362	$1362 - 1600 = -238$	$(-238)^2 = 56,644$
1614	$1614 - 1600 = 14$	$14^2 = 196$
1460	$1460 - 1600 = -140$	$(-140)^2 = 19,600$
1867	$1867 - 1600 = 267$	$267^2 = 71,289$
1439	$1439 - 1600 = -161$	$(-161)^2 = 25,921$
	sum $= 0$	sum $=214,870$

The variance is the sum of the squared deviations divided by one less than the number of observations:

$$s^2 = \frac{214,870}{6} = 35,811.67$$

The standard deviation is the square root of the variance:

$$s = \sqrt{35,811.67} = 189.24 \text{ calories}$$

Notice that the "average" in the variance s^2 divides the sum by one fewer than the number of observations, that is, $n - 1$ rather than n. The reason is that the deviations $x_i - \bar{x}$ always sum to exactly 0, so that knowing $n - 1$ of them determines the last one. Only $n - 1$ of the squared deviations can vary freely,

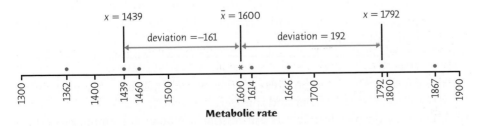

Figure 2.3 Metabolic rates for 7 men, with their mean (*) and the deviations of two observations from the mean.

degrees of freedom

and we average by dividing the total by $n - 1$. The number $n - 1$ is called the **degrees of freedom** of the variance or standard deviation. Many calculators offer a choice between dividing by n and dividing by $n - 1$, so be sure to use $n - 1$.

More important than the details of hand calculation are the properties that determine the usefulness of the standard deviation:

- s measures spread about the mean and should be used only when the mean is chosen as the measure of center.

- $s = 0$ only when there is *no spread*. This happens only when all observations have the same value. Otherwise, $s > 0$. As the observations become more spread out about their mean, s gets larger.

- s has the same units of measurement as the original observations. For example, if you measure metabolic rates in calories, s is also in calories. This is one reason to prefer s to the variance s^2, which is in squared calories.

- Like the mean $\overline{x}$, s is not resistant. Strong skewness or a few outliers can greatly increase s. For example, the standard deviation of the incomes for our 15 college graduates is 24.9. (Use your calculator to verify this.) If we omit the high outlier, the standard deviation drops to 17.7.

You may rightly feel that the importance of the standard deviation is not yet clear. We will see in the next chapter that the standard deviation is the natural measure of spread for an important class of symmetric distributions, the Normal distributions. The usefulness of many statistical procedures is tied to distributions of particular shapes. This is certainly true of the standard deviation.

Choosing measures of center and spread

How do we choose between the five-number summary and $\overline{x}$ and s to describe the center and spread of a distribution? Because the two sides of a strongly skewed distribution have different spreads, no single number such as s describes the spread well. The five-number summary, with its two quartiles and two extremes, does a better job.

CHOOSING A SUMMARY

The five-number summary is usually better than the mean and standard deviation for describing a skewed distribution or a distribution with strong outliers. Use $\overline{x}$ and s only for reasonably symmetric distributions that are free of outliers.

Do remember that a graph gives the best overall picture of a distribution. Numerical measures of center and spread report specific facts about a distribution, but they do not describe its entire shape. Numerical summaries do not disclose the presence of multiple peaks or gaps, for example. Exercise 2.9 shows how misleading numerical summaries can be. **Always plot your data.**

APPLY YOUR KNOWLEDGE

2.8 Blood phosphate. The level of various substances in the blood influences our health. Here are measurements of the level of phosphate in the blood of a patient, in milligrams of phosphate per deciliter of blood, made on 6 consecutive visits to a clinic:

$$5.6 \quad 5.2 \quad 4.6 \quad 4.9 \quad 5.7 \quad 6.4$$

A graph of only 6 observations gives little information, so we proceed to compute the mean and standard deviation.

(a) Find the mean from its definition. That is, find the sum of the 6 observations and divide by 6.

(b) Find the standard deviation from its definition. That is, find the deviations of each observation from the mean, square the deviations, then obtain the variance and the standard deviation. Example 2.6 shows the method.

(c) Now enter the data into your calculator and use the mean and standard deviation buttons to obtain $\bar{x}$ and s. Do the results agree with your hand calculations?

2.9 $\bar{x}$ and s are not enough. The mean $\bar{x}$ and standard deviation s measure center and spread but are not a complete description of a distribution. Data sets with different shapes can have the same mean and standard deviation. To demonstrate this fact, use your calculator to find $\bar{x}$ and s for these two small data sets. Then make a stemplot of each and comment on the shape of each distribution.

Data A	9.14 8.14 8.74 8.77 9.26 8.10 6.13 3.10 9.13 7.26 4.74
Data B	6.58 5.76 7.71 8.84 8.47 7.04 5.25 5.56 7.91 6.89 12.50

2.10 Choose a summary. The shape of a distribution is a rough guide to whether the mean and standard deviation are a helpful summary of center and spread. For which of these distributions would $\bar{x}$ and s be useful? In each case, give a reason for your decision.

(a) Percents of Hispanics in the states, Figure 1.3 (page 11).

(b) Iowa Test scores, Figure 1.4 (page 13).

(c) Annual costs for Massachusetts colleges, Figure 1.5 (page 14).

Using technology

Although a "two-variables statistics" calculator will do the basic calculations we need, more elaborate tools are very helpful. Figure 2.4 displays output describing the percents of Hispanics in the states (Table 1.1, page 10) from three types of tools. Minitab is a statistical software package, Microsoft Excel is spreadsheet software that includes statistical functions, and the Texas

Figure 2.4 Describing the data on the percents of Hispanic residents in the states: output from statistical software, spreadsheet software, and a graphing calculator.

Minitab

```
Session                                                          _ □ ×

Descriptive Statistics: PctHisp


Variable            N        Mean      Median     TrMean      StDev    SE Mean
PctHisp            50        7.73        4.70       6.31       8.91       1.26

Variable      Minimum     Maximum          Q1         Q3
PctHisp          0.70       42.10        2.08       8.77
```

Microsoft Excel

```
Microsoft Excel - ta01-01.dat                                         _ □ ×
        A                  B        C                    D        E      F
 1            Pct Hisp
 2
 3   Mean                  7.73     QUARTILE(B2:B51,1)    2.175
 4   Standard Error        1.2604   QUARTILE(B2:B51,3)    8.525
 5   Median                4.7
 6   Mode                  1.5
 7   Standard Deviation    8.9125
 8   Sample Variance       79.4332
 9   Kurtosis              5.1186
10   Skewness              2.2450
11   Range                 41.4
12   Minimum               0.7
13   Maximum               42.1
14   Sum                   386.5
15   Count                 50

  ► ►│ \ Sheet2 \ Sheet1 / ta01-01 /
```

Texas Instruments TI-83 Plus

```
1-Var Stats                     1-Var Stats
 x̄=7.73                          ↑n=50
 Σx=386.5                         minX=.7
 Σx²=6879.87                      Q₁=2.1
 Sx=8.912528444                   Med=4.7
 σx=8.82295302                    Q₃=8.7
↓n=50                            maxX=42.1
```

Instruments TI-83 Plus is a graphing calculator with statistical functions. All three tools will make statistical graphs, which is very handy. They also do many more statistical calculations than a basic calculator.

Figure 2.4 shows the summary statistics produced by each tool. We are interested in the number of observations n, the five-number summary, and $\bar{x}$ and s. All three displays give this information, but they include other measures that we don't need. This is typical of software—learn to look for what you want and ignore other output. Excel's "Descriptive Statistics" menu item doesn't give the quartiles. We used Excel's separate quartile function to get Q_1 and Q_3.

EXAMPLE 2.7   What is the first quartile?

What is the first quartile of the percents of Hispanics? Arrange the 50 observations in order:

0.7	0.7	0.9	1.2	1.3	1.4	1.5	1.5	1.7	1.9
2.0	2.0	**2.1**	2.4	2.4	2.8	2.8	2.9	3.2	3.3
3.5	3.6	4.1	4.3	**4.7**	**4.7**	4.8	5.2	5.3	5.5
6.4	6.8	7.0	7.2	7.2	7.9	8.0	**8.7**	9.0	9.4
10.7	13.3	15.1	16.8	17.1	19.7	25.3	32.0	32.4	42.1

Applying our rules, the median is midway between the 25th and 26th observations in the list, the two bold **4.7**s. All the outputs agree that M = 4.7.

The first quartile by our rule is the median of the first 25; that is, it is the 13th in the ordered list, $Q_1 = 2.1$. Only the TI-83 gives this answer. Minitab and Excel don't give 2.1, and they differ from each other. The differences are small: the three results are 2.08, 2.1, and 2.175. That's a reminder that there are several rules for exactly determining the quartiles. Our rule is simplest for hand computation. Results from the various rules are always close to each other, so to describe data you should just use the answer your technology gives you.

Chapter 2 SUMMARY

A numerical summary of a distribution should report its **center** and its **spread** or **variability.**

The **mean** $\bar{x}$ and the **median M** describe the center of a distribution in different ways. The mean is the arithmetic average of the observations, and the median is the midpoint of the values.

When you use the median to indicate the center of the distribution, describe its spread by giving the **quartiles.** The **first quartile** Q_1 has one-fourth of the observations below it, and the **third quartile** Q_3 has three-fourths of the observations below it.

The **five-number summary** consisting of the median, the quartiles, and the high and low extremes provides a quick overall description of a distribution. The median describes the center, and the quartiles and extremes show the spread.

Boxplots based on the five-number summary are useful for comparing several distributions. The box spans the quartiles and shows the spread of the central half of the distribution. The median is marked within the box. Lines extend from the box to the extremes and show the full spread of the data.

The **variance** s^2 and especially its square root, the **standard deviation s,** are common measures of spread about the mean as center. The standard deviation s is zero when there is no spread and gets larger as the spread increases.

A **resistant measure** of any aspect of a distribution is relatively unaffected by changes in the numerical value of a small proportion of the total number of observations, no matter how large these changes are. The median and quartiles are resistant, but the mean and the standard deviation are not.

The mean and standard deviation are good descriptions for symmetric distributions without outliers. They are most useful for the Normal distributions introduced in the next chapter. The five-number summary is a better description for skewed distributions.

Numerical summaries do not fully describe the shape of a distribution. Always plot your data.

Chapter 2 EXERCISES

2.11 Incomes of college grads. The Census Bureau reports that the mean and median income of people who worked full-time in 2001 and had at least a bachelor's degree were $53,054 and $72,674. Which of these numbers is the mean and which is the median? Explain your reasoning.

2.12 Internet access. How much do users pay for Internet access? Here are the monthly fees (in dollars) paid by a random sample of 50 users of commercial Internet service providers in August 2000:[4]

```
20   40   22   22   21   21   20   10   20   20
20   13   18   50   20   18   15    8   22   25
22   10   20   22   22   21   15   23   30   12
 9   20   40   22   29   19   15   20   20   20
20   15   19   21   14   22   21   35   20   22
```

(a) Make a stemplot of these data. Briefly describe the pattern you see. About how much do you think America Online and its larger competitors were charging in August 2000? Are there any outliers?

(b) To report a quick summary of how much people pay for Internet service, do you prefer $\bar{x}$ and s or the five-number summary? Why? Calculate your preferred summary.

2.13 Where are the doctors? Table 1.6 (page 29) gives the number of medical doctors per 100,000 people in each state. Exercise 1.29 (page 28) asked you to plot the data. The distribution is right-skewed with several high outliers.

(a) Do you expect the mean to be greater than the median, about equal to the median, or less than the median? Why? Calculate $\bar{x}$ and M and verify your expectation.

(b) The District of Columbia is a high outlier at 758 M.D.'s per 100,000 residents. If you remove D.C. because it is a city rather than a state, do you expect $\bar{x}$ or M to change more? Why? Omitting D.C., calculate both measures for the 50 states and verify your expectation.

2.14 McGwire versus Ruth. Exercises 1.25 and 1.26 (page 27) give the numbers of home runs hit each season by Babe Ruth and Mark McGwire. Find the five-number summaries and make side-by-side boxplots to compare these two home run hitters. What do your plots show? Why are the boxplots less informative than the back-to-back stemplot you made in Exercise 1.26 (page 27)?

2.15 Students' attitudes. Here are the scores of 18 first-year college women on the Survey of Study Habits and Attitudes (SSHA):

154	109	137	115	152	140	154	178	101
103	126	126	137	165	165	129	200	148

(a) Find the mean score from the formula for the mean. Then enter the data into your calculator and use the calculator's $\bar{x}$ button to obtain the mean. Verify that you get the same result.

(b) A stemplot (Exercise 1.8, page 17) suggests that the score 200 is an outlier. Use your calculator to find the mean for the 17 observations that remain when you drop the outlier. How does the outlier change the mean?

2.16 How much fruit do adolescent girls eat? Figure 1.10 (page 23) is a histogram of the number of servings of fruit per day claimed by 74 seventeen-year-old girls. With a little care, you can find the median and the quartiles from the histogram. What are these numbers? How did you find them?

2.17 Students' attitudes. In Exercise 2.15 you found the mean of the SSHA scores of 18 first-year college women. Now find the median of these scores. Is the median smaller or larger than the mean? Explain why this is so.

2.18 Weight of newborns. Exercise 1.18 (page 24) gives the weight class of all babies born in the United States in 1999.

(a) How many babies were there? What are the positions of the median and quartiles in the ordered list of all birth weights?

(b) In which weight classes do the median and quartiles fall?

(Tony Anderson/Taxi/Getty Images)

2.19 Does breast-feeding weaken bones? Breast-feeding mothers secrete calcium into their milk. Some of the calcium may come from their bones, so mothers may lose bone mineral. Researchers compared 47 breast-feeding women with 22 women of similar age who were neither pregnant nor lactating. They measured the percent change in the mineral content of the women's spines over three months. Here are the data:[5]

Breast-feeding women						Other women					
−4.7	−2.5	−4.9	2.7	−0.8	−5.3	2.4	0.0	0.9	−0.2	1.0	1.7
−8.3	−2.1	−6.8	−4.3	2.2	−7.8	2.9	−0.6	1.1	−0.1	−0.4	0.3
−3.1	−1.0	−6.5	−1.8	−5.2	−5.7	1.2	−1.6	−0.1	−1.5	0.7	−0.4
−7.0	−2.2	−6.5	−1.0	−3.0	−3.6	2.2	−0.4	−2.2	−0.1		
−5.2	−2.0	−2.1	−5.6	−4.4	−3.3						
−4.0	−4.9	−4.7	−3.8	−5.9	−2.5						
−0.3	−6.2	−6.8	1.7	0.3	−2.3						
0.4	−5.3	0.2	−2.2	−5.1							

Compare the two distributions using both graphs and numerical summaries. Do the data show distinctly greater bone mineral loss among the breast-feeding women?

2.20 **The density of the earth.** In 1798 the English scientist Henry Cavendish measured the density of the earth with great care. It is common practice to repeat careful measurements several times and use the mean as the final result. Cavendish repeated his work 29 times. Here are his results (the data give the density of the earth as a multiple of the density of water):[6]

5.50	5.61	4.88	5.07	5.26	5.55	5.36	5.29	5.58	5.65
5.57	5.53	5.62	5.29	5.44	5.34	5.79	5.10	5.27	5.39
5.42	5.47	5.63	5.34	5.46	5.30	5.75	5.68	5.85	

Present these measurements with a graph of your choice. Scientists usually give the mean and standard deviation to summarize a set of measurements. Does the shape of this distribution suggest that $\bar{x}$ and s are adequate summaries? Calculate $\bar{x}$ and s.

2.21 **California counties.** You are planning a sample survey of households in California. You decide to select households separately within each county and to choose more households from the more populous counties. To aid in the planning, Table 2.3 gives the populations of California counties from the 2000 census. Examine the distribution

(Claver Carroll/Photolibrary/PictureQuest)

TABLE 2.3 Population of California counties, 2000 census

County	Population	County	Population	County	Population
Alameda	1,443,741	Marin	247,289	San Mateo	707,161
Alpine	1,208	Mariposa	17,130	Santa Barbara	399,347
Amador	35,100	Mendocino	86,265	Santa Clara	1,682,585
Butte	203,171	Merced	210,554	Santa Cruz	255,602
Calaveras	40,554	Modoc	9,449	Shasta	163,256
Colusa	18,804	Mono	12,853	Sierra	3,555
Contra Costa	948,816	Monterey	401,762	Siskiyou	44,301
Del Norte	27,507	Napa	124,279	Solano	394,542
El Dorado	156,299	Nevada	92,033	Sonoma	458,614
Fresno	799,407	Orange	2,846,289	Stanislaus	446,997
Glenn	26,453	Placer	248,399	Sutter	78,930
Humboldt	126,518	Plumas	20,824	Tehama	56,039
Imperial	142,361	Riverside	1,545,387	Trinity	13,022
Inyo	17,945	Sacramento	1,223,499	Tulare	368,021
Kern	661,645	San Benito	53,234	Tuolumne	54,501
Kings	129,461	San Bernardino	1,709,434	Ventura	753,197
Lake	58,309	San Diego	2,813,833	Yolo	168,660
Lassen	33,828	San Francisco	776,733	Yuba	60,219
Los Angeles	9,519,338	San Joaquin	563,598		
Madera	123,109	San Luis Obispo	246,681		

TABLE 2.4 Positions and weights (pounds) for a major college football team

QB	200	QB	209	QB	190	QB	201	RB	210	RB	224	RB	196
RB	218	RB	229	RB	236	OL	281	OL	286	OL	320	OL	369
OL	298	OL	276	OL	293	OL	292	OL	285	OL	286	OL	265
OL	314	OL	318	OL	334	OL	276	OL	300	OL	290	WR	227
WR	178	WR	180	WR	193	WR	190	WR	163	WR	185	WR	200
TE	235	TE	233	TE	225	TE	253	TE	275	TE	200	KP	189
KP	214	KP	185	KP	204	DB	189	DB	173	DB	186	DB	220
DB	170	DB	166	DB	185	DB	194	DB	175	DB	194	DB	193
DB	179	LB	247	LB	208	LB	227	LB	219	LB	219	LB	208
LB	225	LB	203	DL	240	DL	261	DL	302	DL	214	DL	225
DL	250	DL	350	DL	230	DL	271	DL	294	DL	202	DL	297
DL	271	DL	260	DL	257	DL	262						

of county populations both graphically and numerically, using whatever tools are most suitable. Write a brief description of the main features of this distribution. Sample surveys often select households from all of the most populous counties but from only some of the less populous. How would you divide California counties into three groups according to population, with the intent of including all of the first group, half of the second, and a smaller fraction of the third in your survey?

2.22 **A football team.** The University of Miami Hurricanes have been among the more successful teams in college football. Table 2.4 gives the weights in pounds and positions of the players on the 2002 team.[7] The positions are quarterback (QB), running back (RB), offensive line (OL), wide receiver (WR), tight end (TE), kicker/punter (KP), defensive back (DB), linebacker (LB), and defensive line (DL).

(AP/Wide World Photos)

(a) Make side-by-side boxplots of the weights for running backs, wide receivers, offensive linemen, defensive linemen, linebackers, and defensive backs.

(b) Briefly compare the weight distributions. Which position has the heaviest players overall? Which has the lightest?

(c) Are any individual players outliers within their position?

2.23 **Guinea pig survival times.** Here are the survival times in days of 72 guinea pigs after they were injected with infectious bacteria in a medical experiment.[8] Survival times, whether of machines under stress or cancer patients after treatment, usually have distributions that are skewed to the right.

43	45	53	56	56	57	58	66	67	73	74	79
80	80	81	81	81	82	83	83	84	88	89	91
91	92	92	97	99	99	100	100	101	102	102	102
103	104	107	108	109	113	114	118	121	123	126	128
137	138	139	144	145	147	156	162	174	178	179	184
191	198	211	214	243	249	329	380	403	511	522	598

(a) Graph the distribution and describe its main features. Does it show the expected right skew?

(b) Which numerical summary would you choose for these data? Calculate your chosen summary. How does it reflect the skewness of the distribution?

2.24 Never on Sunday: also in Canada? Exercise 1.4 (page 9) gives the number of births in the United States on each day of the week during an entire year. The boxplots in Figure 2.5 are based on more detailed data from Toronto, Canada: the number of births on each of the 365 days in a year, grouped by day of the week.[9] Based on these plots, give a more detailed description of how births depend on the day of the week.

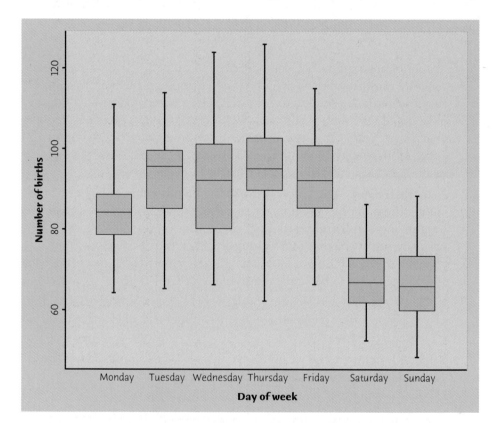

Figure 2.5 Side-by-side boxplots of the distributions of numbers of births in Toronto, Canada, for each day of the week during a year.

2.25 Athletes' salaries. In 2002, the Chicago Cubs failed once again to reach the National League playoffs. Table 2.5 gives the salaries of the players as of opening day of the season. Describe the distribution of salaries both with a graph and with a numerical summary. Then write a brief description of the important features of the distribution.

Identifying suspected outliers. *Whether an observation is an outlier is a matter of judgment: does it appear to clearly stand apart from the rest*

TABLE 2.5 2002 salaries for the Chicago Cubs baseball team

Player	Salary	Player	Salary	Player	Salary
Sammy Sosa	$15,000,000	Jason Bere	$3,400,000	Donovan Osborne	$450,000
Fred McGriff	$7,250,000	Jeff Fassero	$2,800,000	Kyle Farnsworth	$410,000
Todd Hundley	$6,500,000	Tom Gordon	$2,800,000	Robert Machado	$277,500
Moises Alou	$6,000,000	Matt Clement	$2,500,000	Roosevelt Brown	$255,000
Jon Lieber	$5,833,333	Joe Girardi	$1,500,000	Augie Ojeda	$240,000
Alex Gonzalez	$4,250,000	Jesus Sanchez	$1,425,000	Joe Borowski	$235,000
Kerry Wood	$3,695,000	Delino DeShields	$1,250,000	Corey Patterson	$227,500
Antonio Alfonseca	$3,550,000	Chris Stynes	$1,250,000	Mark Bellhorn	$224,000
Bill Mueller	$3,450,000	Darren Lewis	$500,000	Juan Cruz	$215,000
				Will Ohman	$203,500

of the distribution? When large volumes of data are scanned automatically, however, we need a rule to pick out suspected outliers. The most common rule is the **1.5IQR rule**. The **interquartile range** is the distance between the two quartiles, $IQR = Q_3 - Q_1$. This is the spread of the middle half of the data. A point is a suspected outlier if it lies more than $1.5IQR$ below the first quartile Q_1 or above the third quartile Q_3. Exercises 2.26 to 2.28 ask you to apply this rule.

1.5IQR rule
interquartile range

2.26 A high income. In Example 2.1, we noted the influence of one high income of $110,000 among the incomes of a sample of 15 college graduates. Does the 1.5IQR rule identify this income as a suspected outlier?

2.27 Athletes' salaries. Which members of the Chicago Cubs (see Table 2.5) have salaries that are suspected outliers by the 1.5IQR rule?

2.28 Gas guzzlers? The BMW Z8, Ferrari, and Lamborghini have notably low gas mileage among the two-seater cars in Table 1.2 (page 12). Find the five-number summary of the city gas mileages (omitting the Honda Insight). Are any of these cars suspected low outliers by the 1.5IQR rule?

2.29 A standard deviation contest. This is a standard deviation contest. You must choose four numbers from the whole numbers 0 to 10, with repeats allowed.

  (a) Choose four numbers that have the smallest possible standard deviation.

  (b) Choose four numbers that have the largest possible standard deviation.

  (c) Is more than one choice possible in either (a) or (b)? Explain.

2.30 Test your technology. This exercise requires a calculator with a standard deviation button or statistical software on a computer.

(Kim Sayer/Corbis)

The observations

10,001   10,002   10,003

have mean $\overline{x} = 10{,}002$ and standard deviation $s = 1$. Adding a 0 in the center of each number, the next set becomes

100,001   100,002   100,003

The standard deviation remains $s = 1$ as more 0s are added. Use your calculator or software to find the standard deviation of these numbers, adding extra 0s until you get an incorrect answer. How soon did you go wrong? This demonstrates that calculators and software cannot handle an arbitrary number of digits correctly.

2.31 You create the data. Create a set of 5 positive numbers (repeats allowed) that have median 10 and mean 7. What thought process did you use to create your numbers?

2.32 You create the data. Give an example of a small set of data for which the mean is larger than the third quartile.

Chapter 2 MEDIA EXERCISES

2.33 Comparing incomes. Want data on incomes? Go to the Census Bureau Web site, www.census.gov, click on "Income," and open the latest issue of the annual publication *Money Income in the United States.* The tables in this publication contain lots of detail. One set of tables compares the distributions of incomes among households by ethnicity (white, black, Asian and Pacific Islander, and Hispanic origin). What is the median household income for each group? How closely can you find the quartiles from the information given? Make partial boxplots—just the central box—to compare the four distributions. What do the data show?

2.34 Internet access. You can use the *One-Variable Statistical Calculator* applet in place of a calculator or software to do both calculations ($\overline{x}$ and s and the five-number summary) and graphs (histograms and stemplots). The applet is more convenient than most calculators. It is less convenient than good software because, depending on your browser, it may be difficult to enter new data sets in one operation and to print output. Exercise 2.12 (page 48) gives the monthly fees paid for Internet access by a random sample of customers of commercial Internet service providers. This data set is one of those stored in the applet. Use the applet to do Exercise 2.12.

2.35 Mean = median? The *Mean and Median* applet allows you to place observations on a line and see their mean and median visually. Place two observations on the line. Why does only one arrow appear?

2.36 Making resistance visible. In the *Mean and Median* applet, place three observations on the line by clicking below it: two close together near the center of the line and one somewhat to the right of these two.

(a) Pull the single rightmost observation out to the right. (Place the cursor on the point, hold down a mouse button, and drag the point.) How does the mean behave? How does the median behave? Explain briefly why each measure acts as it does.

(b) Now drag the single rightmost point to the left as far as you can. What happens to the mean? What happens to the median as you drag this point past the other two (watch carefully)?

2.37 Behavior of the median. Place five observations on the line in the *Mean and Median* applet by clicking below it.

(a) Add one additional observation *without changing the median*. Where is your new point?

(b) Use the applet to convince yourself that when you add yet another observation (there are now seven in all), the median does not change no matter where you put the seventh point. Explain why this must be true.

2.38 Eruptions of Old Faithful. In Exercise 1.37 (page 31) you used a graph to describe how long eruptions of famous Old Faithful geyser last. Data for 299 eruptions appear in the EESEE story "Is Old Faithful Faithful?" Now you will continue your work.

(a) The shape of the distribution fits the fact that Old Faithful has two reservoirs of water, which can erupt either separately or together. Explain why the shape matches this fact. Also explain why neither $\bar{x}$ and s nor the five-number summary is a good description of this distribution.

(b) Use your judgment, based on the histogram, to divide the eruptions into two clusters. What is the shape of the distribution of eruption lengths in each cluster? Give numerical descriptions of both clusters and compare the two.

2.39 Truth in calories. The EESEE story "Counting Calories" presents data from a study showing that foods advertised as "low calorie" and the like often contain more calories than the label states. One question was whether the degree of understatement of calories is the same for national brands, regional brands, and locally prepared foods. The EESEE story has data on the percent difference between true and label calories for many foods of each type. The differences are greater than zero when the true calories are higher than the label claims. Use both graphs and numerical summaries to compare the three types of food. What are your conclusions?

(Mike Powell/Allsport Concepts/Getty Images)

The Normal Distributions

We now have a kit of graphical and numerical tools for describing distributions. What is more, we have a clear strategy for exploring data on a single quantitative variable:

1. Always plot your data: make a graph, usually a histogram or a stemplot.

2. Look for the overall pattern (shape, center, spread) and for striking deviations such as outliers.

3. Calculate a numerical summary to briefly describe center and spread.

Here is one more step to add to this strategy:

4. Sometimes the overall pattern of a large number of observations is so regular that we can describe it by a smooth curve.

Density curves

Figure 3.1 is a histogram of the scores of all 947 seventh-grade students in Gary, Indiana, on the vocabulary part of the Iowa Test of Basic Skills.[1] Scores of many students on this national test have a quite regular distribution. The histogram is symmetric, and both tails fall off quite smoothly from a single center peak. There are no large gaps or obvious outliers. The smooth curve drawn through the tops of the histogram bars in Figure 3.1 is a good description of the overall pattern of the data. The curve is a **mathematical model** for the distribution.

mathematical model

A mathematical model is an idealized description. It gives a compact picture

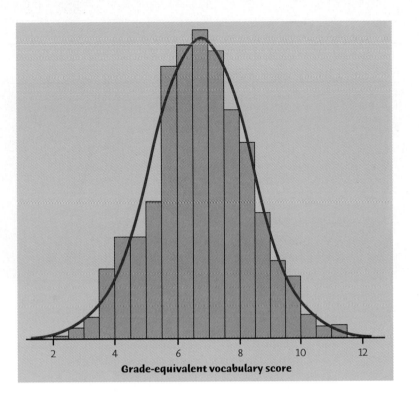

Figure 3.1 Histogram of the vocabulary scores of all seventh-grade students in Gary, Indiana. The smooth curve shows the overall shape of the distribution.

Grade-equivalent vocabulary score

of the overall pattern of the data but ignores minor irregularities as well as any outliers.

We will see that it is easier to work with the smooth curve in Figure 3.1 than with the histogram. The reason is that the histogram depends on our choice of classes, while with a little care we can use a curve that does not depend on any choices we make. Here's how we do it.

EXAMPLE 3.1   From histogram to density curve

Our eyes respond to the *areas* of the bars in a histogram. The bar areas represent proportions of the observations. Figure 3.2(a) is a copy of Figure 3.1 with the leftmost bars shaded. The area of the shaded bars in Figure 3.2(a) represents the students with vocabulary scores 6.0 or lower. There are 287 such students, who make up the proportion $287/947 = 0.303$ of all Gary seventh graders.

Now concentrate on the curve drawn through the bars. In Figure 3.2(b), the area under the curve to the left of 6.0 is shaded. Adjust the scale of the graph so that *the total area under the curve is exactly 1*. This area represents the proportion 1, that is, all the observations. Areas under the curve then represent proportions of the observations. The curve is now a *density curve*. The shaded area under the density curve in Figure 3.2(b) represents the proportion of students with score 6.0 or lower. This area is 0.293, only 0.010 away from the histogram result. You can see that areas under the density curve give quite good approximations of areas given by the histogram.

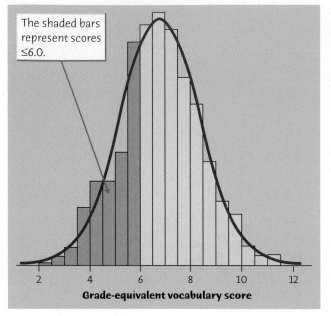

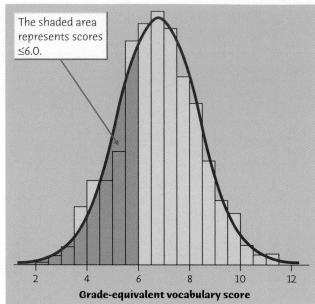

Figure 3.2(a) The proportion of scores less than or equal to 6.0 from the histogram is 0.303.

Figure 3.2(b) The proportion of scores less than or equal to 6.0 from the density curve is 0.293.

> **DENSITY CURVE**
>
> A **density curve** is a curve that
>
> • is always on or above the horizontal axis, and
> • has area exactly 1 underneath it.
>
> A density curve describes the overall pattern of a distribution. The area under the curve and above any range of values is the proportion of all observations that fall in that range.

Normal curve   The density curve in Figures 3.1 and 3.2 is a **Normal curve.** Density curves, like distributions, come in many shapes. Figure 3.3 shows two density curves: a symmetric Normal density curve and a right-skewed curve. A density curve of the appropriate shape is often an adequate description of the overall pattern of a distribution. Outliers, which are deviations from the overall pattern, are not described by the curve. Of course, no set of real data is exactly described by a density curve. The curve is an approximation that is easy to use and accurate enough for practical use.

APPLY YOUR KNOWLEDGE

3.1 **Sketch density curves.** Sketch density curves that might describe distributions with the following shapes:

(a) Symmetric, but with two peaks (that is, two strong clusters of observations).

(b) Single peak and skewed to the left.

The median and mean of a density curve

Our measures of center and spread apply to density curves as well as to actual sets of observations. The median and quartiles are easy. Areas under a density curve represent proportions of the total number of observations. The median is the point with half the observations on either side. So **the median of a density curve is the equal-areas point,** the point with half the area under the curve to its left and the remaining half of the area to its right. The quartiles divide the area under the curve into quarters. One-fourth of the area under the curve is to the left of the first quartile, and three-fourths of the area is to the left of the third quartile. You can roughly locate the median and quartiles of any density curve by eye by dividing the area under the curve into four equal parts.

Because density curves are idealized patterns, a symmetric density curve is exactly symmetric. The median of a symmetric density curve is therefore at its center. Figure 3.3(a) shows the median of a symmetric curve. It isn't so easy to spot the equal-areas point on a skewed curve. There are mathematical ways of finding the median for any density curve. We did that to mark the median on the skewed curve in Figure 3.3(b).

What about the mean? The mean of a set of observations is their arithmetic average. If we think of the observations as weights strung out along a thin rod, the mean is the point at which the rod would balance. This fact is also true of density curves. **The mean is the point at which the curve would balance if made of solid material.** Figure 3.4 illustrates this fact about the mean. A symmetric curve balances at its center because the two sides are identical. **The mean and median of a symmetric density curve are equal,** as in Figure 3.3(a). We know that the mean of a skewed distribution is pulled toward the long tail.

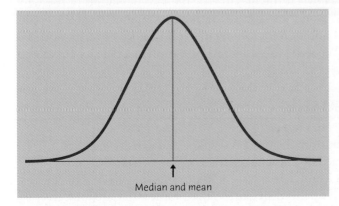

Figure 3.3(a) The median and mean of a symmetric density curve.

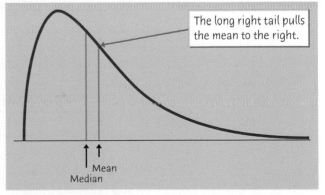

The long right tail pulls the mean to the right.

Figure 3.3(b) The median and mean of a right-skewed density curve.

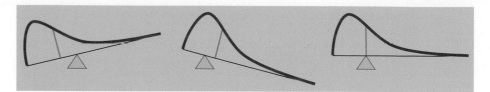

Figure 3.4 The mean is the balance point of a density curve.

Figure 3.3(b) shows how the mean of a skewed density curve is pulled toward the long tail more than is the median. It's hard to locate the balance point by eye on a skewed curve. There are mathematical ways of calculating the mean for any density curve, so we are able to mark the mean as well as the median in Figure 3.3(b).

MEDIAN AND MEAN OF A DENSITY CURVE

The **median** of a density curve is the equal-areas point, the point that divides the area under the curve in half.

The **mean** of a density curve is the balance point, at which the curve would balance if made of solid material.

The median and mean are the same for a symmetric density curve. They both lie at the center of the curve. The mean of a skewed curve is pulled away from the median in the direction of the long tail.

We can roughly locate the mean, median, and quartiles of any density curve by eye. This is not true of the standard deviation. When necessary, we can once again call on more advanced mathematics to learn the value of the standard deviation. The study of mathematical methods for doing calculations with density curves is part of theoretical statistics. Though we are concentrating on statistical practice, we often make use of the results of mathematical study.

Because a density curve is an idealized description of the distribution of data, we need to distinguish between the mean and standard deviation of the density curve and the mean $\bar{x}$ and standard deviation s computed from the actual observations. The usual notation for the mean of an idealized distribution is μ (the Greek letter mu). We write the standard deviation of a density curve as σ (the Greek letter sigma).

mean μ
standard deviation σ

APPLY YOUR KNOWLEDGE

3.2 **A uniform distribution.** Figure 3.5 displays the density curve of a *uniform distribution*. The curve takes the constant value 1 over the interval from 0 to 1 and is zero outside that range of values. This means that data described by this distribution take values that are uniformly spread between 0 and 1. Use areas under this density curve to answer the following questions.

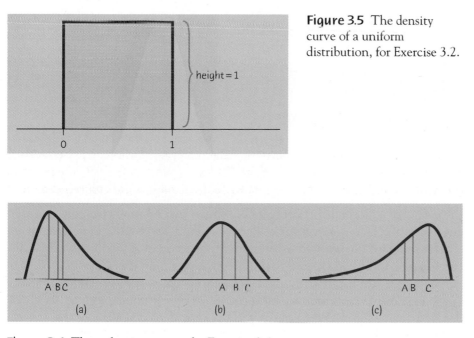

Figure 3.5 The density curve of a uniform distribution, for Exercise 3.2.

Figure 3.6 Three density curves, for Exercise 3.4.

    (a) Why is the total area under this curve equal to 1?

    (b) What percent of the observations lie above 0.8?

    (c) What percent of the observations lie below 0.6?

    (d) What percent of the observations lie between 0.25 and 0.75?

3.3 **Mean and median.** What is the mean μ of the density curve pictured in Figure 3.5? What is the median?

3.4 **Mean and median.** Figure 3.6 displays three density curves, each with three points marked on them. At which of these points on each curve do the mean and the median fall?

Normal distributions

One particularly important class of density curves has already appeared in Figures 3.1 and 3.3(a). These density curves are symmetric, single-peaked, and bell-shaped. They are called *Normal curves*, and they describe **Normal distributions.** Normal distributions play a large role in statistics, but they are rather special and not at all "normal" in the sense of being average or natural. We capitalize Normal to remind you that these curves are special. All Normal distributions have the same overall shape. The exact density curve for a particular Normal distribution is described by giving its mean μ and its standard deviation σ. The mean is located at the center of the symmetric curve and is the same as the median. Changing μ without changing σ moves the Normal curve along the horizontal axis without changing its spread. The standard deviation σ controls

Normal distributions

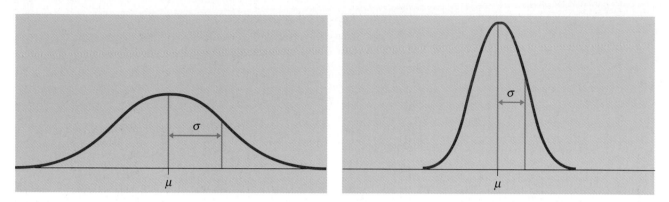

Figure 3.7 Two Normal curves, showing the mean μ and standard deviation σ.

the spread of a Normal curve. Figure 3.7 shows two Normal curves with different values of σ. The curve with the larger standard deviation is more spread out.

The standard deviation σ is the natural measure of spread for Normal distributions. Not only do μ and σ completely determine the shape of a Normal curve, but we can locate σ by eye on the curve. Here's how. Imagine that you are skiing down a mountain that has the shape of a Normal curve. At first, you descend at an ever-steeper angle as you go out from the peak:

Fortunately, before you find yourself going straight down, the slope begins to grow flatter rather than steeper as you go out and down:

The points at which this change of curvature takes place are located at distance σ on either side of the mean μ. You can feel the change as you run a pencil along a Normal curve, and so find the standard deviation. Remember that μ and σ alone do not specify the shape of most distributions, and that the shape of density curves in general does not reveal σ. These are special properties of Normal distributions.

Why are the Normal distributions important in statistics? Here are three reasons. First, Normal distributions are good descriptions for some distributions of *real data*. Distributions that are often close to Normal include scores on tests taken by many people (such as SAT exams and many psychological tests), repeated careful measurements of the same quantity, and characteristics of biological populations (such as lengths of crickets and yields of corn). Second, Normal distributions are good approximations to the results of many kinds of *chance outcomes*, such as tossing a coin many times. Third, and most important, we will see that many *statistical inference* procedures based on Normal distributions work well for other roughly symmetric distributions. However, many sets of data do not follow a Normal distribution. Most income distributions, for example, are skewed to the right and so are not Normal. Non-Normal data, like nonnormal people, not only are common but are sometimes more interesting than their Normal counterparts.

The 68–95–99.7 rule

Although there are many Normal curves, they all have common properties. In particular, all Normal distributions obey the following rule.

THE 68–95–99.7 RULE

In the Normal distribution with mean μ and standard deviation σ:

* **68%** of the observations fall within σ of the mean μ.
* **95%** of the observations fall within 2σ of μ.
* **99.7%** of the observations fall within 3σ of μ.

Figure 3.8 illustrates the 68–95–99.7 rule. By remembering these three numbers, you can think about Normal distributions without constantly making detailed calculations.

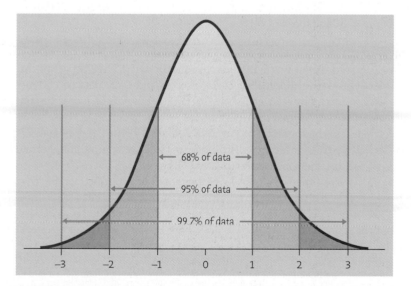

Figure 3.8 The 68–95–99.7 rule for Normal distributions.

EXAMPLE 3.2   Using the 68-95-99.7 rule: SAT scores

The distribution of the scores of the more than 1.3 million high school seniors who took the SAT Reasoning (SAT I) verbal exam in 2002 is close to Normal with mean 504 and standard deviation 111.[2] Figure 3.9 applies the 68–95–99.7 rule to SAT scores.

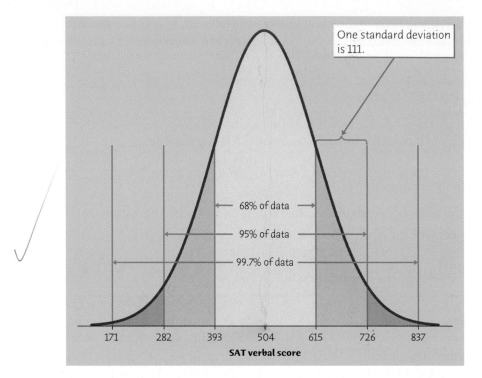

One standard deviation is 111.

68% of data

95% of data

99.7% of data

171   282   393   504   615   726   837

SAT verbal score

Figure 3.9 The 68–95–99.7 rule applied to the distribution of SAT verbal scores for 2002, with $\mu = 504$ and $\sigma = 111$.

Two standard deviations is 222 for this distribution. The 95 part of the 68–95–99.7 rule says that the middle 95% of SAT verbal scores are between $504 - 222$ and $504 + 222$, that is, between 282 and 726. The other 5% of scores lie outside this range. Because the Normal distributions are symmetric, half of these scores are on the high side. So the highest 2.5% of SAT verbal scores are higher than 726. These facts are exactly true for Normal distributions. They are approximately true for SAT scores, whose distribution is approximately Normal.

The 99.7 part of the 68–95–99.7 rule says that almost all scores (99.7% of them) lie between $\mu - 3\sigma$ and $\mu + 3\sigma$. This range of scores is 171 to 837. In fact, SAT scores are only reported between 200 and 800. It is possible to score higher than 800, but the score will be reported as 800. That is, you don't need a perfect paper to score 800 on the SAT.

Because we will mention Normal distributions often, a short notation is helpful. We abbreviate the Normal distribution with mean μ and standard deviation σ as $N(\mu, \sigma)$. For example, the distribution of SAT verbal scores is $N(504, 111)$.

APPLY YOUR KNOWLEDGE

3.5   **Heights of young women.** The distribution of heights of women aged 20 to 29 is approximately Normal with mean 64 inches and standard deviation 2.7 inches.[3] Draw a Normal curve on which this mean and standard deviation are correctly located. (Hint: Draw the curve first, locate the points where the curvature changes, then mark the horizontal axis.)

3.6 **Heights of young women.** The distribution of heights of women aged 20 to 29 is approximately Normal with mean 64 inches and standard deviation 2.7 inches. Use the 68–95–99.7 rule to answer the following questions.

(a) Between what heights do the middle 95% of young women fall?

(b) What percent of young women are taller than 61.3 inches?

3.7 **Length of pregnancies.** The length of human pregnancies from conception to birth varies according to a distribution that is approximately Normal with mean 266 days and standard deviation 16 days. Use the 68–95–99.7 rule to answer the following questions.

(a) Between what values do the lengths of the middle 95% of all pregnancies fall?

(b) How short are the shortest 2.5% of all pregnancies?

The standard Normal distribution

As the 68–95–99.7 rule suggests, all Normal distributions share many common properties. In fact, all Normal distributions are the same if we measure in units of size σ about the mean μ as center. Changing to these units is called *standardizing*. To standardize a value, subtract the mean of the distribution and then divide by the standard deviation.

STANDARDIZING AND z-SCORES

If x is an observation from a distribution that has mean μ and standard deviation σ, the **standardized value** of x is

$$z = \frac{x - \mu}{\sigma}$$

A standardized value is often called a **z-score.**

A z-score tells us how many standard deviations the original observation falls away from the mean, and in which direction. Observations larger than the mean are positive when standardized, and observations smaller than the mean are negative.

EXAMPLE 3.3 Standardizing women's heights

The heights of young women are approximately Normal with $\mu = 64$ inches and $\sigma = 2.7$ inches. The standardized height is

$$z = \frac{\text{height} - 64}{2.7}$$

He said, she said.
The height and weight distributions in this chapter come from actual measurements by a government survey. Good thing that is. When *asked* their weight, almost all women say they weigh less than they really do. Heavier men also underreport their weight—but lighter men claim to weigh more than the scale shows. We leave you to ponder the psychology of the two sexes. Just remember that "say so" is no replacement for measuring.

A woman's standardized height is the number of standard deviations by which her height differs from the mean height of all young women. A woman 70 inches tall, for example, has standardized height

$$z = \frac{70 - 64}{2.7} = 2.22$$

or 2.22 standard deviations above the mean. Similarly, a woman 5 feet (60 inches) tall has standardized height

$$z = \frac{60 - 64}{2.7} = -1.48$$

or 1.48 standard deviations less than the mean height.

If the variable we standardize has a Normal distribution, standardizing does more than give a common scale. It makes all Normal distributions into a single distribution, and this distribution is still Normal. Standardizing a variable that has any Normal distribution produces a new variable that has the *standard Normal distribution*.

STANDARD NORMAL DISTRIBUTION

The **standard Normal distribution** is the Normal distribution $N(0, 1)$ with mean 0 and standard deviation 1.

If a variable x has any Normal distribution $N(\mu, \sigma)$ with mean μ and standard deviation σ, then the standardized variable

$$z = \frac{x - \mu}{\sigma}$$

has the standard Normal distribution.

APPLY YOUR KNOWLEDGE

3.8   **SAT versus ACT.** Eleanor scores 680 on the mathematics part of the SAT. The distribution of SAT math scores in 2002 was Normal with mean 516 and standard deviation 114. Gerald takes the ACT Assessment mathematics test and scores 27. ACT math scores in 2002 were Normally distributed with mean 20.6 and standard deviation 5.0. Find the standardized scores for both students. Assuming that both tests measure the same kind of ability, who has the higher score?

3.9   **Men's and women's heights.** The heights of women aged 20 to 29 are approximately Normal with mean 64 inches and standard deviation 2.7 inches. Men the same age have mean height 69.3 inches with standard deviation 2.8 inches. What are the z-scores for a woman 6 feet tall and a man 6 feet tall? Say in simple language what information the z-scores give that the actual heights do not.

Normal distribution calculations

An area under a density curve is a proportion of the observations in a distribution. Any question about what proportion of observations lie in some range of values can be answered by finding an area under the curve. Because all Normal distributions are the same when we standardize, we can find areas under any Normal curve from a single table, a table that gives areas under the curve for the standard Normal distribution.

EXAMPLE 3.4   Using the standard Normal distribution

What proportion of all young women are less than 70 inches tall? This proportion is the area under the $N(64, 2.7)$ curve to the left of the point 70. Because the standardized height corresponding to 70 inches is

$$z = \frac{x - \mu}{\sigma} = \frac{70 - 64}{2.7} = 2.22$$

this area is the same as the area under the standard Normal curve to the left of the point $z = 2.22$. Figure 3.10(a) shows this area.

Calculators and software often give areas under Normal curves. You can do the work by hand using Table A in the front end covers, which gives areas under the standard Normal curve.

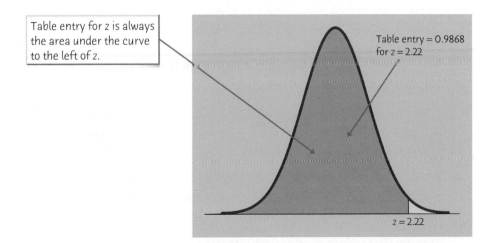

Table entry for z is always the area under the curve to the left of z.

Table entry = 0.9868 for z = 2.22

z = 2.22

Figure 3.10(a) The area under a standard Normal curve to the left of the point $z = 2.22$ is 0.9868. Table A gives areas under the standard Normal curve.

THE STANDARD NORMAL TABLE

Table A is a table of areas under the standard Normal curve. The table entry for each value z is the area under the curve to the left of z.

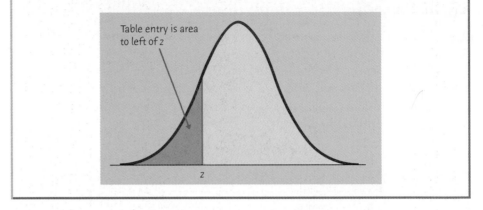

Table entry is area to left of z

z

EXAMPLE 3.5   Using the standard Normal table

Problem: Find the proportion of observations from the standard Normal distribution that are less than 2.22.

Solution: To find the area to the left of 2.22, locate 2.2 in the left-hand column of Table A, then locate the remaining digit 2 as .02 in the top row. The entry opposite 2.2 and under .02 is 0.9868. This is the area we seek. Figure 3.10(a) illustrates the relationship between the value $z = 2.22$ and the area 0.9868. Because $z = 2.22$ is the standardized value of height 70 inches, the proportion of young women who are less than 70 inches tall is 0.9868 (more than 98%).

Problem: Find the proportion of observations from the standard Normal distribution that are greater than −0.67.

Solution: Enter Table A under $z = -0.67$. That is, find −0.6 in the left-hand column and .07 in the top row. The table entry is 0.2514. This is the area to the *left* of −0.67. Because the total area under the curve is 1, the area lying to the *right* of −0.67 is $1 - 0.2514 = 0.7486$. Figure 3.10(b) illustrates these areas.

We can answer any question about proportions of observations in a Normal distribution by standardizing and then using the standard Normal table. Here is an outline of the method for finding the proportion of the distribution in any region.

FINDING NORMAL PROPORTIONS

1. State the problem in terms of the observed variable x.

2. Standardize x to restate the problem in terms of a standard Normal variable z. Draw a picture to show the area under the standard Normal curve.

3. Find the required area under the standard Normal curve, using Table A and the fact that the total area under the curve is 1.

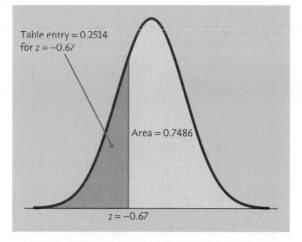

Figure 3.10 (b) Areas under the standard Normal curve to the left and right of $z = -0.67$. Table A gives only areas to the left.

EXAMPLE 3.6   Normal distribution calculations

The level of cholesterol in the blood is important because high cholesterol levels may increase the risk of heart disease. The distribution of blood cholesterol levels in a large population of people of the same age and sex is roughly Normal. For 14-year-old boys, the mean is $\mu = 170$ milligrams of cholesterol per deciliter of blood (mg/dl) and the standard deviation is $\sigma = 30$ mg/dl.[4] Levels above 240 mg/dl may require medical attention. What percent of 14-year-old boys have more than 240 mg/dl of cholesterol?

1. *State the problem.* Call the level of cholesterol in the blood x. The variable x has the $N(170, 30)$ distribution. We want the proportion of boys with $x > 240$.

2. *Standardize.* Subtract the mean, then divide by the standard deviation, to turn x into a standard Normal z:

$$x > 240$$

$$\frac{x - 170}{30} > \frac{240 - 170}{30}$$

$$z > 2.33$$

Figure 3.11 shows the standard Normal curve with the area of interest shaded.

3. *Use the table.* From Table A, we see that the proportion of observations less than 2.33 is 0.9901. About 99% of boys have cholesterol levels less than 240. The area to the right of 2.33 is therefore $1 - 0.9901 = 0.0099$. This is about 0.01, or 1%. Only about 1% of boys have high cholesterol.

In a Normal distribution, the proportion of observations with $x > 240$ is the same as the proportion with $x \geq 240$. There is no area under the curve and

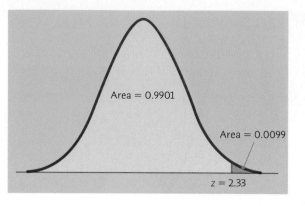

Figure 3.11 Areas under the standard Normal curve for Example 3.6.

exactly over 240, so the areas under the curve with $x > 240$ and $x \geq 240$ are the same. This isn't true of the actual data. There may be a boy with exactly 240 mg/dl of blood cholesterol. The Normal distribution is just an easy-to-use approximation, not a description of every detail in the actual data.

The key to using either software or Table A to do a Normal calculation is to sketch the area you want, then match that area with the areas that the table or software gives you. Here is another example.

EXAMPLE 3.7   More Normal distribution calculations

What percent of 14-year-old boys have blood cholesterol between 170 and 240 mg/dl?

1. *State the problem.* We want the proportion of boys with $170 \leq x \leq 240$.

2. *Standardize:*

$$170 \quad \leq \quad x \quad \leq \quad 240$$
$$\frac{170 - 170}{30} \leq \frac{x - 170}{30} \leq \frac{240 - 170}{30}$$
$$0 \quad \leq \quad z \quad \leq \quad 2.33$$

Figure 3.12 shows the area under the standard Normal curve.

3. *Use the table.* The area between 2.33 and 0 is the area below 2.33 *minus* the area below 0. Look at Figure 3.12 to check this. From Table A,

$$\text{area between 0 and 2.33} = \text{area below 2.33} - \text{area below 0.00}$$
$$= 0.9901 - 0.5000 = 0.4901$$

About 49% of boys have cholesterol levels between 170 and 240 mg/dl.

Sometimes we encounter a value of z more extreme than those appearing in Table A. For example, the area to the left of $z = -4$ is not given directly in the table. The z-values in Table A leave only area 0.0002 in each tail unaccounted

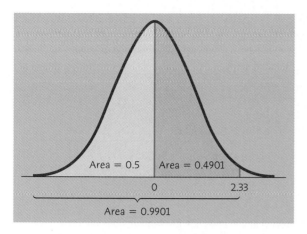

Figure 3.12 Areas under the standard Normal curve for Example 3.7.

for. For practical purposes, we can act as if there is zero area outside the range of Table A.

APPLY YOUR KNOWLEDGE

3.10 Use Table A to find the proportion of observations from a standard Normal distribution that satisfies each of the following statements. In each case, sketch a standard Normal curve and shade the area under the curve that is the answer to the question.

(a) $z < 2.85$

(b) $z > 2.85$

(c) $z > -1.66$

(d) $-1.66 < z < 2.85$

3.11 How hard do locomotives pull? An important measure of the performance of a locomotive is its "adhesion," which is the locomotive's pulling force as a multiple of its weight. The adhesion of one 4400-horsepower diesel locomotive model varies in actual use according to a Normal distribution with mean $\mu = 0.37$ and standard deviation $\sigma = 0.04$.

(a) What proportion of adhesions measured in use are higher than 0.40?

(b) What proportion of adhesions are between 0.40 and 0.50?

(c) Improvements in the locomotive's computer controls change the distribution of adhesion to a Normal distribution with mean $\mu = 0.41$ and standard deviation $\sigma = 0.02$. Find the proportions in (a) and (b) after this improvement.

(CORBIS)

Finding a value given a proportion

Examples 3.6 and 3.7 illustrate the use of Table A to find what proportion of the observations satisfies some condition, such as "blood cholesterol between

170 mg/dl and 240 mg/dl." We may instead want to find the observed value with a given proportion of the observations above or below it. To do this, use Table A backward. Find the given proportion in the body of the table, read the corresponding z from the left column and top row, then "unstandardize" to get the observed value. Here is an example.

EXAMPLE 3.8   "Backward" Normal calculations

Scores on the SAT verbal test in 2002 followed approximately the $N(504, 111)$ distribution. How high must a student score in order to place in the top 10% of all students taking the SAT?

1. *State the problem.* We want to find the SAT score x with area 0.1 to its *right* under the Normal curve with mean $\mu = 504$ and standard deviation $\sigma = 111$. That's the same as finding the SAT score x with area 0.9 to its *left*. Figure 3.13 poses the question in graphical form. Because Table A gives the areas to the left of z-values, always state the problem in terms of the area to the left of x.

2. *Use the table.* Look in the body of Table A for the entry closest to 0.9. It is 0.8997. This is the entry corresponding to $z = 1.28$. So $z = 1.28$ is the standardized value with area 0.9 to its left.

3. *Unstandardize* to transform the solution from the z back to the original x scale. We know that the standardized value of the unknown x is $z = 1.28$.

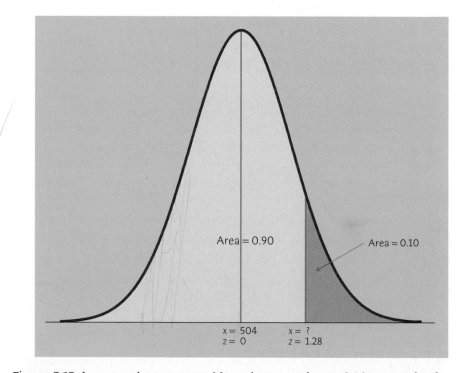

Figure 3.13 Locating the point on a Normal curve with area 0.10 to its right, for Example 3.8.

So x itself satisfies

$$\frac{x - 504}{111} = 1.28$$

Solving this equation for x gives

$$x = 504 + (1.28)(111) = 646.1$$

This equation should make sense: it says that x lies 1.28 standard deviations above the mean on this particular Normal curve. That is the "unstandardized" meaning of $z = 1.28$. We see that a student must score at least 647 to place in the highest 10%.

Here is the general formula for unstandardizing a z-score. To find the value x from the Normal distribution with mean μ and standard deviation σ corresponding to a given standard Normal value z, use

$$x = \mu + z\sigma$$

APPLY YOUR KNOWLEDGE

3.12 Use Table A to find the value z of a standard Normal variable that satisfies each of the following conditions. (Use the value of z from Table A that comes closest to satisfying the condition.) In each case, sketch a standard Normal curve with your value of z marked on the axis.

(a) The point z with 25% of the observations falling below it.

(b) The point z with 40% of the observations falling above it.

3.13 IQ test scores. Scores on the Wechsler Adult Intelligence Scale are approximately Normally distributed with $\mu = 100$ and $\sigma = 15$.

(a) What IQ scores fall in the lowest 25% of the distribution?

(b) How high an IQ score is needed to be in the highest 5%?

The bell curve?

Does the distribution of human intelligence follow the "bell curve" of a Normal distribution? Scores on IQ tests do roughly follow a Normal distribution. That is because a test score is calculated from a person's answers in a way that is designed to produce a Normal distribution. To conclude that intelligence follows a bell curve, we must agree that the test scores directly measure intelligence. Many psychologists don't think there is one human characteristic that we can call "intelligence" and can measure by a single test score.

Chapter 3 SUMMARY

We can sometimes describe the overall pattern of a distribution by a **density curve.** A density curve has total area 1 underneath it. An area under a density curve gives the proportion of observations that fall in a range of values.

A density curve is an idealized description of the overall pattern of a distribution that smooths out the irregularities in the actual data. We write the mean of a density curve as μ and the standard deviation of a density curve as σ to distinguish them from the mean $\overline{x}$ and standard deviation s of the actual data.

The mean, the median, and the quartiles of a density curve can be located by eye. The **mean** μ is the balance point of the curve. The **median** divides the area under the curve in half. The **quartiles** and the median divide the area under the curve into quarters. The **standard deviation** σ cannot be located by eye on most density curves.

The mean and median are equal for symmetric density curves. The mean of a skewed curve is located farther toward the long tail than is the median.

The **Normal distributions** are described by a special family of bell-shaped, symmetric density curves, called **Normal curves.** The mean μ and standard deviation σ completely specify a Normal distribution $N(\mu, \sigma)$. The mean is the center of the curve, and σ is the distance from μ to the change-of-curvature points on either side.

To **standardize** any observation x, subtract the mean of the distribution and then divide by the standard deviation. The resulting **z-score**

$$z = \frac{x - \mu}{\sigma}$$

says how many standard deviations x lies from the distribution mean.

All Normal distributions are the same when measurements are transformed to the standardized scale. In particular, all Normal distributions satisfy the **68–95–99.7 rule,** which describes what percent of observations lie within one, two, and three standard deviations of the mean.

If x has the $N(\mu, \sigma)$ distribution, then the **standardized variable** $z = (x - \mu)/\sigma$ has the **standard Normal distribution** $N(0, 1)$ with mean 0 and standard deviation 1. Table A gives the proportions of standard Normal observations that are less than z for many values of z. By standardizing, we can use Table A for any Normal distribution.

Chapter 3 EXERCISES

3.14 Figure 3.14 shows two Normal curves, both with mean 0. Approximately what is the standard deviation of each of these curves?

> **IQ test scores.** *The Wechsler Adult Intelligence Scale (WAIS) is the most common "IQ test." The scale of scores is set separately for each age group and is approximately Normal with mean 100 and standard deviation 15. Use this distribution and the 68–95–99.7 rule to answer Exercises 3.15 to 3.17.*

3.15 About what percent of people have WAIS scores

    (a) above 100?

    (b) above 145?

    (c) below 85?

3.16 **Retardation.** People with WAIS scores below 70 are considered mentally retarded when, for example, applying for Social Security disability benefits. What percent of adults are retarded by this criterion?

3.17 **MENSA.** The organization MENSA, which calls itself "the high IQ society," requires a WAIS score of 130 or higher for membership.

(Mel Yates/Taxi/Getty Images)

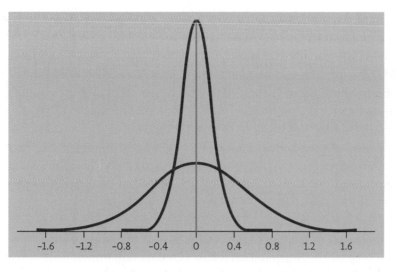

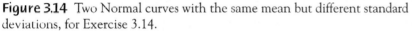

Figure 3.14 Two Normal curves with the same mean but different standard deviations, for Exercise 3.14.

(Similar scores on other tests are also accepted.) What percent of adults would qualify for membership?

3.18 Standard Normal drill. Use Table A to find the proportion of observations from a standard Normal distribution that falls in each of the following regions. In each case, sketch a standard Normal curve and shade the area representing the region.

(a) $z \leq -2.25$

(b) $z \geq -2.25$

(c) $z > 1.77$

(d) $-2.25 < z < 1.77$

3.19 Standard Normal drill.

(a) Find the number z such that the proportion of observations that are less than z in a standard Normal distribution is 0.8.

(b) Find the number z such that 35% of all observations from a standard Normal distribution are greater than z.

3.20 NCAA rules for athletes. The National Collegiate Athletic Association (NCAA) requires Division I athletes to score at least 820 on the combined mathematics and verbal parts of the SAT exam to compete in their first college year. (Higher scores are required for students with poor high school grades.) In 2002, the scores of the 1.3 million students taking the SATs were approximately Normal with mean 1020 and standard deviation 207. What percent of all students had scores less than 820?

3.21 More NCAA rules. The NCAA considers a student a "partial qualifier" eligible to practice and receive an athletic scholarship, but not to compete, if the combined SAT score is at least 720. Use the information in the previous exercise to find the percent of all SAT scores that are less than 720.

3.22 Heights of men and women. The heights of women aged 20 to 29 follow approximately the $N(64, 2.7)$ distribution. Men the same age have heights distributed as $N(69.3, 2.8)$. What percent of young women are taller than the mean height of young men?

3.23 Heights of men and women. The heights of women aged 20 to 29 follow approximately the $N(64, 2.7)$ distribution. Men the same age have heights distributed as $N(69.3, 2.8)$. What percent of young men are shorter than the mean height of young women?

3.24 Length of pregnancies. The length of human pregnancies from conception to birth varies according to a distribution that is approximately Normal with mean 266 days and standard deviation 16 days.

(a) What percent of pregnancies last less than 240 days (that's about 8 months)?

(b) What percent of pregnancies last between 240 and 270 days (roughly between 8 months and 9 months)?

(c) How long do the longest 20% of pregnancies last?

3.25 A surprising calculation. Changing the mean of a Normal distribution by a moderate amount can greatly change the percent of observations in the tails. Suppose that a college is looking for applicants with SAT math scores 750 and above.

(a) In 2002, the scores of men on the math SAT followed the $N(534, 116)$ distribution. What percent of men scored 750 or better?

(b) Women's SAT math scores that year had the $N(500, 110)$ distribution. What percent of women scored 750 or better? You see that the percent of men above 750 is almost three times the percent of women with such high scores.

3.26 Grading managers. Many companies "grade on a bell curve" to compare the performance of their managers and professional workers. This forces the use of some low performance ratings, so that not all workers are listed as "above average." Ford Motor Company's "performance management process" for a time assigned 10% A grades, 80% B grades, and 10% C grades to the company's 18,000 managers. Suppose that Ford's performance scores really are Normally distributed. This year, managers with scores less than 25 received C's and those with scores above 475 received A's. What are the mean and standard deviation of the scores?

3.27 Weights aren't Normal. The heights of people of the same sex and similar ages follow Normal distributions reasonably closely. Weights, on

(Jim McGuire/Index Stock Imagery/PictureQuest)

the other hand, are not Normally distributed. The weights of women aged 20 to 29 have mean 141.7 pounds and median 133.2 pounds. The first and third quartiles are 118.3 pounds and 157.3 pounds. What can you say about the shape of the weight distribution? Why?

3.28 **Quartiles.** The median of any Normal distribution is the same as its mean. We can use Normal calculations to find the quartiles for Normal distributions.

(a) What is the area under the standard Normal curve to the left of the first quartile? Use this to find the value of the first quartile for a standard Normal distribution. Find the third quartile similarly.

(b) Your work in (a) gives the z-scores for the quartiles of any Normal distribution. What are the quartiles for the lengths of human pregnancies? (Use the distribution in Exercise 3.24.)

3.29 **Deciles.** The *deciles* of any distribution are the points that mark off the lowest 10% and the highest 10%. On a density curve, these are the points with areas 0.1 and 0.9 to their left under the curve.
(a) What are the deciles of the standard Normal distribution?
(b) The heights of young women are approximately Normal with mean 64 inches and standard deviation 2.7 inches. What are the deciles of this distribution?

Chapter 3 MEDIA EXERCISES

The *Normal Curve* applet allows you to do Normal calculations quickly. It is somewhat limited by the number of pixels available for use, so that it can't hit every value exactly. In the exercises below, use the closest available values. In each case, *make a sketch* of the curve from the applet marked with the values you used to answer the questions asked.

3.30 **How accurate is 68–95–99.7?** The 68–95–99.7 rule for Normal distributions is a useful approximation. To see how accurate the rule is, drag one flag across the other so that the applet shows the area under the curve between the two flags.

(a) Place the flags one standard deviation on either side of the mean. What is the area between these two values? What does the 68 95 99.7 rule say this area is?

(b) Repeat for locations two and three standard deviations on either side of the mean. Again compare the 68–95–99.7 rule with the area given by the applet.

3.31 **Where are the quartiles?**   How many standard deviations above and below the mean do the quartiles of any Normal distribution lie? (Use the standard Normal distribution to answer this question.)

3.32 Grading managers. In Exercise 3.26, we saw that Ford Motor Company grades its managers in such a way that the top 10% receive an A grade, the bottom 10% a C, and the middle 80% a B. Let's suppose that performance scores follow a Normal distribution. How many standard deviations above and below the mean do the A/B and B/C cutoffs lie? (Use the standard Normal distribution to answer this question.)

3.33 Do the data look Normal? The EESEE stories provide data from many studies in various fields. Make a histogram of each of the following variables. In each case, do the data appear roughly Normal (at least single-peaked, symmetric, tails falling off on both side of the peak) or is their shape clearly not Normal? If the distribution is not Normal, briefly describe its shape.

(a) The number of single owls per square kilometer, from "Habitat of the Spotted Owl."

(b) Pretest Degree of Reading Power scores, from "Checkmating and Reading Skills."

(c) Carbon monoxide emissions from the refinery (only), from "Emissions from an Oil Refinery."

(Richard Hamilton Smith/Corbis)

Scatterplots and Correlation

A medical study finds that short women are more likely to have heart at-tacks than women of average height, while tall women have the fewest heart attacks. An insurance group reports that heavier cars have fewer deaths per 10,000 vehicles registered than do lighter cars. These and many other statisti-cal studies look at the relationship between two variables. To understand such a relationship, we must often examine other variables as well. To conclude that shorter women have higher risk from heart attacks, for example, the researchers had to eliminate the effect of other variables such as weight and exercise habits. In this chapter we begin our study of relationships between variables. One of our main themes is that the relationship between two variables can be strongly influenced by other variables that are lurking in the background.

Explanatory and response variables

Because variation is everywhere, statistical relationships are overall tenden-cies, not ironclad rules. They allow individual exceptions. Although smokers on the average die younger than nonsmokers, some people live to 90 while smoking three packs a day. To study a relationship between two variables, we measure both variables on the same individuals. Often, we think that one of the variables explains or influences the other.

> **RESPONSE VARIABLE, EXPLANATORY VARIABLE**
>
> A **response variable** measures an outcome of a study. An **explanatory variable** explains or influences changes in a response variable.

independent variable
dependent variable

You will often find explanatory variables called **independent variables,** and response variables called **dependent variables.** The idea behind this language is that the response variable depends on the explanatory variable. Because "independent" and "dependent" have other meanings in statistics that are unrelated to the explanatory-response distinction, we prefer to avoid those words.

It is easiest to identify explanatory and response variables when we actually set values of one variable in order to see how it affects another variable.

EXAMPLE 4.1   The effects of alcohol

Alcohol has many effects on the body. One effect is a drop in body temperature. To study this effect, researchers give several different amounts of alcohol to mice, then measure the change in each mouse's body temperature in the 15 minutes after taking the alcohol. Amount of alcohol is the explanatory variable, and change in body temperature is the response variable.

When we don't set the values of either variable but just observe both variables, there may or may not be explanatory and response variables. Whether there are depends on how we plan to use the data.

EXAMPLE 4.2   Voting patterns

A political scientist looks at the Democrats' share of the popular vote for president in each state in the two most recent presidential elections. She does not wish to explain one year's data by the other's, but rather to find a pattern that may shed light on political conditions. The political scientist has two related variables, but she does not use the explanatory-response distinction.

A political consultant looks at the same data with an eye to including the previous election result in a state as one of several explanatory variables in a model to predict how the state will vote in the next election. Now the earlier election result is an explanatory variable and the later result is a response variable.

In Example 4.1, alcohol actually *causes* a change in body temperature. There is no cause-and-effect relationship between votes in different elections in Example 4.2. If there is a close relationship, however, we can use a state's percent Democratic in the last election to help predict the next election. Chapter 5 discusses prediction. Prediction requires that we identify an explanatory variable and a response variable. Some other statistical techniques ignore this distinction. Do remember that calling one variable explanatory and the other response doesn't necessarily mean that changes in one *cause* changes in the other.

Most statistical studies examine data on more than one variable. Fortunately, statistical analysis of several-variable data builds on the tools we used to examine individual variables. The principles that guide our work also remain the same:

- First plot the data, then add numerical summaries.

- Look for overall patterns and deviations from those patterns.

- When the overall pattern is quite regular, use a compact mathematical model to describe it.

After you plot your data, think!

The statistician Abraham Wald (1902–1950) worked on war problems during World War II. Wald invented some statistical methods that were military secrets until the war ended. Here is one of his simpler ideas. Asked where extra armor should be added to airplanes, Wald studied the location of enemy bullet holes in planes returning from combat. He plotted the locations on an outline of the plane. As data accumulated, most of the outline filled up. Put the armor in the few spots with no bullet holes, said Wald. That's where bullets hit the planes that didn't make it back.

APPLY YOUR KNOWLEDGE

4.1 **Explanatory and response variables?** In each of the following situations, is it more reasonable to simply explore the relationship between the two variables or to view one of the variables as an explanatory variable and the other as a response variable? In the latter case, which is the explanatory variable and which is the response variable?

(a) The amount of time a student spends studying for a statistics exam and the grade on the exam.

(b) The weight in kilograms and height in centimeters of a person.

(c) Inches of rain in the growing season and the yield of corn in bushels per acre.

(d) A student's scores on the SAT math exam and the SAT verbal exam.

(e) A family's income and the years of education their eldest child completes.

4.2 **Coral reefs.** How sensitive to changes in water temperature are coral reefs? To find out, measure the growth of corals in aquariums where the water temperature is controlled at different levels. Growth is measured by weighing the coral before and after the experiment. What are the explanatory and response variables? Are they categorical or quantitative?

4.3 **The risks of obesity.** A study observes a large group of people over a 10-year period. The goal is to see if overweight and obese people are more likely to die during the decade than people who weigh less. Such studies can be misleading, because obese people are more likely to be inactive and to be poor. What is the explanatory variable and the response variable? What other variables are mentioned that may influence the relationship between the explanatory variable and the response variable?

Displaying relationships: scatterplots

Relationships between two quantitative variables are best displayed graphically. The most useful graph for this purpose is a *scatterplot*.

EXAMPLE 4.3 State SAT scores

Some people use average SAT scores to rank state or local school systems. This is not proper, because the percent of high school students who take the SAT varies from place to place. Let's examine the relationship between the percent of a state's high school graduates who took the exam in 2002 and the state average SAT verbal score that year.

We think that "percent taking" will help explain "average score." Therefore, "percent taking" is the explanatory variable and "average score" is the response variable. We want to see how average score changes when percent taking changes, so we put percent taking (the explanatory variable) on the horizontal axis. Figure 4.1 is the scatterplot. Each point represents a single state. In Colorado, for example, 28% took the SAT, and the average SAT verbal score was 543. Find 28 on the *x* (horizontal) axis and 543 on the *y* (vertical) axis. Colorado appears as the point (28, 543) above 28 and to the right of 543. Figure 4.1 shows how to locate Colorado's point on the plot.

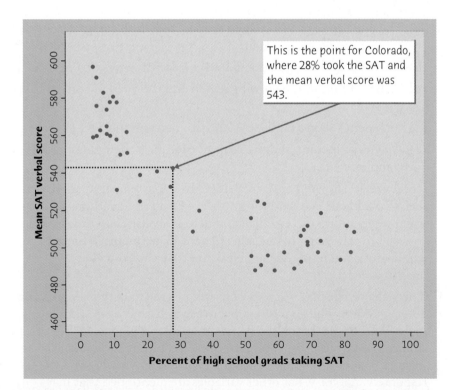

Figure 4.1 Scatterplot of the mean SAT verbal score in each state against the percent of that state's high school graduates who take the SAT. The dotted lines intersect at the point (28, 543), the data for Colorado.

SCATTERPLOT

A **scatterplot** shows the relationship between two quantitative variables measured on the same individuals. The values of one variable appear on the horizontal axis, and the values of the other variable appear on the vertical axis. Each individual in the data appears as the point in the plot fixed by the values of both variables for that individual.

Always plot the explanatory variable, if there is one, on the horizontal axis (the x axis) of a scatterplot. As a reminder, we usually call the explanatory variable x and the response variable y. If there is no explanatory-response distinction, either variable can go on the horizontal axis.

APPLY YOUR KNOWLEDGE

4.4 Bird colonies. One of nature's patterns connects the percent of adult birds in a colony that return from the previous year and the number of new adults that join the colony. Here are data for 13 colonies of sparrowhawks:[1]

Percent returning	New adults	Percent returning	New adults	Percent returning	New adults
74	5	62	15	46	18
66	6	52	16	60	19
81	8	45	17	46	20
52	11	62	18	38	20
73	12				

(Chris Beddall/Papilio/CORBIS)

Plot the count of new birds (response) against the percent of returning birds (explanatory).

Interpreting scatterplots

To interpret a scatterplot, apply the strategies of data analysis learned in Chapters 1 to 3.

EXAMINING A SCATTERPLOT

In any graph of data, look for the **overall pattern** and for striking **deviations** from that pattern.

You can describe the overall pattern of a scatterplot by the **form, direction,** and **strength** of the relationship.

An important kind of deviation is an **outlier,** an individual value that falls outside the overall pattern of the relationship.

---

clusters

EXAMPLE 4.4   Understanding state SAT scores

Figure 4.1 shows a clear *form:* there are two distinct **clusters** of states with a gap between them. In the cluster at the right of the plot, 52% or more of high school graduates take the SAT, and the average scores are low. The states in the cluster at the left have higher SAT scores and no more than 36% of graduates take the test. There are no clear outliers. That is, no points fall clearly outside the clusters.

What explains the clusters? There are two widely used college entrance exams, the SAT and the ACT Assessment exam. Each state favors one or the other. The left cluster in Figure 4.1 contains the ACT states, and the SAT states make up the right cluster. In ACT states, most students who take the SAT are applying to a selective college that requires SAT scores. This select group of students has a higher average score than the much larger group of students who take the SAT in SAT states.

The relationship in Figure 4.1 also has a clear *direction:* states in which a higher percent of students take the SAT tend to have lower average scores. This is a *negative association* between the two variables.

POSITIVE ASSOCIATION, NEGATIVE ASSOCIATION

Two variables are **positively associated** when above-average values of one tend to accompany above-average values of the other, and below-average values also tend to occur together.

Two variables are **negatively associated** when above-average values of one tend to accompany below-average values of the other, and vice versa.

The *strength* of a relationship in a scatterplot is determined by how closely the points follow a clear form. The overall relationship in Figure 4.1 is not strong—states with similar percents taking the SAT show quite a bit of scatter in their average scores. Here is an example of a stronger relationship with a clearer form.

EXAMPLE 4.5   Counting carnivores

Ecologists look at data to learn about nature's patterns. One pattern they have found relates the size of a carnivore (body mass in kilograms) to how many of those carnivores there are in an area. The right measure of "how many" is to count carnivores per 10,000 kilograms of their prey in the area. Table 4.1 gives data for 25 carnivore species.[2]

To see the pattern, plot carnivore abundance (response) against body mass (explanatory). Biologists often find that patterns involving sizes and counts are simpler when we plot the logarithms of the data. Figure 4.2 does that—you can see that 1, 10, 100, and 1000 are equally spaced on the vertical scale.

This scatterplot shows a **moderately strong negative association.** Bigger carnivores are less abundant. The form of the association is **linear.** It is striking that animals from many different parts of the world should fit so simple a pattern. We could use the straight-line pattern to predict the abundance of another carnivore species from its body mass.

TABLE 4.1 Size and abundance of carnivores

Carnivore species	Body mass (kg)	Abundance	Carnivore species	Body mass (kg)	Abundance
Least weasel	0.14	1656.49	Eurasian lynx	20.0	0.46
Ermine	0.16	406.66	Wild dog	25.0	1.61
Small Indian mongoose	0.55	514.84	Dhole	25.0	0.81
Pine marten	1.3	31.84	Snow leopard	40.0	1.89
Kit fox	2.02	15.96	Wolf	46.0	0.62
Channel Island fox	2.16	145.94	Leopard	46.5	6.17
Arctic fox	3.19	21.63	Cheetah	50.0	2.29
Red fox	4.6	32.21	Puma	51.9	0.94
Bobcat	10.0	9.75	Spotted hyena	58.6	0.68
Canadian lynx	11.2	4.79	Lion	142.0	3.40
European badger	13.0	7.35	Tiger	181.0	0.33
Coyote	13.0	11.65	Polar bear	310.0	0.60
Ethiopian wolf	14.5	2.70			

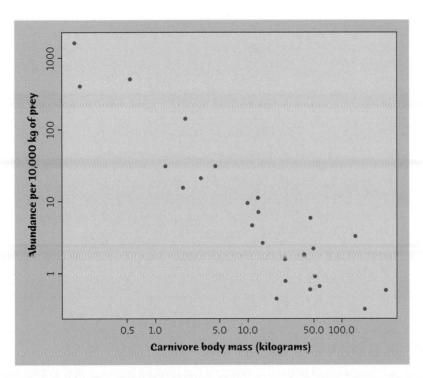

Figure 4.2 Scatterplot of the abundance of 25 species of carnivores against their body mass. Larger carnivores are less abundant. (Logarithmic scales are used for both variables.)

Of course, not all relationships have a simple form and a clear direction that we can describe as positive association or negative association. Exercise 4.6 gives an example that has no clear direction.

APPLY YOUR KNOWLEDGE

4.5 **Bird colonies.** Describe the form, direction, and strength of the relationship between number of new sparrowhawks in a colony and percent of returning adults, as displayed in your plot from Exercise 4.4.

For short-lived birds, the association between these variables is positive: changes in weather and food supply drive the populations of new and returning birds up or down together. For long-lived territorial birds, on the other hand, the association is negative because returning birds claim their territories in the colony and don't leave room for new recruits. Which type of species is the sparrowhawk?

4.6 **Does fast driving waste fuel?** How does the fuel consumption of a car change as its speed increases? Here are data for a British Ford Escort. Speed is measured in kilometers per hour, and fuel consumption is measured in liters of gasoline used per 100 kilometers traveled.[3]

Speed (km/h)	Fuel used (liters/100 km)	Speed (km/h)	Fuel used (liters/100 km)
10	21.00	90	7.57
20	13.00	100	8.27
30	10.00	110	9.03
40	8.00	120	9.87
50	7.00	130	10.79
60	5.90	140	11.77
70	6.30	150	12.83
80	6.95		

(a) Make a scatterplot. (Which is the explanatory variable?)

(b) Describe the form of the relationship. It is not linear. Explain why the form of the relationship makes sense.

(c) It does not make sense to describe the variables as either positively associated or negatively associated. Why?

(d) Is the relationship reasonably strong or quite weak? Explain your answer.

Adding categorical variables to scatterplots

The South long lagged behind the rest of the United States in the performance of its schools. Efforts to improve education have reduced the gap. We wonder if the South stands out in our study of state mean SAT scores.

EXAMPLE 4.6 **The South has risen**

Figure 4.3 enhances the scatterplot in Figure 4.1 by plotting 13 southern states with a different plot color. You see that the southern states blend in with the rest of

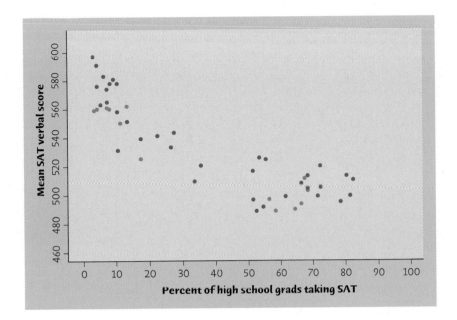

Figure 4.3 Mean SAT verbal score and percent of high school graduates who take the test, by state, with the southern states highlighted.

the country. Some are ACT states and some are SAT states, but neither group has markedly low SAT scores. The South no longer underperforms, at least as measured by SAT scores.

Dividing the states into "southern" and "nonsouthern" introduces a third variable into the scatterplot. This is a categorical variable that has only two values. The two values are displayed by the two different plotting colors.

CATEGORICAL VARIABLES IN SCATTERPLOTS

To add a categorical variable to a scatterplot, use a different plot color or symbol for each category.

APPLY YOUR KNOWLEDGE

4.7 How fast do icicles grow? Japanese researchers measured the growth of icicles in a cold chamber under various conditions of temperature, wind, and water flow.[4] Table 4.2 contains data produced under two sets of conditions. In both cases, there was no wind and the temperature was set at $-11°$ C. Water flowed over the icicle at a higher rate (29.6 milligrams per second) in run 8905 and at a slower rate (11.9 mg/s) in run 8903.

(a) Make a scatterplot of the length of the icicle in centimeters versus time in minutes, using separate symbols or colors for the two runs.

(b) What does your plot show about the pattern of growth of icicles? What does it show about the effect of changing the rate of water flow on icicle growth?

TABLE 4.2 Growth of icicles over time

Run 8903				Run 8905			
Time (min)	Length (cm)	Time (min)	Length (cm)	Time (min)	Length (cm)	Time (min)	Length (cm)
10	0.6	130	18.1	10	0.3	130	10.4
20	1.8	140	19.9	20	0.6	140	11.0
30	2.9	150	21.0	30	1.0	150	11.9
40	4.0	160	23.4	40	1.3	160	12.7
50	5.0	170	24.7	50	3.2	170	13.9
60	6.1	180	27.8	60	4.0	180	14.6
70	7.9			70	5.3	190	15.8
80	10.1			80	6.0	200	16.2
90	10.9			90	6.9	210	17.9
100	12.7			100	7.8	220	18.8
110	14.4			110	8.3	230	19.9
120	16.6			120	9.6	240	21.1

Measuring linear association: correlation

A scatterplot displays the form, direction, and strength of the relationship between two quantitative variables. Linear relations are particularly important because a straight line is a simple pattern that is quite common. We say a linear relation is strong if the points lie close to a straight line, and weak if they are widely scattered about a line. Our eyes are not good judges of how strong a linear relationship is. The two scatterplots in Figure 4.4 depict exactly the same data, but the lower plot is drawn smaller in a large field. The lower plot seems to show a stronger linear relationship. Our eyes can be fooled by changing the plotting scales or the amount of white space around the cloud of points in a scatterplot.[5] We need to follow our strategy for data analysis by using a numerical measure to supplement the graph. *Correlation* is the measure we use.

CORRELATION

The **correlation** measures the direction and strength of the linear relationship between two quantitative variables. Correlation is usually written as r.

Suppose that we have data on variables x and y for n individuals. The values for the first individual are x_1 and y_1, the values for the second individual are x_2 and y_2, and so on. The means and standard deviations of the two variables are $\overline{x}$ and s_x for the x-values, and $\overline{y}$ and s_y for the y-values. The correlation r between x and y is

$$r = \frac{1}{n-1} \sum \left(\frac{x_i - \overline{x}}{s_x} \right) \left(\frac{y_i - \overline{y}}{s_y} \right)$$

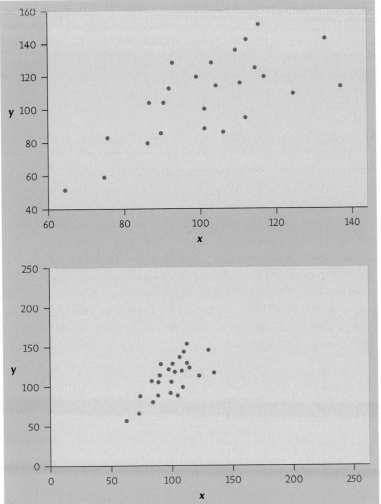

Figure 4.4 Two scatterplots of the same data. The straight-line pattern in the lower plot appears stronger because of the surrounding white space.

As always, the summation sign Σ means "add these terms for all the individuals." The formula for the correlation r is a bit complex. It helps us see what correlation is, but in practice you should use software or a calculator that finds r from keyed-in values of two variables x and y. Exercise 4.8 asks you to calculate a correlation step-by-step from the definition to solidify its meaning.

The formula for r begins by standardizing the observations. Suppose, for example, that x is height in centimeters and y is weight in kilograms and that we have height and weight measurements for n people. Then $\bar{x}$ and s_x are the mean and standard deviation of the n heights, both in centimeters. The value

$$\frac{x_i - \bar{x}}{s_x}$$

Death from superstition?

Is there a relationship between superstitious beliefs and bad things happening? Apparently there is. Chinese and Japanese people think the number 4 is unlucky, because when pronounced it sounds like the word for "death." Sociologists looked at 15 years' of death certificates for Chinese and Japanese Americans and for white Americans. Deaths from heart disease were notably higher on the fourth day of the month among Chinese and Japanese, but not among whites. The sociologists think the explanation is increased stress on "unlucky days."

is the standardized height of the ith person, familiar from Chapter 1. The standardized height says how many standard deviations above or below the mean a person's height lies. Standardized values have no units—in this example, they are no longer measured in centimeters. Standardize the weights also. The correlation r is an average of the products of the standardized height and the standardized weight for the n people.

APPLY YOUR KNOWLEDGE

4.8   **Classifying fossils.** *Archaeopteryx* is an extinct beast having feathers like a bird but teeth and a long bony tail like a reptile. Only six fossil specimens are known. Because these specimens differ greatly in size, some scientists think they belong to different species. We will examine some data. If the specimens belong to the same species and differ in size because some are younger than others, there should be a positive linear relationship between the lengths of a pair of bones from all individuals. An outlier from this relationship would suggest a different species. Here are data on the lengths in centimeters of the femur (a leg bone) and the humerus (a bone in the upper arm) for the five specimens that preserve both bones:[6]

Femur	38	56	59	64	74
Humerus	41	63	70	72	84

(a) Make a scatterplot. Do you think that all five specimens come from the same species?

(b) Find the correlation r step-by-step. That is, find the mean and standard deviation of the femur lengths and of the humerus lengths. (Use your calculator for means and standard deviations.) Then find the five standardized values for each variable and use the formula for r.

(c) Now enter these data into your calculator and use the calculator's correlation function to find r. Check that you get the same result as in (b).

Facts about correlation

The formula for correlation helps us see that r is positive when there is a positive association between the variables. Height and weight, for example, have a positive association. People who are above average in height tend to also be above average in weight. Both the standardized height and the standardized weight are positive. People who are below average in height tend to also have below-average weight. Then both standardized height and standardized weight are negative. In both cases, the products in the formula for r are mostly positive and so r is positive. In the same way, we can see that r is negative when the association between x and y is negative. More detailed study of the formula gives more detailed properties of r. Here is what you need to know in order to interpret correlation.

1. Correlation makes no distinction between explanatory and response variables. It makes no difference which variable you call x and which you call y in calculating the correlation.

2. Correlation requires that both variables be quantitative, so that it makes sense to do the arithmetic indicated by the formula for r. We cannot calculate a correlation between the incomes of a group of people and what city they live in, because city is a categorical variable.

3. Because r uses the standardized values of the observations, r does not change when we change the units of measurement of x, y, or both. Measuring height in inches rather than centimeters and weight in pounds rather than kilograms does not change the correlation between height and weight. The correlation r itself has no unit of measurement; it is just a number.

4. Positive r indicates positive association between the variables, and negative r indicates negative association.

5. The correlation r is always a number between -1 and 1. Values of r near 0 indicate a very weak linear relationship. The strength of the linear relationship increases as r moves away from 0 toward either -1 or 1. Values of r close to -1 or 1 indicate that the points in a scatterplot lie close to a straight line. The extreme values $r = -1$ and $r = 1$ occur only in the case of a perfect linear relationship, when the points lie exactly along a straight line.

6. Correlation measures the strength of only a linear relationship between two variables. Correlation does not describe curved relationships between variables, no matter how strong they are.

7. Like the mean and standard deviation, the correlation is not resistant: r is strongly affected by a few outlying observations. Use r with caution when outliers appear in the scatterplot.

The scatterplots in Figure 4.5 illustrate how values of r closer to 1 or -1 correspond to stronger linear relationships. To make the meaning of r clearer, the standard deviations of both variables in these plots are equal, and the horizontal and vertical scales are the same. In general, it is not so easy to guess the value of r from the appearance of a scatterplot. Remember that changing the plotting scales in a scatterplot may mislead our eyes, but it does not change the correlation. To explore how extreme observations can influence r, use the *Correlation and Regression* applet.

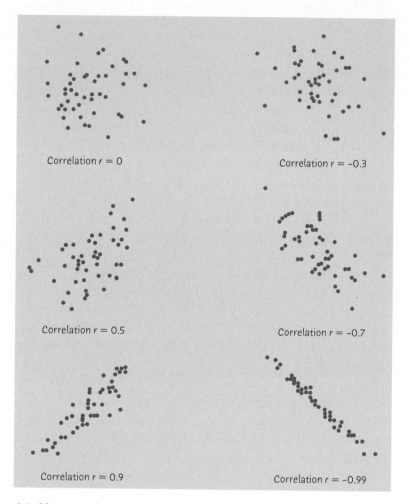

Figure 4.5 How correlation measures the strength of a linear relationship. Patterns closer to a straight line have correlations closer to 1 or −1.

The real data we have examined also illustrate how correlation measures the strength and direction of linear relationships. Figure 4.2 shows a strong negative linear relationship between the logarithms of body mass and abundance for carnivore species. The correlation is $r = -0.9124$. Figure 4.1 shows a weaker but still quite strong negative association between percent of students taking the SAT and the mean SAT verbal score in a state. The correlation is $r = -0.8814$.

Do remember that **correlation is not a complete description of two-variable data,** even when the relationship between the variables is linear. You should give the means and standard deviations of both x and y along with the correlation. (Because the formula for correlation uses the means and standard deviations, these measures are the proper choice to accompany a correlation.) Conclusions based on correlations alone may require rethinking in the light of a more complete description of the data.

EXAMPLE 4.7 Scoring figure skaters

Each phase of a figure-skating competition is scored by a panel of judges who use a scale from 1 to 6. The subjective nature of the scoring often results in controversy. We have the scores awarded by two judges, Marina and Steve, for a large number of performances. How well do they agree? Calculation shows that the correlation between their scores is $r = 0.9$, but the mean of Marina's scores is 0.5 point lower than Steve's mean.

These facts do not contradict each other. They are simply different kinds of information. The mean scores show that Marina awards lower scores than Steve. But because Marina gives *every* skater a score about 0.5 point lower than Steve, the correlation remains high. Adding or subtracting the same number to all values of either x or y does not change the correlation. If Marina and Steve both rate all skaters, the contest is fairly scored because they agree on which performances are better than others. The high r shows their agreement. But if Marina scores one skater and Steve another, we must add 0.5 point to Marina's score to arrive at a fair comparison.

APPLY YOUR KNOWLEDGE

4.9 Changing the correlation.

(a) Use your calculator to find the correlation between the percent of returning birds and the number of new birds from the data in Exercise 4.4.

(b) Make a scatterplot of the data with two new points added. Point A: 10% return, 25 new birds. Point B: 40% return, 5 new birds. Find two new correlations, for the original data plus Point A, and for the original data plus Point B.

(c) Explain in terms of what correlation measures why adding Point A makes the correlation stronger (closer to -1) and adding Point B makes the correlation weaker (closer to 0).

4.10 Wives and husbands. If women always married men who were 2 years older than themselves, what would be the correlation between the ages of husband and wife? (Hint: Draw a scatterplot for several ages.)

4.11 Strong association but no correlation. The gas mileage of an automobile first increases and then decreases as the speed increases. Suppose that this relationship is very regular, as shown by the following data on speed (miles per hour) and mileage (miles per gallon):

Speed	20	30	40	50	60
MPG	24	28	30	28	24

Make a scatterplot of mileage versus speed. Show that the correlation between speed and mileage is $r = 0$. Explain why the correlation is 0 even though there is a strong relationship between speed and mileage.

Chapter 4 SUMMARY

To study relationships between variables, we must measure the variables on the same group of individuals.

If we think that a variable x may explain or even cause changes in another variable y, we call x an **explanatory variable** and y a **response variable.**

A **scatterplot** displays the relationship between two quantitative variables measured on the same individuals. Mark values of one variable on the horizontal axis (x axis) and values of the other variable on the vertical axis (y axis). Plot each individual's data as a point on the graph. Always plot the explanatory variable, if there is one, on the x axis of a scatterplot.

Plot points with different colors or symbols to see the effect of a categorical variable in a scatterplot.

In examining a scatterplot, look for an overall pattern showing the **form, direction,** and **strength** of the relationship, and then for **outliers** or other deviations from this pattern.

Form: Linear relationships, where the points show a straight-line pattern, are an important form of relationship between two variables. Curved relationships and **clusters** are other forms to watch for.

Direction: If the relationship has a clear direction, we speak of either **positive association** (high values of the two variables tend to occur together) or **negative association** (high values of one variable tend to occur with low values of the other variable.)

Strength: The **strength** of a relationship is determined by how close the points in the scatterplot lie to a simple form such as a line.

The **correlation r** measures the strength and direction of the linear association between two quantitative variables x and y. Although you can calculate a correlation for any scatterplot, r measures only straight-line relationships.

Correlation indicates the direction of a linear relationship by its sign: $r > 0$ for a positive association and $r < 0$ for a negative association. Correlation always satisfies $-1 \leq r \leq 1$ and indicates the strength of a relationship by how close it is to -1 or 1. Perfect correlation, $r = \pm 1$, occurs only when the points on a scatterplot lie exactly on a straight line.

Correlation ignores the distinction between explanatory and response variables. The value of r is not affected by changes in the unit of measurement of either variable. Correlation is not resistant, so outliers can greatly change the value of r.

Chapter 4 EXERCISES

4.12 Sports car gas mileage. Table 1.2 (page 12) gives the city and highway gas mileages for two-seater cars. We expect a positive association between the city and highway mileages of a group of vehicles. Make a scatterplot that shows the relationship between city and highway mileage, using city mileage as the explanatory variable. Describe the overall pattern. Does the outlier (the Honda Insight) extend the pattern of the other cars or is it far from the line they form?

4.13 IQ and school grades. Do students with higher IQ test scores tend to do better in school? Figure 4.6 is a scatterplot of IQ and school grade point average (GPA) for all 78 seventh-grade students in a rural Midwest school.[7]

(a) Say in words what a positive association between IQ and GPA would mean. Does the plot show a positive association?

(b) What is the form of the relationship? Is it roughly linear? Is it very strong? Explain your answers.

(c) At the bottom of the plot are several points that we might call outliers. One student in particular has a very low GPA despite an average IQ score. What are the approximate IQ and GPA for this student?

4.14 Sports car gas mileage. Find the correlation between city and highway mileage in Table 1.2, leaving out the Insight. Explain how the correlation matches the pattern of the plot. Based on your plot in Exercise 4.12, will adding the Insight make the correlation stronger

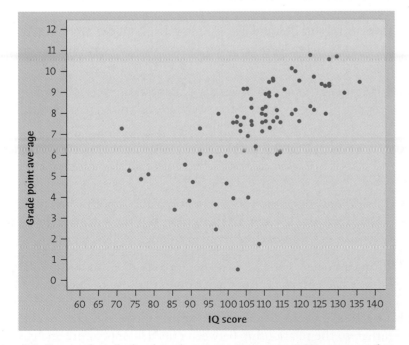

Figure 4.6 Scatterplot of school grade point average versus IQ test score for seventh-grade students, for Exercise 4.13.

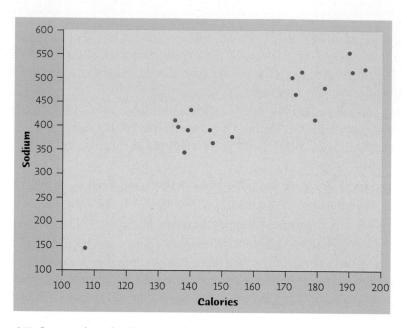

Figure 4.7 Scatterplot of milligrams of sodium and calories in each of 17 brands of meat hot dogs, for Exercise 4.15.

(closer to 1) or weaker? Verify your guess by calculating the correlation for all 22 cars.

4.15 Calories and salt in hot dogs. Are hot dogs that are high in calories also high in salt? Figure 4.7 is a scatterplot of the calories and salt content (measured as milligrams of sodium) in 17 brands of meat hot dogs.[8]

(a) Roughly what are the lowest and highest calorie counts among these brands? Roughly what is the sodium level in the brands with the fewest and with the most calories?

(b) Does the scatterplot show a clear positive or negative association? Say in words what this association means about calories and salt in hot dogs.

(c) Are there any outliers? Is the relationship (ignoring any outliers) roughly linear in form? Still ignoring any outliers, how strong would you say the relationship between calories and sodium is?

4.16 Heating a home. The Sanchez household is about to install solar panels to reduce the cost of heating their house. In order to know how much the solar panels help, they record their consumption of natural gas before the panels are installed. Gas consumption is higher in cold weather, so the relationship between outside temperature and gas consumption is important. Here are data for 16 consecutive months:[9]

Month	Nov.	Dec.	Jan.	Feb.	Mar.	Apr.	May	June
Degree-days	24	51	43	33	26	13	4	0
Gas used	6.3	10.9	8.9	7.5	5.3	4.0	1.7	1.2
Month	July	Aug.	Sept.	Oct.	Nov.	Dec.	Jan.	Feb.
Degree-days	0	1	6	12	30	32	52	30
Gas used	1.2	1.2	2.1	3.1	6.4	7.2	11.0	6.9

Outside temperature is recorded in degree-days, a common measure of demand for heating. A day's degree-days are the number of degrees its average temperature falls below 65° F. Gas used is recorded in hundreds of cubic feet. Make a plot and describe the pattern. Is correlation a reasonable way to describe the pattern? What is the correlation?

4.17 Thinking about correlation. Figure 4.6 is a scatterplot of school grade point average versus IQ score for 78 seventh-grade students.

   (a) Is the correlation r for these data near -1, clearly negative but not near -1, near 0, clearly positive but not near 1, or near 1? Explain your answer.

   (b) Figure 4.7 shows the calories and sodium content in 17 brands of meat hot dogs. Is the correlation here closer to 1 than that for Figure 4.6, or closer to zero? Explain your answer.

4.18 Do solar panels reduce gas usage? After the Sanchez household gathered the information recorded in Exercise 4.16, they added solar panels to their house. They then measured their natural-gas consumption for 23 more months. Here are the data:[10]

Degree-days	19	3	3	0	0	0	8	11	27	46	38	34
Gas	3.2	2.0	1.6	1.0	0.7	0.7	1.6	3.1	5.1	7.7	7.0	6.1
Degree-days	16	9	2	1	0	2	3	18	32	34	40	
Gas	3.0	2.1	1.3	1.0	1.0	1.0	1.2	3.4	6.1	6.5	7.5	

Add the new data to your scatterplot from Exercise 4.16, using a different color or symbol. What do the before-and-after data show about the effect of solar panels?

4.19 Thinking about correlation. Both Figures 4.6 and 4.7 contain outliers. Removing the outliers will *increase* the correlation r in one figure and *decrease* r in the other figure. What happens in each figure, and why?

4.20 Merlins breeding. Often the percent of an animal species in the wild that survive to breed again is lower following a successful breeding season. This is part of nature's self-regulation to keep population size stable. A study of merlins (small falcons) in northern Sweden observed

(Russell Burden/Index Stock Imagery/PictureQuest)

the number of breeding pairs in an isolated area and the percent of males (banded for identification) who returned the next breeding season. Here are data for nine years:[11]

Breeding pairs	Percent of males returning
28	82
29	*7[* 83, 70, 61
30	69
32	58
33	43
38	*49* 50, 47

(a) Why is the response variable the *percent* of males that return rather than the *number* of males that return?

(b) Make a scatterplot. Describe the pattern. Do the data support the theory that a smaller percent of birds survive following a successful breeding season?

4.21 Mutual-fund performance. Many mutual funds compare their performance with that of a benchmark, an index of the returns on all securities of the kind the fund buys. The Vanguard International Growth Fund, for example, takes as its benchmark the Morgan Stanley EAFE (Europe, Australasia, Far East) index of overseas stock market performance. Here are the percent returns for the fund and for the EAFE from 1982 (the first full year of the fund's existence) to 2001:[12]

(Julian Calder/Stone/Getty Images)

Year	Fund	EAFE	Year	Fund	EAFE
1982	5.27	−1.86	1992	−5.79	−12.17
1983	43.08	23.69	1993	44.74	32.56
1984	−1.02	7.38	1994	0.76	7.78
1985	56.94	56.16	1995	14.89	11.21
1986	56.71	69.44	1996	14.65	6.05
1987	12.48	24.63	1997	4.12	1.78
1988	11.61	28.27	1998	16.93	20.00
1989	24.76	10.54	1999	26.34	26.96
1990	−12.05	−23.45	2000	−8.60	−14.17
1991	4.74	12.13	2001	−18.92	−21.44

Make a scatterplot suitable for predicting fund returns from EAFE returns. Is there a clear straight-line pattern? How strong is this pattern? (Give a numerical measure.) Are there any extreme outliers from the straight-line pattern?

4.22 Mice. For a biology project, you measure the tail length (centimeters) and weight (grams) of 12 mice of the same variety.

(a) Explain why you expect the correlation between tail length and weight to be positive.

(b) The mean tail length turns out to be 9.8 centimeters. What is the mean length in inches? (There are 2.54 centimeters in an inch.)

(c) The correlation between tail length and weight turns out to be $r = 0.6$. If you measured length in inches instead of centimeters, what would be the new value of r?

4.23 Statistics for investing. Investment reports now often include correlations. Following a table of correlations among mutual funds, a report adds: "Two funds can have perfect correlation, yet different levels of risk. For example, Fund A and Fund B may be perfectly correlated, yet Fund A moves 20% whenever Fund B moves 10%." Write a brief explanation, for someone who knows no statistics, of how this can happen. Include a sketch to illustrate your explanation.

4.24 Nematodes and tomatoes. To demonstrate the effect of nematodes (microscopic worms) on plant growth, a student introduces different numbers of nematodes into 16 planting pots. He then transplants a tomato seedling into each pot. Here are data on the increase in height of the seedlings (in centimeters) 14 days after planting:[13]

Nematodes	Seedling growth			
0	10.8	9.1	13.5	9.2
1,000	11.1	11.1	8.2	11.3
5,000	5.4	4.6	7.4	5.0
10,000	5.8	5.3	3.2	7.5

(a) Make a scatterplot of the response variable (growth) against the explanatory variable (nematode count). Then compute the mean growth for each group of seedlings, plot the means against the nematode counts, and connect these four points with line segments.

(b) Briefly describe the conclusions about the effects of nematodes on plant growth that these data suggest.

4.25 How many corn plants are too many? How much corn per acre should a farmer plant to obtain the highest yield? Too few plants will give a low yield. On the other hand, if there are too many plants, they will compete with each other for moisture and nutrients, and yields will fall. To find the best planting rate, plant at different rates on several plots of

(Alvis Upitis/The Image Bank/Gerty Images)

ground and measure the harvest. (Be sure to treat all the plots the same except for the planting rate.) Here are data from such an experiment:[14]

Plants per acre	Yield (bushels per acre)			
12,000	150.1	113.0	118.4	142.6
16,000	166.9	120.7	135.2	149.8
20,000	165.3	130.1	139.6	149.9
24,000	134.7	138.4	156.1	
28,000	119.0	150.5		

(a) Is yield or planting rate the explanatory variable?

(b) Make a scatterplot of yield and planting rate. Use a scale of yields from 100 to 200 bushels per acre so that the pattern will be clear.

(c) Describe the overall pattern of the relationship. Is it linear? Is there a positive or negative association, or neither?

(d) Find the mean yield for each of the five planting rates. Plot each mean yield against its planting rate on your scatterplot and connect these five points with lines. This combination of numerical description and graphing makes the relationship clearer. What planting rate would you recommend to a farmer whose conditions were similar to those in the experiment?

4.26 The effect of changing units. Changing the units of measurement can dramatically alter the appearance of a scatterplot. Return to the fossil data from Exercise 4.8:

Femur	38	56	59	64	74
Humerus	41	63	70	72	84

These measurements are in centimeters. Suppose a mad scientist measured the femur in meters and the humerus in millimeters. The data would then be

Femur	0.38	0.56	0.59	0.64	0.74
Humerus	410	630	700	720	840

(a) Draw an x axis from 0 to 75 and a y axis from 0 to 850. Plot the original data on these axes. Then plot the new data on the same axes in a different color. The two plots look very different.

(b) Nonetheless, the correlation is exactly the same for the two sets of measurements. Why do you know that this is true without doing any calculations? Find the two correlations to verify that they are the same.

TABLE 4.3 Wine consumption and heart attacks

Country	Alcohol from wine	Heart disease deaths	Country	Alcohol from wine	Heart disease deaths
Australia	2.5	211	Netherlands	1.8	167
Austria	3.9	167	New Zealand	1.9	266
Belgium	2.9	131	Norway	0.8	227
Canada	2.4	191	Spain	6.5	86
Denmark	2.9	220	Sweden	1.6	207
Finland	0.8	297	Switzerland	5.8	115
France	9.1	71	United Kingdom	1.3	285
Iceland	0.8	211	United States	1.2	199
Ireland	0.7	300	West Germany	2.7	172
Italy	7.9	107			

4.27 Is wine good for your heart? There is some evidence that drinking moderate amounts of wine helps prevent heart attacks. Table 4.3 gives data on yearly wine consumption (liters of alcohol from drinking wine, per person) and yearly deaths from heart disease (deaths per 100,000 people) in 19 developed nations.[15]

(a) Make a scatterplot that shows how national wine consumption helps explain heart disease death rates.

(b) Describe the form of the relationship. Is there a linear pattern? How strong is the relationship?

(c) Is the direction of the association positive or negative? Explain in simple language what this says about wine and heart disease. Do you think these data give good evidence that drinking wine *causes* a reduction in heart disease deaths? Why?

4.28 Teaching and research. A college newspaper interviews a psychologist about student ratings of the teaching of faculty members. The psychologist says, "The evidence indicates that the correlation between the research productivity and teaching rating of faculty members is close to zero." The paper reports this as "Professor McDaniel said that good researchers tend to be poor teachers, and vice versa." Explain why the paper's report is wrong. Write a statement in plain language (don't use the word "correlation") to explain the psychologist's meaning.

4.29 Investment diversification. A mutual-fund company's newsletter says, "A well-diversified portfolio includes assets with low correlations." The newsletter includes a table of correlations between the returns on various classes of investments. For example, the correlation between municipal bonds and large-cap stocks is 0.50, and the correlation between municipal bonds and small-cap stocks is 0.21.

(a) Rachel invests heavily in municipal bonds. She wants to diversify by adding an investment whose returns do not closely follow the returns on her bonds. Should she choose large-cap stocks or small-cap stocks for this purpose? Explain your answer.

(b) If Rachel wants an investment that tends to increase when the return on her bonds drops, what kind of correlation should she look for?

4.30 Sloppy writing about correlation. Each of the following statements contains a blunder. Explain in each case what is wrong.

(a) "There is a high correlation between the gender of American workers and their income."

(b) "We found a high correlation ($r = 1.09$) between students' ratings of faculty teaching and ratings made by other faculty members."

(c) "The correlation between planting rate and yield of corn was found to be $r = 0.23$ bushel."

Chapter 4 MEDIA EXERCISES

4.31 SAT scores. The data plotted in Figure 4.1 came from the Web site of the College Board, www.collegeboard.com. At first glance, this site seems to be of interest mainly to students applying to college and planning to take the SAT. If you can locate it, there is a mine of information about SAT scores. Try clicking on "For Education Professionals" and then on "College Bound Seniors." You can then access a list of data tables. Use data from these to do *one* of these tasks:

(a) Make a time plot of the average SAT verbal score of all students for as many past years as the data table allows. What trend or other pattern do you see? Has the trend changed over time?

(b) Make a scatterplot of the mean SAT verbal and math scores for the states in a recent year. What overall pattern do you see? What is the correlation? Hawaii is often a mild outlier. What might explain Hawaii's position on the plot?

4.32 Match the correlation. You are going to use the *Correlation and Regression* applet to make scatterplots with 10 points that have correlation close to 0.7. The lesson is that many patterns can have the same correlation. Always plot your data before you trust a correlation.

(a) Stop after adding the first two points. What is the value of the correlation? Why does it have this value?

(b) Make a lower-left to upper-right pattern of 10 points with correlation about $r = 0.7$. (You can drag points up or down to adjust r after you have 10 points.) Make a rough sketch of your scatterplot.

(c) Make another scatterplot with 9 points in a vertical stack at the left of the plot. Add one point far to the right and move it until the correlation is close to 0.7. Make a rough sketch of your scatterplot.

(d) Make yet another scatterplot with 10 points in a curved pattern that starts at the lower left, rises to the right, then falls again at the far right. Adjust the points up or down until you have a quite smooth curve with correlation close to 0.7. Make a rough sketch of this scatterplot also.

4.33 Correlation is not resistant. Go to the *Correlation and Regression* applet. Click on the scatterplot to create a group of 10 points in the lower-left corner of the scatterplot with a strong straight-line pattern (correlation about 0.9).

(a) Add one point at the upper right that is in line with the first 10. How does the correlation change?

(b) Drag this last point down until it is opposite the group of 10 points. How small can you make the correlation? Can you make the correlation negative? You see that a single outlier can greatly strengthen or weaken a correlation. Always plot your data to check for outlying points.

4.34 Brain size and IQ score. Do people with larger brains have higher IQ scores? A study looked at 40 volunteer subjects, 20 men and 20 women. Brain size was measured by magnetic resonance imaging. The EESEE story "Brain Size and Intelligence" gives the data. The MRI count is the number of "pixels" the brain covered in the image. IQ was measured by the Wechsler test that we met in Chapter 3.

(a) Make a scatterplot of IQ score versus MRI count, using distinct symbols for men and women. In addition, find the correlation between IQ and MRI for all 40 subjects, for the men alone, and for the women alone.

(b) Men are larger than women on the average, so they have larger brains. How is this size effect visible in your plot? Find the mean MRI count for men and women to verify the difference.

(c) Your result in (b) suggests separating men and women in looking at the relationship between brain size and IQ. Use your work in (a) to comment on the nature and strength of this relationship for women and for men.

4.35 Mercury in bass. Mercury in fish can make eating the fish unsafe. The EESEE story "Mercury in Florida's Bass" contains data from 53 lakes in Florida. We are interested in the relationship between the mercury concentration in the flesh of fish from a lake and various measures of water quality. EESEE asks you to examine several relationships. Start with this one: plot the standardized mercury concentration against the pH (the chemical measure of acidity versus alkalinity) for the 53 lakes. Give a brief verbal and numerical description of the relationship.

(R.Lynn/Photo Researchers)

Regression

Linear (straight-line) relationships between two quantitative variables are easy to understand and quite common. In Chapter 4, we found linear relationships in settings as varied as sparrowhawk colonies, icicle growth, and heating a home. Correlation measures the direction and strength of these relationships. When a scatterplot shows a linear relationship, we would like to summarize the overall pattern by drawing a line on the scatterplot. A *regression line* summarizes the relationship between two variables, but only in a specific setting: one of the variables helps explain or predict the other. That is, regression describes a relationship between an explanatory variable and a response variable.

REGRESSION LINE

A **regression line** is a straight line that describes how a response variable y changes as an explanatory variable x changes. We often use a regression line to predict the value of y for a given value of x.

EXAMPLE 5.1   *Predicting new birds*

We saw in Exercise 4.4 (page 83) that there is a linear relationship between the percent x of adult sparrowhawks that return to a colony from the previous year and the number y of new adult birds that join the colony. The scatterplot in Figure 5.1 displays this relationship.

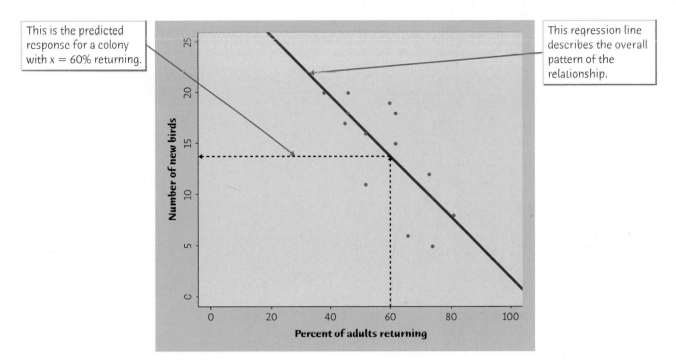

This is the predicted response for a colony with x = 60% returning.

This regression line describes the overall pattern of the relationship.

Figure 5.1 Data on 13 sparrowhawk colonies, with a regression line for predicting number of new birds from percent of returning birds. The dashed lines illustrate how to use the regression line to predict new birds in a colony with 60% returning.

The correlation is $r = -0.7485$, so the straight-line pattern is moderately strong. The line on the plot is a regression line that describes the overall pattern.

An ecologist wants to use the line, based on 13 colonies, to predict how many birds will join another colony, to which 60% of the adults from the previous year return. To **predict** new birds for 60% returning, first locate 60 on the x axis. Then go "up and over" as in the figure to find the y that corresponds to $x = 60$. It appears from the graph that we predict around 13 or 14 new birds.

prediction

The least-squares regression line

Different people will draw different lines by eye on a scatterplot. This is especially true when the points are widely scattered. We need a way to draw a regression line that doesn't depend on our guess as to where the line should go. We will use the line to predict y from x, so the prediction errors we make are errors in y, the vertical direction in the scatterplot. If we predict 14 new birds for a colony with 60% returning birds and in fact 18 new birds join the colony, our prediction error is

$$\text{error} = \text{observed } y - \text{predicted } y$$
$$= 18 - 14 = 4$$

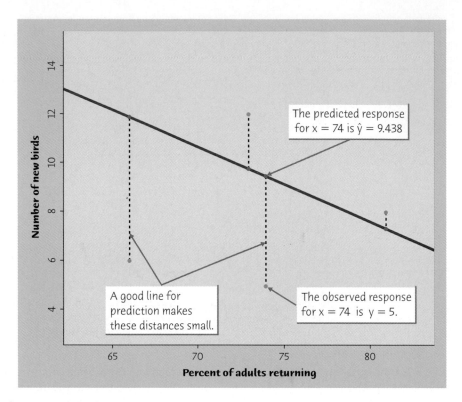

The predicted response
for x = 74 is ŷ = 9.438

A good line for
prediction makes
these distances small.

The observed response
for x = 74 is y = 5.

Percent of adults returning

Number of new birds

Figure 5.2 The least-squares idea. For each observation, find the vertical distance of each point on the scatterplot from a regression line. The least-squares regression line makes the sum of the squares of these distances as small as possible.

No line will pass exactly through all the points in the scatterplot. We want the *vertical* distances of the points from the line to be as small as possible. Figure 5.2 illustrates the idea. This plot shows four of the points from Figure 5.1, along with the line, on an expanded scale. The line passes above two of the points and below two of them. The vertical distances of the data points from the line appear as vertical line segments. There are many ways to make the collection of vertical distances "as small as possible." The most common is the *least-squares* method.

LEAST-SQUARES REGRESSION LINE

The **least-squares regression line** of y on x is the line that makes the sum of the squares of the vertical distances of the data points from the line as small as possible.

One reason for the popularity of the least-squares regression line is that the problem of finding the line has a simple answer. We can give the recipe for the least-squares line in terms of the means and standard deviations of the two variables and their correlation.

EQUATION OF THE LEAST-SQUARES REGRESSION LINE

We have data on an explanatory variable x and a response variable y for n individuals. From the data, calculate the means $\overline{x}$ and $\overline{y}$ and the standard deviations s_x and s_y of the two variables, and their correlation r. The least-squares regression line is the line

$$\hat{y} = a + bx$$

with **slope**

$$b = r\frac{s_y}{s_x}$$

and **intercept**

$$a = \overline{y} - b\overline{x}$$

We write $\hat{y}$ (read "y hat") in the equation of the regression line to emphasize that the line gives a *predicted* response $\hat{y}$ for any x. Because of the scatter of points about the line, the predicted response will usually not be exactly the same as the actually *observed* response y. In practice, you don't need to calculate the means, standard deviations, and correlation first. Software or your calculator will give the slope b and intercept a of the least-squares line from keyed-in values of the variables x and y. You can then concentrate on understanding and using the regression line.

EXAMPLE 5.2   Using a regression line

The line in Figure 5.1 is in fact the least-squares regression line of new birds on percent of returning birds. Enter the data from Exercise 4.4 into your calculator and check that the equation of this line is

$$\hat{y} = 31.9343 - 0.3040x$$

slope

The **slope** of a regression line is usually important for the interpretation of the data. The slope is the rate of change, the amount of change in $\hat{y}$ when x increases by 1. The slope $b = -0.3040$ in this example says that for each additional percent of last year's birds that return we predict about 0.3 fewer new birds.

intercept

The **intercept** of the regression line is the value of $\hat{y}$ when $x = 0$. Although we need the value of the intercept to draw the line, it is statistically meaningful only when x can actually take values close to zero. In our example, $x = 0$ means that a colony disappears because no birds return. The line predicts that on the average 31.9 new birds will appear. This isn't meaningful because a colony disappearing is a different setting than a colony with returning birds.

prediction

The equation of the regression line makes **prediction** easy. Just substitute an x-value into the equation. To predict new birds when 60% return, substitute $x = 60$:

$$\hat{y} = 31.9343 - (0.3040)(60)$$
$$= 31.9343 - 18.24 = 13.69$$

The actual number of new birds must be a whole number. Think of the prediction $\hat{y} = 13.69$ as an "on the average" value for many colonies with 60% returning birds.

plotting a line

To **plot the line** on the scatterplot, use the equation to find $\hat{y}$ for two values of x, one near each end of the range of x in the data. Plot each $\hat{y}$ above its x and draw the line through the two points.

Using technology

Least-squares regression is one of the most common statistical procedures. Any technology you use for statistical calculations will give you the least-squares line and related information. Figure 5.3 displays the regression output for the

Minitab

```
The regression equation is
New birds = 31.9 - 0.304 Pct return

Predictor        Coef      SE Coef           T         P
Constant       31.934        4.838        6.60     0.000
Pct retu     -0.30402      0.08122       -3.74     0.003

S = 3.667        R-Sq = 56.0%     R-Sq(adj) = 52.0%
```

Excel

	A	B	C	D	E	F	G
1	SUMMARY OUTPUT						
2							
3	*Regression Statistics*						
4	Multiple R	0.7485					
5	R Square	0.5602					
6	Adjusted R Square	0.5202					
7	Standard Error	3.6669					
8	Observations	13					
9							
10		Coefficients	Standard Error	t Stat	P-value		
11	Intercept	31.93426	4.83762	6.60124	3.86E-05		
12	Pct return	-0.30402	0.08122	-3.7432	0.00325		
13							

Texas Instruments TI-83 Plus

```
LinReg
 y=a+bx
 a=31.93425919
 b=-.3040229451
 r²=.5602033042
 r=-.7484673034
```

Figure 5.3 Least-squares regression for the sparrowhawk data. Output from statistical software, a spreadsheet, and a graphing calculator.

sparrowhawk data from a statistical software package, a spreadsheet program, and a graphing calculator. Each output records the slope and intercept of the least-squares line. The software also provides information that we do not yet need, although we will use much of it later. (In fact, we left out part of the Minitab and Excel outputs.) Be sure that you can locate the slope and intercept on all three outputs. Once you understand the statistical ideas, you can read and work with almost any software output.

APPLY YOUR KNOWLEDGE

5.1 Verify our claims. Example 5.2 gives the equation of the regression line of new birds y on percent of returning birds x for the data in Exercise 4.4 as

$$\hat{y} = 31.9343 - 0.3040x$$

Enter the data from Exercise 4.4 into your calculator.

(a) Use your calculator's regression function to find the equation of the least-squares regression line.

(b) Use your calculator to find the mean and standard deviation of both x and y and their correlation r. Find the slope b and intercept a of the regression line from these, using the facts in the box Equation of the Least-Squares Regression Line. Verify that in both part (a) and part (b) you get the equation in Example 5.2. (Results may differ slightly because of rounding off.)

5.2 Penguins diving. A study of king penguins looked for a relationship between how deep the penguins dive to seek food and how long they stay under water.[1] For all but the shallowest dives, there is a linear relationship that is different for different penguins. The study report gives a scatterplot for one penguin titled "The relation of dive duration (DD) to depth (D)." Duration DD is measured in minutes and depth D is in meters. The report then says, "The regression equation for this bird is: DD = 2.69 + 0.0138D."

(a) What is the slope of the regression line? Explain in specific language what this slope says about this penguin's dives.

(b) According to the regression line, how long does a typical dive to a depth of 200 meters last?

(c) The dives varied from 40 meters to 300 meters in depth. Plot the regression line from $x = 40$ to $x = 300$.

5.3 Sports car gas mileage. Table 1.2 (page 12) gives the city and highway gas mileages for two-seater cars. A scatterplot (Exercise 4.12) shows a strong positive linear relationship.

(a) Find the least-squares regression line for predicting highway mileage from city mileage, using data from all 22 car models. Make a scatterplot and plot the regression line.

Regression toward the mean

To "regress" means to go backward. Why are statistical methods for predicting a response from an explanatory variable called "regression"? Sir Francis Galton (1822–1911), who was the first to apply regression to biological and psychological data, looked at examples such as the heights of children versus the heights of their parents. He found that the taller-than-average parents tended to have children who were also taller than average but not as tall as their parents. Galton called this fact "regression toward the mean," and the name came to be applied to the statistical method.

(b) What is the slope of the regression line? Explain in words what the slope says about gas mileage for two-seater cars.

(c) Another two-seater is rated at 20 miles per gallon in the city. Predict its highway mileage.

Facts about least-squares regression

One reason for the popularity of least-squares regression lines is that they have many convenient special properties. Here are some facts about least-squares regression lines.

Fact 1. The distinction between explanatory and response variables is essential in regression. Least-squares regression looks at the distances of the data points from the line only in the y direction. If we reverse the roles of the two variables, we get a different least-squares regression line.

EXAMPLE 5.3   The expanding universe

Figure 5.4 is a scatterplot of data that played a central role in the discovery that the universe is expanding. They are the distances from earth of 24 spiral galaxies and the speed at which these galaxies are moving away from us, reported by the astronomer

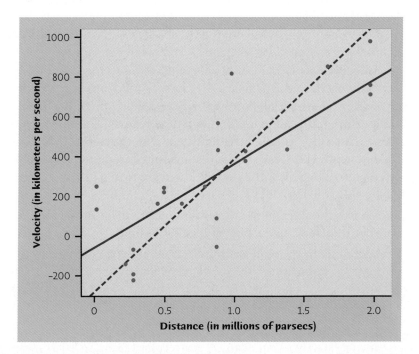

Figure 5.4 Scatterplot of Hubble's data on the distance from earth of 24 galaxies and the velocity at which they are moving away from us. The two lines are the two least-squares regression lines: of velocity on distance (*solid*) and of distance on velocity (*dashed*).

Edwin Hubble in 1929.[2] There is a positive linear relationship, $r = 0.7842$, so that more distant galaxies are moving away more rapidly. Astronomers believe that there is in fact a perfect linear relationship, and that the scatter is caused by imperfect measurements.

The two lines on the plot are the two least-squares regression lines. The regression line of velocity on distance is solid. The regression line of distance on velocity is dashed. *Regression of velocity on distance and regression of distance on velocity give different lines.* In the regression setting you must know clearly which variable is explanatory.

Fact 2. There is a close connection between correlation and the slope of the least-squares line. The slope is

$$b = r\frac{s_y}{s_x}$$

This equation says that along the regression line, **a change of one standard deviation in x corresponds to a change of r standard deviations in y.** When the variables are perfectly correlated ($r = 1$ or $r = -1$), the change in the predicted response $\hat{y}$ is the same (in standard deviation units) as the change in x. Otherwise, because $-1 \leq r \leq 1$, the change in $\hat{y}$ is less than the change in x. As the correlation grows less strong, the prediction $\hat{y}$ moves less in response to changes in x.

Fact 3. **The least-squares regression line always passes through the point $(\overline{x}, \overline{y})$** on the graph of y against x. So the least-squares regression line of y on x is the line with slope rs_y/s_x that passes through the point $(\overline{x}, \overline{y})$.

Fact 4. The correlation r describes the strength of a straight-line relationship. In the regression setting, this description takes a specific form: **the square of the correlation, r^2, is the fraction of the variation in the values of y that is explained by the least-squares regression of y on x.**

The idea is that when there is a linear relationship, some of the variation in y is accounted for by the fact that as x changes it pulls y along with it. Look again at Figure 5.1 on page 105. The number of new birds joining a colony ranges from 5 to 20. Some of this variation in the response y is explained by the fact that the percent x of returning birds varies from 38% to 81%. As x moves from 38% to 81%, it pulls y with it along the line. You would guess a smaller number of new birds for a colony with 80% returning than for a colony with 40% returning. But there is also quite a bit of scatter above and below the line, variation that isn't explained by the straight-line relationship between x and y.

Although we won't do the algebra, it is possible to break the total variation in the observed values of y into two parts. One part is the variation we expect as x moves and $\hat{y}$ moves with it along the regression line. The other measures the variation of the data points about the line. The squared correlation r^2 is the first of these as a fraction of the whole:

$$r^2 = \frac{\text{variation in } \hat{y} \text{ as } x \text{ pulls it along the line}}{\text{total variation in observed values of } y}$$

EXAMPLE 5.4   Using r^2

In Figure 5.1, $r = -0.7485$ and $r^2 = 0.5603$. About 56% of the variation in new birds is accounted for by the linear relationship with percent returning. The other 44% is individual variation among colonies that is not explained by the linear relationship.

Figure 4.2 (page 85) shows a stronger linear relationship in which the points are more tightly concentrated along a line. Here, $r = -0.9124$ and $r^2 = 0.8325$. More than 83% of the variation in carnivore abundance is explained by regression on body mass. Only 17% is variation among species with the same mass.

When you report a regression, give r^2 as a measure of how successful the regression was in explaining the response. All the outputs in Figure 5.3 include r^2, either in decimal form or as a percent. When you see a correlation, square it to get a better feel for the strength of the association. Perfect correlation ($r = -1$ or $r = 1$) means the points lie exactly on a line. Then $r^2 = 1$ and all of the variation in one variable is accounted for by the linear relationship with the other variable. If $r = -0.7$ or $r = 0.7$, $r^2 = 0.49$ and about half the variation is accounted for by the linear relationship. In the r^2 scale, correlation ± 0.7 is about halfway between 0 and ± 1.

Facts 2, 3, and 4 are special properties of least-squares regression. They are not true for other methods of fitting a line to data.

APPLY YOUR KNOWLEDGE

5.4   **Growing corn.** Exercise 4.25 (page 99) gives data from an agricultural experiment. The purpose of the study was to see how the yield of corn changes as we change the planting rate (plants per acre).

   (a) Make a scatterplot of the data. (Use a scale of yields from 100 to 200 bushels per acre.) Find the least-squares regression line for predicting yield from planting rate and add this line to your plot. Why should we *not* use regression for prediction in this setting?

   (b) What is r^2? What does this value say about the success of the regression in predicting yield?

   (c) Even regression lines that make no practical sense obey Facts 1 to 4. Use the equation of the regression line you found in (a) to show that when x is the mean planting rate, the predicted yield $\hat{y}$ is the mean of the observed yields.

5.5   **Sports car gas mileage.** In Exercise 5.3 you found the least-squares regression line for predicting highway mileage from city mileage for the

22 two-seater car models in Table 1.2. Find the mean city mileage and mean highway mileage for these cars. Use your regression line to predict the highway mileage for a car with city mileage equal to the mean for the group. Explain why you knew the answer before doing the prediction.

5.6 **Comparing regressions.** What is the value of r^2 for predicting highway from city mileage in Exercise 5.5? What value did you find for predicting corn yield from planting rate in Exercise 5.4? Explain in simple language why if we knew only these two r^2-values, we would expect predictions using the regression line to be more satisfactory for gas mileage than for corn yield.

Residuals

One of the first principles of data analysis is to look for an overall pattern and also for striking deviations from the pattern. A regression line describes the overall pattern of a linear relationship between an explanatory variable and a response variable. We see deviations from this pattern by looking at the scatter of the data points about the regression line. The vertical distances from the points to the least-squares regression line are as small as possible, in the sense that they have the smallest possible sum of squares. Because they represent "left-over" variation in the response after fitting the regression line, these distances are called *residuals*.

RESIDUALS

A **residual** is the difference between an observed value of the response variable and the value predicted by the regression line. That is,

$$\text{residual} = \text{observed } y - \text{predicted } y$$
$$= y - \hat{y}$$

EXAMPLE 5.5 *Predicting mental ability*

Does the age at which a child begins to talk predict later score on a test of mental ability? A study of the development of young children recorded the age in months at which each of 21 children spoke their first word and their Gesell Adaptive Score, the result of an aptitude test taken much later. The data appear in Table 5.1.[3]

Figure 5.5 is a scatterplot, with age at first word as the explanatory variable x and Gesell score as the response variable y. Children 3 and 13, and also Children 16 and 21, have identical values of both variables. We use a different plotting symbol to show that one point stands for two individuals. The plot shows a negative association. That is, children who begin to speak later tend to have lower test scores than early talkers. The overall pattern is moderately linear. The correlation describes both the direction and the strength of the linear relationship. It is $r = -0.640$.

The line on the plot is the least-squares regression line of Gesell score on age at first word. Its equation is

$$\hat{y} = 109.8738 - 1.1270x$$

TABLE 5.1 Age at first word and Gesell score

Child	Age	Score	Child	Age	Score
1	15	95	11	7	113
2	26	71	12	9	96
3	10	83	13	10	83
4	9	91	14	11	84
5	15	102	15	11	102
6	20	87	16	10	100
7	18	93	17	12	105
8	11	100	18	42	57
9	8	104	19	17	121
10	20	94	20	11	86
			21	10	100

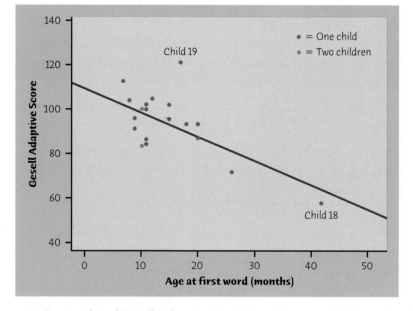

Figure 5.5 Scatterplot of Gesell Adaptive Score versus the age at first word for 21 children, from Table 5.1. The line is the least-squares regression line for predicting Gesell score from age at first word.

For Child 1, who first spoke at 15 months, we predict the score

$$\hat{y} = 109.8738 - (1.1270)(15) = 92.97$$

This child's actual score was 95. The residual is

$$\text{residual} = \text{observed } y - \text{predicted } y$$
$$= 95 - 92.97 = 2.03$$

The residual is positive because the data point lies above the line.

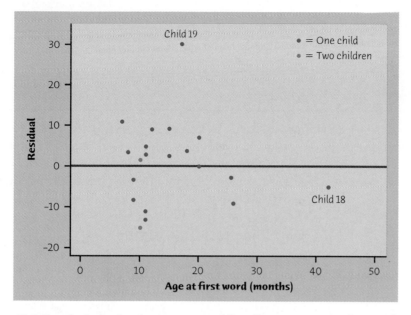

Figure 5.6 Residual plot for the regression of Gesell score on age at first word. Child 19 is an outlier. Child 18 is an influential observation that does not have a large residual.

There is a residual for each data point. Finding the residuals is a bit unpleasant because you must first find the predicted response for every x. Software or a graphing calculator gives you the residuals all at once. Here are the 21 residuals for the Gesell data, from software:

```
residuals:
2.0310 −9.5721 −15.6040 −8.7309   9.0310  −0.3341    3.4120
2.5230  3.1421   6.6659 11.0151  −3.7309 −15.6040 −13.4770
4.5230  1.3960   8.6500 −5.5403  30.2850 −11.4770   1.3960
```

Because the residuals show how far the data fall from our regression line, examining the residuals helps assess how well the line describes the data. Although residuals can be calculated from any model fitted to the data, the residuals from the least-squares line have a special property: **the mean of the least-squares residuals is always zero.**

Compare the scatterplot in Figure 5.5 with the *residual plot* for the same data in Figure 5.6. The horizontal line at zero in Figure 5.6 helps orient us. It corresponds to the regression line in Figure 5.5.

RESIDUAL PLOTS

A **residual plot** is a scatterplot of the regression residuals against the explanatory variable. Residual plots help us assess the fit of a regression line.

By in effect turning the regression line horizontal, a residual plot magnifies the deviations of the points from the line and makes it easier to see unusual observations and patterns.

5.7 Does fast driving waste fuel? Exercise 4.6 (page 86) gives data on the fuel consumption y of a car at various speeds x. Fuel consumption is measured in liters of gasoline per 100 kilometers driven and speed is measured in kilometers per hour. Software tells us that the equation of the least-squares regression line is

$$\hat{y} = 11.058 - 0.01466x$$

The residuals, in the same order as the observations, are

10.09	2.24	−0.62	−2.47	−3.33	−4.28	−3.73	−2.94
−2.17	−1.32	−0.42	0.57	1.64	2.76	3.97	

(a) Make a scatterplot of the observations and draw the regression line on your plot.

(b) Would you use the regression line to predict y from x? Explain your answer.

(c) Check that the residuals have sum zero (up to roundoff error).

(d) Make a plot of the residuals against the values of x. Draw a horizontal line at height zero on your plot. Notice that the residuals show the same pattern about this line as the data points show about the regression line in the scatterplot in (a).

Influential observations

Figures 5.5 and 5.6 show two unusual observations. Children 18 and 19 are unusual in different ways. Child 19 lies far from the regression line. Child 18 is close to the line but far out in the x direction. Child 19 is an *outlier in the y direction*, with a Gesell score so high that we should check for a mistake in recording it. In fact, the score is correct.

Child 18 is an *outlier in the x direction*. This child began to speak much later than any of the other children. Because of its extreme position on the age scale, this point has a strong influence on the position of the regression line. Figure 5.7 adds a second regression line, calculated after leaving out Child 18. You can see that this one point moves the line quite a bit. Least-squares lines make the sum of squares of the vertical distances to the points as small as possible. A point that is extreme in the x direction with no other points near it pulls the line toward itself. We call such points *influential*.

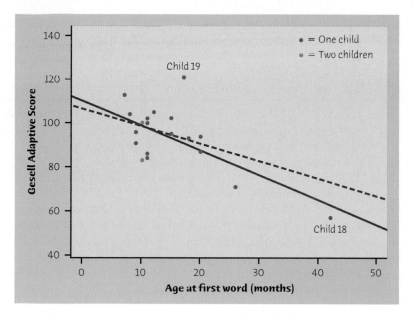

Figure 5.7 Two least-squares regression lines of Gesell score on age at first word. The solid line is calculated from all the data. The dashed line was calculated leaving out Child 18. Child 18 is an influential observation because leaving out this point moves the regression line quite a bit.

OUTLIERS AND INFLUENTIAL OBSERVATIONS IN REGRESSION

An **outlier** is an observation that lies outside the overall pattern of the other observations. Points that are outliers in the y direction of a scatterplot have large regression residuals, but other outliers need not have large residuals.

An observation is **influential** for a statistical calculation if removing it would markedly change the result of the calculation. Points that are outliers in the x direction of a scatterplot are often influential for the least-squares regression line.

We did not need the distinction between outliers and influential observations in Chapter 2. A single large salary that pulls up the mean salary $\overline{x}$ for a group of workers is an outlier because it lies far above the other salaries. It is also influential, because the mean changes when it is removed. In the regression setting, however, not all outliers are influential. The least-squares regression line is most likely to be heavily influenced by observations that are outliers in the x direction. The scatterplot will alert you to observations that are extreme in x and may therefore be influential. The surest way to verify that a point is influential is to find the regression line both with and without the suspect point, as in Figure 5.7. If the line moves more than a small amount when the point is deleted, the point is influential. The *Correlation and Regression* applet allows you to move points and watch how they influence the least-squares line.

EXAMPLE 5.6   An influential observation

The strong influence of Child 18 makes the original regression of Gesell score on age at first word misleading. The original data have $r^2 = 0.41$. That is, the age at which a child begins to talk explains 41% of the variation on a later test of mental ability. This relationship is strong enough to be interesting to parents. If we leave out Child 18, r^2 drops to only 11%. The apparent strength of the association was largely due to a single influential observation.

What should the child development researcher do? She must decide whether Child 18 was so slow to speak that this individual should not be allowed to influence the analysis. If she excludes Child 18, much of the evidence for a connection between the age at which a child begins to talk and later ability score vanishes. If she keeps Child 18, she needs data on other children who were also slow to begin talking, so that the analysis no longer depends so heavily on just one child.

APPLY YOUR KNOWLEDGE

5.8   Influential or not? We have seen that Child 18 in the Gesell data in Table 5.1 (page 114) is an influential observation. Now we will examine the effect of Child 19, who is also an outlier in Figure 5.5.

(a) Find the least-squares regression line of Gesell score on age at first word, leaving out Child 19. Example 5.5 gives the regression line from all the children. Plot both lines on the same graph. (You do not have to make a scatterplot of all the points; just plot the two lines.) Would you call Child 19 very influential? Why?

(b) For all children, $r^2 = 0.41$. How does removing Child 19 change the r^2 for this regression? Explain why r^2 changes in this direction when you drop Child 19.

5.9   Sports car gas mileage. The data on gas mileage of two-seater cars (Table 1.2, page 12) contain an outlier, the Honda Insight. When we predict highway mileage from city mileage, this point is an outlier in *both* the *x* and *y* directions. We wonder if it influences the least-squares line.

(a) Make a scatterplot and draw (again) the least-squares line from all 22 car models.

(b) Find the least-squares line when the Insight is left out of the calculation and draw this line on your plot.

(c) Influence is a matter of degree, not a yes-or-no question. Use both regression lines to predict highway mileages for city mileages of 10, 20, and 25 MPG. (These city mileage values span the range of car models other than the Insight.) Do you think the Insight changes the predictions enough to be important to a car buyer?

Cautions about correlation and regression

Correlation and regression are powerful tools for describing the relationship between two variables. When you use these tools, you must be aware of their limitations, beginning with the fact that **correlation and regression describe only linear relationships.** Also remember that **the correlation r and the least-squares regression line are not resistant.** One influential observation or incorrectly entered data point can greatly change these measures. **Always plot your data** before interpreting regression or correlation. Here are some other cautions to keep in mind when you apply correlation and regression or read accounts of their use.

Beware extrapolation. Suppose that you have data on a child's growth between 3 and 8 years of age. You find a strong linear relationship between age x and height y. If you fit a regression line to these data and use it to predict height at age 25 years, you will predict that the child will be 8 feet tall. Growth slows down and then stops at maturity, so extending the straight line to adult ages is foolish. Few relationships are linear for all values of x. So don't stray far from the range of x that actually appears in your data.

EXTRAPOLATION

Extrapolation is the use of a regression line for prediction far outside the range of values of the explanatory variable x that you used to obtain the line. Such predictions are often not accurate.

Beware the lurking variable. Correlation and regression describe the relationship between two variables. Often the relationship between two variables is strongly influenced by other variables. More advanced statistical methods allow the study of many variables together, so that we can take other variables into account. Sometimes, however, the relationship between two variables is influenced by other variables that we did not measure or even think about. Because these variables are lurking in the background, we call them *lurking variables.*

LURKING VARIABLE

A **lurking variable** is a variable that has an important effect on the relationship among the variables in a study but is not included among the variables studied.

You should always think about possible lurking variables before you draw conclusions based on correlation or regression.

Do left-handers die early?

Yes, said a study of 1000 deaths in California. Left-handed people died at an average age of 66 years; right-handers, at 75 years of age. Should left-handed people fear an early death? No—the lurking variable has struck again. Older people grew up in an era when many natural left-handers were forced to use their right hands. So right-handers are more common among older people, and left-handers are more common among the young. When we look at deaths, the left-handers who die are younger on the average because left-handers in general are younger. Mystery solved.

EXAMPLE 5.7   Magic Mozart?

The Kalamazoo (Michigan) Symphony once advertised a "Mozart for Minors" program with this statement: "Question: Which students scored 51 points higher in verbal skills and 39 points higher in math? Answer: Students who had experience in music."[4]

We could as well answer "Children who played soccer." Why? Children with prosperous and well-educated parents are more likely than poorer children to have experience with music and also to play soccer. They are also likely to attend good schools, get good health care, and be encouraged to study hard. These advantages lead to high test scores. Experience with music and soccer are correlated with high scores just because they go along with the other advantages of having prosperous and educated parents.

APPLY YOUR KNOWLEDGE

5.10 **The declining farm population.** The number of people living on American farms has declined steadily during this century. Here are data on the farm population (millions of persons) from 1935 to 1980:

Year	1935	1940	1945	1950	1955	1960	1965	1970	1975	1980
Population	32.1	30.5	24.4	23.0	19.1	15.6	12.4	9.7	8.9	7.2

(a) Make a scatterplot of these data and find the least-squares regression line of farm population on year.

(b) According to the regression line, how much did the farm population decline each year on the average during this period? What percent of the observed variation in farm population is accounted for by linear change over time?

(c) Use the regression equation to predict the number of people living on farms in 1990. Is this result reasonable? Why?

5.11 **Is math the key to success in college?** A College Board study of 15,941 high school graduates found a strong correlation between how much math minority students took in high school and their later success in college. News articles quoted the head of the College Board as saying that "math is the gatekeeper for success in college."[5] Maybe so, but we should also think about lurking variables. What might lead minority students to take more or fewer high school math courses? Would these same factors influence success in college?

Association does not imply causation

Thinking about lurking variables leads to the most important caution about correlation and regression. When we study the relationship between two variables, we often hope to show that changes in the explanatory variable *cause* changes in the response variable. A strong association between two variables is not enough to draw conclusions about cause and effect. Sometimes an observed association really does reflect cause and effect. The Sanchez household uses more natural gas in colder months because cold weather requires burning more gas to stay warm. In other cases, an association is explained by lurking variables, and the conclusion that x causes y is either wrong or not proved.

EXAMPLE 5.8   Does TV make you live longer?

Measure the number of television sets per person x and the average life expectancy y for the world's nations. There is a high positive correlation: nations with many TV sets have higher life expectancies.

The basic meaning of causation is that by changing x we can bring about a change in y. Could we lengthen the lives of people in Rwanda by shipping them TV sets? No. Rich nations have more TV sets than poor nations. Rich nations also have longer life expectancies because they offer better nutrition, clean water, and better health care. There is no cause-and-effect tie between TV sets and length of life.

Correlations such as that in Example 5.8 are sometimes called "nonsense correlations." The correlation is real. What is nonsense is the conclusion that changing one of the variables causes changes in the other. A lurking variable—such as national wealth in Example 5.8—that influences both x and y can create a high correlation even though there is no direct connection between x and y.

ASSOCIATION DOES NOT IMPLY CAUSATION

An association between an explanatory variable x and a response variable y, even if it is very strong, is not by itself good evidence that changes in x actually cause changes in y.

EXAMPLE 5.9   Obesity in mothers and daughters

Obese parents tend to have obese children. The results of a study of Mexican American girls aged 9 to 12 years are typical. The investigators measured body mass index (BMI), a measure of weight relative to height, for both the girls and their mothers. People with high BMI are overweight or obese. The correlation between the BMI of daughters and the BMI of their mothers was $r = 0.506$.[6]

Body type is in part determined by heredity. Daughters inherit half their genes from their mothers. There is therefore a direct causal link between the BMI of mothers and daughters. But it may be that mothers who are overweight also set an example of little exercise, poor eating habits, and lots of television. Their daughters pick up these habits to some extent, so the influence of heredity is mixed up with influences from the girls' environment. Both contribute to the mother-daughter correlation.

The lesson of Example 5.9 is more subtle than just "association does not imply causation." Even when direct causation is present, it may not be the whole

experiment

explanation for a correlation. You must still worry about lurking variables. Careful statistical studies try to anticipate lurking variables and measure them so that they are no longer "lurking." The mother-daughter study did measure TV viewing, exercise, and diet. Elaborate statistical analysis can remove the effects of these variables to come closer to the direct effect of mother's BMI on daughter's BMI. This remains a second-best approach to causation. The best way to get good evidence that x causes y is to do an **experiment** in which we change x and keep lurking variables under control. We will discuss experiments in Chapter 8.

When experiments cannot be done, finding the explanation for an observed association is often difficult and controversial. Many of the sharpest disputes in which statistics plays a role involve questions of causation that cannot be settled by experiment. Do gun control laws reduce violent crime? Does using cell phones cause brain tumors? Has increased free trade widened the gap between the incomes of more educated and less educated American workers? All of these questions have become public issues. All concern associations among variables. And all have this in common: they try to pinpoint cause and effect in a setting involving complex relations among many interacting variables.

EXAMPLE 5.10   Does smoking cause lung cancer?

Despite the difficulties, it is sometimes possible to build a strong case for causation in the absence of experiments. The evidence that smoking causes lung cancer is about as strong as nonexperimental evidence can be.

Doctors had long observed that most lung cancer patients were smokers. Comparison of smokers and "similar" nonsmokers showed a very strong association between smoking and death from lung cancer. Could the association be explained by lurking variables? Might there be, for example, a genetic factor that predisposes people both to nicotine addiction and to lung cancer? Smoking and lung cancer would then be positively associated even if smoking had no direct effect on the lungs. How were these objections overcome?

Let's answer this question in general terms: What are the criteria for establishing causation when we cannot do an experiment?

- *The association is strong.* The association between smoking and lung cancer is very strong.

- *The association is consistent.* Many studies of different kinds of people in many countries link smoking to lung cancer. That reduces the chance that a lurking variable specific to one group or one study explains the association.

- *Higher doses are associated with stronger responses.* People who smoke more cigarettes per day or who smoke over a longer period get lung cancer more often. People who stop smoking reduce their risk.

- *The alleged cause precedes the effect in time.* Lung cancer develops after years of smoking. The number of men dying of lung cancer rose as smoking became more common, with a lag of about 30 years. Lung cancer kills more men than any other form of cancer. Lung cancer was rare among women until women began to smoke. Lung cancer in women rose along with smoking, again with a lag of about 30 years, and has now passed breast cancer as the leading cause of cancer death among women.

- *The alleged cause is plausible*. Experiments with animals show that tars from cigarette smoke do cause cancer.

Medical authorities do not hesitate to say that smoking causes lung cancer. The U.S. Surgeon General has long stated that cigarette smoking is "the largest avoidable cause of death and disability in the United States."[7] The evidence for causation is overwhelming—but it is not as strong as the evidence provided by well-designed experiments.

APPLY YOUR KNOWLEDGE

5.12 Education and income. There is a strong positive association between the education and income of adults. For example, the Census Bureau reports that the median income of people aged 25 and over increases from $15,800 for those with less than a ninth-grade education, to $24,656 for high school graduates, to $40,939 for holders of a bachelor's degree, and on up for yet more education. In part, this association reflects causation—education helps people qualify for better jobs. Suggest several lurking variables (ask yourself what kinds of people tend to get good educations) that also contribute.

5.13 How's your self-esteem? People who do well tend to feel good about themselves. Perhaps helping people feel good about themselves will help them do better in school and life. Raising self-esteem became for a time a goal in many schools. California even created a state commission to advance the cause. Can you think of explanations for the association between high self-esteem and good school performance other than "Self-esteem causes better work in school"?

5.14 Are big hospitals bad for you? A study shows that there is a positive correlation between the size of a hospital (measured by its number of beds x) and the median number of days y that patients remain in the hospital. Does this mean that you can shorten a hospital stay by choosing a small hospital? Why?

Chapter 5 SUMMARY

A **regression line** is a straight line that describes how a response variable y changes as an explanatory variable x changes. You can use a regression line to **predict** the value of y for any value of x by substituting this x into the equation of the line.

The **slope** b of a regression line $\hat{y} = a + bx$ is the rate at which the predicted response $\hat{y}$ changes along the line as the explanatory variable x changes. Specifically, b is the change in $\hat{y}$ when x increases by 1.

The **intercept** a of a regression line $\hat{y} = a + bx$ is the predicted response $\hat{y}$ when the explanatory variable $x = 0$. This prediction is of no statistical use unless x can actually take values near 0.

The most common method of fitting a line to a scatterplot is least squares. The **least-squares regression line** is the straight line $\hat{y} = a + bx$ that minimizes the sum of the squares of the vertical distances of the observed points from the line.

The least-squares regression line of y on x is the line with slope $r s_y/s_x$ and intercept $a = \overline{y} - b\overline{x}$. This line always passes through the point $(\overline{x}, \overline{y})$.

Correlation and regression are closely connected. The correlation r is the slope of the least-squares regression line when we measure both x and y in standardized units. The **square of the correlation** r^2 is the fraction of the variance of one variable that is explained by least-squares regression on the other variable.

Correlation and regression must be **interpreted with caution. Plot the data** to be sure the relationship is roughly linear and to detect outliers and influential observations. A plot of the **residuals** makes these effects easier to see.

Look for **influential observations,** individual points that substantially change the regression line. Influential observations are often outliers in the x direction.

Avoid **extrapolation,** the use of a regression line for prediction for values of the explanatory variable far outside the range of the data from which the line was calculated.

Lurking variables that you did not measure may explain the relations between the variables you did measure. Correlation and regression can be misleading if you ignore important lurking variables.

Most of all, be careful not to conclude that there is a cause-and-effect relationship between two variables just because they are strongly associated. **High correlation does not imply causation.** The best evidence that an association is due to causation comes from an **experiment** in which the explanatory variable is directly changed and other influences on the response are controlled.

Chapter 5 EXERCISES

5.15 Sisters and brothers. How strongly do physical characteristics of sisters and brothers correlate? Here are data on the heights (in inches) of 11 adult pairs:[8]

Brother	71	68	66	67	70	71	70	73	72	65	66
Sister	69	64	65	63	65	62	65	64	66	59	62

(a) Verify using your calculator or software that the least-squares line for predicting sister's height from brother's height is $\hat{y} = 27.64$

$+\,0.527x$. What is the correlation between sister's height and brother's height?

(b) Damien is 70 inches tall. Predict the height of his sister Tonya.

5.16 Husbands and wives. The mean height of American women in their twenties is about 64 inches, and the standard deviation is about 2.7 inches. The mean height of men the same age is about 69.3 inches, with standard deviation about 2.8 inches. If the correlation between the heights of husbands and wives is about $r = 0.5$, what is the slope of the regression line of the husband's height on the wife's height in young couples? Draw a graph of this regression line. Predict the height of the husband of a woman who is 67 inches tall.

(Superstock/Superstock/PictureQuest)

5.17 Measuring water quality. Biochemical oxygen demand (BOD) measures organic pollutants in water by measuring the amount of oxygen consumed by microorganisms that break down these compounds. BOD is hard to measure accurately. Total organic carbon (TOC) is easy to measure, so it is common to measure TOC and use regression to predict BOD. A typical regression equation for water entering a municipal treatment plant is[9]

$$BOD = -55.43 + 1.507\ TOC$$

Both BOD and TOC are measured in milligrams per liter of water.

(a) What does the slope of this line say about the relationship between BOD and TOC?

(b) What is the predicted BOD when TOC = 0? Values of BOD less than 0 are impossible. Why does the prediction give an impossible value?

5.18 IQ and school GPA. Figure 4.6 (page 95) plots school grade point average (GPA) against IQ test score for 78 seventh-grade students. Calculation shows that the mean and standard deviation of the IQ scores are

$$\bar{x} = 108.9 \qquad s_x = 13.17$$

For the grade point averages,

$$\bar{y} = 7.447 \qquad s_y = 2.10$$

The correlation between IQ and GPA is $r = 0.6337$.

(a) Find the equation of the least-squares line for predicting GPA from IQ.

(b) What percent of the observed variation in these students' GPAs can be explained by the linear relationship between GPA and IQ?

(c) One student has an IQ of 103 but a very low GPA of 0.53. What is the predicted GPA for a student with IQ = 103? What is the residual for this particular student?

5.19 A growing child. Sarah's parents are concerned that she seems short for her age. Their doctor has the following record of Sarah's height:

Age (months)	36	48	51	54	57	60
Height (cm)	86	90	91	93	94	95

(a) Make a scatterplot of these data. Note the strong linear pattern.

(b) Using your calculator, find the equation of the least-squares regression line of height on age.

(c) Predict Sarah's height at 40 months and at 60 months. Use your results to draw the regression line on your scatterplot.

(d) What is Sarah's rate of growth, in centimeters per month? Normally growing girls gain about 6 cm in height between ages 4 (48 months) and 5 (60 months). What rate of growth is this in centimeters per month? Is Sarah growing more slowly than normal?

5.20 Heating a home. Exercise 4.16 (page 96) gives data on degree-days and natural gas consumed by the Sanchez home for 16 consecutive months. There is a very strong linear relationship. Mr. Sanchez asks, "If a month averages 20 degree-days per day (that's 45° F), how much gas will we use?" Use your calculator or software to find the least-squares regression line and answer his question.

5.21 A nonsense prediction. Use the least-squares regression line for the data in Exercise 5.19 to predict Sarah's height at age 40 years (480 months). Your prediction is in centimeters. Convert it to inches using the fact that a centimeter is 0.3937 inch. The data have r^2 almost 0.99. Why is the prediction clearly silly?

5.22 Merlins breeding. Exercise 4.20 (page 97) gives data on the number of breeding pairs of merlins in an isolated area in each of nine years and the percent of males who returned the next year. The data show that the percent returning is lower after successful breeding seasons and that the relationship is roughly linear. Use your calculator or software to find the least-squares regression line and predict the percent of returning males after a season with 30 breeding pairs.

5.23 Keeping water clean. Keeping water supplies clean requires regular measurement of levels of pollutants. The measurements are indirect—a typical analysis involves forming a dye by a chemical reaction with the dissolved pollutant, then passing light through the solution and measuring its "absorbence." To calibrate such measurements, the laboratory measures known standard solutions and uses regression to relate absorbence to pollutant concentration. This is usually done every day. Here is one series of data on the absorbence for different levels of nitrates. Nitrates are measured in milligrams per liter of water.[10]

Nitrates	50	50	100	200	400	800	1200	1600	2000	2000
Absorbence	7.0	7.5	12.8	24.0	47.0	93.0	138.0	183.0	230.0	226.0

(a) Chemical theory says that these data should lie on a straight line. If the correlation is not at least 0.997, something went wrong and the calibration procedure is repeated. Plot the data and find the correlation. Must the calibration be done again?

(b) What is the equation of the least-squares line for predicting absorbence from concentration? If the lab analyzed a specimen with 500 milligrams of nitrates per liter, what do you expect the absorbence to be? Based on your plot and the correlation, do you expect your predicted absorbence to be very accurate?

5.24 Comparing regressions. What are the correlations between the explanatory and response variables in Exercises 5.20 and 5.22? What does r^2 say about the two regressions? Which of the two predictions do you expect to be more accurate? Explain why.

5.25 Is wine good for your heart? Table 4.3 (page 101) gives data on wine consumption and heart disease death rates in 19 countries. A scatterplot (Exercise 4.27) shows a moderately strong relationship.

(a) The correlation for these variables is $r = -0.843$. What does a negative correlation say about wine consumption and heart disease deaths? About what percent of the variation among countries in heart disease death rates is explained by the straight-line relationship with wine consumption?

(b) The least-squares regression line for predicting heart disease death rate from wine consumption is

$$\hat{y} = 260.56 - 22.969x$$

Use this equation to predict the heart disease death rate in another country where adults average 4 liters of alcohol from wine each year.

(Digital Vision/Getty Images)

(c) The correlation in (a) and the slope of the least-squares line in (b) are both negative. Is it possible for these two quantities to have opposite signs? Explain your answer.

5.26 Always plot your data! Table 5.2 presents four sets of data prepared by the statistician Frank Anscombe to illustrate the dangers of calculating without first plotting the data.[11]

(a) Without making scatterplots, find the correlation and the least-squares regression line for all four data sets. What do you notice? Use the regression line to predict y for $x = 10$.

(b) Make a scatterplot for each of the data sets and add the regression line to each plot.

TABLE 5.2 Four data sets for exploring correlation and regression

Data Set A

x	10	8	13	9	11	14	6	4	12	7	5
y	8.04	6.95	7.58	8.81	8.33	9.96	7.24	4.26	10.84	4.82	5.68

Data Set B

x	10	8	13	9	11	14	6	4	12	7	5
y	9.14	8.14	8.74	8.77	9.26	8.10	6.13	3.10	9.13	7.26	4.74

Data Set C

x	10	8	13	9	11	14	6	4	12	7	5
y	7.46	6.77	12.74	7.11	7.81	8.84	6.08	5.39	8.15	6.42	5.73

Data Set D

x	8	8	8	8	8	8	8	8	8	8	19
y	6.58	5.76	7.71	8.84	8.47	7.04	5.25	5.56	7.91	6.89	12.50

(c) In which of the four cases would you be willing to use the regression line to describe the dependence of y on x? Explain your answer in each case.

5.27 **Lots of wine.** Exercise 5.25 gives the least-squares line for predicting a nation's heart disease death rate from its wine consumption. What is the predicted heart disease death rate for a country that drinks enough wine to supply 150 liters of alcohol per person? Explain why this result can't be true. Explain why using the regression line for this prediction is not intelligent.

5.28 **What's my grade?** In Professor Friedman's economics course the correlation between the students' total scores prior to the final examination and their final examination scores is $r = 0.6$. The pre-exam totals for all students in the course have mean 280 and standard deviation 30. The final exam scores have mean 75 and standard deviation 8. Professor Friedman has lost Julie's final exam but knows that her total before the exam was 300. He decides to predict her final exam score from her pre-exam total.

(a) What is the slope of the least-squares regression line of final exam scores on pre-exam total scores in this course? What is the intercept?

(b) Use the regression line to predict Julie's final exam score.

(c) Julie doesn't think this method accurately predicts how well she did on the final exam. Use r^2 to argue that her actual score could have been much higher (or much lower) than the predicted value.

5.29 Going to class. A study of class attendance and grades among first-year students at a state university showed that in general students who attended a higher percent of their classes earned higher grades. Class attendance explained 16% of the variation in grade index among the students. What is the numerical value of the correlation between percent of classes attended and grade index?

5.30 Will I bomb the final? We expect that students who do well on the midterm exam in a course will usually also do well on the final exam. Gary Smith of Pomona College looked at the exam scores of all 346 students who took his statistics class over a 10-year period.[12] The least-squares line for predicting final exam score from midterm exam score was $\hat{y} = 46.6 + 0.41x$.

Octavio scores 10 points above the class mean on the midterm. How many points above the class mean do you predict that he will score on the final? (Hint: Use the fact that the least-squares line passes through the point $(\bar{x}, \bar{y})$ and the fact that Octavio's midterm score is $\bar{x} + 10$. This is an example of the phenomenon that gave "regression" its name: students who do well on the midterm will on the average do less well, but still above average, on the final.)

5.31 Height and reading score. A study of elementary school children, ages 6 to 11, finds a high positive correlation between height x and score y on a test of reading comprehension. What explains this correlation?

5.32 Do artificial sweeteners cause weight gain? People who use artificial sweeteners in place of sugar tend to be heavier than people who use sugar. Does this mean that artificial sweeteners cause weight gain? Give a more plausible explanation for this association.

5.33 What explains grade inflation? Students at almost all colleges and universities get higher grades than was the case 10 or 20 years ago. Is grade inflation caused by lower grading standards? Suggest some lurking variables that might affect the distribution of grades even if standards have remained the same.

5.34 The benefits of foreign language study. Members of a high school language club believe that study of a foreign language improves a student's command of English. From school records, they obtain the scores on an English achievement test given to all seniors. The mean score of seniors who studied a foreign language for at least two years is much higher than the mean score of seniors who studied no foreign language. These data are not good evidence that language study strengthens English skills. Identify the explanatory and response variables in this study. Then explain what lurking variable prevents the conclusion that language study improves students' English scores.

5.35 Beware correlations based on averages. The variables used for regression and correlation are sometimes averages of a number of individual values. For example, both degree-days and gas consumption for the Sanchez household (Exercise 4.16) are averages over the days of a month. The values for individual days vary about the monthly average. If you calculated the correlation for the 485 days in these 16 months, would r be closer to 1 or closer to 0 than the r for the 16 monthly averages? Why?

5.36 Beavers and beetles. Ecologists sometimes find rather strange relationships in our environment. One study seems to show that beavers benefit beetles. The researchers laid out 23 circular plots, each 4 meters in diameter, in an area where beavers were cutting down cottonwood trees. In each plot, they counted the number of stumps from trees cut by beavers and the number of clusters of beetle larvae. Here are the data:[13]

(Daniel J. Cox/Natural Exposures)

Stumps	2	2	1	3	3	4	3	1	2	5	1	3
Beetle larvae	10	30	12	24	36	40	43	11	27	56	18	40

Stumps	2	1	2	2	1	1	4	1	2	1	4
Beetle larvae	25	8	21	14	16	6	54	9	13	14	50

(a) Make a scatterplot that shows how the number of beaver-caused stumps influences the number of beetle larvae clusters. What does your plot show? (Ecologists think that the new sprouts from stumps are more tender than other cottonwood growth, so that beetles prefer them.)

(b) Find the least-squares regression line and draw it on your plot.

(c) What percent of the observed variation in beetle larvae counts can be explained by straight-line dependence on stump counts?

5.37 A computer game. A multimedia statistics learning system includes a test of skill in using the computer's mouse. The software displays a circle at a random location on the computer screen. The subject tries to click in the circle with the mouse as quickly as possible. A new circle appears as soon as the subject clicks the old one. Table 5.3 gives data for one subject's trials, 20 with each hand. Distance is the distance from the cursor location to the center of the new circle, in units whose actual size depends on the size of the screen. Time is the time required to click in the new circle, in milliseconds.[14]

(a) We suspect that time depends on distance. Make a scatterplot of time against distance, using separate symbols for each hand.

(b) Describe the pattern. How can you tell that the subject is right-handed?

TABLE 5.3 Reaction times in a computer game

Time	Distance	Hand	Time	Distance	Hand
115	190.70	right	240	190.70	left
96	138.52	right	190	138.52	left
110	165.08	right	170	165.08	left
100	126.19	right	125	126.19	left
111	163.19	right	315	163.19	left
101	305.66	right	240	305.66	left
111	176.15	right	141	176.15	left
106	162.78	right	210	162.78	left
96	147.87	right	200	147.87	left
96	271.46	right	401	271.46	left
95	40.25	right	320	40.25	left
96	24.76	right	113	24.76	left
96	104.80	right	176	104.80	left
106	136.80	right	211	136.80	left
100	308.60	right	238	308.60	left
113	279.80	right	316	279.80	left
123	125.51	right	176	125.51	left
111	329.80	right	173	329.80	left
95	51.66	right	210	51.66	left
108	201.95	right	170	201.95	left

(c) Find the regression line of time on distance separately for each hand. Draw these lines on your plot. Which regression does a better job of predicting time from distance? Give numerical measures that describe the success of the two regressions.

5.38 Using residuals. It is possible that the subject in Exercise 5.37 got better in later trials due to learning. It is also possible that he got worse due to fatigue. Plot the residuals from each regression against the time order of the trials (down the columns in Table 5.3). Is either of these systematic effects of time visible in the data?

5.39 How residuals behave. Return to the merlin data regression of Exercise 5.22. Use your calculator or software to obtain the residuals. The residuals are the part of the response left over after the straight-line tie to the explanatory variable is removed. Find the correlation between the residuals and the explanatory variable. Your result should not be a surprise.

5.40 Using residuals. Make a residual plot (residual against explanatory variable) for the merlin regression of Exercise 5.22. Use a y scale from -20 to 20 or wider to better see the pattern. Add a horizontal line at $y = 0$, the mean of the residuals.

(a) Describe the pattern if we ignore the two years with $x = 38$. Do the $x = 38$ years fit this pattern?

(b) Return to the original data. Make a scatterplot with two least-squares lines: with all nine years and without the two $x = 38$ years. Although the original regression in Exercise 5.22 seemed satisfactory, the two $x = 38$ years are influential. We would like more data for years with x greater than 33.

5.41 Using residuals. Return to the regression of highway mileage on city mileage in Exercise 5.3 (page 109). Use your calculator or software to obtain the residuals. Make a residual plot (residuals against city mileage) and add a horizontal line at $y = 0$ (the mean of the residuals).

(a) Which car has the largest positive residual? The largest negative residual?

(b) The Honda Insight, an extreme outlier, does not have the largest residual in either direction. Why is this not surprising?

(c) Explain briefly what a large positive residual says about a car. What does a large negative residual say?

Chapter 5 MEDIA EXERCISES

5.42 Influence in regression. The *Correlation and Regression* applet allows you to create a scatterplot and to move points by dragging with the mouse. Click to create a group of 10 points in the lower-left corner of the scatterplot with a strong straight-line pattern (correlation about 0.9). Click the "Show least-squares line" box to display the regression line.

(a) Add one point at the upper right that is far from the other 10 points but exactly on the regression line. Why does this outlier have no effect on the line even though it changes the correlation?

(b) Now drag this last point down until it is opposite the group of 10 points. You see that one end of the least-squares line chases this single point, while the other end remains near the middle of the original group of 10. What makes the last point so influential?

5.43 Is regression useful? In Exercise 4.32 (page 102) you used the *Correlation and Regression* applet to create three scatterplots having correlation about $r = 0.7$ between the horizontal variable x and the vertical variable y. Create three similar scatterplots again, and click the "Show least-squares line" box to display the regression lines. Correlation $r = 0.7$ is considered reasonably strong in many areas of work. Because there is a reasonably strong correlation, we might use a regression line to predict y from x. In which of your three scatterplots does it make sense to use a straight line for prediction?

5.44 Guessing a regression line. Click on the scatterplot to create a group of 15 to 20 points from lower left to upper right with a clear positive straight-line pattern (correlation around 0.7). Click the "Draw line" button and use the mouse (right-click and drag) to draw a line through the middle of the cloud of points from lower left to upper right. Note the "thermometer" above the plot. The red portion is the sum of the squared vertical distances from the points in the plot to the least-squares line. The green portion is the "extra" sum of squares for your line—it shows by how much your line misses the smallest possible sum of squares.

(a) You drew a line by eye through the middle of the pattern. Yet the right-hand part of the bar is probably almost entirely green. What does that tell you?

(b) Now click the "Show least-squares line" box. Is the slope of the least-squares line smaller (the new line is less steep) or larger (line is steeper) than that of your line? If you repeat this exercise several times, you will consistently get the same result. The least-squares line minimizes the *vertical* distances of the points from the line. It is *not* the line through the "middle" of the cloud of points. This is one reason why it is hard to draw a good regression line by eye.

5.45 **An influenza epidemic.** In 1918 and 1919 a worldwide outbreak of influenza killed more than 25 million people. The EESEE story "Influenza Outbreak of 1918" includes the following data on the number of new influenza cases and the number of deaths from the epidemic in San Francisco week by week from October 5, 1918, to January 25, 1919. The date given is the last day of the week.

EESEE

Date	Oct. 5	Oct. 12	Oct. 19	Oct. 26	Nov. 2	Nov. 9	Nov. 16	Nov. 23	Nov. 30
Cases	36	531	4233	8682	7164	2229	600	164	57
Deaths	0	0	130	552	738	414	198	90	56

Date	Dec. 7	Dec. 14	Dec. 21	Dec. 28	Jan. 4	Jan. 11	Jan. 18	Jan. 25
Cases	722	1517	1828	1539	2416	3148	3465	1440
Deaths	50	71	137	178	194	290	310	149

We expect the number of deaths to lag behind the number of new cases because the disease takes some time to kill its victims.

(a) Make three scatterplots of deaths (the response variable) against each of new cases the same week, new cases one week earlier, and new cases two weeks earlier. Describe and compare the patterns you see.

(b) Find the correlations that go with your three plots.

(c) What do you conclude? Do the cases data predict deaths best with no lag, a one-week lag, or a two-week lag?

(d) Find the least-squares line for predicting weekly deaths for the choice of explanatory variable that gives the best predictions.

(Digital Vision/PictureQuest)

Two-Way Tables*

We have concentrated on relationships in which at least the response variable is quantitative. Now we will shift to describing relationships between two or more categorical variables. Some variables—such as gender, race, and occupation—are categorical by nature. Other categorical variables are created by grouping values of a quantitative variable into classes. Published data often appear in grouped form to save space. To analyze categorical data, we use the *counts* or *percents* of individuals that fall into various categories.

> **EXAMPLE 6.1** Age and education
>
> *two-way table*
> *row and column variables*
>
> Table 6.1 presents Census Bureau data for the year 2000 on the level of education reached by Americans of different ages.[1] Many people under 25 years of age have not completed their education, so they are left out of the table. Both variables, age and education, are grouped into categories. This is a **two-way table** because it describes two categorical variables. Education is the **row variable** because each row in the table describes people with one level of education. Age is the **column variable** because each column describes one age group. The entries in the table are the counts of persons in each age-by-education class. Although both age and education in this table are categorical variables, both have a natural order from least to most. The order of the rows and the columns in Table 6.1 reflects the order of the categories.

---

*This material is important in statistics, but it is needed later in this book only for Chapter 20. You may omit it if you do not plan to read Chapter 20 or delay reading it until you reach Chapter 20.

TABLE 6.1 Years of school completed, by age (thousands of persons)

Education	Age group 25 to 34	Age group 35 to 54	Age group 55 and over	Total
Did not complete high school	4,459	9,174	14,226	27,859
Completed high school	11,562	26,455	20,060	58,077
College, 1 to 3 years	10,693	22,647	11,125	44,465
College, 4 or more years	11,071	23,160	10,597	44,828
Total	37,786	81,435	56,008	175,230

Marginal distributions

How can we best grasp the information contained in Table 6.1? First, *look at the distribution of each variable separately*. The distribution of a categorical variable says how often each outcome occurred. The "Total" column at the right of the table contains the totals for each of the rows. These row totals give the distribution of education level (the row variable) among all people 25 years of age and older: 27,859,000 did not complete high school, 58,077,000 finished high school but did not attend college, and so on. In the same way, the "Total" row at the bottom of the table gives the age distribution. If the row and column totals are missing, the first thing to do in studying a two-way table is to calculate them. The distributions of education alone and age alone are called **marginal distributions** because they appear at the right and bottom margins of the two-way table.

marginal distribution

If you check the row and column totals in Table 6.1, you will notice some discrepancies. For example, the sum of the entries in the "25 to 34" column is 37,785. The entry in the "Total" row for that column is 37,786. The explanation is **roundoff error.** The table entries are in thousands of persons, and each is rounded to the nearest thousand. The Census Bureau obtained the "Total" entry by rounding the exact number of people aged 25 to 34 to the nearest thousand. The result was 37,786,000. Adding the row entries, each of which is already rounded, gives a slightly different result.

roundoff error

Percents are often more informative than counts. We can display the marginal distribution of education level in terms of percents by dividing each row total by the table total and converting to a percent.

EXAMPLE 6.2   Calculating a marginal distribution

The percent of people 25 years of age and older who have at least 4 years of college is

$$\frac{\text{total with 4 years of college}}{\text{table total}} = \frac{44,828}{175,230} = 0.256 = 25.6\%$$

Do three more such calculations to obtain the marginal distribution of education level in percents. Here it is:

	Did not complete high school	Completed high school	1 to 3 years of college	≥4 years of college
Percent	15.9	33.1	25.4	25.6

The total is 100% because everyone is in one of the four education categories.

Each marginal distribution from a two-way table is a distribution for a single categorical variable. As we saw in Chapter 1 (page 8), we can use a bar graph or a pie chart to display such a distribution. Figure 6.1 is a bar graph of the distribution of years of schooling. We see that people with at least some college education make up about half of the 25 and over population.

In working with two-way tables, you must calculate lots of percents. Here's a tip to help decide what fraction gives the percent you want. Ask, "What group represents the total that I want a percent of?" The count for that group is the denominator of the fraction that leads to the percent. In Example 6.2, we wanted a percent "of people aged 25 years and over," so the count of people aged 25 and over (the table total) is the denominator.

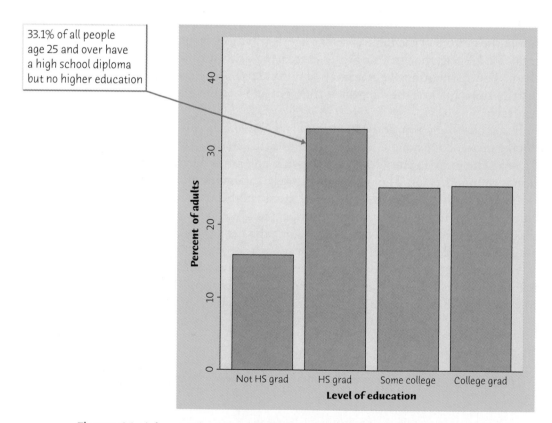

33.1% of all people age 25 and over have a high school diploma but no higher education

Figure 6.1 A bar graph of the distribution of years of schooling completed among people aged 25 years and over. This is one of the marginal distributions for Table 6.1.

APPLY YOUR KNOWLEDGE

6.1   Age and education. The counts in the "Total" column at the right of Table 6.1 are the counts of people in each education group. Explain why the sum of these counts is not exactly equal to 175,230, the table total that appears at the lower-right corner of the table.

6.2   Age and education. Give the marginal distribution of age (in percents) among people 25 years of age and older, starting from the counts in Table 6.1.

6.3   Risks of playing soccer. A study in Sweden looked at former elite soccer players, people who had played soccer but not at the elite level, and people of the same age who did not play soccer. Here is a two-way table that classifies these subjects by whether or not they had arthritis of the hip or knee by their mid-50s:[2]

	Elite	Non-elite	Did not play
Arthritis	10	9	24
No arthritis	61	206	548

(a) How many people do these data describe?

(b) How many of these people have arthritis of the hip or knee?

(c) Give the marginal distribution of participation in soccer, both as counts and as percents.

Relationships between categorical variables

Table 6.1 contains much more information than the two marginal distributions of age alone and education alone. The nature of the relationship between age and education cannot be deduced from the separate distributions but requires the full table. **Relationships between categorical variables are described by calculating appropriate percents from the counts given** We use percents because counts are often hard to compare. For example, 23,160,000 persons aged 35 to 54 have completed college, and only 10,597,000 persons in the 55-and-up age group have done so. These counts do not accurately describe the association, however, because there are many more people in the younger age group.

EXAMPLE 6.3   How common is college education?

What percent of people aged 25 to 34 have completed 4 years of college? This is the count of those who are 25 to 34 and have 4 years of college as a percent of the age group total:

$$\frac{11,071}{37,786} = 0.293 = 29.3\%$$

"People aged 25 to 34" is the group we want a percent of, so the count for that group is the denominator. In the same way, find the percent of people in each age group who have completed college. The comparison of all three groups is

	25 to 34	35 to 54	55 and over
Percent with 4 years of college	29.3	28.4	18.9

These percentages make it clear that a college education is less common among Americans over age 55 than among younger adults. This is an important aspect of the association between age and education.

These percents do *not* add to 100% because they are not the parts of a whole. They describe three different age groups.

APPLY YOUR KNOWLEDGE

6.4 **Age and education.** Using the counts in Table 6.1, find the percent of people in each age group who did not complete high school. Draw a bar graph that compares these percents. State briefly what the data show.

6.5 **Risks of playing soccer.** Find the percent of each group in the soccer-risk data of Exercise 6.3 who have arthritis. What do these percents say about the association between playing soccer and later arthritis?

Conditional distributions

Example 6.3 does not compare the complete distributions of years of schooling in the three age groups. It compares only the percents who finished college. Let's look at the complete picture.

EXAMPLE 6.4 Calculating a conditional distribution

Information about the 25 to 34 age group occupies the first column in Table 6.1. To find the complete distribution of education in this age group, look only at that column. Compute each count as a percent of the column total, which is 37,786. Here is the distribution:

	Did not complete high school	Completed high school	1 to 3 years of college	≥4 years of college
Percent	11.8	30.6	28.3	29.3

conditional distribution

These percents do add to 100% because all 25- to 34-year-olds fall in one of the educational categories. The four percents together are the **conditional distribution** of education, given that a person is 25 to 34 years of age. We use the term "conditional" because the distribution refers only to people who satisfy the condition that they are 25 to 34 years old.

```
Session                                              _ □ X
              25 to 34     35 to 54     55 up        All

1:NotHS           4459         9174     14226        27859
                 11.80        11.27     25.40        15.90

2:HSgrad         11562        26455     20060        58077
                 30.60        32.49     35.82        33.14

3:SomeCo         10693        22647     11125        44465
                 28.30        27.81     19.86        25.38

4:CollGr         11071        23160     10597        44828
                 29.30        28.44     18.92        25.58

All              37785        81436     56008       175229
                100.00       100.00    100.00       100.00

  Cell Contents-
            Count
            % of Col
```

Figure 6.2 Minitab output of the two-way table of education by age with the three conditional distributions of education in each age group. The percents in each column add to 100%.

Now focus in turn on the second column (people aged 35 to 54) and then the third column (people 55 and over) of Table 6.1 in order to find two more conditional distributions. Statistical software can speed the task of finding each entry in a two-way table as a percent of its column total. Figure 6.2 displays the result. The software found the row and column totals from the table entries, so they differ slightly from those in Table 6.1. The software produces a legend at the bottom to remind us that the entries for each cell are the count and the count as a percent of its column total.

Each cell in this table contains a count from Table 6.1 along with that count as a percent of the column total. The percents in each column form the conditional distribution of years of schooling for one age group. The percents in each column add to 100% because everyone in the age group is accounted for. Comparing the conditional distributions reveals the nature of the association between age and education. The distributions of education in the two younger groups are quite similar, but higher education is less common in the 55-and-over group.

Bar graphs can help make the association visible. We could make three side-by-side bar graphs, each resembling Figure 6.1, to present the three conditional distributions. Figure 6.3 shows an alternative form of bar graph. Each set of three bars compares the percents in the three age groups who have reached a specific educational level. We see at once that the "25 to 34" and "35 to 54" bars are similar for all four levels of education, and that the "55 and over" bars show that many more people in this group did not finish high school and that many fewer have any college.

No single graph (such as a scatterplot) portrays the form of the relationship between categorical variables. No single numerical measure (such as the correlation) summarizes the strength of the association. Bar graphs are flexible enough to be helpful, but you must think about what comparisons you want

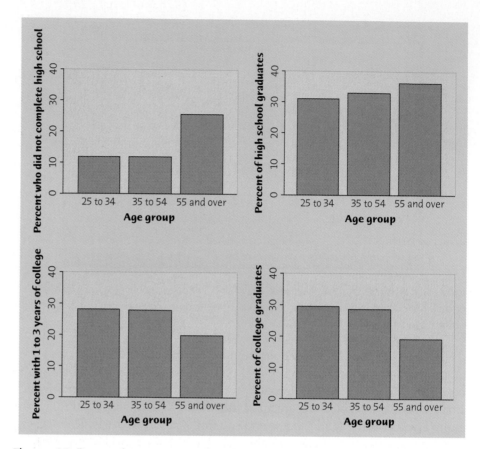

Figure 6.3 Bar graphs to compare the education levels of three age groups. Each graph compares the percents of three groups who fall in one of the four education levels.

to display. For numerical measures, we rely on well-chosen percents. You must decide which percents you need. Here is a hint: compare the conditional distributions of the response variable (education) for the separate values of the explanatory variable (age). That's what we did in Figure 6.3.

APPLY YOUR KNOWLEDGE

6.6 Education among older people. Show the calculations to find the conditional distribution of education among persons 55 years old and over. Your results should agree with the "55 up" column in Figure 6.2 and with the heights of the "55 and over" bars in the four bar graphs in Figure 6.3.

6.7 Majors for men and women in business. A study of the career plans of young women and men sent questionnaires to all 722 members of the senior class in the College of Business Administration at the University of Illinois. One question asked which major within the business program the student had chosen. Here are the data from the students who responded:[3]

	Female	Male
Accounting	68	56
Administration	91	40
Economics	5	6
Finance	61	59

(a) Find the two conditional distributions of major, one for women and one for men. Based on your calculations, describe the differences between women and men with a graph and in words.

(b) What percent of the students did not respond to the questionnaire? The nonresponse weakens conclusions drawn from these data.

6.8 Marginal distributions aren't the whole story. Here are the row and column totals for a two-way table with two rows and two columns:

a	b	50
c	d	50
60	40	100

Find *two different* sets of counts a, b, c, and d for the body of the table that give these same totals. This shows that the relationship between two variables cannot be obtained from the two individual distributions of the variables.

Simpson's paradox

As is the case with quantitative variables, the effects of lurking variables can change or even reverse relationships between two categorical variables. Here is an example that demonstrates the surprises that can await the unsuspecting user of data.

Attack of the killer TVs!

Despite occasional media frenzy, you won't find shark attacks high in any count of causes of death and injury. In fact, the 28 children killed by falling TV sets in the United States between 1990 and 1997 is four times the total number of people killed by great white sharks in the twentieth century.

EXAMPLE 6.5 Do medical helicopters save lives?

Accident victims are sometimes taken by helicopter from the accident scene to a hospital. The helicopter may save time and also brings medical care to the accident scene. Does the use of helicopters save lives? We might compare the percents of accident victims who die with helicopter evacuation and with the usual transport to a hospital by road. Here are hypothetical data that illustrate a practical difficulty:[4]

	Helicopter	Road
Victim died	64	260
Victim survived	136	840
Total	200	1100

We see that 32% (64 out of 200) helicopter patients died, compared with only 24% (260 out of 1100) of the others. That seems discouraging. The explanation is that the helicopter is sent mostly to serious accidents, so that the victims transported by helicopter are more often seriously injured than other victims. They are more likely to die with or without helicopter evacuation. Here are the same data broken down by the seriousness of the accident:

Serious Accidents	Helicopter	Road		Less Serious Accidents	Helicopter	Road
Died	48	60		Died	16	200
Survived	52	40		Survived	84	800
Total	100	100		Total	100	1000

(Ashley/Cooper/PICIMPACT/CORBIS)

Inspect these tables to convince yourself that they describe the same 1300 accidents as the original two-way table. For example, 200 were moved by helicopter, and 64 (48 + 16) of these died.

Among victims of serious accidents, the helicopter saves 52% (52 out of 100) compared with 40% for road transport. If we look only at less serious accidents, 84% of those transported by helicopter survive, versus 80% of those transported by road. Both groups of victims have a higher survival rate when evacuated by helicopter.

At first, it seems paradoxical that the helicopter does better for both groups of victims but worse when all victims are lumped together. Examining the data makes the explanation clear. Half the helicopter transport patients are from serious accidents, compared with only 100 of the 1100 road transport patients. So the helicopter carries patients who are more likely to die. The seriousness of the accident was a lurking variable that, until we uncovered it, made the relationship between survival and mode of transport to a hospital hard to interpret. Example 6.5 illustrates *Simpson's paradox*.

SIMPSON'S PARADOX

An association or comparison that holds for all of several groups can reverse direction when the data are combined to form a single group. This reversal is called **Simpson's paradox.**

The lurking variable in Simpson's paradox is categorical. That is, it breaks the individuals into groups, as when accident victims are classified as injured in a "serious accident" or a "less serious accident." Simpson's paradox is just an extreme form of the fact that observed associations can be misleading when there are lurking variables.

APPLY YOUR KNOWLEDGE

6.9 **Airline flight delays.** Here are the numbers of flights on time and delayed for two airlines at five airports in one month. Overall on-time percentages for each airline are often reported in the news. The airport that flights serve is a lurking variable that can make such reports misleading.[5]

	Alaska Airlines		America West	
	On time	Delayed	On time	Delayed
Los Angeles	497	62	694	117
Phoenix	221	12	4840	415
San Diego	212	20	383	65
San Francisco	503	102	320	129
Seattle	1841	305	201	61

(a) What percent of all Alaska Airlines flights were delayed? What percent of all America West flights were delayed? These are the numbers usually reported.

(b) Now find the percent of delayed flights for Alaska Airlines at each of the five airports. Do the same for America West.

(c) America West does worse at *every one* of the five airports, yet does better overall. That sounds impossible. Explain carefully, referring to the data, how this can happen. (The weather in Phoenix and Seattle lies behind this example of Simpson's paradox.)

6.10 **Race and the death penalty.** Whether a convicted murderer gets the death penalty seems to be influenced by the race of the victim. Here are data on 326 cases in which the defendant was convicted of murder:[6]

White Defendant	White victim	Black victim		Black Defendant	White victim	Black victim
Death	19	0		Death	11	6
Not	132	9		Not	52	97

(a) Use these data to make a two-way table of defendant's race (white or black) versus death penalty (yes or no).

(b) Show that Simpson's paradox holds: a higher percent of white defendants are sentenced to death overall, but for both black and white victims a higher percent of black defendants are sentenced to death.

(c) Use the data to explain why the paradox holds in language that a judge could understand.

Chapter 6 SUMMARY

A **two-way table** of counts organizes data about two categorical variables. Values of the **row variable** label the rows that run across the table, and values of the **column variable** label the columns that run down the table. Two-way tables are often used to summarize large amounts of information by grouping outcomes into categories.

The **row totals** and **column totals** in a two-way table give the **marginal distributions** of the two individual variables. It is clearer to present these distributions as percents of the table total. Marginal distributions tell us nothing about the relationship between the variables.

To find the **conditional distribution** of the row variable for one specific value of the column variable, look only at that one column in the table. Find each entry in the column as a percent of the column total.

There is a conditional distribution of the row variable for each column in the table. Comparing these conditional distributions is one way to describe the association between the row and the column variables. It is particularly useful when the column variable is the explanatory variable.

Bar graphs are a flexible means of presenting categorical data. There is no single best way to describe an association between two categorical variables.

A comparison between two variables that holds for each individual value of a third variable can be changed or even reversed when the data for all values of the third variable are combined. This is **Simpson's paradox.** Simpson's paradox is an example of the effect of lurking variables on an observed association.

Chapter 6 EXERCISES

Marital status and job level. *We sometimes hear that getting married is good for your career. Table 6.2 presents data from one of the studies behind this generalization. To avoid gender effects, the investigators*

TABLE 6.2 Marital status and job level

Job grade	Marital status				Total
	Single	Married	Divorced	Widowed	
1	58	874	15	8	955
2	222	3927	70	20	4239
3	50	2396	34	10	2490
4	7	533	7	4	551
Total	337	7730	126	42	8235

looked only at men. The data describe the marital status and the job level of all 8235 male managers and professionals employed by a large manufacturing firm.[7] The firm assigns each position a grade that reflects the value of that particular job to the company. The authors of the study grouped the many job grades into quarters. Grade 1 contains jobs in the lowest quarter of the job grades, and Grade 4 contains those in the highest quarter. Exercises 6.11 to 6.15 are based on these data.

6.11 Marginal distributions. Give (in percents) the two marginal distributions, for marital status and for job grade. Do each of your two sets of percents add to exactly 100%? If not, why not?

6.12 Percents. What percent of single men hold Grade 1 jobs? What percent of Grade 1 jobs are held by single men?

6.13 Conditional distribution. Give (in percents) the conditional distribution of job grade among single men. Should your percents add to 100% (up to roundoff error)?

6.14 Marital status and job grade. One way to see the relationship is to look at who holds Grade 1 jobs.

(a) There are 874 married men with Grade 1 jobs, and only 58 single men with such jobs. Explain why these counts by themselves don't describe the relationship between marital status and job grade.

(b) Find the percent of men in each marital status group who have Grade 1 jobs. Then find the percent in each marital group who have Grade 4 jobs. What do these percents say about the relationship?

6.15 Association is not causation. The data in Table 6.2 show that single men are more likely to hold lower-grade jobs than are married men. We should not conclude that single men can help their career by getting married. What lurking variables might help explain the association between marital status and job grade?

6.16 Smoking by students and their parents. Here are data from eight high schools on smoking among students and among their parents:[8]

	Neither parent smokes	One parent smokes	Both parents smoke
Student does not smoke	1168	1823	1380
Student smokes	188	416	400

(a) How many students do these data describe?

(b) What percent of these students smoke?

(c) Calculate and compare percents to show how parents' smoking influences students' smoking. Briefly state your conclusions about the relationship.

6.17 Python eggs. How is the hatching of water python eggs influenced by the temperature of the snake's nest? Researchers assigned newly laid eggs to one of three temperatures: hot, neutral, or cold. Hot duplicates the extra warmth provided by the mother python, and cold duplicates the absence of the mother. Here are the data on the number of eggs and the number that hatched:[9]

	Cold	Neutral	Hot
Number of eggs	27	56	104
Number hatched	16	38	75

(a) Make a two-way table of temperature by outcome (hatched or not).

(b) Calculate the percent of eggs in each group that hatched. The researchers anticipated that eggs would not hatch at cold temperatures. Do the data support that anticipation?

6.18 Firearm deaths. Firearms are second to motor vehicles as a cause of nondisease deaths in the United States. Here are counts from a study of all firearm-related deaths in Milwaukee, Wisconsin, between 1990 and 1994.[10] We want to compare the types of firearms used in homicides and in suicides. We suspect that long guns (shotguns and rifles) will more often be used in suicides because many people keep them at home for hunting. Make a careful comparison of homicides and suicides with a bar graph. What do you find about long guns versus handguns?

	Handgun	Shotgun	Rifle	Unknown	Total
Homicides	468	28	15	13	524
Suicides	124	22	24	5	175

6.19 Helping cocaine addicts. Cocaine addiction is hard to break. Addicts need cocaine to feel any pleasure, so perhaps giving them an antidepressant drug will help. A 3-year study with 72 chronic cocaine users compared an antidepressant drug called desipramine with lithium and a placebo. (Lithium is a standard drug to treat cocaine addiction. A placebo is a dummy drug, used so that the effect of being in the study but not taking any drug can be seen.) One-third of the subjects, chosen at random, received each drug. Here are the results:[11]

	Desipramine	Lithium	Placebo
Relapse	10	18	20
No relapse	14	6	4
Total	24	24	24

(a) Compare the effectiveness of the three treatments in preventing relapse. Use percents and draw a bar graph.

(b) Do you think that this study gives good evidence that desipramine actually *causes* a reduction in relapses?

6.20 Female college professors. Purdue University is a Big Ten university that emphasizes engineering, scientific, and technical fields. University faculty start as assistant professors, then are promoted to associate professor and eventually to professor. Here is a two-way table that breaks down Purdue's 1621 faculty members in the 1998–1999 academic year by gender and academic rank:

	Female	Male	Total
Assistant professors	126	213	339
Associate professors	149	411	560
Professors	60	662	722
Total	335	1286	1621

(a) Describe the relationship between rank and gender by finding and commenting on several percents.

(b) One possible explanation for the association might be discrimination (women find it harder to win promotion to higher ranks). Suggest other possible explanations.

6.21 Do angry people have more heart disease? People who get angry easily tend to have more heart disease. That's the conclusion of a study that followed a random sample of 12,986 people from three locations for about four years. All subjects were free of heart disease at the beginning of the study. The subjects took the Spielberger Trait Anger Scale test, which measures how prone a person is to sudden anger. Here are data for the 8474 people in the sample who had normal blood pressure.[12] CHD stands for "coronary heart disease." This includes people who had heart attacks and those who needed medical treatment for heart disease.

(Henryk Kaiser/eStock Photography/PictureQuest)

	Low anger	Moderate anger	High anger	Total
CHD	53	110	27	190
No CHD	3057	4621	606	8284
Total	3110	4731	633	8474

Present evidence from this two-way table that backs up the study's conclusion about the relationship between anger and heart disease.

6.22 Which hospital is safer? To help consumers make informed decisions about health care, the government releases data about patient outcomes in hospitals. You want to compare Hospital A and Hospital B, which serve your community. Here are data on all patients undergoing surgery in a recent time period. The data include the condition of the patient ("good" or "poor") before the surgery. "Survived" means that the patient lived at least 6 weeks following surgery.

Good Condition

	Hospital A	Hospital B
Died	6	8
Survived	594	592
Total	600	600

Poor Condition

	Hospital A	Hospital B
Died	57	8
Survived	1443	192
Total	1500	200

(a) Compare percents to show that Hospital A has a higher survival rate for both groups of patients.

(b) Combine the data into a single two-way table of outcome ("survived" or "died") by hospital (A or B). The local paper reports just these overall survival rates. Which hospital has the higher rate?

(c) Explain from the data, in language that a reporter can understand, how Hospital B can do better overall even though Hospital A does better for both groups of patients.

6.23 Discrimination? Wabash Tech has two professional schools, business and law. Here are two-way tables of applicants to both schools, categorized by gender and admission decision. (Although these data are made up, similar situations occur in reality.)[13]

Business

	Admit	Deny
Male	480	120
Female	180	20

Law

	Admit	Deny
Male	10	90
Female	100	200

(a) Make a two-way table of gender by admission decision for the two professional schools together by summing entries in these tables.

(b) From the two-way table, calculate the percent of male applicants who are admitted and the percent of female applicants who are admitted. Wabash admits a higher percent of male applicants.

(c) Now compute separately the percents of male and female applicants admitted by the business school and by the law school. Each school admits a higher percent of female applicants.

(d) This is Simpson's paradox: both schools admit a higher percent of the women who apply, but overall Wabash admits a lower percent of female applicants than of male applicants. Explain carefully, as if speaking to a skeptical reporter, how it can happen that Wabash appears to favor males when each school individually favors females.

6.24 Obesity and health. Recent studies have shown that earlier reports underestimated the health risks associated with being overweight. The error was due to overlooking lurking variables. In particular, smoking tends both to reduce weight and to lead to earlier death. Illustrate Simpson's paradox by a simplified version of this situation. That is, make up tables of overweight (yes or no) by early death (yes or no) by smoker (yes or no) such that

- Overweight smokers and overweight nonsmokers both tend to die earlier than those not overweight.

- But when smokers and nonsmokers are combined into a two-way table of overweight by early death, persons who are not overweight tend to die earlier.

(Aneal Vohra/Index Stock Imagery/PictureQuest)

Chapter 6 MEDIA EXERCISES

6.25 Baldness and heart attacks. The EESEE story "Baldness and Heart Attacks" reports results from comparing men under the age of 55 who survived a first heart attack with men admitted to the same hospitals for other reasons. As part of the study, each patient was asked to rate his own degree of baldness, from 1 (no baldness) to 5 (extreme baldness). Here are the counts:

Baldness score	Heart attack patients	Other patients
1	251	331
2	165	221
3	195	185
4	50	34
5	2	1

Do these data show a relationship between degree of baldness and having a heart attack? What kind of relationship? (The EESEE story points out that there are lurking variables—in particular, the ages of the men.)

(Tom Stewart/CORBIS)

Exploring Data
Part I Review

Data analysis is the art of describing data using graphs and numerical summaries. The purpose of data analysis is to help us see and understand the most important features of a set of data. Chapter 1 commented on graphs to display distributions: pie charts and bar graphs for categorical variables, histograms and stemplots for quantitative variables. In addition, time plots show how a quantitative variable changes over time. Chapter 2 presented numerical tools for describing the center and spread of the distribution of one variable. Chapter 3 discussed density curves for describing the overall pattern of a distribution, with emphasis on the Normal distributions.

The first STATISTICS IN SUMMARY figure on the following page organizes the big ideas for exploring a quantitative variable. Plot your data, then describe their center and spread using either the mean and standard deviation or the five-number summary. The last step, which makes sense only for some data, is to summarize the data in compact form by using a Normal curve as a model for the overall pattern. The question marks at the last two stages remind us that the usefulness of numerical summaries and Normal distributions depends on what we find when we examine graphs of our data. No short summary does justice to irregular shapes or to data with several distinct clusters.

Chapters 4 and 5 applied the same ideas to relationships between two quantitative variables. The second STATISTICS IN SUMMARY figure retraces the big ideas, with details that fit the new setting. Always begin by making graphs

STATISTICS IN SUMMARY
Analyzing Data for One Variable

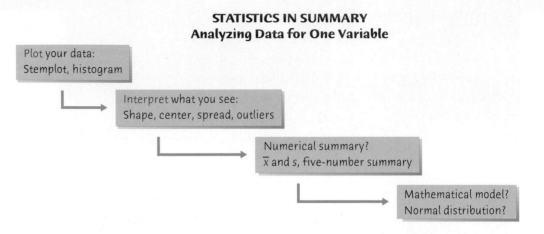

of your data. In the case of a scatterplot, we have learned a numerical summary only for data that show a roughly linear pattern on the scatterplot. The summary is then the means and standard deviations of the two variables and their correlation. A regression line drawn on the plot gives a compact model of the overall pattern that we can use for prediction. Once again there are question marks at the last two stages to remind us that correlation and regression describe only straight-line relationships.

STATISTICS IN SUMMARY
Analyzing Data for Two Variables

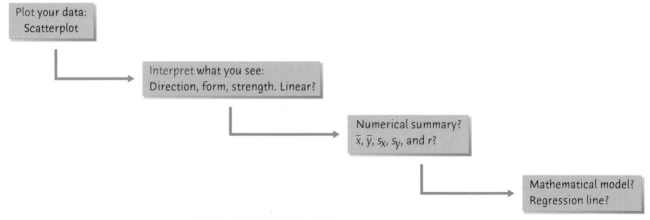

Part I SUMMARY

Here are the most important skills you should have acquired from reading Chapters 1 to 6.

A. DATA

1. Identify the individuals and variables in a set of data.

2. Identify each variable as categorical or quantitative. Identify the units in which each quantitative variable is measured.

3. Identify the explanatory and response variables in situations where one variable explains or influences another.

B. DISPLAYING DISTRIBUTIONS

1. Recognize when a pie chart can and cannot be used.
2. Make a bar graph of the distribution of a categorical variable, or in general to compare related quantities.
3. Interpret pie charts and bar graphs.
4. Make a time plot of a quantitative variable over time.
5. Recognize patterns such as trends and seasonal variation in time plots.
6. Make a histogram of the distribution of a quantitative variable.
7. Make a stemplot of the distribution of a small set of observations. Round leaves or split stems as needed to make an effective stemplot.

C. DESCRIBING DISTRIBUTIONS (QUANTITATIVE VARIABLE)

1. Look for the overall pattern and for major deviations from the pattern.
2. Assess from a histogram or stemplot whether the shape of a distribution is roughly symmetric, distinctly skewed, or neither. Assess whether the distribution has one or more major peaks.
3. Describe the overall pattern by giving numerical measures of center and spread in addition to a verbal description of shape.
4. Decide which measures of center and spread are more appropriate: the mean and standard deviation (especially for symmetric distributions) or the five-number summary (especially for skewed distributions).
5. Recognize outliers and give plausible explanations for them.

D. NUMERICAL SUMMARIES OF DISTRIBUTIONS

1. Find the median M and the quartiles Q_1 and Q_3 for a set of observations.
2. Give the five-number summary and draw a boxplot; assess center, spread, symmetry, and skewness from a boxplot.
3. Find the mean $\bar{x}$ and the standard deviation s for a set of observations.
4. Understand that the median is more resistant than the mean. Recognize that skewness in a distribution moves the mean away from the median toward the long tail.
5. Know the basic properties of the standard deviation: $s \geq 0$ always; $s = 0$ only when all observations are identical and increases as the spread increases; s has the same units as the original measurements; s is pulled strongly up by outliers or skewness.

E. DENSITY CURVES AND NORMAL DISTRIBUTIONS

1. Know that areas under a density curve represent proportions of all observations and that the total area under a density curve is 1.

2. Approximately locate the median (equal-areas point) and the mean (balance point) on a density curve.

3. Know that the mean and median both lie at the center of a symmetric density curve and that the mean moves farther toward the long tail of a skewed curve.

4. Recognize the shape of Normal curves and estimate by eye both the mean and standard deviation from such a curve.

5. Use the 68–95–99.7 rule and symmetry to state what percent of the observations from a Normal distribution fall between two points when both points lie at the mean or one, two, or three standard deviations on either side of the mean.

6. Find the standardized value (z-score) of an observation. Interpret z-scores and understand that any Normal distribution becomes standard Normal $N(0, 1)$ when standardized.

7. Given that a variable has a Normal distribution with a stated mean μ and standard deviation σ, calculate the proportion of values above a stated number, below a stated number, or between two stated numbers.

8. Given that a variable has a Normal distribution with a stated mean μ and standard deviation σ, calculate the point having a stated proportion of all values above it. Also calculate the point having a stated proportion of all values below it.

F. SCATTERPLOTS AND CORRELATION

1. Make a scatterplot to display the relationship between two quantitative variables measured on the same subjects. Place the explanatory variable (if any) on the horizontal scale of the plot.

2. Add a categorical variable to a scatterplot by using a different plotting symbol or color.

3. Describe the form, direction, and strength of the overall pattern of a scatterplot. In particular, recognize positive or negative association and linear (straight-line) patterns. Recognize outliers in a scatterplot.

4. Judge whether it is appropriate to use correlation to describe the relationship between two quantitative variables. Find the correlation r.

5. Know the basic properties of correlation: r measures the strength and direction of only straight-line relationships; r is always a number between -1 and 1; $r = \pm1$ only for perfect straight-line relationships; r moves away from 0 toward ±1 as the straight-line relationship gets stronger.

G. REGRESSION LINES

1. Explain what the slope b and the intercept a mean in the equation $y = a + bx$ of a straight line.

2. Draw a graph of the straight line when you are given its equation.

3. Find the least-squares regression line of a response variable y on an explanatory variable x from data.

4. Find the slope and intercept of the least-squares regression line from the means and standard deviations of x and y and their correlation.

5. Use the regression line to predict y for a given x. Recognize extrapolation and be aware of its dangers.

6. Use r^2, the square of the correlation, to describe how much of the variation in one variable can be accounted for by a straight-line relationship with another variable.

7. Recognize outliers and potentially influential observations from a scatterplot with the regression line drawn on it.

8. Calculate the residuals and plot them against the explanatory variable x or against other variables. Recognize unusual patterns.

H. CAUTIONS ABOUT CORRELATION AND REGRESSION

1. Understand that both r and the least-squares regression line can be strongly influenced by a few extreme observations.

2. Recognize possible lurking variables that may explain the observed association between two variables x and y.

3. Understand that even a strong correlation does not mean that there is a cause-and-effect relationship between x and y.

4. Give plausible explanations for an observed association between two variables: direct cause and effect, the influence of lurking variables, or both.

I. CATEGORICAL DATA (Optional)

1. From a two-way table of counts, find the marginal distributions of both variables by obtaining the row sums and column sums.

2. Express any distribution in percents by dividing the category counts by their total.

3. Describe the relationship between two categorical variables by computing and comparing percents. Often this involves comparing the conditional distributions of one variable for the different categories of the other variable.

4. Recognize Simpson's paradox and be able to explain it.

Review EXERCISES

Review exercises are short and straightforward exercises that help you solidify the basic ideas and skills in each part of this book.

I.1 **Chimps and humans.** Some chimpanzee populations in West Africa use stones as tools to crack nuts. Archaeologists compared the sizes of stone

(Peter Davey/Bruce Coleman, Inc.)

flakes from chimpanzees with those left by hominids (ancestors of humans) in Africa. Similar distributions may indicate that the hominids were busy cracking nuts. Here are data from two sites.[1] The archaeologists conclude that "stone by-products of chimpanzee nut cracking fall within the [size and shape range] observed in a subset of the earliest known hominim technological repertoires."

Size (mm)	1–10	11–20	21–30	31–40	41–50	51–60
Chimps	235	144	34	19	19	14
Hominids	80	102	35	5	1	0

Size (mm)	61–70	71–80	81–90	91–100	101–110
Chimps	5	5	2	1	1
Hominids	0	0	0	0	0

(a) How many flakes were measured at each site? To compare the two size distributions, we prefer a histogram of the percent of flakes in each size range rather than a histogram of the counts. Why?

(b) Make histograms, using the same scale for ease of comparison. Write a brief comparison. Do you agree with the archaeologists' conclusion?

I.2 **Remember what you ate.** How well do people remember their past diet? Data are available for 91 people who were asked about their diet when they were 18 years old. Researchers asked them at about age 55 to describe their eating habits at age 18. The study looked at the correlations between actual intake of many foods at age 18 and the intake the subjects now remember for age 18. The median correlation was $r = 0.217$. The authors say, "We conclude that memory of food intake in the distant past is fair to poor."[2] Explain why $r = 0.217$ points to this conclusion.

I.3 **College students studying.** Do women study more than men? We asked the students in a large first-year college class how many minutes they studied on a typical weeknight. Here are the responses of random samples of 30 women and 30 men from the class:

	Women					Men			
180	120	180	360	240	90	120	30	90	200
120	180	120	240	170	90	45	30	120	75
150	120	180	180	150	150	120	60	240	300
200	150	180	150	180	240	60	120	60	30
120	60	120	180	180	30	230	120	95	150
90	240	180	115	120	0	200	120	120	180

Driving in Canada

Canada is a civilized and restrained nation, at least in the eyes of Americans. A survey cosponsored by the Canada Safety Council suggests that driving in Canada may be more adventurous than expected. Of the Canadian drivers surveyed, 71% said road rage was on the rise, and 90% admitted to falling asleep at the wheel. What really alarms us is the name of the annual survey: the Nerves of Steel Aggressive Driving Study.

(a) Examine the data. Why are you not surprised that most responses are multiples of 10 minutes? We eliminated one student who claimed to study 30,000 minutes per night. Are there any other responses you consider suspicious?

(b) Make a back-to-back stemplot of these data. Does it appear that women study more than men (or at least claim that they do)? Give numerical summaries that back up your conclusion.

I.4 **Who sells soft drinks?** Here are data on the market share of the leading producers of soft drinks in 2001:

Company	Market share
Coca-Cola	43.7%
Pepsi-Cola	31.6%
Cadbury (Dr Pepper, Seven Up)	15.6%
Cott Corp. (store brands)	3.8%
National Beverage (Shasta, Faygo)	2.2%
Big Red	0.4%
Seagram	0.3%

(Steve Raymer/CORBIS)

Display these data in a graph. What percent of the soft drink market is held by other producers (mainly store labels)?

I.5 **More on study times.** In Exercise I.3 you examined the nightly study time claimed by first-year college men and women. The most common methods for formal comparison of two groups use $\bar{x}$ and s to summarize the data.

(a) What kinds of distributions are best summarized by $\bar{x}$ and s?

(b) Each set of study times appears to contain a high outlier. How much does removing the outlier change $\bar{x}$ and s for each group? The presence of outliers makes us reluctant to use the mean and standard deviation for these data unless we remove the outliers on the grounds that these students were exaggerating.

I.6 **A big toe problem.** Hallux abducto valgus (call it HAV) is a deformation of the big toe that is not common in youth and often requires surgery. Doctors used X-rays to measure the angle (in degrees) of deformity in 38 consecutive patients under the age of 21 who came to a medical center for surgery to correct HAV.[3] The angle is a measure of the seriousness of the deformity. The data appear in Table I.1 as "HAV angle." Make a graph and give a numerical description of this distribution. Are there any outliers? Write a brief discussion of the shape, center, and spread of the angle of deformity among young patients needing surgery for this condition.

I.7 **More on a big toe problem.** The HAV angle data in the previous exercise contain one high outlier. Calculate the median, the mean, and

TABLE I.1 Angle of deformity (degrees) for two types of foot deformity

HAV angle	MA angle	HAV angle	MA angle	HAV angle	MA angle
28	18	21	15	16	10
32	16	17	16	30	12
25	22	16	10	30	10
34	17	21	7	20	10
38	33	23	11	50	12
26	10	14	15	25	25
25	18	32	12	26	30
18	13	25	16	28	22
30	19	21	16	31	24
26	10	22	18	38	20
28	17	20	10	32	37
13	14	18	15	21	23
20	20	26	16		

the standard deviation for the full data set and also for the 37 observations remaining when you remove the outlier. How strongly does the outlier affect each of these measures?

I.8 Predicting foot problems. Metatarsus adductus (call it MA) is a turning in of the front part of the foot that is common in adolescents and usually corrects itself. Table I.1 gives the severity of MA ("MA angle") as well. Doctors speculate that the severity of MA can help predict the severity of HAV.

(a) Make a scatterplot of the data. (Which is the explanatory variable?)

(b) Describe the form, direction, and strength of the relationship between MA angle and HAV angle. Are there any clear outliers in your graph?

(c) Do you think the data confirm the doctors' speculation?

I.9 Predicting foot problems, continued.

(a) Find the equation of the least-squares regression line for predicting HAV angle from MA angle. Add this line to the scatterplot you made in the previous exercise.

(b) A new patient has MA angle 25 degrees. What do you predict this patient's HAV angle to be?

(c) Does knowing MA angle allow doctors to predict HAV angle accurately? Explain your answer from the scatterplot, then calculate a numerical measure to support your finding.

I.10 How much oil? How much oil wells in a given field will ultimately produce is key information in deciding whether to drill more wells. Here are the estimated total amounts of oil recovered from 64 wells in the

Devonian Richmond Dolomite area of the Michigan basin, in thousands of barrels:[4]

21.71	53.2	46.4	42.7	50.4	97.7	103.1	51.9
43.4	69.5	156.5	34.6	37.9	12.9	2.5	31.4
79.5	26.9	18.5	14.7	32.9	196	24.9	118.2
82.2	35.1	47.6	54.2	63.1	69.8	57.4	65.6
56.4	49.4	44.9	34.6	92.2	37.0	58.8	21.3
36.6	64.9	14.8	17.6	29.1	61.4	38.6	32.5
12.0	28.3	204.9	44.5	10.3	37.7	33.7	81.1
12.1	20.1	30.5	7.1	10.1	18.0	3.0	2.0

Describe the distribution with both a graph and suitable numerical summaries. What are the main features of the distribution that might be of interest to a landowner in this area thinking about drilling for oil?

I.11 Data on mice. For a biology project, you measure the tail length (centimeters) and weight (grams) of 12 mice of the same variety. What units of measurement do each of the following have?

(a) The mean length of the tails.

(b) The first quartile of the tail lengths.

(c) The standard deviation of the tail lengths.

(d) The correlation between tail length and weight.

I.12 Voting. Here are the U.S. resident population of voting age and the votes cast for president, both in thousands, for presidential elections between 1960 and 2000:

(Susan Steinkamp/CORBIS)

Year	Population	Votes	Year	Population	Votes
1960	109,672	68,838	1984	173,995	92,653
1964	114,090	70,645	1988	181,956	91,595
1968	120,285	73,212	1992	189,524	104,425
1972	140,777	77,719	1996	196,511	96,456
1976	152,308	81,556	2000	209,128	105,363
1980	163,945	86,515			

(a) For each year compute the percent of people who voted. Make a time plot of the percent who voted. Describe the change over time in participation in presidential elections.

(b) Before proposing political explanations for this change, we should examine possible lurking variables. The minimum voting age in presidential elections dropped from 21 to 18 years in 1970. Use this fact to propose a partial explanation for the trend you saw in (a).

I.13 How are schools doing? The nonprofit group Public Agenda conducted telephone interviews with randomly selected parents of high school children. There were 202 black parents, 202 Hispanic parents, and 201 white parents. One question asked was, "Are the high schools in your state doing an excellent, good, fair or poor job, or don't you know enough to say?" Here are the survey results:[5]

	Black parents	Hispanic parents	White parents
Excellent	12	34	22
Good	69	55	81
Fair	75	61	60
Poor	24	24	24
Don't know	22	28	14
Total	202	202	201

Write a brief analysis of these results that focuses on the relationship between parent group and opinions about schools.

I.14 Regulating guns. The 1998 National Gun Policy Survey, conducted by the National Opinion Research Center at the University of Chicago, asked many questions about regulation of guns in the United States. One of the questions was "Do you think there should be a law that would ban possession of handguns except for the police and other authorized persons?" Here are the responses, broken down by the respondents' level of education:[6]

	Yes	No
Less than high school	58	58
High school graduate	84	129
Some college	169	294
College graduate	98	135
Postgraduate degree	77	99

How does the proportion of the sample who favor banning possession of handguns differ among people with different levels of education? Make a bar graph that compares the proportions and briefly describe the relationship between education and opinion about a handgun ban.

Brains and bodies. Figure I.1 plots the average brain weight in grams versus average body weight in kilograms for 96 species of mammals. There are many small mammals whose points at the lower left overlap. Exercises I.15 to I.20 are based on this scatterplot.[7]

I.15 Dolphins and hippos. The points for the dolphin and hippopotamus are labeled in Figure I.1. Read from the graph the approximate body weight and brain weight for these two species.

(Superstock/Superstock/PictureQuest)

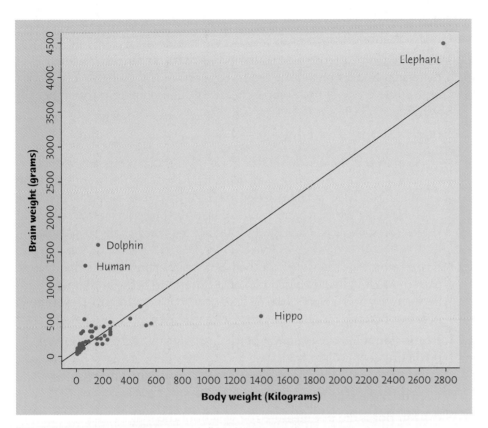

Figure I.1 Brain weight and body weight for 96 species of mammals, with the least-squares regression line.

I.16 Dolphins and hippos, continued. One reaction to this scatterplot is "Dolphins are smart, hippos are dumb." What feature of the plot lies behind this reaction?

I.17 Outliers. The African elephant is much larger than any other mammal in the data set but lies roughly in the overall straight-line pattern. Dolphins, humans, and hippos lie outside the overall pattern. The correlation between body weight and brain weight for the entire data set is $r = 0.86$.

   (a) If we removed the elephant, would this correlation increase or decrease or not change much? Explain your answer.

   (b) If we removed dolphins, hippos, and humans, would this correlation increase or decrease or not change much? Explain your answer.

I.18 Brain and body. The correlation between body weight and brain weight is $r = 0.86$. How well does body weight explain brain weight for mammals? Give a number to answer this question, and briefly explain what the number tells us.

I.19 Prediction. The line on the scatterplot in Figure I.1 is the least-squares regression line for predicting brain weight from body weight. Suppose that a new mammal species is discovered hidden in the rain forest with body weight 600 kilograms. Predict the brain weight for this species.

I.20 **Slope.** The line on the scatterplot in Figure I.1 is the least-squares regression line for predicting brain weight from body weight. The slope of this line is one of the numbers below. Which number is the slope? Why?

(a) $b = 0.5$. (b) $b = 1.3$. (c) $b = 3.2$.

> *Soap in the shower. From Rex Boggs in Australia comes an unusual data set: before showering in the morning, he weighed the bar of soap in his shower stall. The weight goes down as the soap is used. The data appear in Table I.2 (weights in grams). Notice that Mr. Boggs forgot to weigh the soap on some days. Exercises I.21 to I.23 are based on the soap data set.*

I.21 **Scatterplot.** Plot the weight of the bar of soap against day. Is the overall pattern roughly linear? Based on your scatterplot, is the correlation between day and weight close to 1, positive but not close to 1, close to 0, negative but not close to -1, or close to -1? Explain your answer.

I.22 **Regression.** Find the equation of the least-squares regression line for predicting soap weight from day.

(a) What is the equation? Explain what it tells us about the rate at which the soap lost weight.

(b) Mr. Boggs did not measure the weight of the soap on day 4. Use the regression equation to predict that weight.

(c) Draw the regression line on your scatterplot from the previous exercise.

I.23 **Prediction?** Use the regression equation in the previous exercise to predict the weight of the soap after 30 days. Why is it clear that your answer makes no sense? What's wrong with using the regression line to predict weight after 30 days?

I.24 **Prey attract predators.** Here is one way in which nature regulates the size of animal populations: high population density attracts predators, who remove a higher proportion of the population than when the density of the prey is low. One study looked at kelp perch and their common predator, the kelp bass. The researcher set up four large circular

TABLE I.2 Weight (grams) of a bar of soap used to shower

Day	Weight	Day	Weight	Day	Weight
1	124	8	84	16	27
2	121	9	78	18	16
5	103	10	71	19	12
6	96	12	58	20	8
7	90	13	50	21	6

pens on sandy ocean bottom in southern California. He chose young perch at random from a large group and placed 10, 20, 40, and 60 perch in the four pens. Then he dropped the nets protecting the pens, allowing bass to swarm in, and counted the perch left after 2 hours. Here are data on the proportions of perch eaten in four repetitions of this setup:[8]

Perch	Proportion killed			
10	0.0	0.1	0.3	0.3
20	0.2	0.3	0.3	0.6
40	0.075	0.3	0.6	0.725
60	0.517	0.55	0.7	0.817

(a) Why is the response variable the *proportion* of perch in a pen who are killed rather than the *count* of perch killed?

(b) Make a graph to examine how proportion killed changes with the number of perch in the pen. The relationship is roughly linear. Give a numerical measure of its strength.

(c) Do the data support the principle that "more prey attract more predators, who drive down the number of prey"?

I.25 **Statistics for investing.** Joe's retirement plan invests in stocks through an "index fund" that follows the behavior of the stock market as a whole, as measured by the Standard & Poor's 500 Index. Joe wants to buy a mutual fund that does not track the index closely. He reads that monthly returns from Fidelity Technology Fund have correlation $r = 0.77$ with the S&P 500 Index and that Fidelity Real Estate Fund has correlation $r = 0.37$ with the index.

(a) Which of these funds has the closer relationship to returns from the stock market as a whole? How do you know?

(b) Does the information given tell Joe anything about which fund has had higher returns?

I.26 **Initial public offerings.** The business magazine *Forbes* reports that 4567 companies sold their first stock to the public between 1990 and 2000. The *mean* change in the stock price of these companies since the first stock was issued was +111%. The *median* change was −31%.[9] Explain how this could happen. (Hint: Start with the fact that Cisco Systems stock went up 60,600%.)

I.27 **Highly paid athletes.** A news article reports that of the 411 players on National Basketball Association rosters in February 1998, only 139 "made more than the league average salary" of $2.36 million. Is $2.36 million the mean or median salary for NBA players? How do you know?

I.28 **Mean versus median.** Last year a small accounting firm paid each of its five clerks $22,000, two junior accountants $50,000 each, and the firm's

Beer in South Dakota

Take a break from doing exercises to apply your math to beer cans in South Dakota. A newspaper there reported that every year an average of 650 beer cans per mile are tossed onto the state's highways. South Dakota has about 83,000 miles of roads. How many beer cans is that in all? The 2000 census found 756,600 people in South Dakota. How many beer cans does each person in the state toss on the road per year? Either South Dakota is home to champion (and sloppy) beer drinkers or the newspaper got its numbers wrong.

owner $270,000. What is the mean salary paid at this firm? How many of the employees earn less than the mean? What is the median salary?

I.29 **Nonresponse in a survey.** A business school conducted a survey of companies in its state. They mailed a questionnaire to 200 small companies, 200 medium-sized companies, and 200 large companies. The rate of nonresponse is important in deciding how reliable survey results are. Here are the data on response to this survey:

	Small	Medium	Large
Response	125	81	40
No response	75	119	160
Total	200	200	200

(a) What was the overall percent of nonresponse?

(b) Describe how nonresponse is related to the size of the business. (Use percents to make your statements precise.)

(c) Draw a bar graph to compare the nonresponse percents for the three size categories.

I.30 **Aspirin and heart attacks.** Does taking aspirin regularly help prevent heart attacks? The Physicians' Health Study tried to find out. The subjects were 22,071 healthy male doctors at least 40 years old. Half the subjects, chosen at random, took aspirin every other day. The other half took a placebo, a dummy pill that looked and tasted like aspirin. Here are the results.[10] (The row for "None of these" is left out of the two-way table.)

	Aspirin group	Placebo group
Fatal heart attacks	10	26
Other heart attacks	129	213
Strokes	119	98
Total	11,037	11,034

What do the data show about the association between taking aspirin and heart attacks and stroke? Use percents to make your statements precise. Do you think the study provides evidence that aspirin actually reduces heart attacks (cause and effect)?

I.31 **Moving in step?** One reason to invest abroad is that markets in different countries don't move in step. When American stocks go down, foreign stocks may go up. So an investor who holds both bears less risk. That's the theory. Now we read: "The correlation between changes in American and European share prices has risen from 0.4 in the mid-1990s to 0.8 in 2000."[11] Explain to an investor who knows no statistics why this fact reduces the protection provided by buying European stocks.

I.32 **Interpreting correlation.** The same article that claims that the correlation between changes in stock prices in Europe and the United States was 0.8 in 2000 goes on to say: "Crudely, that means that movements on Wall Street can explain 80% of price movements in Europe." Is this true? What is the correct percent explained if $r = 0.8$?

I.33 **Coaching for the SATs.** A study finds that high school students who take the SAT, enroll in an SAT coaching course, and then take the SAT a second time raise their SAT mathematics scores from a mean of 521 to a mean of 561.[12] What factors other than "taking the course causes higher scores" might explain this improvement?

Supplementary EXERCISES

Supplementary exercises apply the skills you have learned in ways that require more thought or more elaborate use of technology.

Falling through the ice. *The Nenana Ice Classic is an annual contest to guess the exact time in the spring thaw when a tripod erected on the frozen Tanana River near Nenana, Alaska, will fall through the ice. The 2002 jackpot prize was $304,000. The contest has been run since 1917. Table I.3 gives simplified data that record only the date on which the*

TABLE I.3 Days from April 20 for the Tanana River tripod to fall

Year	Day	Year	Day	Year	Day	Year	Day	Year	Day	Year	Day
1917	11	1932	12	1947	14	1962	23	1977	17	1992	25
1918	22	1933	19	1948	24	1963	16	1978	11	1993	4
1919	14	1934	11	1949	25	1964	31	1979	11	1994	10
1920	22	1935	26	1950	17	1965	18	1980	10	1995	7
1921	22	1936	11	1951	11	1966	19	1981	11	1996	16
1922	23	1937	23	1952	23	1967	15	1982	21	1997	11
1923	20	1938	17	1953	10	1968	19	1983	10	1998	1
1924	22	1939	10	1954	17	1969	9	1984	20	1999	10
1925	16	1940	1	1955	20	1970	15	1985	23	2000	12
1926	7	1941	14	1956	12	1971	19	1986	19	2001	19
1927	23	1942	11	1957	16	1972	21	1987	16	2002	18
1928	17	1943	9	1958	10	1973	15	1988	8		
1929	16	1944	15	1959	19	1974	17	1989	12		
1930	19	1945	27	1960	13	1975	21	1990	5		
1931	21	1946	16	1961	16	1976	13	1991	12		

tripod fell each year. The earliest date so far is April 20. To make the data easier to use, the table gives the date each year in days starting with April 20. That is, April 20 is 1, April 21 is 2, and so on. You will need software or a graphing calculator to analyze these data in Exercises I.34 to I.36.[13]

I.34 When does the ice break up? We have 86 years of data on the date of ice breakup on the Tanana River. Describe the distribution of the breakup date with both a graph or graphs and appropriate numerical summaries. What is the median date (month and day) for ice breakup?

I.35 Global warming? Because of the high stakes, the falling of the tripod has been carefully observed for many years. If the date the tripod falls has been getting earlier, that may be evidence for the effects of global warming.

(a) Make a time plot of the date the tripod falls against year.

(b) There is a great deal of year-to-year variation. Fitting a regression line to the data may help us see the trend. Fit the least-squares line and add it to your time plot. What do you conclude?

(c) There is much variation about the line. Give a numerical description of how much of the year-to-year variation in ice breakup time is accounted for by the time trend represented by the regression line.

I.36 More on global warming. Side-by-side boxplots offer a different look at the data. Group the data into periods of roughly equal length: 1917 to 1939, 1940 to 1959, 1960 to 1979, and 1980 to 2002. Make boxplots to compare ice breakup dates in these four time periods. Write a brief description of what the plots show.

I.37 Save the eagles. The pesticide DDT was especially threatening to bald eagles. Here are data on the productivity of the eagle population in northwestern Ontario, Canada.[14] The eagles nest in an area free of DDT but migrate south and eat prey contaminated with the pesticide. DDT was banned at the end of 1972. The researcher observed every nesting area he could reach every year between 1966 and 1981. He measured productivity by the count of young eagles per nesting area.

(Lamm/Premium Stock/PictureQuest)

Year	Count	Year	Count	Year	Count	Year	Count
1966	1.26	1970	0.54	1974	0.46	1978	0.82
1967	0.73	1971	0.60	1975	0.77	1979	0.98
1968	0.89	1972	0.54	1976	0.86	1980	0.93
1969	0.84	1973	0.78	1977	0.96	1981	1.12

(a) Make a time plot of the data. Does the plot support the claim that banning DDT helped save the eagles?

(b) It appears that the overall pattern might be described by *two* straight lines. Find the least-squares line for 1966 to 1972 (pre-ban) and also the least-squares line for 1975 to 1981 (allowing a few years for DDT to leave the environment after the ban). Draw these lines on your plot. Would you use the second line to predict young per nesting area in the several years after 1981?

I.38 **Thin monkeys, fat monkeys.** Animals and people that take in more energy than they expend will get fatter. Here are data on 12 rhesus monkeys: 6 lean monkeys (4% to 9% body fat) and 6 obese monkeys (13% to 44% body fat). The data report the energy expended in 24 hours (kilojoules per minute) and the lean body mass (kilograms, leaving out fat) for each monkey.[15]

Lean		Obese	
Mass	Energy	Mass	Energy
6.6	1.17	7.9	0.93
7.8	1.02	9.4	1.39
8.9	1.46	10.7	1.19
9.8	1.68	12.2	1.49
9.7	1.06	12.1	1.29
9.3	1.16	10.8	1.31

(a) What is the mean lean body mass of the lean monkeys? Of the obese monkeys? Because animals with higher lean mass usually expend more energy, we can't directly compare energy expended.

(b) Instead, look at how energy expended is related to body mass. Make a scatterplot of energy versus mass, using different plot symbols for lean and obese monkeys. Then add to the plot two regression lines, one for lean monkeys and one for obese monkeys. What do these lines suggest about the monkeys?

I.39 **Casting aluminum.** In casting metal parts, molten metal flows through a "gate" into a die that shapes the part. The gate velocity (the speed at which metal is forced through the gate) plays a critical role in die casting. A firm that casts cylindrical aluminum pistons examined 12 types formed from the same alloy. What is the relationship between the cylinder wall thickness (inches) and the gate velocity (feet per second) chosen by the skilled workers who do the casting? If there is a clear pattern, it can be used to direct new workers or to automate the process. Analyze these data and report your findings.[16]

Thickness	Velocity	Thickness	Velocity	Thickness	Velocity
0.248	123.8	0.524	228.6	0.697	145.2
0.359	223.9	0.552	223.8	0.752	263.1
0.366	180.9	0.628	326.2	0.806	302.4
0.400	104.8	0.697	302.4	0.821	302.4

I.40 **Age and income.** How do the incomes of working-age people change with age? Because many older women have been out of the labor force for much of their lives, we look only at men between the ages of 25 and 65. Because education strongly influences income, we look only at men who have a bachelor's degree but no higher degree. A government sample survey tells us the age and income of a random sample of 5712 such men.[17] Figure I.2 is a scatterplot of these data. Here is software output for regressing income on age. The line in the scatterplot is the least-squares regression line from this output.

	Coefficients	Standard Error	t Stat	P-value
Intercept	24874.3745	2637.4198	9.4313	5.75E-21
AGE	892.1135	61.7639	14.4439	1.79E-46

(a) The scatterplot in Figure I.2 has a distinctive form. Why do the points fall into vertical stacks?

(b) Give some reasons why older men in this population might earn more than younger men. Give some reasons why younger men might earn more than older men. What do the data show about the

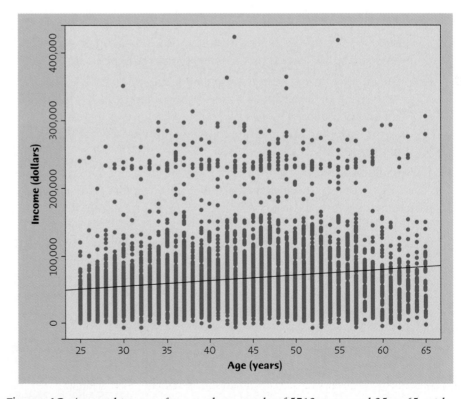

Figure I.2 Age and income for a random sample of 5712 men aged 25 to 65, with the least-squares regression line.

relationship between age and income in the sample? Is the relationship very strong?

(c) What is the equation of the least-squares line for predicting income from age? What specifically does the slope of this line tell us?

I.41 Weeds among the corn. Lamb's-quarter is a common weed that interferes with the growth of corn. An agriculture researcher planted corn at the same rate in 16 small plots of ground, then weeded the plots by hand to allow a fixed number of lamb's-quarter plants to grow in each meter of corn row. No other weeds were allowed to grow. Here are the yields of corn (bushels per acre) in each of the plots:[18]

Weeds per meter	Corn yield	Weeds per meter	Corn yield	Weeds per meter	Corn yield	Weeds per meter	Corn yield
0	166.7	1	166.2	3	158.6	9	162.8
0	172.2	1	157.3	3	176.4	9	142.4
0	165.0	1	166.7	3	153.1	9	162.8
0	176.9	1	161.1	3	156.0	9	162.4

(a) What are the explanatory and response variables in this experiment?

(b) Make side-by-side stemplots of the yields, after rounding to the nearest bushel. Give the median yield for each group (using the unrounded data). What do you conclude about the effect of this weed on corn yield?

I.42 Weeds among the corn, continued. We can also use regression to analyze the data on weeds and corn yield. The advantage of regression over the side-by-side comparison in the previous exercise is that we can use the fitted model to draw conclusions for counts of weeds other than the ones the researcher actually used.

(a) Make a scatterplot of corn yield against weeds per meter. Find the least-squares regression line and add it to your plot. What does the slope of the fitted line tell us about the effect of lamb's-quarter on corn yield?

(b) Predict the yield for corn grown under these conditions with 6 lamb's-quarter plants per meter of row.

I.43 A fact about the mean. Use the definition of the mean $\bar{x}$ to show that the sum of the deviations $x_i - \bar{x}$ of the observations from their mean is always zero. This is one reason why the variance and standard deviation use squared deviations.

I.44 A fact about regression. Use the equation for the least-squares regression line to show that this line always passes through the point $(\bar{x}, \bar{y})$. That is, set $x = \bar{x}$ and show that the line predicts that $y = \bar{y}$.

TABLE I.4 Count and weight of seeds produced by common tree species

Tree species	Seed count	Seed weight (mg)	Tree species	Seed count	Seed weight (mg)
Paper birch	27,239	0.6	American beech	463	247
Yellow birch	12,158	1.6	American beech	1,892	247
White spruce	7,202	2.0	Black oak	93	1,851
Engelmann spruce	3,671	3.3	Scarlet oak	525	1,930
Red spruce	5,051	3.4	Red oak	411	2,475
Tulip tree	13,509	9.1	Red oak	253	2,475
Ponderosa pine	2,667	37.7	Pignut hickory	40	3,423
White fir	5,196	40.0	White oak	184	3,669
Sugar maple	1,751	48.0	Chestnut oak	107	4,535
Sugar pine	1,159	216.0			

I.45 Transforming data. Table I.4 gives data on the mean number of seeds produced in a year by several common tree species and the mean weight (in milligrams) of the seeds produced. (Some species appear twice because their seeds were counted in two locations.) We might expect that trees with heavy seeds produce fewer of them, but what is the form of the relationship?[19]

(a) Make a scatterplot showing how the weight of tree seeds helps explain how many seeds the tree produces. Describe the form, direction, and strength of the relationship.

(b) When dealing with sizes and counts, the logarithms of the original data are often the "natural" variables. Use your calculator or software to obtain the logarithms of both the seed weights and the seed counts in Table I.4. Make a new scatterplot using these new variables. Now what are the form, direction, and strength of the relationship?

I.46 The computing revolution. Gordon Moore, one of the founders of Intel Corporation, predicted in 1965 that the number of transistors on an integrated circuit chip would double every 18 months. This is "Moore's law," one way to measure the revolution in computing. Here are data on the dates and number of transistors for Intel microprocessors:[20]

Processor	Date	Transistors	Processor	Date	Transistors
4004	1971	2,250	486 DX	1989	1,180,000
8008	1972	2,500	Pentium	1993	3,100,000
8080	1974	5,000	Pentium II	1997	7,500,000
8086	1978	29,000	Pentium III	1999	24,000,000
286	1982	120,000	Pentium 4	2000	42,000,000
386	1985	275,000			

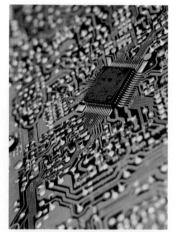

(Andre Kudyusov/Photo Disc/Getty Images)

Make a time plot of these data. The growth is certainly rapid, but it's hard to say if Moore was right in predicting regular doubling. The way to verify that pattern is to plot the logarithms of transistor counts against time. If the original data double at fixed time intervals, the logarithms will lie on a straight line. Plot the logarithms against time. Find the correlation between the logarithms and time and report r^2. What is your verdict: has chip capacity doubled at regular intervals?

From Exploration to Inference

The purpose of statistics is to gain understanding from data. We can seek understanding in different ways, depending on the circumstances. We have studied one approach to data, *exploratory data analysis*, in some detail. Now we move from data analysis toward *statistical inference*. Both types of reasoning are essential to effective work with data. Here is a brief sketch of the differences between them.

EXPLORATORY DATA ANALYSIS	STATISTICAL INFERENCE
Purpose is unrestricted exploration of the data, searching for interesting patterns.	Purpose is to answer specific questions, posed before the data were produced.
Conclusions apply only to the individuals and circumstances for which we have data in hand.	Conclusions apply to a larger group of individuals or a broader class of circumstances.
Conclusions are informal, based on what we see in the data.	Conclusions are formal, backed by a statement of our confidence in them.

Our journey toward inference begins in Chapters 7 and 8 with *data production*, statistical ideas for producing data to answer specific questions. Chapters 9 and 10 (and the optional Chapters 11 and 12) concern *probability*, the language of formal statistical conclusions. Finally, Chapters 13, 14, and 15 present the core concepts of inference. In practice, data analysis and inference cooperate. Successful inference requires good data production, data analysis to ensure that the data are regular enough, and the language of probability to state conclusions.

"I was an eight, too, until the number crunchers got hold of me."

(AP/Wide World Photos)

Producing Data: Sampling

Exploratory data analysis seeks to discover and describe what data say by using graphs and numerical summaries. The conclusions we draw from data analysis apply to the specific data that we examine. Often, however, we want to answer questions about some large group of individuals. To get sound answers, we must produce data in a way that is designed to answer our questions.

Suppose our question is "What percent of American adults agree that the United Nations should continue to have its headquarters in the United States?" To answer the question, we interview American adults. We can't afford to ask all adults, so we put the question to a **sample** chosen to represent the entire adult population. How shall we choose a sample that truly represents the opinions of the entire population? Statistical designs for choosing samples are the topic of this chapter.

sample

Observation versus experiment

Our goal in choosing a sample is a picture of the population, disturbed as little as possible by the act of gathering information. Sample surveys are one kind of *observational study*. In other settings, we gather data from an *experiment*. In doing an experiment, we don't just observe individuals or ask them questions. We actively impose some treatment in order to observe the response. Experiments can answer questions such as "Does aspirin reduce the chance of a heart attack?" and "Does a majority of college students prefer Pepsi to Coke when they taste both without knowing which they are drinking?" Experiments, like samples, provide useful data only when properly designed. We will discuss statistical design of experiments in Chapter 8. The distinction between experiments and observational studies is one of the most important ideas in statistics.

OBSERVATION VERSUS EXPERIMENT

An **observational study** observes individuals and measures variables of interest but does not attempt to influence the responses.

An **experiment,** on the other hand, deliberately imposes some treatment on individuals in order to observe their responses.

Observational studies are essential sources of data about topics from the opinions of voters to the behavior of animals in the wild. But an observational study, even one based on a statistical sample, is a poor way to gauge the effect of an intervention. To see the response to a change, we must actually impose the change. When our goal is to understand cause and effect, experiments are the only source of fully convincing data.

EXAMPLE 7.1 *The rise and fall of hormone replacement*

Should women take hormones such as estrogen after menopause, when natural production of these hormones ends? In 1992, several major medical organizations said "Yes." In particular, women who took hormones seemed to reduce their risk of a heart attack by 35% to 50%. The risks of taking hormones appeared small compared with the benefits.

The evidence in favor of hormone replacement came from a number of observational studies that compared women who were taking hormones with others who were not. But women who choose to take hormones are very different from women who do not: they are richer and better educated and see doctors more often. These women do many things to maintain their health. It isn't surprising that they have fewer heart attacks.

Experiments don't let women decide what to do. They assign women to either hormone replacement or to dummy pills that look and taste the same as the hormone pills. The assignment is done by a coin toss, so that all kinds of women are equally likely to get either treatment. By 2002, several experiments with women of different ages agreed that hormone replacement does *not* reduce the risk of heart attacks. The National Institutes of Health, after reviewing the evidence, concluded that the observational studies were wrong. Taking hormones after menopause quickly fell out of favor.[1]

When we simply observe women, the effects of actually taking hormones are *confounded* with (mixed up with) the characteristics of women who choose to take hormones.

CONFOUNDING

Two variables (explanatory variables or lurking variables) are **confounded** when their effects on a response variable cannot be distinguished from each other.

Observational studies of the effect of one variable on another often fail because the explanatory variable is confounded with lurking variables. We will see that well-designed experiments take steps to defeat confounding.

EXAMPLE 7.2   Wine, beer, or spirits?

Moderate use of alcohol is associated with better health. Some studies suggest that drinking wine rather than beer or spirits confers added health benefits. But people who prefer wine are different from those who drink mainly beer or stronger stuff. Wine drinkers eat more fruit and vegetables and less fried food. Their diets contain less fat, less cholesterol, and also less alcohol. They are less likely to smoke. Finally, wine drinkers as a group are richer and better educated than people who prefer beer or spirits. A large study therefore concludes: "The apparent health benefits of wine compared with other alcoholic beverages, as described by others, may be a result of confounding by dietary habits and other lifestyle factors."[2]

APPLY YOUR KNOWLEDGE

7.1   **Cell phones and brain cancer.** A study of cell phones and the risk of brain cancer looked at a group of 469 people who have brain cancer. The investigators matched each cancer patient with a person of the same sex, age, and race who did not have brain cancer, then asked about use of cell phones.[3] Result: "Our data suggest that use of handheld cellular telephones is not associated with risk of brain cancer." Is this an observational study or an experiment? Why? What are the explanatory and response variables?

7.2   **Teaching economics.** An educational software company wants to compare the effectiveness of its computer animation for teaching about supply, demand, and market clearing with that of a textbook presentation. The company tests the economic knowledge of each of a group of first-year college students, then divides them into two groups. One group uses the animation, and the other studies the text. The company retests all the students and compares the increase in economic understanding in the two groups. Is this an experiment? Why or why not? What are the explanatory and response variables?

7.3   **TV viewing and aggression.** A typical hour of prime-time television shows three to five violent acts. Research shows that there is a clear association between time spent watching TV and aggressive behavior by adolescents. Nonetheless, it is hard to conclude that watching TV *causes* aggression. Suggest several lurking variables describing an adolescent's home life that may be confounded with how much TV he or she watches.[4]

sample design

Sampling

An opinion poll wants to know what fraction of the public approves of the president's performance in office. A quality engineer must estimate what fraction of the bearings rolling off an assembly line are defective. Government economists inquire about household income. In all these situations, we want to gather information about a large group of people or things. Time, cost, and inconvenience usually forbid inspecting every bearing or contacting every household. In such cases, we gather information about only part of the group in order to draw conclusions about the whole.

POPULATION, SAMPLE

The entire group of individuals that we want information about is called the **population.**

A **sample** is a part of the population that we actually examine in order to gather information.

Notice that the population is defined in terms of our desire for knowledge. If we wish to draw conclusions about all U.S. college students, that group is our population even if only local students are available for questioning. The sample is the part from which we draw conclusions about the whole. The **design** of a sample refers to the method used to choose the sample from the population. Poor sample designs can produce misleading conclusions. Here are some examples of badly designed samples.

EXAMPLE 7.3 A "good" sample isn't a good idea

A mill produces large coils of thin steel for use in manufacturing home appliances. A quality engineer wants to submit a sample of 5-centimeter squares to detailed laboratory examination. She asks a technician to cut a sample of 10 such squares. Wanting to provide "good" pieces of steel, the technician carefully avoids the visible defects in the steel when cutting the sample. The laboratory results are wonderful, but the customers complain about the material they are receiving.

EXAMPLE 7.4 Call-in opinion polls

Television news programs like to conduct call-in polls of public opinion. The program announces a question and asks viewers to call one telephone number to respond "Yes" and another for "No." Telephone companies charge for these calls. The ABC network program *Nightline* once asked whether the United Nations should continue to have its headquarters in the United States. More than 186,000 callers responded, and 67% said "No."

People who spend the time and money to respond to call-in polls are not representative of the entire adult population. People who feel strongly, especially those with strong negative opinions, are more likely to call. It is not surprising that a properly designed sample showed that 72% of adults want the UN to stay.[5]

Call-in opinion polls are an example of *voluntary response sampling*. A voluntary response sample can easily produce 67% "No" when the truth about the population is close to 72% "Yes."

VOLUNTARY RESPONSE SAMPLE

A **voluntary response sample** consists of people who choose themselves by responding to a general appeal. Voluntary response samples are biased because people with strong opinions, especially negative opinions, are most likely to respond.

Voluntary response is one common type of bad sample design. Another is **convenience sampling** which chooses the individuals easiest to reach. Here is an example of convenience sampling.

convenience sampling

EXAMPLE 7.5   Interviewing at the mall

Manufacturers and advertising agencies often use interviews at shopping malls to gather information about the habits of consumers and the effectiveness of ads. A sample of mall shoppers is fast and cheap. But people contacted at shopping malls are not representative of the entire U.S. population. They are richer, for example, and more likely to be teenagers or retired. Moreover, mall interviewers tend to select neat, safe-looking individuals from the stream of customers. Decisions based on mall interviews may not reflect the preferences of all consumers.[6]

Samples chosen by either voluntary response or convenience sampling are almost guaranteed not to represent the entire population. These sampling methods display *bias*, or systematic error, in favoring some parts of the population over others.

BIAS

The design of a study is **biased** if it systematically favors certain outcomes.

APPLY YOUR KNOWLEDGE

What is the population?   In each of Exercises 7.4 to 7.6, identify the population as exactly as possible. That is, say what kind of individuals the population consists of and say exactly which individuals fall in the population. If the information given is not sufficient, complete the description of the population in a reasonable way.

7.4   **An opinion poll.** An opinion poll contacts 1161 adults and asks, "Which political party do you think has better ideas for leading the country in the twenty-first century?"

7.5 **The census.** The 2000 census tried to gather basic information from every household in the United States. In addition, a "long form" requesting much additional information was sent to a sample of about 17% of households.

7.6 **Is the quality OK?** A machinery manufacturer purchases voltage regulators from a supplier. There are reports that variation in the output voltage of the regulators is affecting the performance of the finished products. To assess the quality of the supplier's production, the manufacturer sends a sample of 5 regulators from the last shipment to a laboratory for study.

7.7 **An online poll.** The Excite Poll can be found online at poll.excite.com. The question appears on the screen, and you simply click to choose a response. On October 14, 2002, the question was

> It is now possible for school students to log on to Internet sites and download homework. Everything from book reports to doctoral dissertations can be downloaded free or for a fee. Do you believe giving a student who is caught plagiarizing an "F" for their assignment is the right punishment?

In all, 14,793 people clicked "Yes," 1778 clicked "No, it's too harsh," 2566 clicked "No, it's not harsh enough," and 988 clicked don't know or don't care.

(a) What is the sample size for this poll?

(b) That's a much larger sample than standard sample surveys. In spite of this, we can't trust the result to give good information about any clearly defined population. Why?

Simple random samples

In a voluntary response sample, people choose whether to respond. In a convenience sample, the interviewer makes the choice. In both cases, personal choice produces bias. The statistician's remedy is to allow impersonal chance to choose the sample. A sample chosen by chance allows neither favoritism by the sampler nor self-selection by respondents. Choosing a sample by chance attacks bias by giving all individuals an equal chance to be chosen. Rich and poor, young and old, black and white, all have the same chance to be in the sample.

   The simplest way to use chance to select a sample is to place names in a hat (the population) and draw out a handful (the sample). This is the idea of *simple random sampling*.

SIMPLE RANDOM SAMPLE

A **simple random sample (SRS)** of size n consists of n individuals from the population chosen in such a way that every set of n individuals has an equal chance to be the sample actually selected.

An SRS not only gives each individual an equal chance to be chosen (thus avoiding bias in the choice) but also gives every possible sample an equal chance to be chosen. There are other random sampling designs that give each individual, but not each sample, an equal chance. Exercise 7.36 describes one such design, called systematic random sampling.

The idea of an SRS is to choose our sample by drawing names from a hat. In practice, statistical software can choose an SRS almost instantly from a list of the individuals in the population. (Spreadsheet programs and graphing calculators do this less well.) The *Simple Random Sample* applet also speeds the choosing of an SRS. If you don't use software, you can randomize by using a *table of random digits*.

RANDOM DIGITS

A **table of random digits** is a long string of the digits 0, 1, 2, 3, 4, 5, 6, 7, 8, 9 with these two properties:

1. Each entry in the table is equally likely to be any of the 10 digits 0 through 9.

2. The entries are independent of each other. That is, knowledge of one part of the table gives no information about any other part.

Table B at the back of the book is a table of random digits. You can think of Table B as the result of asking an assistant (or a computer) to mix the digits 0 to 9 in a hat, draw one, then replace the digit drawn, mix again, draw a second digit, and so on. The assistant's mixing and drawing save us the work of mixing and drawing when we need to randomize. Table B begins with the digits 19223950340575628713. To make the table easier to read, the digits appear in groups of five and in numbered rows. The groups and rows have no meaning—the table is just a long list of randomly chosen digits. Because the digits in Table B are random:

*   Each entry is equally likely to be any of the 10 possibilities 0, 1,..., 9.

*   Each pair of entries is equally likely to be any of the 100 possible pairs 00, 01,..., 99.

*   Each triple of entries is equally likely to be any of the 1000 possibilities 000, 001,..., 999; and so on.

These "equally likely" facts make it easy to use Table B to choose an SRS. Here is an example that shows how.

Are these random digits really random?

Not a chance. The random digits in Table B were produced by a computer program. Computer programs do exactly what you tell them to do. Give the program the same input and it will produce exactly the same "random" digits. Of course, clever people have devised computer programs that produce output that *looks* like random digits. These are called "pseudo-random numbers," and that's what Table B contains. Pseudo-random numbers work fine for statistical randomizing, but they have hidden nonrandom patterns that can mess up more refined uses.

EXAMPLE 7.6   How to choose an SRS

Joan's small accounting firm serves 30 business clients. Joan wants to interview a sample of 5 clients in detail to find ways to improve client satisfaction. To avoid bias, she chooses an SRS of size 5.

Step 1.   Label. Give each client a numerical label, using as few digits as possible. Two digits are needed to label 30 clients, so we use labels

$$01, 02, 03, \ldots, 29, 30$$

It is also correct to use labels 00 to 29 or even another choice of 30 two-digit labels. Here is the list of clients, with labels attached:

01	A-1 Plumbing	16	JL Records
02	Accent Printing	17	Johnson Commodities
03	Action Sport Shop	18	Keiser Construction
04	Anderson Construction	19	Liu's Chinese Restaurant
05	Bailey Trucking	20	MagicTan
06	Balloons Inc.	21	Peerless Machine
07	Bennett Hardware	22	Photo Arts
08	Best's Camera Shop	23	River City Books
09	Blue Print Specialties	24	Riverside Tavern
10	Central Tree Service	25	Rustic Boutique
11	Classic Flowers	26	Satellite Services
12	Computer Answers	27	Scotch Wash
13	Darlene's Dolls	28	Sewer's Center
14	Fleisch Realty	29	Tire Specialties
15	Hernandez Electronics	30	Von's Video Store

Step 2.   Table. Enter Table B anywhere and read two-digit groups. Suppose we enter at line 130, which is

69051   64817   87174   09517   84534   06489   87201   97245

The first 10 two-digit groups in this line are

69   05   16   48   17   87   17   40   95   17

Each successive two-digit group is a label. The labels 00 and 31 to 99 are not used in this example, so we ignore them. The first 5 labels between 01 and 30 that we encounter in the table choose our sample. Of the first 10 labels in line 130, we ignore 5 because they are too high (over 30). The others are 05, 16, 17, 17, and 17. The clients labeled 05, 16, and 17 go into the sample. Ignore the second and third 17s because that client is already in the sample. Now run your finger across line 130 (and continue to line 131 if needed) until 5 clients are chosen.

The sample is the clients labeled 05, 16, 17, 20, 19. These are Bailey Trucking, JL Records, Johnson Commodities, MagicTan, and Liu's Chinese Restaurant.

CHOOSING AN SRS

Choose an SRS in two steps:

Step 1.   Label. Assign a numerical label to every individual in the population.

Step 2.   Table. Use Table B to select labels at random.

You can assign labels in any convenient manner, such as alphabetical order for names of people. Be certain that all labels have the same number of digits. Only then will all individuals have the same chance to be chosen. Use the shortest possible labels: one digit for a population of up to 10 members, two digits for 11 to 100 members, three digits for 101 to 1000 members, and so on. As standard practice, we recommend that you begin with label 1 (or 01 or 001, as needed). You can read digits from Table B in any order—across a row, down a column, and so on—because the table has no order. As standard practice, we recommend reading across rows.

APPLY YOUR KNOWLEDGE

7.8 Apartment living. You are planning a report on apartment living in a college town. You decide to select three apartment complexes at random for in-depth interviews with residents. Use Table B, starting at line 117, to select a simple random sample of three of the following apartment complexes.

01 Ashley Oaks	Country View	Mayfair Village
02 Bay Pointe	Country Villa	Nobb Hill
03 Beau Jardin	Crestview	Pemberly Courts
04 Bluffs	Del-Lynn	Peppermill
Brandon Place	Fairington	Pheasant Run
Briarwood	Fairway Knolls	River Walk
Brownstone	Fowler	Sagamore Ridge
Burberry Place	Franklin Park	Salem Courthouse
Cambridge	Georgetown	Village Square
Chauncey Village	Greenacres	Waterford Court
Country Squire	Lahr House	Williamsburg

7.9 Minority managers. A firm wants to understand the attitudes of its minority managers toward its system for assessing management performance. Below is a list of all the firm's managers who are members of minority groups. Use Table B at line 139 to choose six to be interviewed in detail about the performance appraisal system.

(Bill Lai/Index Stock Imagery/PictureQuest)

01 Abdulhamid	08 Duncan	15 Huang	22 Puri
02 Agarwal	09 Fernandez	16 Kim	23 Richards
03 Baxter	10 Fleming	17 Liao	24 Rodriguez
04 Bonds	11 Gates	18 Mourning	25 Santiago
05 Brown	12 Goel	19 Naber	26 Shen
06 Castillo	13 Gomez	20 Peters	27 Vega
07 Cross	14 Hernandez	21 Pliego	28 Wang

7.10 Sampling retail outlets. You must choose an SRS of 10 of the 440 retail outlets in New York that sell your company's products. How would you label this population? Use Table B, starting at line 105, to choose your sample.

The telemarketer's pause

People who do sample surveys hate telemarketing. We all get so many unwanted sales pitches by phone that many people hang up before learning that the caller is conducting a survey rather than selling vinyl siding. Here's a tip. Both sample surveys and telemarketers dial telephone numbers at random. Telemarketers automatically dial many numbers, and their sellers come on the line only after you pick up the phone. Once you know this, the telltale "telemarketer's pause" gives you a chance to hang up before the seller arrives. Sample surveys have a live interviewer on the line when you answer.

Other sampling designs

The general framework for statistical sampling is a *probability sample*.

> **PROBABILITY SAMPLE**
>
> A **probability sample** is a sample chosen by chance. We must know what samples are possible and what chance, or probability, each possible sample has.

Some probability sampling designs (such as an SRS) give each member of the population an equal chance to be selected. This may not be true in more elaborate sampling designs. In every case, however, the use of chance to select the sample is the essential principle of statistical sampling.

Designs for sampling from large populations spread out over a wide area are usually more complex than an SRS. For example, it is common to sample important groups within the population separately, then combine these samples. This is the idea of a *stratified random sample*.

> **STRATIFIED RANDOM SAMPLE**
>
> To select a **stratified random sample,** first divide the population into groups of similar individuals, called **strata.** Then choose a separate SRS in each stratum and combine these SRSs to form the full sample.

Choose the strata based on facts known before the sample is taken. For example, a population of election districts might be divided into urban, suburban, and rural strata. A stratified design can produce more exact information than an SRS of the same size by taking advantage of the fact that individuals in the same stratum are similar to one another.

EXAMPLE 7.7   Seat belt use in Hawaii

Each state conducts an annual survey of seat belt use by drivers, following guidelines set by the federal government. The guidelines require probability samples. Seat belt use is observed at randomly chosen road locations at random times during daylight hours. The locations are not an SRS of all locations in the state but rather a stratified sample using the state's counties as strata.

In Hawaii, the counties are the islands that make up the state's territory. The 2002 sample consisted of 135 road locations in the four most populated islands: 66 in Oahu, 24 in Maui, 23 in Hawaii, and 22 in Kauai. The sample sizes on the islands are proportional to the amount of road traffic.[7]

multistage sample   Seat belt surveys in larger states often use **multistage samples.** Counties are grouped into strata by population. At the first stage, choose a stratified sample of counties that includes all of the most populated counties and a selection of smaller counties. The second stage selects locations at random within each county chosen at the first stage. These are also stratified samples with locations grouped into strata by, for example, high, medium and low traffic volume.

Most large-scale sample surveys use multistage samples. The samples at individual stages may be SRSs but are often stratified. Analysis of data from sampling designs more complex than an SRS takes us beyond basic statistics. But the SRS is the building block of more elaborate designs, and analysis of other designs differs more in complexity of detail than in fundamental concepts.

APPLY YOUR KNOWLEDGE

7.11 A stratified sample. A club has 30 student members and 10 faculty members. The students are

01 Abel	07 Fisher	13 Huber	19 Miranda	25 Reinmann
02 Carson	08 Ghosh	14 Jimenez	20 Moskowitz	26 Santos
03 Chen	09 Griswold	15 Jones	21 Neyman	27 Shaw
04 David	10 Hein	16 Kim	22 O'Brien	28 Thompson
05 Deming	11 Hernandez	17 Klotz	23 Pearl	29 Utts
06 Elashoff	12 Holland	18 Liu	24 Potter	30 Varga

The faculty members are

01 Andrews	03 Fernandez	05 Kim	07 Moore	09 West
02 Besicovitch	04 Gupta	06 Lightman	08 Vicario	10 Yang

The club can send 4 students and 2 faculty members to a convention. It decides to choose those who will go by random selection. Use Table B to choose a stratified random sample of 4 students and 2 faculty members.

7.12 Sampling by accountants. Accountants use stratified samples during audits to verify a company's records of such things as accounts receivable. The stratification is based on the dollar amount of the item and often includes 100% sampling of the largest items. One company reports 5000 accounts receivable. Of these, 100 are in amounts over $50,000; 500 are in amounts between $1000 and $50,000; and the remaining 4400 are in amounts under $1000. Using these groups as strata, you decide to verify all of the largest accounts and to sample 5% of the midsize accounts and 1% of the small accounts. How would you label the two strata from which you will sample? Use Table B, starting at line 115, to select *only the first 5* accounts from each of these strata.

Cautions about sample surveys

Random selection eliminates bias in the choice of a sample from a list of the population. When the population consists of human beings, however, accurate information from a sample requires much more than a good sampling design. To begin, we need an accurate and complete list of the population. Because such a list is rarely available, most samples suffer from some degree of *undercoverage*. A sample survey of households, for example, will miss not only homeless people but prison inmates and students in dormitories. An opinion poll conducted by telephone will miss the 6% of American households without residential phones. The results of national sample surveys therefore have some bias if the people not covered—who most often are poor people—differ from the rest of the population.

A more serious source of bias in most sample surveys is *nonresponse,* which occurs when a selected individual cannot be contacted or refuses to cooperate. Nonresponse to sample surveys often reaches 50% or more, even with careful planning and several callbacks. Because nonresponse is higher in urban areas, most sample surveys substitute other people in the same area to avoid favoring rural areas in the final sample. If the people contacted differ from those who are rarely at home or who refuse to answer questions, some bias remains.

UNDERCOVERAGE AND NONRESPONSE

Undercoverage occurs when some groups in the population are left out of the process of choosing the sample.

Nonresponse occurs when an individual chosen for the sample can't be contacted or refuses to cooperate.

Counting Nigerians

Speaking of census undercounts, how many Nigerians are there? Nigeria is Africa's most populous nation, but that's all we can agree on. The World Bank says there are 127 million Nigerians. The State Department says 120 million. The United Nations says 114 million. Nigeria's last official estimate was 105 million, but that was more than ten years ago.

EXAMPLE 7.8 How bad is nonresponse?

The government's monthly Current Population Survey (CPS) produces data on employment, unemployment, and many other topics. The CPS has the lowest nonresponse rate of any poll we know: only about 6% or 7% of the households in the sample don't respond. People are more likely to respond to a government survey, and the CPS contacts its sample in person before doing later interviews by phone.

The University of Chicago's General Social Survey (GSS) is the nation's most important social science survey. (See Figure 7.1.) The GSS also contacts its sample in person, and it is run by a university. Despite these advantages, its most recent survey had a 30% rate of nonresponse.

What about opinion polls by the media and opinion-polling firms? We don't know their rates of nonresponse, because they won't say. That itself is a bad sign. The Pew Research Center imitated a careful telephone survey and published the results: out of 2879 households called, 1658 were never at home, refused, or would not finish the interview. That's a nonresponse rate of 58%.[8]

EXAMPLE 7.9 The census undercount

Even the U.S. census suffers from undercoverage and nonresponse. The census begins by mailing forms to every household in the country. There is some undercoverage due to incomplete lists of addresses and the difficulty of counting homeless

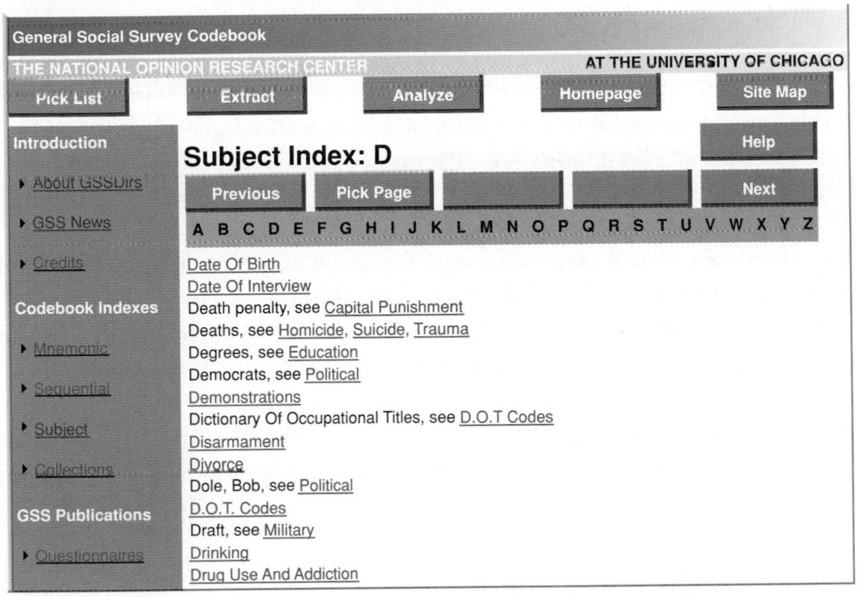

Figure 7.1 A small part of the subject index of the General Social Survey. The GSS has tracked opinions about a wide variety of issues since 1972.

people. As with most surveys, the big problem is nonresponse: in 2000, 28% of the households that received census forms did not mail them back. The nonresponse rate was higher in large cities, among young people, and among minorities.

The Census Bureau sent interviewers to every household that did not respond to the mail questionnaire. The follow-up in 2000 was more intense than in the past, and all agree that the 2000 census was the most accurate ever. Nonetheless, some people were missed and others were counted twice. (For example, retired people who spend part of the year in Florida may be counted both there and in their home state.)

How accurate was the 2000 census? Based on an elaborate follow-up sample survey of 300,000 households, the Census Bureau estimated that overall it undercounted the nation by 1.15%. But the bureau also estimated that it missed 2.07% of blacks and about 2.8% of Hispanics and that it slightly overcounted people over age 50. (For comparison, the 1990 census was estimated to have missed about 4.6% of blacks and 5% of Hispanics.)

Alas, the story is more complicated than this. A careful look at records of births, deaths, and immigration didn't agree with the follow-up survey. It suggested an overall undercount of only 0.12% and (more seriously) bigger differences among ethnic groups. Blacks, said the new analysis, were undercounted by 2.78%, and nonblacks were actually overcounted, by 0.29%. An expert panel concluded that the follow-up survey had missed too many overcounts to be completely trustworthy. We are left with a census that was very accurate but that did undercount minorities by an amount that we can't pin down precisely.[9]

In addition, the behavior of the respondent or of the interviewer can cause **response bias** in sample results. Respondents may lie, especially if asked about illegal or unpopular behavior. The sample then underestimates the presence of

response bias

such behavior in the population. An interviewer whose attitude suggests that some answers are more desirable than others will get these answers more often. The race or sex of the interviewer can influence responses to questions about race relations or attitudes toward feminism. Answers to questions that ask respondents to recall past events are often inaccurate because of faulty memory. For example, many people "telescope" events in the past, bringing them forward in memory to more recent time periods. "Have you visited a dentist in the last 6 months?" will often draw a "Yes" from someone who last visited a dentist 8 months ago.[10] Careful training of interviewers and careful supervision to avoid variation among the interviewers can greatly reduce response bias. Good interviewing technique is another aspect of a well-done sample survey.

wording effects   The **wording of questions** is the most important influence on the answers given to a sample survey. Confusing or leading questions can introduce strong bias, and even minor changes in wording can change a survey's outcome. Here are some examples.[11]

EXAMPLE 7.10   Help the poor?

How do Americans feel about government help for the poor? Only 13% think we are spending too much on "assistance to the poor," but 44% think we are spending too much on "welfare."

EXAMPLE 7.11   Independence for Scotland?

How do the Scots feel about the movement to become independent from England? Well, 51% would vote for "independence for Scotland," but only 34% support "an independent Scotland separate from the United Kingdom."

It seems that "assistance to the poor" and "independence" are nice, hopeful words. "Welfare" and "separate" are negative words. Never trust the results of a sample survey until you have read the exact questions posed. The amount of nonresponse and the date of the survey are also important. Good statistical design is a part, but only a part, of a trustworthy survey.

APPLY YOUR KNOWLEDGE

7.13 Ring-no-answer. A common form of nonresponse in telephone surveys is "ring-no-answer." That is, a call is made to an active number but no one answers. The Italian National Statistical Institute looked at nonresponse to a government survey of households in Italy during the periods January 1 to Easter and July 1 to August 31. All calls were made between 7 and 10 p.m., but 21.4% gave "ring-no-answer" in one period versus 41.5% "ring-no-answer" in the other period.[12] Which period do you think had the higher rate of no answers? Why? Explain why a high rate of nonresponse makes sample results less reliable.

7.14 Question wording. During the 2000 presidential campaign, the candidates debated what to do with the large government surplus. The Pew Research Center asked two questions of random samples of adults. Both questions stated that Social Security would be "fixed." Here are the uses suggested for the remaining surplus:

> *Should the money be used for a tax cut, or should it be used to fund new government programs?*

> *Should the money be used for a tax cut, or should it be spent on programs for education, the environment, health care, crime-fighting and military defense?*

One of these questions drew 60% favoring a tax cut. The other drew only 22%. Which wording pulls respondents toward a tax cut? Why?

Inference about the population

Despite the many practical difficulties in carrying out a sample survey, using chance to choose a sample does eliminate bias in the actual selection of the sample from the list of available individuals. But it is unlikely that results from a sample are exactly the same as for the entire population. Sample results, like the official unemployment rate obtained from the monthly Current Population Survey, are only estimates of the truth about the population. If we select two samples at random from the same population, we will draw different individuals. So the sample results will almost certainly differ somewhat. Properly designed samples avoid systematic bias, but their results are rarely exactly correct and they vary from sample to sample.

How accurate is a sample result like the monthly unemployment rate? We can't say for sure, because the result would be different if we took another sample. But the results of random sampling don't change haphazardly from sample to sample. Because we deliberately use chance, the results obey the laws of probability that govern chance behavior. We can say how large an error we are likely to make in drawing conclusions about the population from a sample. Results from a sample survey usually come with a margin of error that sets bounds on the size of the likely error. How to do this is part of the business of statistical inference. We will describe the reasoning in Chapter 13.

One point is worth making now: **larger random samples give more accurate results than smaller samples.** By taking a very large sample, you can be confident that the sample result is very close to the truth about the population. The Current Population Survey contacts about 55,000 households, so it estimates the national unemployment rate very accurately. Opinion polls that contact 1000 or 1500 people give less accurate results. Of course, only probability samples carry this guarantee. *Nightline*'s voluntary response sample is worthless even though 186,000 people called in. Using a probability sampling design and taking care to deal with practical difficulties reduce bias in a sample. The size of the sample then determines how close to the population truth the sample result is likely to fall.

APPLY YOUR KNOWLEDGE

7.15 Ask more people. Just before a presidential election, a national opinion-polling firm increases the size of its weekly sample from the usual 1500 people to 4000 people. Why do you think the firm does this?

Chapter 7 SUMMARY

We can produce data intended to answer specific questions by **observational studies** or **experiments. Sample surveys** that select a part of a population of interest to represent the whole are one type of observational study. **Experiments,** unlike observational studies, actively impose some treatment on the subjects of the experiment.

Observational studies often fail to show that changes in an explanatory variable actually cause changes in a response variable, because the explanatory variable is **confounded** with lurking variables. Variables are confounded when their effects on a response can't be distinguished from each other.

A sample survey selects a **sample** from the **population** of all individuals about which we desire information. We base conclusions about the population on data about the sample.

The **design** of a sample refers to the method used to select the sample from the population. **Probability sampling** designs use chance to select a sample.

The basic probability sample is a **simple random sample (SRS).** An SRS gives every possible sample of a given size the same chance to be chosen.

Choose an SRS by labeling the members of the population and using a **table of random digits** to select the sample. Software can automate this process.

To choose a **stratified random sample,** divide the population into **strata,** groups of individuals that are similar in some way that is important to the response. Then choose a separate SRS from each stratum.

Failure to use probability sampling often results in **bias,** or systematic errors in the way the sample represents the population. **Voluntary response samples,** in which the respondents choose themselves, are particularly prone to large bias.

In human populations, even probability samples can suffer from bias due to **undercoverage** or **nonresponse,** from **response bias,** or from misleading results due to **poorly worded questions.** Sample surveys must deal expertly with these potential problems in addition to using a probability sampling design.

Chapter 7 EXERCISES

7.16 Alcohol and heart attacks. Many studies have found that people who drink alcohol in moderation have lower risk of heart attacks than either nondrinkers or heavy drinkers. Does alcohol consumption also improve survival after a heart attack? One study followed 1913 people who were hospitalized after severe heart attacks. In the year before their heart attack, 47% of these people did not drink, 36% drank moderately, and 17% drank heavily. After four years, fewer of the moderate drinkers had died.[13] Is this an observational study or an experiment? Why? What are the explanatory and response variables?

7.17 Term limits? There is strong public support for "term limits" that restrict the number of terms that legislators can serve. One possible explanation for this support is that voters are dissatisfied with the performance of Congress and other legislative bodies. A political scientist asks a sample of voters if they support term limits for members of Congress and also asks several questions that gauge their satisfaction with Congress. He finds no relationship between approval of Congress and support for term limits. Is this an observational study or an experiment? Why? What are the explanatory and response variables?

(Mark Wilson/Getty Images)

7.18 Safety of anesthetics. The National Halothane Study was a major investigation of the safety of anesthetics used in surgery. Records of over 850,000 operations performed in 34 major hospitals showed the following death rates for four common anesthetics:[14]

Anesthetic	A	B	C	D
Death rate	1.7%	1.7%	3.4%	1.9%

There is a clear association between the anesthetic used and the death rate of patients. Anesthetic C appears dangerous.

(a) Explain why we call the National Halothane Study an observational study rather than an experiment, even though it compared the results of using different anesthetics in actual surgery.

(b) When the study looked at other variables that are confounded with a doctor's choice of anesthetic, it found that Anesthetic C was not causing extra deaths. Suggest several variables that are mixed up with what anesthetic a patient receives.

7.19 Sampling students. A political scientist wants to know how college students feel about the Social Security system. She obtains a list of the 3456 undergraduates at her college and mails a questionnaire to 250 students selected at random. Only 104 questionnaires are returned. What is the population in this study? What is the sample from which information was actually obtained? What is the rate (percent) of nonresponse?

7.20 Sampling insurance claims. An insurance company wants to monitor the quality of its procedures for handling loss claims from its auto insurance policyholders. Each month the company selects an SRS from all auto insurance claims filed that month to examine them for accuracy and promptness. What is the population in this study? What is the sample?

7.21 Sampling small businesses. A business school researcher wants to know what factors affect the survival and success of small businesses. She selects a sample of 150 eating-and-drinking establishments from those listed in the telephone directory Yellow Pages for a large city. What is the population in this study? What is the sample?

7.22 Rating the president. A newspaper article about an opinion poll says that "43% of Americans approve of the president's overall job performance." Toward the end of the article, you read: "The poll is based on telephone interviews with 1210 adults from around the United States, excluding Alaska and Hawaii." What variable did this poll measure? What population do you think the poll wants information about? What was the sample? Are there any sources of bias in the sampling method used?

7.23 Ann Landers takes a sample. Advice columnist Ann Landers once asked her female readers whether they would be content with affectionate treatment by men, with no sex ever. Over 90,000 women wrote in, with 72% answering "Yes." Many of the letters described unfeeling treatment at the hands of men. Explain why this sample is certainly biased. What is the likely direction of the bias? That is, is 72% probably higher or lower than the truth about the population of all adult women?

(Ryan McVay/Photo Disc/Getty Images)

7.24 Seat belt use. A study in El Paso, Texas, looked at seat belt use by drivers. Drivers were observed at randomly chosen convenience stores. After they left their cars, they were invited to answer questions that included questions about seat belt use. In all, 75% said they always used seat belts, yet only 61.5% were wearing seat belts when they pulled into the store parking lots.[15] Explain the reason for the bias observed in responses to the survey. Do you expect bias in the same direction in most surveys about seat belt use?

7.25 Rating the police. The Miami Police Department wants to know how black residents of Miami feel about police service. A sociologist prepares several questions about the police. A sample of 300 mailing addresses in predominantly black neighborhoods is chosen, and a uniformed black police officer goes to each address to ask the questions of an adult living there. What are the population and the sample? Why are the results likely to be biased?

7.26 Do you trust the Internet? You want to ask a sample of college students the question "How much do you trust information about health that you find on the Internet—a great deal, somewhat, not much, or not at all?" You try out this and other questions on a pilot group of 10 students chosen from your class. The class members are

Anderson	Eckstein	Johnson	Puri
Arroyo	Fernandez	Kim	Richards
Batista	Fullmer	Molina	Rodriguez
Bell	Gandhi	Morgan	Samuels
Burke	Garcia	Nguyen	Shen
Calloway	Glaus	Palmiero	Velasco
Delluci	Helling	Percival	Washburn
Drasin	Husain	Prince	Zhao

Use Table B, starting at line 117, to choose your sample.

7.27 Using random digits. In using Table B repeatedly to choose random samples, you should not always begin at the same place, such as line 101. Why not?

7.28 Do you trust the Internet, continued. After testing the wording of your questions on the pilot group you chose in Exercise 7.26, you are ready to contact an SRS of students at your college. The online student directory lists 2968 students. You will attempt to call an SRS of 200 of them. Explain how you will choose your sample, and use Table B, starting at line 128, to choose the first 6 members of the sample.

7.29 Sampling from a census tract. The Census Bureau divides the entire country into "census tracts" that contain about 4000 people. Each tract is in turn divided into small "blocks," which in urban areas are bounded by local streets. An SRS of blocks from a census tract is often the next-to-last stage in a multistage sample. Figure 7.2 shows part of census tract 8051.12, in Cook County, Illinois, west of Chicago. The 44 blocks in this tract are divided into three "block groups." Group 1 contains 6 blocks numbered 1000 to 1005; group 2 (outlined in Figure 7.2) contains 12 blocks numbered 2000 to 2011; group 3 contains 26 blocks numbered 3000 to 3025. Use Table B, beginning at line 125, to choose an SRS of 5 of the 44 blocks in this census tract. Explain carefully how you labeled the blocks.

7.30 Nonresponse. Academic sample surveys, unlike commercial polls, often discuss nonresponse. A survey of drivers began by randomly sampling all listed residential telephone numbers in the United States. Of 45,956 calls to these numbers, 5029 were completed.[16] What was the rate of nonresponse for this sample? (Only one call was made to each number. Nonresponse would be lower if more calls were made.)

7.31 Sampling drivers. The sample described in the previous exercise produced a list of 5024 licensed drivers. The investigators then chose an

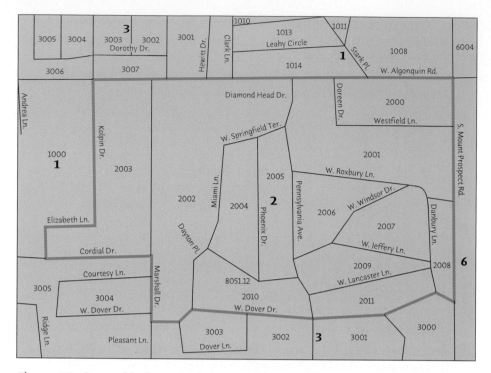

Figure 7.2 Census blocks in Cook County, Illinois, for Exercise 7.29. The outlined area is a block group.

SRS of 880 of these drivers to answer a 58-item questionnaire about their driving habits. How would you assign labels to the 5024 drivers? Use Table B, starting at line 104, to choose the first 5 drivers in the sample.

7.32 **Sampling at a party.** At a party there are 30 students over age 21 and 20 students under age 21. You choose at random 3 of those over 21 and separately choose at random 2 of those under 21 to interview about attitudes toward alcohol. You have given every student at the party the same chance to be interviewed: what is that chance? Why is your sample not an SRS?

7.33 **Random digits.** Which of the following statements are true of a table of random digits, and which are false? Briefly explain your answers.

(a) There are exactly four 0s in each row of 40 digits.

(b) Each pair of digits has chance 1/100 of being 00.

(c) The digits 0000 can never appear as a group, because this pattern is not random.

7.34 **Sampling at a party.** At a large block party there are 290 men and 110 women. You want to ask opinions about how to improve the next party. To be sure that women's opinions are adequately represented, you decide to choose a stratified random sample of 20 men and 20 women.

Explain how you will assign labels to the names of the people at the party. Enter Table B at line 130 and give the labels of the first 3 men and the first 3 women in your sample.

7.35 Sampling college faculty. A labor organization wants to study the attitudes of college faculty members toward collective bargaining. These attitudes appear to differ depending on the type of college. The American Association of University Professors classifies colleges as follows:

Class I.   Offer doctorate degrees and award at least 15 per year.
Class IIA. Award degrees above the bachelor's but are not in Class I.
Class IIB. Award no degrees beyond the bachelor's.
Class III. Two-year colleges.

Discuss the design of a sample of faculty from colleges in your state, with total sample size about 200.

7.36 Systematic random samples. *Systematic random samples* are often used   *systematic random sample*
to choose a sample of apartments in a large building or dwelling units in a block at the last stage of a multistage sample. An example will illustrate the idea of a systematic sample. Suppose that we must choose 4 addresses out of 100. Because $100/4 = 25$, we can think of the list as four lists of 25 addresses. Choose 1 of the first 25 at random, using Table B. The sample contains this address and the addresses 25, 50, and 75 places down the list from it. If 13 is chosen, for example, then the systematic random sample consists of the addresses numbered 13, 38, 63, and 88.

(a) Use Table B to choose a systematic random sample of 5 addresses from a list of 200. Enter the table at line 120.

(b) Like an SRS, a systematic sample gives all individuals the same chance to be chosen. Explain why this is true, then explain carefully why a systematic sample is nonetheless *not* an SRS.

7.37 Random-digit dialing. The list of individuals from which a sample is   *sampling frame*
actually selected is called the **sampling frame.** Ideally, the frame should list every individual in the population, but in practice this is often difficult. A frame that leaves out part of the population is a common source of undercoverage.

(a) Suppose that a sample of households in a community is selected at random from the telephone directory. What households are omitted from this frame? What types of people do you think are likely to live in these households? These people will probably be underrepresented in the sample.

(b) It is usual in telephone surveys to use random-digit dialing equipment that selects the last four digits of a telephone number at random after being given the exchange (the first three digits). Which of the households you mentioned in your answer to (a) will be included in the sampling frame by random-digit dialing?

7.38 Wording survey questions. Comment on each of the following as a potential sample survey question. Is the question clear? Is it slanted toward a desired response?

(a) "Some cell phone users have developed brain cancer. Should all cell phones come with a warning label explaining the danger of using cell phones?"

(b) "Do you agree that a national system of health insurance should be favored because it would provide health insurance for everyone and would reduce administrative costs?"

(c) "In view of the negative externalities in parent labor force participation and pediatric evidence associating increased group size with morbidity of children in day care, do you support government subsidies for day care programs?"

7.39 A survey on regulating guns. The 1998 National Gun Policy Survey, carried out by the University of Chicago's National Opinion Research Center, asked respondents' opinions about government regulation of firearms. A report from the survey says, "Participating households were identified through random-digit dialing; the respondent in each household was selected by the most-recent-birthday method."[17]

(a) What is "random-digit dialing?" Why is it a practical method for obtaining (almost) an SRS of households?

(b) The survey wants the opinion of an individual adult. Several adults may live in a household. In that case, the survey interviewed the adult with the most recent birthday. Why is this preferable to simply interviewing the person who answers the phone?

7.40 Roll your own bad questions. Write your own examples of bad sample survey questions.

(a) Write a biased question designed to get one answer rather than another.

(b) Write a question that is confusing, so that it is hard to answer.

7.41 Canada's national health care. The Ministry of Health in the Canadian Province of Ontario wants to know whether the national health care system is achieving its goals in the province. Much information about health care comes from patient records, but that source doesn't allow us to compare people who use health services with those who don't. So the Ministry of Health conducted the Ontario Health Survey, which interviewed a random sample of 61,239 people who live in the Province of Ontario.[18]

(a) What is the population for this sample survey? What is the sample?

(b) The survey found that 76% of males and 86% of females in the sample had visited a general practitioner at least once in the past year. Do you think these estimates are close to the truth about the entire population? Why?

7.42 Errors in a poll. A *New York Times* opinion poll on women's issues contacted a sample of 1025 women and 472 men by randomly selecting telephone numbers. The *Times* publishes descriptions of its polling methods. Here is part of the description for this poll:

> *In theory, in 19 cases out of 20 the results based on the entire sample will differ by no more than three percentage points in either direction from what would have been obtained by seeking out all adult Americans.*

> *The potential sampling error for smaller subgroups is larger. For example, for men it is plus or minus five percentage points.*[19]

Explain why the margin of error is larger for conclusions about men alone than for conclusions about all adults.

Chapter 7 MEDIA EXERCISES

7.43 The Gallup Poll. Go to the Web site of the Gallup Organization, www.gallup.com. Reports of several recent opinion polls appear on the home page. (You will have to avoid "premium content," which is available only to those who pay for it.) Choose a topic of interest and summarize key information: the sample size, the data, one of the questions asked, and the response percents.

7.44 Choose a stratified sample. Go to the Census Bureau Web site, www.census.gov. You want the population of all counties in your home state. This may take some searching. At the end of 2002, you could choose "Estimates" under "People" on the home page, then choose "Population change for counties," and pick your state. (Canadians: Go to Statistics Canada, www.statcan.ca, choose "Canadian Statistics," then "Population" and "Data tables." Get the populations of the Census Divisions in your home province or territory.)

You are planning a sample survey. The first stage will be a stratified sample of half the counties in your state. You will fill half of this sample with the most populated counties, then take a random sample of the remaining counties. Do this. Print a list of all the counties with their populations, and mark the ones that are in your sample.

7.45 Sampling retail outlets. Exercise 7.10 asks you to choose an SRS of 10 from the 440 retail outlets in New York that sell your product. Use the *Simple Random Sample* applet to choose this sample. Which outlets were chosen? (That was faster than using Table B.)

7.46 What do schoolkids want? The EESEE story "What Makes a Pre-Teen Popular?" examines the most important goals of schoolchildren. Do girls and boys have different goals? Are goals different in urban, suburban, and rural areas? To find out, researchers wanted to ask children in the fourth, fifth, and sixth grades this question:

What would you most like to do at school?

   A. *Make good grades.*

   B. *Be good at sports.*

   C. *Be popular.*

Because most children live in heavily populated urban and suburban areas, an SRS might contain few rural children. Moreover, it is too expensive to choose children at random from a large region—we must start by choosing schools rather than children. Describe a suitable sample design for this study and explain the reasoning behind your choice of design.

(Paul A. Souders/CORBIS)

Producing Data: Experiments

A study is an experiment when we actually do something to people, animals, or objects in order to observe the response. Because the purpose of an experiment is to reveal the response of one variable to changes in other variables, the distinction between explanatory and response variables is essential. Here is the basic vocabulary of experiments.

SUBJECTS, FACTORS, TREATMENTS

The individuals studied in an experiment are often called **subjects,** particularly when they are people.

The explanatory variables in an experiment are often called **factors.**

A **treatment** is any specific experimental condition applied to the subjects. If an experiment has several factors, a treatment is a combination of specific values of each factor.

EXAMPLE 8.1 Effects of good day care

Does day care help low-income children stay in school and hold good jobs later in life? The Carolina Abecedarian Project (the name suggests the ABCs) has followed a group of children since 1972. The *subjects* are 111 people who in 1972 were healthy but low-income black infants in Chapel Hill, North Carolina. All the infants received nutritional supplements and help from social workers. Half, chosen at random, were also placed in an intensive preschool program. The experiment compares these two *treatments*. There is a single *factor*, "preschool, yes or no." There are many *response variables*, recorded over more than 20 years, including academic test scores, college attendance, and employment.[1]

EXAMPLE 8.2 Effects of TV advertising

What are the effects of repeated exposure to an advertising message? The answer may depend both on the length of the ad and on how often it is repeated. An experiment investigated this question using undergraduate students as *subjects*. All subjects viewed a 40-minute television program that included ads for a digital camera. Some subjects saw a 30-second commercial; others, a 90-second version. The same commercial was shown either 1, 3, or 5 times during the program.

This experiment has two *factors*: length of the commercial, with 2 values, and repetitions, with 3 values. The 6 combinations of one value of each factor form 6 *treatments*. Figure 8.1 shows the layout of the treatments. After viewing, all of the subjects answered questions about their recall of the ad, their attitude toward the camera, and their intention to purchase it. These are the *response variables*.[2]

Examples 8.1 and 8.2 illustrate the advantages of experiments over observational studies. In an experiment, we can study the effects of the specific treatments we are interested in. By assigning subjects to treatments, we can avoid confounding. If, for example, we simply compare children whose parents did and did not choose an intensive preschool program, we may find that children in the program come from richer and better-educated parents. Example 8.1 avoids that. Moreover, we can control the environment of the subjects to hold

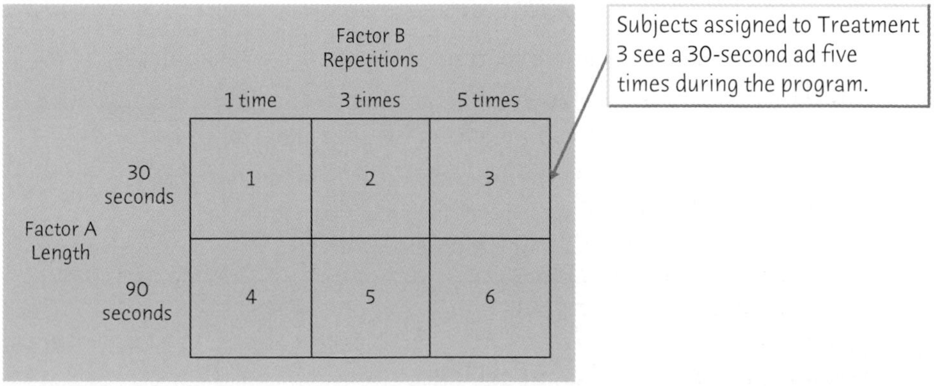

Figure 8.1 The treatments in the experimental design of Example 8.2. Combinations of values of the two factors form six treatments.

constant factors that are of no interest to us, such as the specific product adver-tised in Example 8.2.

Another advantage of experiments is that we can study the combined effects of several factors simultaneously. The interaction of several factors can produce effects that could not be predicted from looking at the effect of each factor alone. Perhaps longer commercials increase interest in a product, and more commercials also increase interest, but if we both make a commercial longer and show it more often, viewers get annoyed and their interest in the product drops. The two-factor experiment in Example 8.2 will help us find out.

APPLY YOUR KNOWLEDGE

8.1 Internet telephone calls. You can use your computer to make long-distance telephone calls over the Internet. How will the cost affect the use of this service? A university plans an experiment to find out. It will offer the service to all 350 students in one of its dormitories. Some students will pay a low flat rate. Others will pay higher rates at peak periods and very low rates off-peak. The university is interested in the amount and time of use and in the effect on the congestion of the network. What are the subjects, the factors, the treatments, and the response variables in this experiment?

8.2 Sealing food packages. A manufacturer of food products uses package liners that are sealed at the top by applying heated jaws after the package is filled. The customer peels the sealed pieces apart to open the package. What effect does the temperature of the jaws have on the force needed to peel the liner? To answer this question, engineers obtain 20 pairs of pieces of package liner. They seal five pairs at each of 250° F, 275° F, 300° F, and 325° F. Then they measure the force needed to peel each seal.

(a) What are the individuals?

(b) There is one factor (explanatory variable). What is it, and what are its values?

(c) What is the response variable?

8.3 An industrial experiment. A chemical engineer is designing the production process for a new product. The chemical reaction that produces the product may have higher or lower yield, depending on the temperature and the stirring rate in the vessel in which the reaction takes place. The engineer decides to investigate the effects of combinations of two temperatures (50° C and 60° C) and three stirring rates (60 rpm, 90 rpm, and 120 rpm) on the yield of the process. She will process two batches of the product at each combination of temperature and stirring rate.

(a) What are the individuals and the response variable in this experiment?

(b) How many factors are there? How many treatments? Use a diagram like that in Figure 8.1 to lay out the treatments.

(c) How many individuals are required for the experiment?

Comparative experiments

Experiments in the science laboratory often have a simple design: impose the treatment and see what happens. We can outline that design like this:

Subjects $\longrightarrow$ **Treatment** $\longrightarrow$ **Response**

In the laboratory, we try to avoid confounding by rigorously controlling the environment of the experiment so that nothing except the experimental treatment influences the response. Once we get out of the laboratory, however, there are almost always lurking variables waiting to confound us. When our subjects are people or animals rather than electrons or chemical compounds, confounding can happen even in the controlled environment of a laboratory or medical clinic. Here are typical examples.

EXAMPLE 8.3   How to experiment badly

Plant a new variety of seed corn at an agricultural research station. The yield is very high. But this growing season had ideal weather, with rain in the right amounts and at the right times. The high yield may just reflect great growing conditions.

Offer a college course online. The students do much better than past groups have done with classroom instruction. But the students in the online course were older and better prepared than past classroom students. No wonder they performed well.

"Gastric freezing" is a clever treatment for stomach ulcers. The patient swallows tubes through which a cold solution is pumped to cool the stomach and perhaps reduce acid production. The patients say they have less ulcer pain. But many patients respond to attention from doctors and the promise of a new treatment. Perhaps high hopes rather than gastric freezing explain the patients' experiences.

All of these experiments are *biased*. They systematically favor the experimental treatments because other variables that lead to favorable outcomes are confounded with the treatment. Fortunately, the remedy is simple. Experiments should *compare* treatments rather than attempt to assess a single treatment in isolation. Plant a standard variety of corn as well as the new variety, so that the good weather acts on both. Assign some of a semester's students to a classroom presentation and the rest to an online version. Give some ulcer patients gastric freezing and give others a **placebo,** a dummy treatment that will bring out the same high hopes. In this case, the dummy treatment pumps a solution at body temperature through the tubes so that there is no cooling of the stomach. The comparison group of corn plantings or students or ulcer patients is called a **control group,** because it enables us to control the effects of lurking variables on the outcome.

placebo

control group

Randomized comparative experiments

The design of an experiment first describes the response variables, the factors (explanatory variables), and the layout of the treatments, with *comparison* as the leading principle. The second aspect of design is the rule used to assign the subjects to the treatments. Comparison of the effects of several treatments is valid only when all treatments are applied to similar groups of individuals. If one corn variety is planted on more fertile ground, or if one method of teaching is tried on better students, comparisons among treatments are biased. How can we assign individuals to treatments in a way that is fair to all the treatments?

Our answer is the same as in sampling: let impersonal chance make the assignment. The use of chance to divide experimental subjects into groups is called **randomization.** Groups formed by randomization don't depend on any characteristic of the subjects or on the judgment of the experimenter. An experiment that uses both comparison and randomization is a **randomized comparative experiment.** Here is an example.

randomization

randomized comparative experiment

EXAMPLE 8.4   Testing a breakfast food

A food company assesses the nutritional quality of a new "instant breakfast" product by feeding it to newly weaned male white rats. The response variable is a rat's weight gain over a 28-day period. A control group of rats eats a standard diet but otherwise receives exactly the same treatment as the experimental group.

This experiment has one factor (the diet) with two values. The researchers use 30 rats for the experiment and so must divide them into two groups of 15. To do this in an unbiased fashion, put the cage numbers of the 30 rats in a hat, mix them up, and draw 15. These rats form the experimental group and the remaining 15 make up the control group. That is, *each group is an SRS of the available rats*. Figure 8.2 outlines the design of this experiment.

We can use software or the table of random digits to randomize. Label the rats 01 to 30. Enter Table B at (say) line 130. Run your finger along this line (and continue to lines 131 and 132 as needed) until 15 rats are chosen. They are the rats labeled

05, 16, 17, 20, 19, 04, 25, 29, 18, 07, 13, 02, 23, 27, 21

These rats form the experimental group. The remaining 15 are the control group

Golfing at random

Random drawings give all the same chance to be chosen, so they offer a fair way to decide who gets a scarce good—like a round of golf. Lots of golfers want to play the famous Old Course at St. Andrews, Scotland. A few can reserve in advance. Most must hope that chance favors them in the daily random drawing for tee times. At the height of the summer season, only 1 in 6 wins the right to pay $120 for a round.

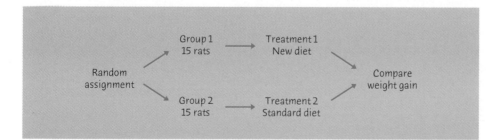

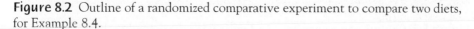

Figure 8.2 Outline of a randomized comparative experiment to compare two diets, for Example 8.4.

Completely randomized designs

The design in Figure 8.2 combines comparison and randomization to arrive at the simplest statistical design for an experiment. This "flowchart" outline presents all the essentials: randomization, the sizes of the groups and which treatment they receive, and the response variable. There are, as we will see later, statistical reasons for generally using treatment groups about equal in size. We call designs like that in Figure 8.2 *completely randomized*.

COMPLETELY RANDOMIZED DESIGN

In a **completely randomized** experimental design, all the subjects are allocated at random among all the treatments.

Completely randomized designs can compare any number of treatments. Here is an example that compares three treatments.

EXAMPLE 8.5   Conserving energy

Many utility companies have introduced programs to encourage energy conservation among their customers. An electric company considers placing electronic meters in households to show what the cost would be if the electricity use at that moment continued for a month. Will meters reduce electricity use? Would cheaper methods work almost as well? The company decides to design an experiment.

One cheaper approach is to give customers a chart and information about monitoring their electricity use. The experiment compares these two approaches (meter, chart) and also a control. The control group of customers receives information about energy conservation but no help in monitoring electricity use. The response variable is total electricity used in a year. The company finds 60 single-family residences in the same city willing to participate, so it assigns 20 residences at random to each of the three treatments. Figure 8.3 outlines the design.

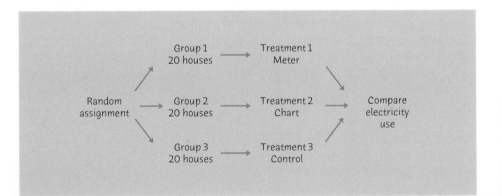

Figure 8.3 Outline of a completely randomized design comparing three energy-saving programs, for Example 8.5.

To carry out the random assignment, label the 60 households 01 to 60. Enter Table B (or use software) to select an SRS of 20 to receive the meters. Continue in Table B, selecting 20 more to receive charts. The remaining 20 form the control group.

Examples 8.4 and 8.5 describe completely randomized designs that compare values of a single factor. In Example 8.4, the factor is the diet fed to the rats. In Example 8.5, it is the method used to encourage energy conservation. Completely randomized designs can have more than one factor. The advertising experiment of Example 8.2 has two factors: the length and the number of repetitions of a television commercial. Their combinations form the six treatments outlined in Figure 8.1. A completely randomized design assigns subjects at random to these six treatments. Once the layout of treatments is set, the randomization needed for a completely randomized design is tedious but straightforward.

APPLY YOUR KNOWLEDGE

8.4 **Does ginkgo improve memory?** The law allows marketers of herbs and other natural substances to make health claims that are not supported by evidence. Brands of ginkgo extract claim to "improve memory and concentration." A randomized comparative experiment found no evidence for such effects.[3] The subjects were 230 healthy people over 60 years old. They were randomly assigned to ginkgo or a placebo pill that looked and tasted the same. All the subjects took a battery of tests for learning and memory before treatment started and again after six weeks.

Outline the design of this experiment. (When you outline the design of an experiment, indicate the size of the treatment groups and the response variable. Figure 8.2 is a model.) Use Table B, starting at line 103, to choose the first 5 members of the ginkgo group.

8.5 **Can tea prevent cataracts?** Eye cataracts are responsible for over 40% of blindness around the world. Can drinking tea regularly slow the growth of cataracts? We can't experiment on people, so we use rats as subjects. Researchers injected 18 young rats with a substance that causes cataracts. One group of the rats also received black tea extract; a second group received green tea extract; and a third got a placebo, a substance with no effect on the body. The response variable was the growth of cataracts over the next six weeks. Yes, both tea extracts did slow cataract growth.

Outline the design of this experiment. (When you outline the design of an experiment, indicate the size of the treatment groups and the response variable. Figure 8.3 is a model.) Use Table B, starting at line 142, to assign rats to treatments.[4]

8.6 **Sealing food packages.** Use a diagram to describe a completely randomized experimental design for the package liner experiment of Exercise 8.2. Use software or Table B (starting at line 120) to do the randomization required by your design.

The logic of randomized comparative experiments

Randomized comparative experiments are designed to give good evidence that differences in the treatments actually *cause* the differences we see in the response. The logic is as follows:

- Random assignment of subjects forms groups that should be similar in all respects before the treatments are applied. Exercise 8.44 uses the *Simple Random Sample* applet to demonstrate this.

- Comparative design ensures that influences other than the experimental treatments operate equally on all groups.

- Therefore, differences in average response must be due either to the treatments or to the play of chance in the random assignment of subjects to the treatments.

That "either-or" deserves more thought. In Example 8.4, we cannot say that *any* difference in the average weight gains of rats fed the two diets must be caused by a difference between the diets. There would be some difference even if both groups received the same diet, because the natural variability among rats means that some grow faster than others. Chance assigns the faster-growing rats to one group or the other, and this creates a chance difference between the groups. We would not trust an experiment with just one rat in each group, for example. The results would depend too much on which group got lucky and received the faster-growing rat. If we assign many rats to each diet, however, the effects of chance will average out, and there will be little difference in the average weight gains in the two groups unless the diets themselves cause a difference. "Use enough subjects to reduce chance variation" is the third big idea of statistical design of experiments.

What's news?

Randomized comparative experiments provide the best evidence for medical advances. Do newspapers care? Maybe not. University researchers looked at 1192 articles in medical journals, of which 7% were turned into stories by the two newspapers examined. Of the journal articles, 37% concerned observational studies and 25% described randomized experiments. Among the articles publicized by the newspapers, 58% were observational studies and only 6% were randomized experiments. Conclusion: the newspapers want exciting stories, especially bad-news stories, whether or not the evidence is good.

PRINCIPLES OF EXPERIMENTAL DESIGN

The basic principles of statistical design of experiments are

1. **Control** the effects of lurking variables on the response, most simply by comparing several treatments.
2. **Randomize**—use impersonal chance to assign subjects to treatments.
3. **Use enough subjects** in each group to reduce chance variation in the results.

We hope to see a difference in the responses so large that it is unlikely to happen just because of chance variation. We can use the laws of probability, which give a mathematical description of chance behavior, to learn if the treatment effects are larger than we would expect to see if only chance were operating. If they are, we call them *statistically significant*.

STATISTICAL SIGNIFICANCE

An observed effect so large that it would rarely occur by chance is called **statistically significant.**

If we observe statistically significant differences among the groups in a comparative randomized experiment, we have good evidence that the treatments actually caused these differences. You will often see the phrase "statistically significant" in reports of investigations in many fields of study. The great advantage of randomized comparative experiments is that they can produce data that give good evidence for a cause-and-effect relationship between the explanatory and response variables. We know that in general a strong association does not imply causation. A statistically significant association in data from a well-designed experiment does imply causation.

APPLY YOUR KNOWLEDGE

8.7 **Conserving energy.** Example 8.5 describes an experiment to learn whether providing households with electronic meters or charts will reduce their electricity consumption. An executive of the electric company objects to including a control group. He says: "It would be simpler to just compare electricity use last year (before the meter or chart was provided) with consumption in the same period this year. If households use less electricity this year, the meter or chart must be working." Explain clearly why this design is inferior to that in Example 8.5.

8.8 **Exercise and heart attacks.** Does regular exercise reduce the risk of a heart attack? Here are two ways to study this question. Explain clearly why the second design will produce more trustworthy data.

   1. A researcher finds 2000 men over 40 who exercise regularly and have not had heart attacks. She matches each with a similar man who does not exercise regularly, and she follows both groups for 5 years.

   2. Another researcher finds 4000 men over 40 who have not had heart attacks and are willing to participate in a study. She assigns 2000 of the men to a regular program of supervised exercise. The other 2000 continue their usual habits. The researcher follows both groups for 5 years.

8.9 **The Monday effect.** Puzzling but true: stocks tend to go down on Mondays. There is no convincing explanation for this fact. A study looked at this "Monday effect" in more detail, using data on the daily returns of stocks over a 30-year period. Here are some of the findings:

   To summarize, our results indicate that the well-known Monday effect is caused largely by the Mondays of the last two weeks of the month. The mean Monday return of the first three weeks of the month is, in general, not significantly different from zero and is generally significantly higher than the mean Monday return of the last two weeks. Our finding seems to make it more difficult to explain the Monday effect.[5]

A friend thinks that "significantly" in this article has its plain English meaning, roughly "I think this is important." Explain in simple language what "significantly higher" and "not significantly different from zero" tell us here.

Scratch my furry ears

Rats and rabbits, specially bred to be uniform in their inherited characteristics, are the subjects in many experiments. It turns out that animals, like people, are quite sensitive to how they are treated. This creates opportunities for hidden bias. For example, human affection can change the cholesterol level of rabbits. Choose some rabbits at random and regularly remove them from their cages to have their heads scratched by friendly people. Leave other rabbits unloved. All the rabbits eat the same diet, but the rabbits that receive affection have lower cholesterol.

Cautions about experimentation

The logic of a randomized comparative experiment depends on our ability to treat all the subjects identically in every way except for the actual treatments being compared. Good experiments therefore require careful attention to details. For example, the subjects in both groups of the experiment on the effects of ginkgo (Exercise 8.4) all got the same attention before and during the treatment period. Moreover, the study was *double-blind*. The subjects didn't know what was in the pills they took. Neither did the investigators who worked with them. Pills for each day—ginkgo or a placebo that looked and tasted the same—were placed in sealed envelopes whose contents were known only to the one investigator who did the randomizing. The double-blind method avoids unconscious bias by, for example, a doctor who doesn't think that "just a placebo" can benefit a patient. The report of the ginkgo study says that the study design was a "randomized, double-blind, placebo-controlled, parallel-group trial." Doctors are supposed to know what this means. Now you also know.

DOUBLE-BLIND EXPERIMENTS

In a **double-blind** experiment, neither the subjects nor the people who work with them know which treatment each subject is receiving.

---

lack of realism   The most serious potential weakness of experiments is **lack of realism.** The subjects or treatments or setting of an experiment may not realistically duplicate the conditions we really want to study. Here are two examples.

EXAMPLE 8.6   Response to advertising

The study of television advertising in Example 8.2 showed a 40-minute videotape to students who knew an experiment was going on. We can't be sure that the results apply to everyday television viewers. Many behavioral science experiments use as subjects students who know they are subjects in an experiment. That's not a realistic setting.

EXAMPLE 8.7   *Center brake lights*

Do those high center brake lights, required on all cars sold in the United States since 1986, really reduce rear-end collisions? Randomized comparative experiments with fleets of rental and business cars, done before the lights were required, showed that the third brake light reduced rear-end collisions by as much as 50%. Alas, requiring the third light in all cars led to only a 5% drop.

What happened? Most cars did not have the extra brake light when the experiments were carried out, so it caught the eye of following drivers. Now that almost all cars have the third light, they no longer capture attention.

Lack of realism can limit our ability to apply the conclusions of an experiment to the settings of greatest interest. Most experimenters want to generalize their conclusions to some setting wider than that of the actual experiment. Statistical analysis of the original experiment cannot tell us how far the results will generalize. Nonetheless, the randomized comparative experiment, because of its ability to give convincing evidence for causation, is one of the most important ideas in statistics.

APPLY YOUR KNOWLEDGE

8.10 **Dealing with pain.** Health care providers are giving more attention to relieving the pain of cancer patients. An article in the journal *Cancer* surveyed a number of studies and concluded that controlled-release morphine tablets, which release the painkiller gradually over time, are more effective than giving standard morphine when the patient needs it.[6] The "methods" section of the article begins: "Only those published studies that were controlled (i.e., randomized, double blind, and comparative), repeated-dose studies with CR morphine tablets in cancer pain patients were considered for this review." Explain the terms in parentheses to someone who knows nothing about medical trials.

8.11 **Does meditation reduce anxiety?** An experiment that claimed to show that meditation reduces anxiety proceeded as follows. The experimenter interviewed the subjects and rated their level of anxiety. Then the subjects were randomly assigned to two groups. The experimenter taught one group how to meditate and they meditated daily for a month. The other group was simply told to relax more. At the end of the month, the experimenter interviewed all the subjects again and rated their anxiety level. The meditation group now had less anxiety. Psychologists said that the results were suspect because the ratings were not blind. Explain what this means and how lack of blindness could bias the reported results.

Matched pairs designs

Completely randomized designs are the simplest statistical designs for experiments. They illustrate clearly the principles of control, randomization, and adequate number of subjects. However, completely randomized designs are often inferior to more elaborate statistical designs. In particular, matching the subjects in various ways can produce more precise results than simple randomization.

matched pairs design

One common design that combines matching with randomization is the **matched pairs design.** A matched pairs design compares just two treatments. Choose pairs of subjects that are as closely matched as possible. Assign one of the treatments to one of the subjects in a pair by tossing a coin or reading odd and even digits from Table B. The other subject gets the remaining treatment. Sometimes each "pair" in a matched pairs design consists of just one subject, who gets both treatments one after the other. Each subject serves as his or her own control. The *order* of the treatments can influence the subject's response, so we randomize the order for each subject, again by a coin toss.

EXAMPLE 8.8 Coke versus Pepsi

Pepsi wanted to demonstrate that Coke drinkers prefer Pepsi when they taste both colas blind. The subjects, all people who said they were Coke drinkers, tasted both colas from glasses without brand markings and said which they liked better. This is a matched pairs design in which each subject compares the two colas. Because responses may depend on which cola is tasted first, the order of tasting should be chosen at random for each subject.

When more than half the Coke drinkers chose Pepsi, Coke claimed that the experiment was biased. The Pepsi glasses were marked M and Coke glasses were marked Q. Aha, said Coke, this just shows that people like the letter M better than the letter Q. A careful experiment would in fact take care to avoid any distinction other than the actual treatments.[7]

Block designs

Matched pairs designs use the principles of comparison of treatments and randomization. However, the randomization is not complete—we do not randomly assign all the subjects at once to the two treatments. Instead, we randomize only within each matched pair. This allows matching to reduce the effect of variation among the subjects. Matched pairs are an example of *block designs*.

BLOCK DESIGN

A **block** is a group of individuals that are known before the experiment to be similar in some way that is expected to affect the response to the treatments.

In a **block design,** the random assignment of individuals to treatments is carried out separately within each block.

A block design combines the idea of creating equivalent treatment groups by matching with the principle of forming treatment groups at random. Blocks are another form of *control*. They control the effects of some outside variables

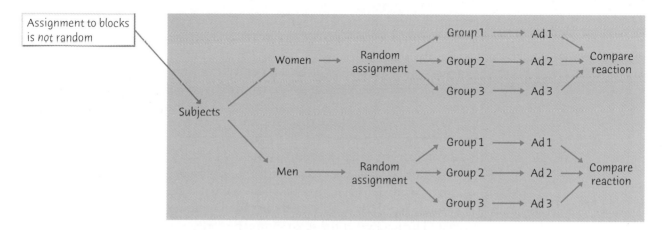

Figure 8.4 Outline of a block design, for Example 8.9. The blocks consist of male and female subjects. The treatments are three advertisements for the same product.

by bringing those variables into the experiment to form the blocks. Here are some typical examples of block designs.

EXAMPLE 8.9 Men, women, and advertising

Women and men respond differently to advertising. An experiment to compare the effectiveness of three television commercials for the same product will want to look separately at the reactions of men and women, as well as assess the overall response to the ads.

A completely randomized design considers all subjects, both men and women, as a single pool. The randomization assigns subjects to three treatment groups without regard to their sex. This ignores the differences between men and women. A better design considers women and men separately. Randomly assign the women to three groups, one to view each commercial. Then separately assign the men at random to three groups. Figure 8.4 outlines this improved design.

EXAMPLE 8.10 Comparing welfare policies

A social policy experiment will assess the effect on family income of several proposed new welfare systems and compare them with the present welfare system. Because the future income of a family is strongly related to its present income, the families who agree to participate are divided into blocks of similar income levels. The families in each block are then allocated at random among the welfare systems.

Blocks allow us to draw separate conclusions about each block, for example, about men and women in Example 8.9. Blocking also allows more precise overall conclusions, because the systematic differences between men and women can be removed when we study the overall effects of the three commercials. The idea of blocking is an important additional principle of statistical design of experiments. A wise experimenter will form blocks based on the most important unavoidable sources of variability among the subjects. Randomization will then average out the effects of the remaining variation and allow an unbiased comparison of the treatments.

Like the design of samples, the design of complex experiments is a job for experts. Now that we have seen a bit of what is involved, we will concentrate for the most part on completely randomized experiments.

APPLY YOUR KNOWLEDGE

8.12 Comparing hand strength. Is the right hand generally stronger than the left in right-handed people? You can crudely measure hand strength by placing a bathroom scale on a shelf with the end protruding, then squeezing the scale between the thumb below and the four fingers above it. The reading of the scale shows the force exerted. Describe the design of a matched pairs experiment to compare the strength of the right and left hands, using 10 right-handed people as subjects. (You need not actually do the randomization.)

8.13 Does charting help investors? Some investment advisors believe that charts of past trends in the prices of securities can help predict future prices. Most economists disagree. In an experiment to examine the effects of using charts, business students trade (hypothetically) a foreign currency at computer screens. There are 20 student subjects available, named for convenience A, B, C, . . . , T. Their goal is to make as much money as possible, and the best performances are rewarded with small prizes. The student traders have the price history of the foreign currency in dollars in their computers. They may or may not also have software that highlights trends. Describe two designs for this experiment, a completely randomized design and a matched pairs design in which each student serves as his or her own control. In both cases, carry out the randomization required by the design.

(Scott Markewitz/Taxi/Getty Images)

8.14 Protecting ultramarathon runners. An ultramarathon, as you might guess, is a foot race longer than the 26.2 miles of a marathon. Runners commonly develop respiratory infections after an ultramarathon. Will taking 600 milligrams of vitamin C daily reduce these infections? Researchers randomly assigned ultramarathon runners to receive either vitamin C or a placebo. Separately, they also randomly assigned these treatments in a group of nonrunners the same age as the runners. All subjects were watched for 14 days after the big race to see if infections developed.[8]

(a) What is the name for this experimental design?

(b) Use a diagram to outline the design.

8.15 Technology for teaching statistics. The Brigham Young University statistics department is performing randomized comparative experiments to compare teaching methods. Response variables include students' final-exam scores and a measure of their attitude toward statistics. One study compares two levels of technology for large lectures: standard (overhead projectors and chalk) and multimedia. The individuals in the study are the 8 lectures in a basic statistics course. There are four instructors, each of whom teaches two lectures. Because the lecturers differ, their lectures form four blocks.[9] Suppose the lectures and lecturers are as follows:

Lecture	Lecturer	Lecture	Lecturer
1	Hilton	5	Tolley
2	Christensen	6	Hilton
3	Hadfield	7	Tolley
4	Hadfield	8	Christensen

Outline a randomized block design and do the randomization that your design requires.

Chapter 8 SUMMARY

In an experiment, we impose one or more **treatments** on individuals, often called **subjects.** Each treatment is a combination of values of the explanatory variables, which we call **factors.**

The **design** of an experiment describes the choice of treatments and the manner in which the subjects are assigned to the treatments.

The basic principles of statistical design of experiments are **control** and **randomization** to combat bias and **using enough subjects** to reduce chance variation.

The simplest form of control is **comparison.** Experiments should compare two or more treatments in order to avoid **confounding** of the effect of a treatment with other influences, such as lurking variables.

Randomization uses chance to assign subjects to the treatments. Randomization creates treatment groups that are similar (except for chance variation) before the treatments are applied. Randomization and comparison together prevent **bias,** or systematic favoritism, in experiments.

You can carry out randomization by giving numerical labels to the subjects and using a **table of random digits** to choose treatment groups.

Applying each treatment to many subjects reduces the role of chance variation and makes the experiment more sensitive to differences among the treatments.

Good experiments require attention to detail as well as good statistical design. Many behavioral and medical experiments are **double-blind. Lack of realism** in an experiment can prevent us from generalizing its results.

In addition to comparison, a second form of control is to restrict randomization by forming **blocks** of individuals that are similar in some way that is important to the response. Randomization is then carried out separately within each block.

Matched pairs are a common form of blocking for comparing just two treatments. In some matched pairs designs, each subject receives both treatments in a random order. In others, the subjects are matched in pairs as closely as possible, and one subject in each pair receives each treatment.

Chapter 8 EXERCISES

8.16 Wine, beer, or spirits? Example 7.2 (page 177) describes a study that compared three groups of people: the first group drinks mostly wine, the second drinks mostly beer, and the third drinks mostly spirits. This study is comparative, but it is not an experiment. Why not?

8.17 Reducing risky sex. The National Institute of Mental Health (NIMH) wants to know whether intense education about the risks of AIDS will help change the behavior of people who report sexual activities that put them at risk of infection. NIMH investigators screened 38,893 people and identified 3706 suitable subjects. The subjects were assigned to a control group (1855 people) or an intervention group (1851 people). The control group attended a one-hour AIDS education session; the intervention group attended seven single-sex discussion sessions, each lasting 90 to 120 minutes. After 12 months, 64% of the intervention group and 52% of the control group said they used condoms. (None of the subjects used condoms regularly before the study began.)[10]

(a) Because none of the subjects used condoms when the study started, we might just offer the intervention sessions and find that 64% used condoms 12 months after the sessions. Explain why this greatly overstates the effectiveness of the intervention.

(b) Outline the design of this experiment.

(c) You must randomly assign 3706 subjects. How would you label them? Use line 119 of Table B to choose the first 5 subjects for the intervention group.

8.18 Wine, beer, or spirits? You have recruited 300 adults aged 45 to 65 who are willing to follow your orders about alcohol consumption over the next five years. You want to compare the effects of moderate drinking of just wine, just beer, or just spirits on heart disease. Outline the design of a completely randomized experiment to do this. (No such experiment has been done because subjects aren't willing to have their drinking regulated for years.)

8.19 Response to TV ads. You decide to use a completely randomized design in the two-factor experiment on response to advertising described in Example 8.2 (page 200). The 36 students named below will serve as subjects. (Ignore the asterisks.) Outline the design. Then use Table B at line 130 to randomly assign the subjects to the 6 treatments.

Alomar	Denman	Han	Liang	Padilla*	Valasco
Asihiro*	Durr*	Howard*	Maldonado	Plochman	Vaughn
Bennett	Edwards*	Hruska	Marsden	Rosen*	Wei
Bikalis	Farouk	Imrani	Montoya*	Solomon	Wilder*
Chao*	Fleming	James	O'Brian	Trujillo	Willis
Clemente	George	Kaplan*	Ogle*	Tullock	Zhang*

8.20 Wine, beer, or spirits? Women as a group develop heart disease much later than men. We can improve the completely randomized design of Exercise 8.18 by using women and men as blocks. Your 300 subjects include 120 women and 180 men. Outline a block design for comparing wine, beer, and spirits. Be sure to say how many subjects you will put in each group in your design.

8.21 Response to TV ads, continued. We can improve on the completely randomized design you outlined in Exercise 8.19. The 36 subjects include 24 women and 12 men. Men and women often react differently to advertising. You therefore decide to use a block design with the two genders as blocks. You must assign the 6 treatments at random within each block separately.

(a) Outline the design with a diagram.

(b) The 12 men are marked with asterisks in the list in Exercise 8.19. Use Table B, beginning at line 140, to do the randomization. Report your result in a table that lists the 24 women and 12 men and the treatment you assigned to each.

8.22 Fabric finishing. A maker of fabric for clothing is setting up a new line to "finish" the raw fabric. The line will use either metal rollers or natural bristle rollers to raise the surface of the fabric; a dyeing cycle time of either 30 minutes or 40 minutes; and a temperature of either 150° or 175° Celsius. An experiment will compare all combinations of these choices. Three specimens of fabric will be subjected to each treatment and scored for quality.

(a) What are the factors and the treatments? How many individuals (fabric specimens) does the experiment require?

(b) Outline a completely randomized design for this experiment. (You need not actually do the randomization.)

8.23 Aspirin and heart attacks. Can aspirin help prevent heart attacks? The Physicians' Health Study, a large medical experiment involving 22,000 male physicians, attempted to answer this question. One group of about 11,000 physicians took an aspirin every second day, while the rest took a placebo. After several years the study found that subjects in

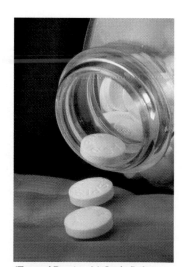

(Tom and Dee Ann McCarthy/Index Stock Imagery/PictureQuest)

the aspirin group had significantly fewer heart attacks than subjects in the placebo group.

(a) Identify the experimental subjects, the factor and its values, and the response variable in the Physicians' Health Study.

(b) Use a diagram to outline a completely randomized design for the Physicians' Health Study.

(Steve Raymer/CORBIS)

8.24 Prayer and meditation. You read in a magazine that "nonphysical treatments such as meditation and prayer have been shown to be effective in controlled scientific studies for such ailments as high blood pressure, insomnia, ulcers, and asthma." Explain in simple language what the article means by "controlled scientific studies" and why such studies might show that meditation and prayer are effective treatments for some medical problems.

8.25 Treating breast cancer. The most common treatment for breast cancer discovered in its early stages was once removal of the breast. It is now usual to remove only the tumor and nearby lymph nodes, followed by radiation. To study whether these treatments differ in their effectiveness, a medical team examines the records of 25 large hospitals and compares the survival times after surgery of all women who have had either treatment.

(a) What are the explanatory and response variables?

(b) Explain carefully why this study is not an experiment.

(c) Explain why confounding will prevent this study from discovering which treatment is more effective. (The current treatment was in fact recommended after several large randomized comparative experiments.)

8.26 Treating drunk drivers. Once a person has been convicted of drunk driving, one purpose of court-mandated treatment or punishment is to prevent future offenses of the same kind. Suggest three different treatments that a court might require. Then outline the design of an experiment to compare their effectiveness. Be sure to specify the response variables you will measure.

8.27 Reducing health care spending. Will people spend less on health care if their health insurance requires them to pay some part of the cost themselves? An experiment on this issue asked if the percent of medical costs that are paid by health insurance has an effect either on the amount of medical care that people use or on their health. The treatments were four insurance plans. Each plan paid all medical costs above a ceiling. Below the ceiling, the plans paid 100%, 75%, 50%, or 0% of costs incurred.

(a) Outline the design of a randomized comparative experiment suitable for this study.

(b) Describe briefly the practical and ethical difficulties that might arise in such an experiment.

8.28 Sickle cell disease. Sickle cell disease is an inherited disorder of the red blood cells that in the United States affects mostly blacks. It can cause severe pain and many complications. Can the drug hydroxyurea reduce the severe pain caused by sickle cell disease? A study by the National Institutes of Health gave the drug to 150 sickle cell sufferers and a placebo (a dummy medication) to another 150. The researchers then counted the episodes of pain reported by each subject. What are the subjects, the factor, the treatments, and the response variable?

8.29 The benefits of red wine. Does red wine protect moderate drinkers from heart disease better than other alcoholic beverages? Red wine contains substances called polyphenols that may change blood chemistry in a desirable way. This calls for a randomized comparative experiment. The subjects were healthy men aged 35 to 65. They were randomly assigned to drink red wine (9 subjects), drink white wine (9 subjects), drink white wine and also take polyphenols from red wine (6 subjects), take polyphenols alone (9 subjects), or drink vodka and lemonade (6 subjects).[11] Outline the design of the experiment. Use Table B, starting at line 107, or the *Simple Random Sample* applet to assign the 39 subjects to the 5 groups. (See Exercise 8.43 for directions on using the applet.)

8.30 Sickle cell disease, continued. Exercise 8.28 describes a medical study of a new treatment for sickle cell disease.

(a) Outline the design of this experiment.

(b) Use of a placebo is considered ethical if there is no effective standard treatment to give the control group. It might seem humane to give all the subjects hydroxyurea in the hope that it will help them. Explain clearly why this would not provide information about the effectiveness of the drug. (In fact, the experiment was stopped ahead of schedule because the hydroxyurea group had only half as many pain episodes as the control group. Ethical standards required stopping the experiment as soon as statistically significant evidence became available.)

8.31 Growing trees faster. The concentration of carbon dioxide (CO_2) in the atmosphere is increasing rapidly due to our use of fossil fuels. Because plants use CO_2 to fuel photosynthesis, more CO_2 may cause trees and other plants to grow faster. An elaborate apparatus allows researchers to pipe extra CO_2 to a 30-meter circle of forest. We want to compare the growth in base area of trees in treated and untreated areas to see if extra CO_2 does in fact increase growth. We can afford to treat three circular areas.[12]

(a) Describe the design of a completely randomized experiment using 6 well-separated 30-meter circular areas in a pine forest. Sketch the circles and carry out the randomization your design calls for.

(b) Areas within the forest may differ in soil fertility. Describe a matched pairs design using three pairs of circles that will reduce the

(Greg Ryan & Sally Beyer/Stock Connection/PictureQuest)

extra variation due to different fertility. Sketch the circles and carry out the randomization your design calls for.

8.32 Temperature and work performance. An industrial psychologist is interested in the effect of room temperature on the performance of tasks requiring manual dexterity. She chooses temperatures of 70° F and 90° F as treatments. The response variable is the number of correct insertions, during a 30-minute period, in a peg-and-hole apparatus that requires the use of both hands simultaneously. Each subject is trained on the apparatus and then asked to make as many insertions as possible in 30 minutes of continuous effort.

(a) Outline a completely randomized design to compare dexterity at 70° and 90°. Twenty subjects are available.

(b) Because individuals differ greatly in dexterity, the wide variation in individual scores may hide the systematic effect of temperature unless there are many subjects in each group. Describe in detail the design of a matched pairs experiment in which each subject serves as his or her own control.

8.33 Daytime running lights. Canada requires that cars be equipped with "daytime running lights," headlights that automatically come on at a low level when the car is started. Many manufacturers are now equipping cars sold in the United States with running lights. Will running lights reduce accidents by making cars more visible?

(a) Briefly discuss the design of an experiment to help answer this question. In particular, what response variables will you examine?

(b) Example 8.7 (page 209) discusses center brake lights. What cautions do you draw from that example that apply to an experiment on the effects of running lights?

8.34 Do antioxidants prevent cancer? People who eat lots of fruits and vegetables have lower rates of colon cancer than those who eat little of these foods. Fruits and vegetables are rich in "antioxidants" such as vitamins A, C, and E. Will taking antioxidants help prevent colon cancer? A clinical trial studied this question with 864 people who were at risk of colon cancer. The subjects were divided into four groups: daily beta-carotene, daily vitamins C and E, all three vitamins every day, and daily placebo. After four years, the researchers were surprised to find no significant difference in colon cancer among the groups.[13]

(a) What are the explanatory and response variables in this experiment?

(b) Outline the design of the experiment. Use your judgment in choosing the group sizes.

(c) Assign labels to the 864 subjects and use Table B, starting at line 118, to choose the *first* 5 subjects for the beta-carotene group.

(d) The study was double-blind. What does this mean?

(e) What does "no significant difference" mean in describing the outcome of the study?

(f) Suggest some lurking variables that could explain why people who eat lots of fruits and vegetables have lower rates of colon cancer. The experiment suggests that these variables, rather than the antioxidants, may be responsible for the observed benefits of fruits and vegetables.

8.35 **Comparing corn varieties.** New varieties of corn with altered amino acid content may have higher nutritional value than standard corn, which is low in the amino acid lysine. An experiment compares two new varieties, called opaque-2 and floury-2, with normal corn. The researchers mix corn-soybean meal diets using each type of corn at each of three protein levels: 12% protein, 16% protein, and 20% protein. They feed each diet to 10 one-day-old male chicks and record their weight gains after 21 days. The weight gain of the chicks is a measure of the nutritional value of their diet.

(a) What are the individuals and the response variable in this experiment?

(b) How many factors are there? How many treatments? Use a diagram like Figure 8.1 to describe the treatments. How many individuals does the experiment require?

(c) Use a diagram to describe a completely randomized design for this experiment. (You do not need to actually do the randomization.)

8.36 **An industrial experiment.** A chemical engineer is designing the production process for a new product. The chemical reaction that produces the product may have higher or lower yield, depending on the temperature and the stirring rate in the vessel in which the reaction takes place. The engineer decides to investigate the effects of all combinations of two temperatures (50° C and 60° C) and three stirring rates (60 rpm, 90 rpm, and 120 rpm) on the yield of the process. She will process two batches of the product at each combination of temperature and stirring rate. In Exercise 8.3 you identified the treatments.

(a) Outline in graphic form the design of an appropriate experiment.

(b) The randomization in this experiment determines the order in which batches of the product will be processed according to each treatment. Use Table B, starting at line 128, to carry out the randomization and state the result.

8.37 **McDonald's versus Wendy's.** Do consumers prefer the taste of a cheeseburger from McDonald's or from Wendy's in a blind test in which neither burger is identified? Describe briefly the design of a matched pairs experiment to investigate this question.

8.38 Athletes taking oxygen. We often see players on the sidelines of a football game inhaling oxygen. Their coaches think this will speed their recovery. We might measure recovery from intense exercise as follows: Have a football player run 100 yards three times in quick succession. Then allow three minutes to rest before running 100 yards again. Time the final run. Because players vary greatly in speed, you plan a matched pairs experiment using 25 football players as subjects. Discuss the design of such an experiment to investigate the effect of inhaling oxygen during the rest period.

8.39 An herb for depression? Does the herb Saint-John's-wort relieve major depression? Here are some excerpts from the report of a study of this issue.[14] The study concluded that the herb is no more effective than a placebo.

(a) "Design: Randomized, double-blind, placebo-controlled clinical trial...." Explain the meaning of each of the terms in this description.

(b) "Participants ... were randomly assigned to receive either Saint-John's-wort extract ($n = 98$) or placebo ($n = 102$).... The primary outcome measure was the rate of change in the Hamilton Rating Scale for Depression over the treatment period." Based on this information, use a diagram to outline the design of this clinical trial.

8.40 Reading a medical journal. The article in the *New England Journal of Medicine* that presents the final results of the Physicians' Health Study begins with these words: "The Physicians' Health Study is a randomized, double-blind, placebo-controlled trial designed to determine whether low-dose aspirin (325 mg every other day) decreases cardiovascular mortality and whether beta carotene reduces the incidence of cancer."[15] Doctors are expected to understand this. Explain to a doctor who knows no statistics what "randomized," "double-blind," and "placebo-controlled" mean.

Chapter 8 MEDIA EXERCISES

8.41 Medical journals. Randomized comparative experiments are so important in medicine that the *Journal of the American Medical Association* now regularly includes the words "a randomized controlled trial" in the title of articles to alert readers. The Web sites of major medical journals offer an endless supply of experiments as well as observational studies. Some sites are *British Medical Journal*, www.bmj.com/; *Journal of the American Medical Association*, jama.ama-assn.org/; *New England Journal of Medicine*, www.nejm.org. Go to one of these Web sites. From the contents of a recent journal, choose a paper describing an experiment on a topic that interests you.

(*JAMA* makes only abstracts and selected papers available without charge; *NEJM* makes papers available six months after publication; *BMJ* makes all papers available. An abstract is enough for this exercise.) Briefly describe the nature and findings of the study. *Quote* the abstract's description of the design of the experiment, and explain each statistical term in this description.

8.42 Testing a breakfast food. Because experimental randomization chooses SRSs of the subjects, we can use the *Simple Random Sample* applet here as well as for sampling problems. Example 8.4 (page 203) describes a randomized comparative experiment in which 30 rats are assigned at random to a treatment group of 15 and a control group of 15. Use the applet to choose the 15 rats for the treatment group. Which rats did you choose? The remaining 15 rats make up the control group.

8.43 Conserving energy. The *Simple Random Sample* applet allows you to randomly assign subjects to more than two groups. Example 8.5 (page 204) describes a randomized comparative experiment in which 60 houses are randomly assigned to three groups of 20.

(a) Use the applet to choose an SRS of 20 out of 60 houses to form the first group. Which houses are in this group?

(b) The "Population hopper" now contains the 40 houses that were not chosen, in scrambled order. Click "Sample" again to choose an SRS of 20 of these remaining houses to make up the second group. Which houses were chosen?

(c) The 20 houses remaining in the "Population hopper" form the third group. Which houses are these?

8.44 Randomization avoids bias. Suppose that the 15 even-numbered rats among the 30 rats available in the setting of Exercise 8.42 are (unknown to the experimenters) a fast-growing variety. We hope that these rats will be roughly equally distributed between the two groups. Use the *Simple Random Sample* applet to take 20 samples of size 15 from the 30 rats. (Be sure to click "Reset" after each sample.) Record the counts of even-numbered rats in each of your 20 samples. You see that there is considerable chance variation but no systematic bias in favor of one or the other group in assigning the fast-growing rats. Larger samples from a larger population will on the average do a better job of making the two groups equivalent.

8.45 Repairing knees. The EESEE story "Blinded Knee Doctors" describes an experiment on ways to reduce patient discomfort after knee surgery. Will a nonsteroidal anti-inflammatory drug (NSAID) reduce pain? Eighty-three patients were placed in three groups. Group A received the NSAID both before and after surgery. Group B was given a placebo before and the NSAID after. Group C received a placebo both before and after surgery. The patients recorded a pain score by answering questions one day after the surgery.

(a) Outline the design of this experiment. You do not need to do the randomization that your design requires.

(b) You read that "the patients, physicians, and physical therapists were blinded" during the study. What does this mean?

(c) You also read that "the pain scores for Group A were significantly lower than Group C but not significantly lower than Group B." What does this mean? What does this finding lead you to conclude about the use of NSAIDs?

(Matthias Kulka/CORBIS)

Introducing Probability

Why is probability, the science of chance behavior, needed to understand statistics, the science of data? Let's look at a typical sample survey.

EXAMPLE 9.1   Do you lotto?

What proportion of all adults bought a lottery ticket in the past 12 months? We don't know, but we do have the results of a Gallup Poll. Gallup took a sample of 1523 adults. It happens that 868 of the people in the sample bought tickets. The proportion who bought tickets was

$$\text{sample proportion} = \frac{868}{1523} = 0.57 \ \ (\text{that is, } 57\%)$$

Because all adults had the same chance to be among the chosen 1523, it seems reasonable to use this 57% as an estimate of the unknown proportion in the population. It's a *fact* that 57% of the sample bought lottery tickets—we know because Gallup asked them. We don't know what percent of all adults bought tickets, but we *estimate* that about 57% did. This is a basic move in statistics: use a result from a sample to estimate something about a population.

What if Gallup took a second random sample of 1523 adults? The new sample would have different people in it. It is almost certain that there would not be exactly 868 positive responses. That is, Gallup's estimate of the proportion of adults who bought a lottery ticket will vary from sample to sample. Could it happen that one random sample finds that 57% of adults recently bought a lottery ticket and a second random sample finds that only 37% had done so? Random samples eliminate *bias* from the act of choosing a sample, but they can still be wrong because of the *variability* that results when we choose at random.

If the variation when we take repeat samples from the same population is too great, we can't trust the results of any one sample.

This is where we need facts about probability to make progress in statistics. Because Gallup uses impersonal chance to choose its samples, the laws of probability govern the behavior of the samples. Gallup says that when the Gallup Poll takes a sample, the probability is 0.95 that the estimate from the sample comes within ±3 percentage points of the truth about the population of all adults. The first step toward understanding this statement is to understand what "probability 0.95" means. Our purpose in this chapter is to understand the language of probability, but without going into the mathematics of probability theory.

The idea of probability

To understand why we can trust random samples and randomized comparative experiments, we must look closely at chance behavior. The big fact that emerges is this: **chance behavior is unpredictable in the short run but has a regular and predictable pattern in the long run.**

Toss a coin, or choose an SRS. The result can't be predicted in advance, because the result will vary when you toss the coin or choose the sample repeatedly. But there is still a regular pattern in the results, a pattern that emerges clearly only after many repetitions. This remarkable fact is the basis for the idea of probability.

(Super Stock)

EXAMPLE 9.2 **Coin tossing**

When you toss a coin, there are only two possible outcomes, heads or tails. Figure 9.1 shows the results of tossing a coin 5000 times twice. For each number of tosses from 1 to 5000, we have plotted the proportion of those tosses that gave a head. Trial A (solid line) begins tail, head, tail, tail. You can see that the proportion of heads for Trial A starts at 0 on the first toss, rises to 0.5 when the second toss gives a head, then falls to 0.33 and 0.25 as we get two more tails. Trial B, on the other hand, starts with five straight heads, so the proportion of heads is 1 until the sixth toss.

The proportion of tosses that produce heads is quite variable at first. Trial A starts low and Trial B starts high. As we make more and more tosses, however, the proportion of heads for both trials gets close to 0.5 and stays there. If we made yet a third trial at tossing the coin a great many times, the proportion of heads would again settle down to 0.5 in the long run. We say that 0.5 is the *probability* of a head. The probability 0.5 appears as a horizontal line on the graph.

The *Probability* applet animates Figure 9.1. It allows you to choose the probability of a head and simulate any number of tosses of a coin with that probability. Experience shows that the proportion of heads gradually settles down close to the probability. Equally important, it also shows that the proportion in a small or moderate number of tosses can be far from the probability. Probability describes *only* what happens in the long run.

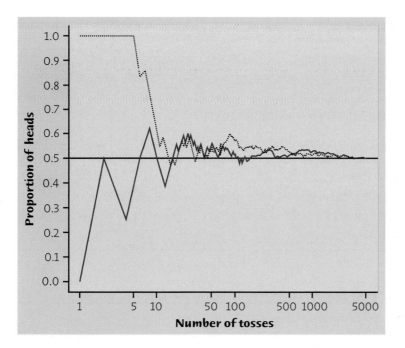

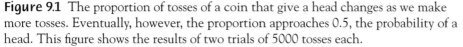

Figure 9.1 The proportion of tosses of a coin that give a head changes as we make more tosses. Eventually, however, the proportion approaches 0.5, the probability of a head. This figure shows the results of two trials of 5000 tosses each.

"Random" in statistics is not a synonym for "haphazard" but a description of a kind of order that emerges only in the long run. We often encounter the unpredictable side of randomness in our everyday experience, but we rarely see enough repetitions of the same random phenomenon to observe the long-term regularity that probability describes. You can see that regularity emerging in Figure 9.1. In the very long run, the proportion of tosses that give a head is 0.5. This is the intuitive idea of probability. Probability 0.5 means "occurs half the time in a very large number of trials."

We might suspect that a coin has probability 0.5 of coming up heads just because the coin has two sides. As Exercises 9.3 and 9.4 illustrate, such suspicions are not always correct. The idea of probability is empirical. That is, it is based on observation rather than theorizing. Probability describes what happens in very many trials, and we must actually observe many trials to pin down a probability. In the case of tossing a coin, some diligent people have in fact made thousands of tosses.

EXAMPLE 9.3   Some coin tossers

The French naturalist Count Buffon (1707–1788) tossed a coin 4040 times. Result: 2048 heads, or proportion 2048/4040 = 0.5069 for heads.

Around 1900, the English statistician Karl Pearson heroically tossed a coin 24,000 times. Result: 12,012 heads, a proportion of 0.5005.

While imprisoned by the Germans during World War II, the South African mathematician John Kerrich tossed a coin 10,000 times. Result: 5067 heads, a proportion of 0.5067.

RANDOMNESS AND PROBABILITY

We call a phenomenon **random** if individual outcomes are uncertain but there is nonetheless a regular distribution of outcomes in a large number of repetitions.

The **probability** of any outcome of a random phenomenon is the proportion of times the outcome would occur in a very long series of repetitions.

APPLY YOUR KNOWLEDGE

9.1   Deal a straight. A "straight" in poker is a hand of five cards whose ranks are consecutive when you arrange them in order, such as 8, 9, 10, Jack, Queen. You read in a book on poker that the probability of being dealt a straight is 4/1000. Explain carefully what this means.

9.2   Three of a kind. "Three of a kind" in poker is a hand of five cards that contains three of the same rank, such as three 7's, along with two other cards that are not 7's and that differ from each other. You read in a book on poker that the probability of being dealt three of a kind is 1/50. Explain why this does *not* mean that if you deal 50 hands, exactly one will contain three of a kind.

9.3   Nickels spinning. Hold a nickel upright on its edge under your forefinger on a hard surface, then snap it with your other forefinger so that it spins for some time before falling. Based on 50 spins, estimate the probability of heads.

9.4   Tossing a thumbtack. Toss a thumbtack on a hard surface 100 times. How many times did it land with the point up? What is the approximate probability of landing point up?

Thinking about randomness

That some things are random is an observed fact about the world. The outcome of a coin toss, the time between emissions of particles by a radioactive source, and the sexes of the next litter of lab rats are all random. So is the outcome of a random sample or a randomized experiment. Probability theory is the branch of mathematics that describes random behavior. Of course, we can never observe a probability exactly. We could always continue tossing the coin, for example. Mathematical probability is an idealization based on imagining what would happen in an indefinitely long series of trials.

Does God play dice?

Few things in the world are truly random in the sense that no amount of information will allow us to predict the outcome. We could in principle apply the laws of physics to a tossed coin, for example, and calculate whether it will land heads or tails. But randomness does rule events inside individual atoms. Albert Einstein didn't like this feature of the new quantum theory. "I shall never believe that God plays dice with the world," said the great scientist. Eighty years later, it appears that Einstein was wrong.

The best way to understand randomness is to observe random behavior—not only the long-run regularity but the unpredictable results of short runs. You can do this with physical devices, as in Exercises 9.3 and 9.4, but computer simulations (imitations) of random behavior allow faster exploration. Exercises 9.5 and 9.6 suggest some simulations of random behavior using the *Probability* applet. There are more applet simulations in the Media Exercises at the end of this chapter. As you explore randomness, remember:

independence

- You must have a long series of **independent** trials. That is, the outcome of one trial must not influence the outcome of any other. Imagine a crooked gambling house where the operator of a roulette wheel can stop it where she chooses—she can prevent the proportion of "red" from settling down to a fixed number. These trials are not independent.

- The idea of probability is empirical. Computer simulations start with given probabilities and imitate random behavior, but we can estimate a real-world probability only by actually observing many trials.

- Nonetheless, computer simulations are very useful because we need long runs of trials. In situations such as coin tossing, the proportion of an outcome often requires several hundred trials to settle down to the probability of that outcome. The kinds of physical random devices suggested in the exercises are too slow for this. Short runs give only rough estimates of a probability.

APPLY YOUR KNOWLEDGE

9.5 Random digits. The table of random digits (Table B) was produced by a random mechanism that gives each digit probability 0.1 of being a 0.

(a) What proportion of the first 50 digits in the table are 0s? This proportion is an estimate, based on 50 repetitions, of the true probability, which in this case is known to be 0.1.

(b) The *Probability* applet can imitate random digits. Set the probability of heads in the applet to 0.1. Check "Show true probability" to show this value on the graph. A head stands for a 0 in the random digit table and a tail stands for any other digit. Simulate 200 digits (40 at a time—don't click "Reset"). If you kept going forever, presumably you would get 10% heads. What was the result of your 200 tosses?

9.6 How many tosses to get a head? When we toss a coin, experience shows that the probability (long-term proportion) of a head is close to 1/2. Suppose now that we toss the coin repeatedly until we get a head. What is the probability that the first head comes up in an odd number of tosses (1, 3, 5, and so on)?

(a) Start with an actual coin. Repeat this experiment 10 times, and keep a record of the number of tosses needed to get a head on each of your 10 trials. Use your results to estimate the probability that the first head appears on an odd-numbered toss.

(b) Now use the *Probability* applet. Set the probability of heads to 0.5. Toss coins one at a time until the first head appears. Do this 50 times (click "Reset" after each trial). What is your estimate of the probability that the first head appears on an odd toss?

9.7 **Probability says . . . .** Probability is a measure of how likely an event is to occur. Match one of the probabilities that follow with each statement of likelihood given. (The probability is usually a more exact measure of likelihood than is the verbal statement.)

$$0 \quad 0.01 \quad 0.3 \quad 0.6 \quad 0.99 \quad 1$$

(a) This event is impossible. It can never occur.

(b) This event is certain. It will occur on every trial.

(c) This event is very unlikely, but it will occur once in a while in a long sequence of trials.

(d) This event will occur more often than not.

Probability models

Gamblers have known for centuries that the fall of coins, cards, and dice displays clear patterns in the long run. The idea of probability rests on the observed fact that the average result of many thousands of chance outcomes can be known with near certainty. How can we give a mathematical description of long-run regularity?

To see how to proceed, think first about a very simple random phenomenon, tossing a coin once. When we toss a coin, we cannot know the outcome in advance. What do we know? We are willing to say that the outcome will be either heads or tails. We believe that each of these outcomes has probability 1/2. This description of coin tossing has two parts:

- A list of possible outcomes.
- A probability for each outcome.

Such a description is the basis for all probability models. Here is the basic vocabulary we use.

PROBABILITY MODELS

The **sample space** *S* of a random phenomenon is the set of all possible outcomes.

An **event** is an outcome or a set of outcomes of a random phenomenon. That is, an event is a subset of the sample space.

A **probability model** is a mathematical description of a random phenomenon consisting of two parts: a sample space *S* and a way of assigning probabilities to events.

The sample space *S* can be very simple or very complex. When we toss a coin once, there are only two outcomes, heads and tails. The sample space is

(Super Stock)

Figure 9.2 The 36 possible outcomes in rolling two dice. If the dice are carefully made, all of these outcomes have the same probability.

$S = \{H, T\}$. When Gallup draws a random sample of 1523 adults, the sample space contains all possible choices of 1523 of the 210 million adults in the country. This S is extremely large. Each member of S is a possible sample, which explains the term *sample space*.

EXAMPLE 9.4   Rolling dice

Rolling two dice is a common way to lose money in casinos. There are 36 possible outcomes when we roll two dice and record the up-faces in order (first die, second die). Figure 9.2 displays these outcomes. They make up the sample space S. "Roll a 5" is an event, call it A, that contains four of these 36 outcomes:

Gamblers care only about the number of pips on the up-faces of the dice. The sample space for rolling two dice and counting the pips is

$$S = \{2, 3, 4, 5, 6, 7, 8, 9, 10, 11, 12\}$$

Comparing this S with Figure 9.2 reminds us that we can change S by changing the detailed description of the random phenomenon we are describing.

APPLY YOUR KNOWLEDGE

9.8 **Sample space.** Choose a student at random from a large statistics class. Describe a sample space S for each of the following. (In some cases you may have some freedom in specifying S.)

(a) Ask how much time the student spent studying during the past 24 hours.

(b) Ask how much money in coins (not bills) the student is carrying.

(c) Record the student's letter grade at the end of the course.

(d) Ask whether the student did or did not take a math class in each of the two previous years of school.

9.9 Sample space. In each of the following situations, describe a sample space S for the random phenomenon.

(a) A basketball player shoots four free throws. You record the sequence of hits and misses.

(b) A basketball player shoots four free throws. You record the number of baskets she makes.

9.10 Sample space. In each of the following situations, describe a sample space S for the random phenomenon. In some cases you have some freedom in specifying S, especially in setting the largest and the smallest value in S.

(a) A study of healthy adults measures the weights of adult women.

(b) The Physicians' Health Study asked 11,000 physicians to take an aspirin every other day and observed how many of them had a heart attack in a five-year period.

(c) In a test of a new package design, you drop a carton of a dozen eggs from a height of 1 foot and count the number of broken eggs.

Probability rules

The true probability of any outcome—say, "roll a 5 when we toss two dice"—can be found only by actually tossing two dice many times, and then only approximately. How then can we describe probability mathematically? Rather than try to give "correct" probabilities, we start by laying down facts that must be true for any assignment of probabilities. These facts follow from the idea of probability as "the long-run proportion of repetitions on which an event occurs."

1. **Any probability is a number between 0 and 1.** Any proportion is a number between 0 and 1, so any probability is also a number between 0 and 1. An event with probability 0 never occurs, and an event with probability 1 occurs on every trial. An event with probability 0.5 occurs in half the trials in the long run.

2. **All possible outcomes together must have probability 1.** Because some outcome must occur on every trial, the sum of the probabilities for all possible outcomes must be exactly 1.

3. **The probability that an event does not occur is 1 minus the probability that the event does occur.** If an event occurs in (say) 70% of all trials, it fails to occur in the other 30%. The probability that an event occurs and the probability that it does not occur always add to 100%, or 1.

4. **If two events have no outcomes in common, the probability that one or the other occurs is the sum of their individual probabilities.** If one event occurs in 40% of all trials, a different event occurs in 25% of all trials, and the two can never occur together, then one or the other occurs on 65% of all trials because 40% + 25% = 65%.

We can use mathematical notation to state Facts 1 to 4 more concisely. Capital letters near the beginning of the alphabet denote events. If A is any event, we write its probability as $P(A)$. Here are our probability facts in formal language. As you apply these rules, remember that they are just another form of intuitively true facts about long-run proportions.

PROBABILITY RULES

Rule 1. The probability $P(A)$ of any event A satisfies $0 \leq P(A) \leq 1$.

Rule 2. If S is the sample space in a probability model, then $P(S) = 1$.

Rule 3. For any event A,

$$P(A \text{ does not occur}) = 1 - P(A)$$

Rule 4. Two events A and B are **disjoint** if they have no outcomes in common and so can never occur simultaneously. If A and B are disjoint,

$$P(A \text{ or } B) = P(A) + P(B)$$

This is the **addition rule for disjoint events.**

The probability of rain is …

You work all week. Then it rains on the weekend. Can there really be a statistical truth behind our perception that the weather is against us? At least on the east coast of the United States, the answer is "Yes." Going back to 1946, it seems that Sundays receive 22% more precipitation than Mondays. The likely explanation is that the pollution from all those workday cars and trucks forms the seeds for raindrops—with just enough delay to cause rain on the weekend.

EXAMPLE 9.5 Benford's law

Faked numbers in tax returns, payment records, invoices, expense account claims, and many other settings often display patterns that aren't present in legitimate records. Some patterns, like too many round numbers, are obvious and easily avoided by a clever crook. Others are more subtle. It is a striking fact that the first digits of numbers in legitimate records often follow a distribution known as **Benford's law.** Here it is:[1]

First digit	1	2	3	4	5	6	7	8	9
Proportion	0.301	0.176	0.125	0.097	0.079	0.067	0.058	0.051	0.046

The probabilities assigned to first digits by Benford's law are all between 0 and 1. They add to 1 because the first digits listed make up the sample space S. (Note that a first digit can't be 0.)

Benford's law

The probability that a first digit is anything other than 1 is, by Rule 3,

$$P(\text{not a } 1) = 1 - P(\text{first digit is } 1)$$
$$= 1 - 0.301 = 0.699$$

That is, if 30.1% of first digits are 1s, the other 69.9% are not 1s.

"First digit is a 1" and "first digit is a 2" are disjoint events. So the addition rule (Rule 4) says

$$P(1 \text{ or } 2) = P(1) + P(2)$$
$$= 0.301 + 0.176 = 0.477$$

About 48% of first digits in data governed by Benford's law are 1s or 2s. Fraudulent records generally have many fewer 1s and 2s.

EXAMPLE 9.6 Probabilities for rolling dice

Figure 9.2 (page 229) displays the 36 possible outcomes of rolling two dice. What probabilities should we assign to these outcomes?

Casino dice are carefully made. Their spots are not hollowed out, which would give the faces different weights, but are filled with white plastic of the same density as the colored plastic of the body. For casino dice it is reasonable to assign the same probability to each of the 36 outcomes in Figure 9.2. Because all 36 outcomes together must have probability 1 (Rule 2), each outcome must have probability 1/36.

Gamblers are often interested in the sum of the pips on the up-faces. What is the probability of rolling a 5? Because the event "roll a 5" contains the four outcomes displayed in Example 9.4, the addition rule (Rule 4) says that its probability is

$$P(\text{roll a 5}) = P(\boxed{\cdot}\ \boxed{::}) + P(\boxed{\because}\ \boxed{:\cdot}) + P(\boxed{\cdot\cdot}\ \boxed{\because}) + P(\boxed{::}\ \boxed{\cdot})$$

$$= \frac{1}{36} + \frac{1}{36} + \frac{1}{36} + \frac{1}{36}$$

$$= \frac{4}{36} = 0.111$$

What about the probability of rolling a 7? In Figure 9.2 you will find six outcomes for which the sum of the pips is 7. The probability is 6/36, or about 0.167.

APPLY YOUR KNOWLEDGE

9.11 Rolling a soft 4. A "soft 4" in rolling two dice is a roll of 1 on one die and 3 on the other. If you roll two dice, what is the probability of rolling a soft 4? Of rolling a 4?

9.12 Preparing for the GMAT. A company that offers courses to prepare would-be MBA students for the Graduate Management Admission Test (GMAT) has the following information about its customers: 20% are currently undergraduate students in business; 15% are undergraduate students in other fields of study; 60% are college graduates who are currently employed; and 5% are college graduates who are not employed.

(a) Is this a legitimate assignment of probabilities to customer backgrounds? Why?

(b) What percent of customers are currently undergraduates?

9.13 Forests in Missouri. What happens to trees over a five-year period? A study lasting more than 30 years found these probabilities for a randomly chosen 12-inch-diameter tree in the Ozark Highlands of Missouri: stay in the 12-inch class, 0.686; move to the 14-inch class, 0.256.[2] The remaining trees die during the five-year period. What is the probability that a tree dies?

Assigning probabilities: finite number of outcomes

Examples 9.5 and 9.6 illustrate one way to assign probabilities to events: assign a probability to every individual outcome, then add these probabilities to find the probability of any event. If such an assignment is to satisfy the rules of probability, the probabilities of all the individual outcomes must sum to exactly 1.

PROBABILITIES IN A FINITE SAMPLE SPACE

Assign a probability to each individual outcome. These probabilities must be numbers between 0 and 1 and must have sum 1.

The probability of any event is the sum of the probabilities of the outcomes making up the event.

EXAMPLE 9.7   Random digits versus Benford's law

You might think that first digits in financial records are distributed "at random" among the digits 1 to 9. The 9 possible outcomes would then be equally likely. The sample space for a single digit is

$$S = \{1, 2, 3, 4, 5, 6, 7, 8, 9\}$$

Call a randomly chosen first digit X for short. The probability model for X is completely described by this table:

First digit X	1	2	3	4	5	6	7	8	9
Probability	1/9	1/9	1/9	1/9	1/9	1/9	1/9	1/9	1/9

The probability that a first digit is equal to or greater than 6 is

$$P(X \geq 6) = P(X = 6) + P(X = 7) + P(X = 8) + P(X = 9)$$
$$= \frac{1}{9} + \frac{1}{9} + \frac{1}{9} + \frac{1}{9} = \frac{4}{9} = 0.444$$

Note that this is not the same as the probability that a random digit is strictly greater than 6, $P(X > 6)$. The outcome $X = 6$ is included in the event $\{X \geq 6\}$ and is omitted from $\{X > 6\}$.

Compare this with the probability of the same event among first digits from financial records that obey Benford's law. A first digit V chosen from such records has this distribution:

First digit V	1	2	3	4	5	6	7	8	9
Probability	0.301	0.176	0.125	0.097	0.079	0.067	0.058	0.051	0.046

The probability that this digit is equal to or greater than 6 is

$$P(V \geq 6) = 0.067 + 0.058 + 0.051 + 0.046 = 0.222$$

Benford's law allows easy detection of phony financial records that have been based on randomly generated numbers. Such records tend to have too few first digits of 1 and 2 and too many first digits of 6 or greater.

Figure 9.3 Four assignments of probabilities to the six faces of a die, for Exercise 9.14.

Outcome	Probability			
	Model 1	Model 2	Model 3	Model 4
⚀	1/7	1/3	1/3	1
⚁	1/7	1/6	1/6	1
⚂	1/7	1/6	1/6	2
⚃	1/7	0	1/6	1
⚄	1/7	1/6	1/6	1
⚅	1/7	1/6	1/6	2

APPLY YOUR KNOWLEDGE

9.14 Rolling a die. Figure 9.3 displays several assignments of probabilities to the six faces of a die. We can learn which assignment is actually *accurate* for a particular die only by rolling the die many times. However, some of the assignments are not *legitimate* assignments of probability. That is, they do not obey the rules. Which are legitimate and which are not? In the case of the illegitimate models, explain what is wrong.

9.15 Benford's law. If first digits follow Benford's law (Examples 9.5 and 9.7), consider these events:

$$A = \{\text{first digit is 7 or greater}\}$$
$$B = \{\text{first digit is odd}\}$$

(a) What outcomes make up event A? What is $P(A)$?

(b) What outcomes make up event B? What is $P(B)$?

(c) What outcomes make up the event "A or B"? What is $P(A \text{ or } B)$? Why is this probability not equal to $P(A) + P(B)$?

9.16 Race and ethnicity. The 2000 census allowed each person to choose from a long list of races. That is, in the eyes of the Census Bureau, you belong to whatever race you say you belong to. "Hispanic/Latino" is a separate category; Hispanics may be of any race. If we choose a resident of the United States at random, the 2000 census gives these probabilities:

	Hispanic	Not Hispanic
Asian	0.000	0.036
Black	0.003	0.121
White	0.060	0.691
Other	0.062	0.027

(a) Verify that this is a legitimate assignment of probabilities.

(b) What is the probability that a randomly chosen American is Hispanic?

(c) Non-Hispanic whites are the historical majority in the United States. What is the probability that a randomly chosen American is not a member of this group?

Assigning probabilities: intervals of outcomes

A software random number generator is designed to produce a number between 0 and 1 chosen at random. It is only a slight idealization to consider *any* number in this range as a possible outcome. The sample space is then

$$S = \{\text{all numbers between 0 and 1}\}$$

Call the outcome of the random number generator Y for short. How can we assign probabilities to such events as $\{0.3 \leq Y \leq 0.7\}$? As in the case of selecting a random digit, we would like all possible outcomes to be equally likely. But we cannot assign probabilities to each individual value of Y and then add them, because there are infinitely many possible values.

We use a new way of assigning probabilities directly to events—as *areas under a density curve*. Any density curve has area exactly 1 underneath it, corresponding to total probability 1. We first met density curves as models for data in Chapter 3 (page 58).

EXAMPLE 9.8   Random numbers

The random number generator will spread its output uniformly across the entire interval from 0 to 1 as we allow it to generate a long sequence of numbers. The results of many trials are represented by the uniform density curve shown in Figure 9.4. This density curve has height 1 over the interval from 0 to 1. The area under the curve is 1, and the probability of any event is the area under the curve and above the event in question.

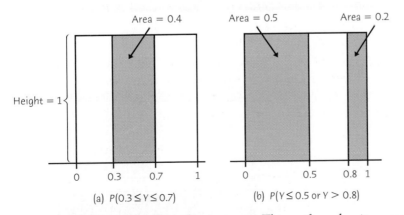

Figure 9.4 Probability as area under a density curve. This uniform density curve spreads probability evenly between 0 and 1.

As Figure 9.4(a) illustrates, the probability that the random number generator produces a number between 0.3 and 0.7 is

$$P(0.3 \leq Y \leq 0.7) = 0.4$$

because the area under the density curve and above the interval from 0.3 to 0.7 is 0.4. The height of the curve is 1 and the area of a rectangle is the product of height and length, so the probability of any interval of outcomes is just the length of the interval. Similarly,

$$P(Y \leq 0.5) = 0.5$$
$$P(Y > 0.8) = 0.2$$
$$P(Y \leq 0.5 \text{ or } Y > 0.8) = 0.7$$

Notice that the last event consists of two nonoverlapping intervals, so the total area above the event is found by adding two areas, as illustrated by Figure 9.4(b). This assignment of probabilities obeys all of our rules for probability.

APPLY YOUR KNOWLEDGE

9.17 Random numbers. Let X be a random number between 0 and 1 produced by the idealized random number generator described in Example 9.8 and Figure 9.4. Find the following probabilities:

(a) $P(0 \leq X \leq 0.4)$

(b) $P(0.4 \leq X \leq 1)$

(c) $P(0.3 \leq X \leq 0.5)$

9.18 The probability that $X = 0.5$. Let X be a random number between 0 and 1 produced by the idealized random number generator described in Example 9.8 and Figure 9.4. Find the following probabilities as areas under the density function:

(a) $P(X < 0.5)$

(b) $P(X \leq 0.5)$

(c) Compare your two answers: what must be the probability that X is exactly equal to 0.5? This is always true when we assign probabilities as areas under a density curve.

9.19 Adding random numbers. Generate two random numbers between 0 and 1 and take Y to be their sum. The sum Y can take any value between 0 and 2. The density curve of Y is the triangle shown in Figure 9.5.

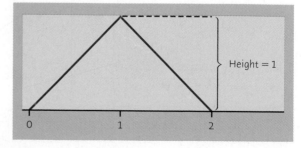

Figure 9.5 The density curve for the sum of two random numbers, for Exercise 9.19. This density curve spreads probability between 0 and 2.

(a) Verify by geometry that the area under this curve is 1.

(b) What is the probability that Y is less than 1? (Sketch the density curve, shade the area that represents the probability, then find that area. Do this for (c) also.)

(c) What is the probability that Y is less than 0.5?

Normal probability models

Any density curve can be used to assign probabilities. The density curves that are most familiar to us are the Normal curves. So **Normal distributions are probability models.** There is a close connection between a Normal distribution as an idealized description for data and a Normal probability model. If we look at the heights of all young women, we find that they closely follow the Normal distribution with mean $\mu = 64$ inches and standard deviation $\sigma = 2.7$ inches. That is a distribution for a large set of data. Now choose one young woman at random. Call her height X. If we repeat the random choice very many times, the distribution of values of X is the same Normal distribution.

EXAMPLE 9.9   The height of young women

What is the probability that a randomly chosen young woman has height between 68 and 70 inches?

The height X of the woman we choose has the $N(64, 2.7)$ distribution. Find the probability by standardizing and using Table A, the table of standard Normal probabilities. We will reserve capital Z for a standard Normal variable.

$$P(68 \leq X \leq 70) = P\left(\frac{68-64}{2.7} \leq \frac{X-64}{2.7} \leq \frac{70-64}{2.7}\right)$$
$$= P(1.48 \leq Z \leq 2.22)$$
$$= 0.9868 - 0.9306 = 0.0562$$

Figure 9.6 shows the areas under the standard Normal curve. The calculation is the same as those we did in Chapter 3. Only the language of probability is new.

Random variables

Examples 9.7 to 9.9 use a shorthand notation that is often convenient. In Example 9.9, we let X stand for the result of choosing a woman at random and measuring her height. We know that X would take a different value if we made another random choice. Because its value changes from one random choice to another, we call the height X a *random variable*.

RANDOM VARIABLE

A **random variable** is a variable whose value is a numerical outcome of a random phenomenon.

The **probability distribution** of a random variable X tells us what values X can take and how to assign probabilities to those values.

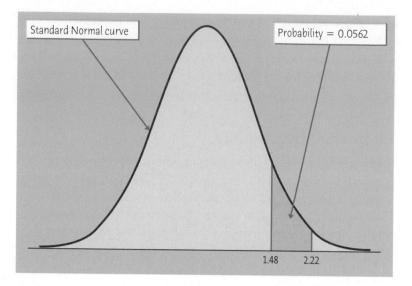

Figure 9.6 The probability in Example 9.9 as an area under the standard Normal curve.

We usually denote random variables by capital letters near the end of the alphabet, such as X or Y. Of course, the random variables of greatest interest to us are outcomes such as the mean $\bar{x}$ of a random sample, for which we will keep the familiar notation. There are two main types of random variables, corresponding to two ways of assigning probability: *discrete* and *continuous*.

EXAMPLE 9.10 Discrete and continuous random variables

The random variable X in Example 9.7 (page 233) has as its possible values the whole numbers {1, 2, 3, 4, 5, 6, 7, 8, 9}. The distribution of X assigns a probability to each of these outcomes. Random variables that have a finite list of possible outcomes are called **discrete**.

discrete random variable

Compare the random variable Y in Example 9.8 (page 235). The values of Y fill the entire interval of numbers between 0 and 1. The probability distribution of Y is given by its density curve, shown in Figure 9.4. Random variables that can take on any value in an interval, with probabilities given as areas under a density curve, are called **continuous**.

continuous random variable

APPLY YOUR KNOWLEDGE

9.20 **Grades in an accounting course.** North Carolina State University posts the grade distributions for its courses online.[3] Students in Accounting 210 in the spring 2001 semester received 18% A's, 32% B's, 34% C's, 9% D's, and 7% F's. Choose an Accounting 210 student at random. To "choose at random" means to give every student the same chance to be chosen. The student's grade on a four-point scale (with A = 4) is a discrete random variable X.

The value of X changes when we repeatedly choose students at random, but it is always one of 0, 1, 2, 3, or 4. Here is the distribution of X:

Value of X	0	1	2	3	4
Probability	0.07	0.09	0.34	0.32	0.18

(a) Say in words what the meaning of $P(X \geq 3)$ is. What is this probability?

(b) Write the event "the student got a grade poorer than C" in terms of values of the random variable X. What is the probability of this event?

9.21 Rolling the dice. Example 9.6 describes the assignment of probabilities to the 36 possible outcomes of tossing two dice. Take the random variable X to be the sum of the values on the two up-faces. Example 9.6 shows that $P(X = 5) = 4/36$ and that $P(X = 7) = 6/36$. Give the complete probability distribution of the discrete random variable X. That is, list the possible values of X and say what the probability of each value is. Start from the 36 outcomes displayed in Figure 9.2.

9.22 Iowa Test scores. The Normal distribution with mean $\mu = 6.8$ and standard deviation $\sigma = 1.6$ is a good description of the Iowa Test vocabulary scores of seventh-grade students in Gary, Indiana. If we use this model, the score Y of a randomly chosen student is a continuous random variable. Figure 3.1 (page 57) pictures the density function.

(a) Write the event "the student chosen has a score of 10 or higher" in terms of Y.

(b) Find the probability of this event.

Personal probability*

We began our discussion of probability with one idea: the probability of an outcome of a random phenomenon is the proportion of times that outcome would occur in a very long series of repetitions. This idea ties probability to actual outcomes. It allows us, for example, to estimate probabilities by simulating random phenomena. Yet we often meet another, quite different, idea of probability.

EXAMPLE 9.11   Joe and the Chicago Cubs

Joe sits staring into his beer as his favorite baseball team, the Chicago Cubs, loses another game. The Cubbies have some good young players, so let's ask Joe, "What's the chance that the Cubs will go to the World Series next year?" Joe brightens up. "Oh, about 10%," he says.

Does Joe assign probability 0.10 to the Cubs' appearing in the World Series? The outcome of next year's pennant race is certainly unpredictable, but we can't

*This section is optional.

What are the odds?

Gamblers often express chance in terms of *odds* rather than probability. Odds of A to B against an outcome means that the probability of that outcome is $B/(A + B)$. So "odds of 5 to 1" is another way of saying "probability 1/6." A probability is always between 0 and 1, but odds range from 0 to infinity. Although odds are mainly used in gambling, they give us a way to make very small probabilities clearer. "Odds of 999 to 1" may be easier to understand than "probability 0.001."

reasonably ask what would happen in many repetitions. Next year's baseball season will happen only once and will differ from all other seasons in players, weather, and many other ways. If probability measures "what would happen if we did this many times," Joe's 0.10 is not a probability. Probability is based on data about many repetitions of the same random phenomenon. Joe is giving us something else, his personal judgment.

Although Joe's 0.10 isn't a probability in our usual sense, it gives useful information about Joe's opinion. More seriously, a company asking, "How likely is it that building this plant will pay off within five years?" can't employ an idea of probability based on many repetitions of the same thing. The opinions of company officers and advisors are nonetheless useful information, and these opinions can be expressed in the language of probability. These are *personal probabilities*.

PERSONAL PROBABILITY

A **personal probability** of an outcome is a number between 0 and 1 that expresses an individual's judgment of how likely the outcome is.

Just as Rachel's opinion about the Cubs may differ from Joe's, the opinions of several company officers about the new plant may differ. Personal probabilities are indeed personal: they vary from person to person. Moreover, a personal probability can't be called right or wrong. If we say, "In the long run, this coin will come up heads 60% of the time," we can find out if we are right by actually tossing the coin several thousand times. If Joe says, "I think the Cubs have a 10% chance of going to the World Series next year," that's just Joe's opinion. Why think of personal probabilities as probabilities? Because any set of personal probabilities that makes sense obeys the same basic Rules 1 to 4 that describe any legitimate assignment of probabilities to events. If Joe thinks there's a 10% chance that the Cubs will go to the World Series, he must also think that there's a 90% chance that they won't go. There is just one set of rules of probability, even though we now have two interpretations of what probability means.

APPLY YOUR KNOWLEDGE

9.23 Will you have an accident? The probability that a randomly chosen driver will be involved in an accident in the next year is about 0.2. This is based on the proportion of millions of drivers who have accidents. "Accident" includes things like crumpling a fender in your own driveway, not just highway accidents.

(a) What do you think is your own probability of being in an accident in the next year? This is a personal probability.

(b) Give some reasons why your personal probability might be a more accurate prediction of your "true chance" of having an accident than the probability for a random driver.

(c) Almost everyone says their personal probability is lower than the random driver probability. Why do you think this is true?

Chapter 9 SUMMARY

A **random phenomenon** has outcomes that we cannot predict but that nonetheless have a regular distribution in very many repetitions.

The **probability** of an event is the proportion of times the event occurs in many repeated trials of a random phenomenon.

A **probability model** for a random phenomenon consists of a sample space S and an assignment of probabilities P.

The **sample space** S is the set of all possible outcomes of the random phenomenon. Sets of outcomes are called **events.** P assigns a number $P(A)$ to an event A as its probability.

Any assignment of probability must obey the rules that state the basic properties of probability:

1. $0 \leq P(A) \leq 1$ for any event A.
2. $P(S) = 1$.
3. For any event A, $P(A \text{ does not occur}) = 1 - P(A)$.
4. **Addition rule:** Events A and B are **disjoint** if they have no outcomes in common. If A and B are disjoint, then $P(A \text{ or } B) = P(A) + P(B)$.

When a sample space S contains finitely many possible values, a probability model assigns each of these values a probability between 0 and 1 such that the sum of all the probabilities is exactly 1. The probability of any event is the sum of the probabilities of all the values that make up the event.

A sample space can contain all values in some interval of numbers. A probability model assigns probabilities as **areas under a density curve.** The probability of any event is the area under the curve above the values that make up the event.

A **random variable** is a variable taking numerical values determined by the outcome of a random phenomenon. The **probability distribution** of a random variable X tells us what the possible values of X are and how probabilities are assigned to those values.

A random variable X and its distribution can be **discrete** or **continuous.** A **discrete random variable** has finitely many possible values. Its distribution gives the probability of each value. A **continuous random variable** takes all values in some interval of numbers. A **density curve** describes the probability distribution of a continuous random variable.

Chapter 9 EXERCISES

9.24 Nickels falling over. You may feel that it is obvious that the probability of a head in tossing a coin is about 1/2 because the coin has two faces.

Such opinions are not always correct. Exercise 9.3 asked you to spin a nickel rather than toss it—that changes the probability of a head. Now try another variation. Stand a nickel on edge on a hard, flat surface. Pound the surface with your hand so that the nickel falls over. What is the probability that it falls with heads upward? Make at least 50 trials to estimate the probability of a head.

9.25 Got money? Choose a student at random and record the number of dollars in bills (ignore change) that he or she is carrying. Give a reasonable sample space S for this random phenomenon. (We don't know the largest amount that a student could reasonably carry, so you will have to make a choice in stating the sample space.)

(Galen Rowell/CORBIS)

9.26 Land in Canada. Statistics Canada says that the land area of Canada is 9,094,000 square kilometers. Of this land, 4,176,000 square kilometers are forested. Choose a square kilometer of land in Canada at random.

(a) What is the probability that the area you choose is forested?

(b) What is the probability that it is not forested?

9.27 Deaths on the job. Government data on job-related deaths assign a single occupation for each such death that occurs in the United States. The data show that the probability is 0.134 that a randomly chosen death was agriculture-related, and 0.119 that it was manufacturing-related. What is the probability that a death was either agriculture-related or manufacturing-related? What is the probability that the death was related to some other occupation?

9.28 Foreign language study. Choose a student in grades 9 to 12 at random and ask if he or she is studying a language other than English. Here is the distribution of results:

Language	Spanish	French	German	All others	None
Probability	0.26	0.09	0.03	0.03	0.59

(a) Explain why this is a legitimate probability model.

(b) What is the probability that a randomly chosen student is studying a language other than English?

(c) What is the probability that a randomly chosen student is studying French, German, or Spanish?

9.29 Car colors. Choose a new car or light truck at random and note its color. Here are the probabilities of the most popular colors for vehicles made in North America in 2001:[4]

Color	Silver	White	Black	Blue	Red	Green
Probability	0.210	0.156	0.112	0.112	0.099	0.076

(a) What is the probability that the vehicle you choose has any color other than the six listed?

(b) What is the probability that a randomly chosen vehicle is either silver or white?

9.30 Colors of M&M's. If you draw an M&M candy at random from a bag of the candies, the candy you draw will have one of seven colors. The probability of drawing each color depends on the proportion of each color among all candies made. Here is the distribution for milk chocolate M&M's:[5]

Color	Purple	Yellow	Red	Orange	Brown	Green	Blue
Probability	0.2	0.2	0.2	0.1	0.1	0.1	?

(a) What must be the probability of drawing a blue candy?

(b) What is the probability that you do not draw a brown candy?

(c) What is the probability that the candy you draw is either yellow, orange, or red?

9.31 More M&M's. Caramel M&M candies are equally likely to be brown, red, yellow, green, orange, or blue. What is the probability for each color?

9.32 Legitimate probabilities? In each of the following situations, state whether or not the given assignment of probabilities to individual outcomes is legitimate, that is, satisfies the rules of probability. If not, give specific reasons for your answer.

(a) When a coin is spun, $P(H) = 0.55$ and $P(T) = 0.45$.

(b) When two coins are tossed, $P(HH) = 0.4$, $P(HT) = 0.4$, $P(TH) = 0.4$, and $P(TT) = 0.4$.

(c) When a die is rolled, the number of spots on the up-face has $P(1) = 1/2$, $P(4) = 1/6$, $P(5) = 1/6$, and $P(6) = 1/6$.

9.33 Who goes to Paris? Abby, Deborah, Mei-Ling, Sam, and Roberto work in a firm's public relations office. Their employer must choose two of them to attend a conference in Paris. To avoid unfairness, the choice will be made by drawing two names from a hat. (This is an SRS of size 2.)

(a) Write down all possible choices of two of the five names. This is the sample space.

(b) The random drawing makes all choices equally likely. What is the probability of each choice?

(c) What is the probability that Mei-Ling is chosen?

(d) What is the probability that neither of the two men (Sam and Roberto) is chosen?

(Neil Beer/Photo Disc/PictureQuest)

9.34 Roulette. A roulette wheel has 38 slots, numbered 0, 00, and 1 to 36. The slots 0 and 00 are colored green, 18 of the others are red, and 18 are black. The dealer spins the wheel and at the same time rolls a small ball along the wheel in the opposite direction. The wheel is carefully balanced so that the ball is equally likely to land in any slot when the wheel slows. Gamblers can bet on various combinations of numbers and colors.

(a) What is the probability of any one of the 38 possible outcomes? Explain your answer.

(b) If you bet on "red," you win if the ball lands in a red slot. What is the probability of winning?

(c) The slot numbers are laid out on a board on which gamblers place their bets. One column of numbers on the board contains all multiples of 3, that is, 3, 6, 9,..., 36. You place a "column bet" that wins if any of these numbers comes up. What is your probability of winning?

9.35 Birth order. A couple plans to have three children. There are 8 possible arrangements of girls and boys. For example, GGB means the first two children are girls and the third child is a boy. All 8 arrangements are (approximately) equally likely.

(a) Write down all 8 arrangements of the sexes of three children. What is the probability of any one of these arrangements?

(b) Let X be the number of girls the couple has. What is the probability that $X = 2$?

(c) Starting from your work in (a), find the distribution of X. That is, what values can X take, and what are the probabilities for each value?

9.36 Living in San Jose. Let the random variable X be the number of rooms in a randomly chosen owner-occupied housing unit in San Jose, California. Here is the distribution of X:[6]

Rooms X	1	2	3	4	5	6	7	8	9	10
Probability	0.003	0.002	0.023	0.104	0.210	0.224	0.197	0.149	0.053	0.035

(a) Is the random variable X discrete or continuous? Why?

(b) Express "the unit has no more than 2 rooms" in terms of X. What is the probability of this event?

(c) Express the event $\{X > 5\}$ in words. What is its probability?

9.37 Living in San Jose, continued. The previous exercise gives the distribution of the number of rooms in owner-occupied housing in San Jose, California. Here is the distribution for rented housing:

Rooms Y	1	2	3	4	5	6	7	8	9	10
Probability	0.008	0.027	0.287	0.363	0.164	0.093	0.039	0.013	0.003	0.003

What are the most important differences between the distributions of the random variables X and Y (owner-occupied and rented housing)? Compare at least two probabilities for X and Y to justify your answer.

9.38 How many cars in the garage? Choose an American household at random and let the random variable X be the number of cars (including SUVs and light trucks) they own. Here is the probability model if we ignore the few households that own more than 5 cars:

Number of cars X	0	1	2	3	4	5
Probability	0.09	0.36	0.35	0.13	0.05	0.02

(a) Verify that this is a legitimate discrete distribution.

(b) Say in words what the event $\{X \geq 1\}$ is. Find $P(X \geq 1)$.

(c) A housing company builds houses with two-car garages. What percent of households have more cars than the garage can hold?

9.39 Unusual dice. Nonstandard dice can produce interesting distributions of outcomes. You have two balanced, six-sided dice. One is a standard die, with faces having 1, 2, 3, 4, 5, and 6 spots. The other die has three faces with 0 spots and three faces with 6 spots. Find the probability distribution for the total number of spots Y on the up-faces when you roll these two dice. (Hint: Start with a picture like Figure 9.2 for the possible up-faces. Label the three 0 faces on the second die 0a, 0b, 0c in your picture, and similarly distinguish the three 6 faces.)

9.40 Random numbers. Many random number generators allow users to specify the range of the random numbers to be produced. Suppose that you specify that the random number Y can take any value between 0 and 2. Then the density curve of the outcomes has constant height between 0 and 2, and height 0 elsewhere.

(a) Is the random variable Y discrete or continuous? Why?

(b) What is the height of the density curve between 0 and 2? Draw a graph of the density curve.

(c) Use your graph from (b) and the fact that probability is area under the curve to find $P(Y \leq 1)$.

9.41 Polling women. Suppose that 47% of all adult women think they do not get enough time for themselves. An opinion poll interviews 1025 randomly chosen women and records the sample proportion who don't feel they get enough time for themselves. Call this proportion V. It will vary from sample to sample if the poll is repeated. The distribution of the random variable V is approximately Normal with mean 0.47 and standard deviation about 0.016. Sketch this Normal curve and use it to answer the following questions.

(a) The truth about the population is 0.47. In what range will the middle 95% of all sample results fall?

(b) What is the probability that the poll gets a sample in which fewer than 45% say they do not get enough time for themselves?

9.42 More random numbers. Find these probabilities as areas under the density curve you sketched in Exercise 9.40.

(a) $P(0.5 < Y < 1.3)$.

(b) $P(Y \geq 0.8)$.

9.43 NAEP math scores. Scores on the National Assessment of Educational Progress 12th-grade mathematics test for the year 2000 were approximately Normal with mean 300 points (out of 500 possible) and standard deviation 35 points. Let Y stand for the score of a randomly chosen student. Express each of the following events in terms of Y and use the 68–95–99.7 rule to give the approximate probability.

(a) The student has a score above 300.

(b) The student's score is above 370.

Figure 9.7 A tetrahedron. Exercises 9.44 and 9.46 ask about probabilities for dice having this shape.

9.44 Tetrahedral dice. Psychologists sometimes use tetrahedral dice to study our intuition about chance behavior. A tetrahedron (Figure 9.7) is a pyramid with 4 faces, each a triangle with all sides equal in length. Label the 4 faces of a tetrahedral die with 1, 2, 3, and 4 spots. Give a probability model for rolling such a die and recording the number of spots on the down-face. Explain why you think your model is at least close to correct.

9.45 Playing Pick 4. The Pick 4 games in many state lotteries announce a four-digit winning number each day. The winning number is essentially a four-digit group from a table of random digits. You win if your choice matches the winning digits. Suppose your chosen number is 5974.

(a) What is the probability that your number matches the winning number exactly?

(b) What is the probability that your number matches the digits in the winning number *in any order*?

9.46 More tetrahedral dice. Tetrahedral dice are described in Exercise 9.44. Give a probability model for rolling two such dice. That is, write down all possible outcomes and give a probability to each. (Figure 9.2 may help you.) What is the probability that the sum of the down-faces is 5?

9.47 Playing Pick 4, continued. The Wisconsin version of Pick 4 pays out $5000 on a $1 bet if your number matches the winning number exactly. It pays $200 on a $1 bet if the digits in your number match those of the winning number in any order. You choose which of these two bets to make. On the average over many bets, your winnings will be

mean amount won = payout amount × probability of winning

What are the mean payout amounts for these two bets? Is one of the two bets a better choice?

9.48 An edge in Pick 4. Exercise 9.45 describes Pick 4 lottery games. Some states (New Jersey, for example) use the "pari-mutuel system" in which the total winnings are divided among all players who matched the winning digits. That suggests a way to get an edge. Suppose you choose to try to match the winning number exactly.

(a) The winning number might be, for example, either 2873 or 8888. Explain why these two outcomes have exactly the same probability.

(b) It is likely that fewer people will choose one of these numbers than the other, because it "doesn't look random." You prefer the less popular number because you will win more if fewer people share a winning number. Which of these two numbers do you prefer?

Chapter 9 MEDIA EXERCISES

9.49 What probability doesn't say. The idea of probability is that the *proportion* of heads in many tosses of a balanced coin eventually gets close to 0.5. But does the actual *count* of heads get close to one-half the number of tosses? Let's find out. Set the "Probability of heads" in the *Probability* applet to 0.5 and the number of tosses to 40. You can extend the number of tosses by clicking "Toss" again to get 40 more. Don't click "Reset" during this exercise.

(a) After 40 tosses, what is the proportion of heads? What is the count of heads? What is the difference between the count of heads and 20 (one-half the number of tosses)?

(b) Keep going to 120 tosses. Again record the proportion and count of heads and the difference between the count and 60 (half the number of tosses).

(c) Keep going. Stop at 240 tosses and again at 480 tosses to record the same facts. Although it may take a long time, the laws of probability say that the proportion of heads will always get close to 0.5 and also that the difference between the count of heads and half the number of tosses will always grow without limit.

9.50 Shaq's free throws. The basketball player Shaquille O'Neal makes about half of his free throws over an entire season. Use the *Probability* applet or software to simulate 100 free throws shot independently by a player who has probability 0.5 of making each shot. (In most software, the key phrase to look for is "Bernoulli trials." This is the technical term for independent trials with Yes/No outcomes. Our outcomes here are "Hit" and "Miss.")

(a) What percent of the 100 shots did he hit?

(b) Examine the sequence of hits and misses. How long was the longest run of shots made? Of shots missed? (Sequences of random outcomes often show runs longer than our intuition thinks likely.)

9.51 Simulating an opinion poll. A recent opinion poll showed that about 65% of the American public have a favorable opinion of the software company Microsoft. Suppose that this is exactly true. Choosing a person at random then has probability 0.65 of getting one who has a favorable opinion of Microsoft. Use the *Probability* applet or your statistical software to simulate choosing many people independently. (In most software, the key phrase to look for is "Bernoulli trials." This is the technical term for independent trials with Yes/No outcomes. Our outcomes here are "Favorable" or not.)

(a) Simulate drawing 20 people, then 80 people, then 320 people. What proportion has a favorable opinion of Microsoft in each case? We expect (but because of chance variation we can't be sure) that the proportion will be closer to 0.65 in longer runs of trials.

(b) Simulate drawing 20 people 10 times and record the percents in each trial who have a favorable opinion of Microsoft. Then simulate drawing 320 people 10 times and again record the 10 percents. Which set of 10 results is less variable? We expect the results of 320 trials to be more predictable (less variable) than the results of 20 trials. That is "long-run regularity" showing itself.

Sampling Distributions

What is the mean income of households in the United States? The government's Current Population Survey contacted a sample of 55,000 households in March 2002. Their mean income in 2001 was $x = \$58{,}208$.[1] That $58,208 describes the sample, but we use it to estimate the mean income of all households. This is an example of statistical inference: we use information from a sample to infer something about a wider population.

Because the results of random samples and randomized comparative experiments include an element of chance, we can't guarantee that our inferences are correct. What we can do is guarantee that the methods we use usually give correct answers. We will see that the reasoning of statistical inference rests on asking, "How often would this method give a correct answer if I used it very many times?" If our data come from random sampling or randomized comparative experiments, the laws of probability answer the question "What would happen if we did this many times?" This chapter presents some facts about probability that help answer this question.

Parameters and statistics

Once we begin to use sample data to draw conclusions about a wider population, we must take care to keep straight whether a number describes a sample or a population. Here is the vocabulary we use.

PARAMETER, STATISTIC

A **parameter** is a number that describes the population. In statistical practice, the value of a parameter is not known because we cannot examine the entire population.

A **statistic** is a number that can be computed from the sample data without making use of any unknown parameters. In practice, we often use a statistic to estimate an unknown parameter.

EXAMPLE 10.1   Household income

The mean income of the sample of households contacted by the Current Population Survey was $\overline{x} = \$58,208$. The number $58,208 is a *statistic* because it describes this one Current Population Survey sample. The population that the poll wants to draw conclusions about is all 110 million U.S. households. The *parameter* of interest is the mean income of all of these households. We don't know the value of this parameter.

Remember: **s**tatistics come from **s**amples, and **p**arameters come from **p**opulations. As long as we were just doing data analysis, the distinction between population and sample was not important. Now, however, it is essential. The notation we use must reflect this distinction. We write μ (the Greek letter mu) for the **mean of a population.** This is a fixed parameter that is unknown when we use a sample for inference. The **mean of the sample** is the familiar $\overline{x}$, the average of the observations in the sample. This is a statistic that would almost certainly take a different value if we chose another sample from the same population. The sample mean $\overline{x}$ from a sample or an experiment is an estimate of the mean μ of the underlying population.

population mean μ
sample mean $\overline{x}$

APPLY YOUR KNOWLEDGE

State whether each boldface number in Exercises 10.1 to 10.3 is a parameter or a statistic.

10.1 Apartment rents. Your local newspaper contains a large number of advertisements for unfurnished one-bedroom apartments. You choose 10 at random and calculate that their mean monthly rent is **$540** and that the standard deviation of their rents is **$80.**

10.2 Indianapolis voters. Voter registration records show that **68%** of all voters in Indianapolis are registered as Republicans. To test a random-digit dialing device, you use the device to call 150 randomly chosen residential telephones in Indianapolis. Of the registered voters contacted, **73%** are registered Republicans.

10.3 Inspecting bearings. A carload lot of ball bearings has mean diameter **2.5003** centimeters (cm). This is within the specifications for acceptance of the lot by the purchaser. By chance, an inspector chooses

/ 5

100 bearings from the lot that have mean diameter **2.5009** cm. Because this is outside the specified limits, the lot is mistakenly rejected.

Statistical estimation and the law of large numbers

Statistical inference uses sample data to draw conclusions about the entire population. Because good samples are chosen randomly, statistics such as $\overline{x}$ are random variables. We can describe the behavior of a sample statistic by a probability model that answers the question "What would happen if we did this many times?" Here is an example that will lead us toward the probability ideas most important for statistical inference.

EXAMPLE 10.2   Does this wine smell bad?

Sulfur compounds such as dimethyl sulfide (DMS) are sometimes present in wine. DMS causes "off-odors" in wine, so winemakers want to know the odor threshold, the lowest concentration of DMS that the human nose can detect. Different people have different thresholds, so we start by asking about the mean threshold μ in the population of all adults. The number μ is a parameter that describes this population.

To estimate μ, we present tasters with both natural wine and the same wine spiked with DMS at different concentrations to find the lowest concentration at which they identify the spiked wine. Here are the odor thresholds (measured in micrograms of DMS per liter of wine) for 10 randomly chosen subjects:

$$28 \quad 40 \quad 28 \quad 33 \quad 20 \quad 31 \quad 29 \quad 27 \quad 17 \quad 21$$

The mean threshold for these subjects is $\overline{x} = 27.4$. It seems reasonable to use the sample result $\overline{x} = 27.4$ to estimate the unknown μ. An SRS should fairly represent the population, so the mean $\overline{x}$ of the sample should be somewhere near the mean μ of the population. Of course, we don't expect $\overline{x}$ to be exactly equal to μ, and we realize that if we choose another SRS, the luck of the draw will probably produce a different $\overline{x}$.

If $\overline{x}$ is rarely exactly right and varies from sample to sample, why is it nonetheless a reasonable estimate of the population mean μ? Here is one answer: if we keep on taking larger and larger samples, the statistic $\overline{x}$ is *guaranteed* to get closer and closer to the parameter μ. We have the comfort of knowing that if we can afford to keep on measuring more subjects, eventually we will estimate the mean odor threshold of all adults very accurately. This remarkable fact is called the *law of large numbers*. It is remarkable because it holds for *any* population, not just for some special class such as Normal distributions.

LAW OF LARGE NUMBERS

Draw observations at random from any population with finite mean μ. As the number of observations drawn increases, the mean $\overline{x}$ of the observed values gets closer and closer to the mean μ of the population.

High-tech gambling

There are more than 450,000 slot machines in the United States. Once upon a time, you put in a coin and pulled the lever to spin three wheels, each with 20 symbols. No longer. Now the machines are video games with flashy graphics and outcomes produced by random number generators. Machines can accept many coins at once, can pay off on a bewildering variety of outcomes, and can be networked to allow common jackpots. Gamblers still search for systems, but in the long run the law of large numbers guarantees the house its 5% profit.

The law of large numbers can be proved mathematically starting from the basic laws of probability. The behavior of $\bar{x}$ is similar to the idea of probability. In the long run, the *proportion* of outcomes taking any value gets close to the probability of that value, and the *average* outcome gets close to the population mean. Figure 9.1 (page 225) shows how proportions approach probability in one example. Here is an example of how sample means approach the population mean.

EXAMPLE 10.3   **The law of large numbers in action**

In fact, the distribution of odor thresholds among all adults has mean 25. The mean $\mu = 25$ is the true value of the parameter we seek to estimate. Figure 10.1 shows how the sample mean $\bar{x}$ of an SRS drawn from this population changes as we add more subjects to our sample.

The first subject in Example 10.2 had threshold 28, so the line in Figure 10.1 starts there. The mean for the first two subjects is

$$\bar{x} = \frac{28 + 40}{2} = 34$$

This is the second point on the graph. At first, the graph shows that the mean of the sample changes as we take more observations. Eventually, however, the mean of the observations gets close to the population mean $\mu = 25$ and settles down at that value.

If we started over, again choosing people at random from the population, we would get a different path from left to right in Figure 10.1. The law of large numbers

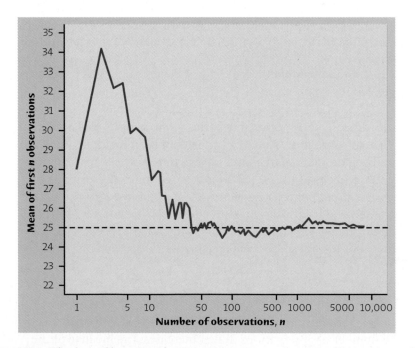

Figure 10.1 The law of large numbers in action: as we take more observations, the sample mean $\bar{x}$ always approaches the mean μ of the population.

says that whatever path we get will always settle down at 25 as we draw more and more people.

The *Law of Large Numbers* applet animates Figure 10.1 in a different setting. You can use the applet to watch $\bar{x}$ change as you average more observations until it eventually settles down at the mean μ. The law of large numbers is the foundation of such business enterprises as gambling casinos and insurance companies. The winnings (or losses) of a gambler on a few plays are uncertain—that's why gambling is exciting. In Figure 10.1, the mean of even 100 observations is not yet very close to μ. It is only *in the long run* that the mean outcome is predictable. The house plays tens of thousands of times. So the house, unlike individual gamblers, can count on the long-run regularity described by the law of large numbers. The average winnings of the house on tens of thousands of plays will be very close to the mean of the distribution of winnings. Needless to say, this mean guarantees the house a profit. That's why gambling can be a business.

APPLY YOUR KNOWLEDGE

10.4 Means in action. Figure 10.1 shows how the mean of n observations behaves as we keep adding more observations to those already in hand. The first 10 observations are given in Example 10.2. Demonstrate that you grasp the idea of Figure 10.1: find the mean of the first one, then two, three, four, and five of these observations and plot the successive means against n. Verify that your plot agrees with the first part of the plot in Figure 10.1.

10.5 Insurance. The idea of insurance is that we all face risks that are unlikely but carry high cost. Think of a fire destroying your home. Insurance spreads the risk: we all pay a small amount, and the insurance policy pays a large amount to those few of us whose homes burn down. An insurance company looks at the records for millions of homeowners and sees that the mean loss from fire in a year is $\mu = \$250$ per person. (Most of us have no loss, but a few lose their homes. The $250 is the average loss.) The company plans to sell fire insurance for $250 plus enough to cover its costs and profit. Explain clearly why it would be unwise to sell only 12 policies. Then explain why selling thousands of such policies is a safe business.

Sampling distributions

The law of large numbers assures us that if we measure enough subjects, the statistic $\bar{x}$ will eventually get very close to the unknown parameter μ. But our study in Example 10.2 had just 10 subjects. What can we say about $\bar{x}$ from 10 subjects as an estimate of μ? We ask: "What would happen if we took many samples of 10 subjects from this population?" Here's how to answer this question:

- Take a large number of samples of size 10 from the population.
- Calculate the sample mean $\bar{x}$ for each sample.
- Make a histogram of the values of $\bar{x}$.
- Examine the distribution displayed in the histogram for shape, center, and spread, as well as outliers or other deviations.

simulation

In practice it is too expensive to take many samples from a large population such as all adult U.S. residents. But we can imitate many samples by using software. Using software to imitate chance behavior is called **simulation.**

EXAMPLE 10.4   **What would happen in many samples?**

Extensive studies have found that the DMS odor threshold of adults follows roughly a Normal distribution with mean $\mu = 25$ micrograms per liter and standard deviation $\sigma = 7$ micrograms per liter. With this information, we can simulate many repetitions of Example 10.2 with different subjects drawn at random from the population.

Figure 10.2 illustrates the process of choosing many samples and finding the sample mean threshold $\bar{x}$ for each one. Follow the flow of the figure from the population at the left, to choosing an SRS and finding the $\bar{x}$ for this sample, to collecting together the $\bar{x}$'s from many samples. The first sample has $\bar{x} = 26.42$. The second sample contains a different 10 people, with $\bar{x} = 24.28$, and so on. The histogram at the right of the figure shows the distribution of the values of $\bar{x}$ from 1000 separate SRSs of size 10. This histogram displays the *sampling distribution* of the statistic $\bar{x}$.

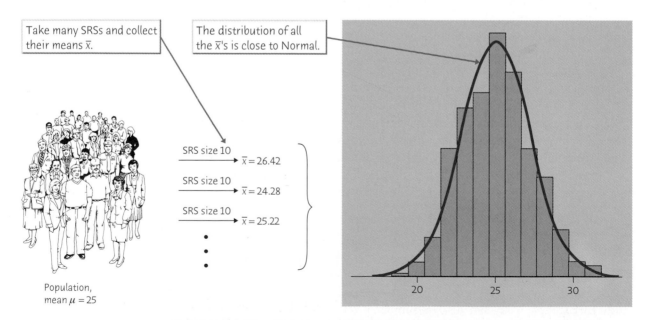

Figure 10.2 The idea of a sampling distribution: take many samples from the same population, collect the $\bar{x}$'s from all the samples, and display the distribution of the $\bar{x}$'s. The histogram shows the results of 1000 samples.

> ## SAMPLING DISTRIBUTION
>
> The **sampling distribution** of a statistic is the distribution of values taken by the statistic in all possible samples of the same size from the same population.

Strictly speaking, the sampling distribution is the ideal pattern that would emerge if we looked at all possible samples of size 10 from our population. A distribution obtained from a fixed number of trials, like the 1000 trials in Figure 10.2, is only an approximation to the sampling distribution. One of the uses of probability theory in statistics is to obtain exact sampling distributions without simulation. The interpretation of a sampling distribution is the same, however, whether we obtain it by simulation or by the mathematics of probability.

We can use the tools of data analysis to describe any distribution. Let's apply those tools to Figure 10.2. What can we say about the shape, center, and spread of this distribution?

- **Shape:** It looks Normal! Detailed examination confirms that the distribution of $\bar{x}$ from many samples does have a distribution that is very close to Normal.

- **Center:** The mean of the 1000 $\bar{x}$'s is 24.95. That is, the distribution is centered very close to the population mean $\mu = 25$.

- **Spread:** The standard deviation of the 1000 $\bar{x}$'s is 2.217, notably smaller than the standard deviation $\sigma = 7$ of the population of individual subjects.

Although these results describe just one simulation of a sampling distribution, they reflect facts that are true whenever we use random sampling.

APPLY YOUR KNOWLEDGE

10.6 **Generating a sampling distribution.** Let us illustrate the idea of a sampling distribution in the case of a very small sample from a very small population. The population is the scores of 10 students on an exam:

Student	0	1	2	3	4	5	6	7	8	9
Score	82	62	80	58	72	73	65	66	74	62

The parameter of interest is the mean score μ in this population. The sample is an SRS of size $n = 4$ drawn from the population. Because the students are labeled 0 to 9, a single random digit from Table B chooses one student for the sample.

Rigging the lottery

We have all seen televised lottery drawings in which numbered balls bubble about and are randomly popped out by air pressure. How might we rig such a drawing? In 1980, when the Pennsylvania lottery used just three balls, a drawing was rigged by the host and several stagehands. They injected paint into all balls bearing 8 of the 10 digits. This weighed them down and guaranteed that all three balls for the winning number would have the remaining 2 digits. The perps then bet on all combinations of these digits. When 6-6-6 popped out, they won $1.2 million. Yes, they were caught.

unbiased estimator

(a) Find the mean of the 10 scores in the population. This is the population mean μ.

(b) Use Table B to draw an SRS of size 4 from this population. Write the four scores in your sample and calculate the mean $\bar{x}$ of the sample scores. This statistic is an estimate of μ.

(c) Repeat this process 10 times using different parts of Table B. Make a histogram of the 10 values of $\bar{x}$. You are constructing the sampling distribution of $\bar{x}$. Is the center of your histogram close to μ?

The sampling distribution of $\bar{x}$

Figure 10.2 suggests that when we choose many SRSs from a population, the sampling distribution of the sample means is centered at the mean of the original population and less spread out than the distribution of individual observations. Here are the facts.

MEAN AND STANDARD DEVIATION OF A SAMPLE MEAN[2]

Suppose that $\bar{x}$ is the mean of an SRS of size n drawn from a large population with mean μ and standard deviation σ. Then the **mean** of the sampling distribution of $\bar{x}$ is μ and its **standard deviation** is $\sigma/\sqrt{n}$.

Both the mean and the standard deviation of the sampling distribution of $\bar{x}$ have important implications for statistical inference.

• The mean of the statistic $\bar{x}$ is always the same as the mean μ of the population. That is, the sampling distribution of $\bar{x}$ is centered at μ. In repeated sampling, $\bar{x}$ will sometimes fall above the true value of the parameter μ and sometimes below, but there is no systematic tendency to overestimate or underestimate the parameter. This makes the idea of lack of bias in the sense of "no favoritism" more precise. Because the mean of $\bar{x}$ is equal to μ, we say that the statistic $\bar{x}$ is an **unbiased estimator** of the parameter μ.

• An unbiased estimator is "correct on the average" in many samples. How close the estimator falls to the parameter in most samples is determined by the spread of the sampling distribution. If individual observations have standard deviation σ, then sample means $\bar{x}$ from samples of size n have standard deviation $\sigma/\sqrt{n}$. **Averages are less variable than individual observations.**

We have described the center and spread of the sampling distribution of a sample mean $\bar{x}$, but not its shape. The shape of the distribution of $\bar{x}$ depends on the shape of the population. Here is one important case: if measurements in the population follow a Normal distribution, then so does the sample mean.

SAMPLING DISTRIBUTION OF A SAMPLE MEAN

If individual observations have the $N(\mu, \sigma)$ distribution, then the sample mean $\bar{x}$ of n independent observations has the $N(\mu, \sigma/\sqrt{n})$ distribution.

EXAMPLE 10.5   Population distribution, sampling distribution

If we measure the DMS odor thresholds of individual adults, the values follow the Normal distribution with mean $\mu = 25$ micrograms per liter and standard deviation $\sigma = 7$ micrograms per liter. We call this the **population distribution** because it shows how measurements vary within the population.

Take many SRSs of size 10 from this population and find the sample mean $\bar{x}$ for each sample, as in Figure 10.2. The *sampling distribution* describes how the values of $\bar{x}$ vary among samples. That sampling distribution is also Normal, with mean $\mu = 25$ and standard deviation

$$\frac{\sigma}{\sqrt{n}} = \frac{7}{\sqrt{10}} = 2.2136$$

Figure 10.3 contrasts these two Normal distributions. Both are centered at the population mean, but sample means are much less variable than individual observations.

Not only is the standard deviation of the distribution of $\bar{x}$ smaller than the standard deviation of individual observations, but it gets smaller as we take

[handwritten notes in margin:]
$M = 25$
$\sigma = 7$

population distribution

$M = 25$

$\dfrac{\sigma}{\sqrt{n}} = \dfrac{7}{\sqrt{10}} = 2.2136$

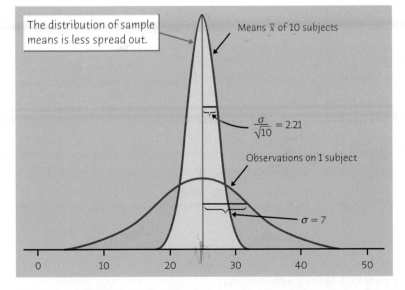

Figure 10.3 The distribution of single observations compared with the distribution of the means $\bar{x}$ of 10 observations. Averages are less variable than individual observations.

larger samples. **The results of large samples are less variable than the results of small samples.** If n is large, the standard deviation of $\bar{x}$ is small, and almost all samples will give values of $\bar{x}$ that lie very close to the true parameter μ. That is, the sample mean from a large sample can be trusted to estimate the population mean accurately. Notice, however, that the standard deviation of the sampling distribution gets smaller only at the rate $\sqrt{n}$. To cut the standard deviation of $\bar{x}$ in half, we must take four times as many observations, not just twice as many.

APPLY YOUR KNOWLEDGE

10.7 Measurements in the lab. Juan makes a measurement in a chemistry laboratory and records the result in his lab report. The standard deviation of students' lab measurements is $\sigma = 10$ milligrams. Juan repeats the measurement 3 times and records the mean $\bar{x}$ of his 3 measurements.

(a) What is the standard deviation of Juan's mean result? (That is, if Juan kept on making 3 measurements and averaging them, what would be the standard deviation of all his $\bar{x}$'s?)

(b) How many times must Juan repeat the measurement to reduce the standard deviation of $\bar{x}$ to 5? Explain to someone who knows no statistics the advantage of reporting the average of several measurements rather than the result of a single measurement.

10.8 Measuring blood cholesterol. The distribution of blood cholesterol level in the population of young men aged 20 to 34 years is close to Normal, with mean $\mu = 188$ milligrams per deciliter (mg/dl) and standard deviation $\sigma = 41$ mg/dl. You measure the cholesterol level of 100 young men chosen at random and calculate the mean $\bar{x}$.

(a) If you did this many times, what would be the mean and standard deviation of the distribution of all the $\bar{x}$-values?

(b) What is the probability that your sample has mean $\bar{x}$ less than 180?

10.9 National math scores. The scores of 12th-grade students on the National Assessment of Educational Progress year 2000 mathematics test have a distribution that is approximately Normal with mean $\mu = 300$ and standard deviation $\sigma = 35$.

(a) Choose one 12th-grader at random. What is the probability that his or her score is higher than 300? Higher than 335?

(b) Now choose an SRS of four 12th-graders. What is the probability that their mean score is higher than 300? Higher than 335?

The central limit theorem

The facts about the mean and standard deviation of $\bar{x}$ are true no matter what the shape of the population distribution may be. But what is the shape of the sampling distribution when the population distribution is not Normal? It is a remarkable fact that as the sample size increases, the distribution of $\bar{x}$ changes shape: it looks less like that of the population and more like a Normal distribution. When the sample is large enough, the distribution of $\bar{x}$ is very close to Normal. This is true no matter what shape the population distribution has, as long as the population has a finite standard deviation σ. This famous fact of probability theory is called the *central limit theorem*. It is much more useful than the fact that the distribution of $\bar{x}$ is exactly Normal if the population is exactly Normal.

CENTRAL LIMIT THEOREM

Draw an SRS of size n from any population with mean μ and finite standard deviation σ. When n is large, the sampling distribution of the sample mean $\bar{x}$ is approximately Normal:

$$\bar{x} \text{ is approximately } N\left(\mu, \frac{\sigma}{\sqrt{n}}\right)$$

More general versions of the central limit theorem say that the distribution of a sum or average of many small random quantities is close to Normal. This is true even if the quantities are not independent (as long as they are not too highly correlated) and even if they have different distributions (as long as no one random quantity is so large that it dominates the others). The central limit theorem suggests why the Normal distributions are common models for observed data. Any variable that is a sum of many small influences will have approximately a Normal distribution.

How large a sample size n is needed for $\bar{x}$ to be close to Normal depends on the population distribution. More observations are required if the shape of the population distribution is far from Normal.

EXAMPLE 10.6   The central limit theorem in action

Figure 10.4 shows how the central limit theorem works for a very non-Normal population. Figure 10.4(a) displays the density curve of a single observation, that is, of the population. The distribution is strongly right-skewed, and the most probable outcomes are near 0. The mean μ of this distribution is 1, and its standard deviation σ is also 1. This particular distribution is called an *exponential distribution*. Exponential distributions are used as models for the lifetime in service of electronic components and for the time required to serve a customer or repair a machine.

Figures 10.4(b), (c), and (d) are the density curves of the sample means of 2, 10, and 25 observations from this population. As n increases, the shape becomes more Normal. The mean remains at $\mu = 1$, and the standard deviation decreases, taking the value $1/\sqrt{n}$. The density curve for 10 observations is still somewhat skewed to the right but already resembles a Normal curve having $\mu = 1$ and $\sigma = 1/\sqrt{10} = 0.32$. The density curve for $n = 25$ is yet more Normal. The contrast between the shapes of the population distribution and of the distribution of the mean of 10 or 25 observations is striking.

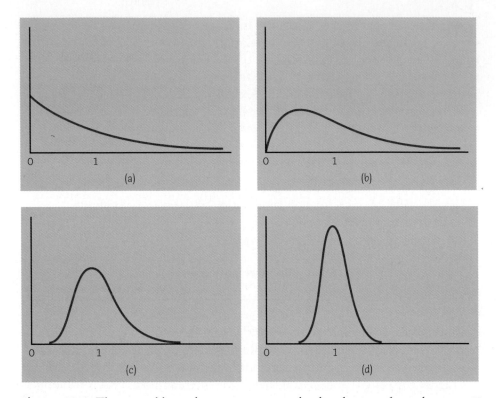

Figure 10.4 The central limit theorem in action: the distribution of sample means $\bar{x}$ from a strongly non-Normal population becomes more Normal as the sample size increases. (a) The distribution of one observation. (b) The distribution of $\bar{x}$ for 2 observations. (c) The distribution of $\bar{x}$ for 10 observations. (d) The distribution of $\bar{x}$ for 25 observations.

The central limit theorem allows us to use Normal probability calculations to answer questions about sample means from many observations even when the population distribution is not Normal.

EXAMPLE 10.7   Maintaining air conditioners

The time X that a technician requires to perform preventive maintenance on an air-conditioning unit is governed by the exponential distribution whose density curve appears in Figure 10.4(a). The mean time is $\mu = 1$ hour and the standard deviation is $\sigma = 1$ hour. Your company operates 70 of these units. What is the probability that their average maintenance time exceeds 50 minutes?

The central limit theorem says that the sample mean time $\bar{x}$ (in hours) spent working on 70 units has approximately the Normal distribution with mean equal to the population mean $\mu = 1$ hour and standard deviation

$$\frac{\sigma}{\sqrt{70}} = \frac{1}{\sqrt{70}} = 0.12 \text{ hour}$$

The distribution of $\bar{x}$ is therefore approximately $N(1, 0.12)$. This Normal curve is the solid curve in Figure 10.5.

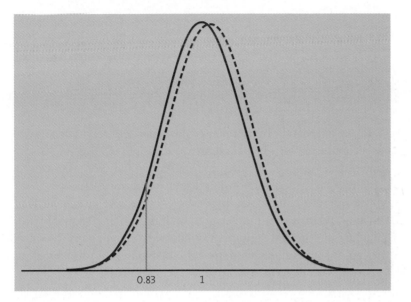

Figure 10.5 The exact distribution (*dashed*) and the Normal approximation from the central limit theorem (*solid*) for the average time needed to maintain an air conditioner, for Example 10.7.

Because 50 minutes is 50/60 of an hour, or 0.83 hour, the probability we want is $P(\bar{x} > 0.83)$. A Normal distribution calculation gives this probability as 0.9222. This is the area to the right of 0.83 under the solid Normal curve in Figure 10.5.

Using more mathematics, we could start with the exponential distribution and find the actual density curve of $\bar{x}$ for 70 observations. This is the dashed curve in Figure 10.5. You can see that the solid Normal curve is a good approximation. The exactly correct probability is the area under the dashed density curve. It is 0.9294. The central limit theorem Normal approximation is off by only about 0.007.

APPLY YOUR KNOWLEDGE

10.10 More on insurance. The insurance company of Exercise 10.5 sees that in the entire population of homeowners, the mean loss from fire is $\mu = \$250$ and the standard deviation of the loss is $\sigma = \$1000$. The distribution of losses is strongly right-skewed: many policies have $0 loss, but a few have large losses. If the company sells 10,000 policies, what is the approximate probability that the average loss will be greater than $275?

10.11 ACT scores. The scores of students on the ACT college entrance examination in 2001 had mean $\mu = 21.0$ and standard deviation $\sigma = 4.7$. The distribution of scores is only roughly Normal.

   (a) What is the approximate probability that a single student randomly chosen from all those taking the test scores 23 or higher?

(b) Now take an SRS of 50 students who took the test. What are the mean and standard deviation of the sample mean score $\overline{x}$ of these 50 students? What is the approximate probability that the mean score $\overline{x}$ of these students is 23 or higher?

(c) Which of your two Normal probability calculations in (a) and (b) is more accurate? Why?

10.12 **Flaws in carpets.** The number of flaws per square yard in a type of carpet material varies with mean 1.6 flaws per square yard and standard deviation 1.2 flaws per square yard. The population distribution cannot be Normal, because a count takes only whole-number values. An inspector samples 200 square yards of the material, records the number of flaws found in each square yard, and calculates $\overline{x}$, the mean number of flaws per square yard inspected. Use the central limit theorem to find the approximate probability that the mean number of flaws exceeds 2 per square yard.

Statistical process control*

The sampling distribution of the sample mean $\overline{x}$ has an immediate application to *statistical process control*. The goal of statistical process control is to make a process stable over time and then keep it stable unless planned changes are made. You might want, for example, to keep your weight constant over time. A manufacturer of machine parts wants the critical dimensions to be the same for all parts. "Constant over time" and "the same for all" are not realistic requirements. They ignore the fact that *all processes have variation.* Your weight fluctuates from day to day; the critical dimension of a machined part varies a bit from item to item; the time to process a college admission application is not the same for all applications. Variation occurs in even the most precisely made product due to small changes in the raw material, the adjustment of the machine, the behavior of the operator, and even the temperature in the plant. Because variation is always present, we can't expect to hold a variable exactly constant over time. The statistical description of stability over time requires that the *pattern of variation* remain stable, not that there be no variation in the variable measured.

> ### STATISTICAL CONTROL
>
> A variable that continues to be described by the same distribution when observed over time is said to be in statistical control, or simply **in control.**
>
> **Control charts** are statistical tools that monitor a process and alert us when the process has been disturbed so that it is now **out of control.** This is a signal to find and correct the cause of the disturbance.

*The rest of this chapter is optional. A more complete treatment of process control appears in Companion Chapter 24.

Control charts work by distinguishing the natural variation in the process from the additional variation that suggests that the process has changed. A control chart sounds an alarm when it sees too much variation. The most common application of control charts is to monitor the performance of an industrial process. The same methods, however, can be used to check the stability of quantities as varied as the ratings of a television show, the level of ozone in the atmosphere, and the gas mileage of your car. Control charts combine graphical and numerical descriptions of data with use of sampling distributions. They therefore provide a natural bridge between exploratory data analysis and formal statistical inference.

x̄ charts*

The population in the control chart setting is all items that would be produced by the process if it ran on forever in its present state. The items actually produced form samples from this population. We generally speak of the process rather than the population. Choose a quantitative variable, such as a diameter or a voltage, that is an important measure of the quality of an item. The process mean μ is the long-term average value of this variable; μ describes the center or aim of the process. The sample mean $\bar{x}$ of several items estimates μ and helps us judge whether the center of the process has moved away from its proper value. The most common control chart plots the means $\bar{x}$ of small samples taken from the process at regular intervals over time.

When you first apply control charts to a process, the process may not be in control. Even if it is in control, you don't yet understand its behavior. You will have to collect data from the process, establish control by uncovering and removing special causes, and then set up control charts to maintain control. To quickly explain the main ideas, we'll assume that you know the usual behavior of the process from long experience. Here are the conditions we will work with.

PROCESS-MONITORING CONDITIONS

Measure a quantitative variable x that has a **Normal distribution.** The process has been operating in control for a long period, so that we know the **process mean** μ and the **process standard deviation** σ that describe the distribution of x as long as the process remains in control.

EXAMPLE 10.8   *Making computer monitors*

A manufacturer of computer monitors must control the tension on the mesh of fine wires that lies behind the surface of the viewing screen. Too much tension will tear the mesh, and too little will allow wrinkles. Tension is measured by an electrical device with output readings in millivolts (mV). The proper tension is 275 mV. Some variation is always present in the production process. When the process is operating properly, the standard deviation of the tension readings is $\sigma = 43$ mV.

TABLE 10.1 Twenty control chart samples of mesh tension

Sample	Tension measurements				$\bar{x}$
1	234.5	272.3	234.5	272.3	253.4
2	311.1	305.8	238.5	286.2	285.4
3	247.1	205.3	252.6	316.1	255.3
4	215.4	296.8	274.2	256.8	260.8
5	327.9	247.2	283.3	232.6	272.7
6	304.3	236.3	201.8	238.5	245.2
7	268.9	276.2	275.6	240.2	265.2
8	282.1	247.7	259.8	272.8	265.6
9	260.8	259.9	247.9	345.3	278.5
10	329.3	231.8	307.2	273.4	285.4
11	266.4	249.7	231.5	265.2	253.2
12	168.8	330.9	333.6	318.3	287.9
13	349.9	334.2	292.3	301.5	319.5
14	235.2	283.1	245.9	263.1	256.8
15	257.3	218.4	296.2	275.2	261.8
16	235.1	252.7	300.6	297.6	271.5
17	286.3	293.8	236.2	275.3	272.9
18	328.1	272.6	329.7	260.1	297.6
19	316.4	287.4	373.0	286.0	315.7
20	296.8	350.5	280.6	259.8	296.9

The operator measures the tension on a sample of 4 monitors each hour. The mean $\bar{x}$ of each sample estimates the mean tension μ for the process at the time of the sample. Table 10.1 shows the samples and their means $\bar{x}$'s for 20 consecutive hours of production. How can we use these data to keep the process in control?

A time plot helps us see whether or not the process is stable. Figure 10.6 is a plot of the successive sample means against the order in which the samples were taken. We have plotted each sample mean from the table against its sample number. For example, the mean of the first sample is 253.4 mV, and this is the value plotted for sample 1. Because the target value for the process mean is $\mu = 275$ mV, we draw a *center line* at that level across the plot. How much variation about this center line do we expect to see? For example, are samples 13 and 19 so high that they suggest lack of control?

The tension measurements are roughly Normal, and the central limit theorem effect implies that sample means will be closer to Normal than individual measurements. So the $\bar{x}$-values from successive samples will follow a Normal distribution. If the standard deviation of the individual screens remains at $\sigma = 43$ mV, the standard deviation of $\bar{x}$ from 4 screens is

$$\frac{\sigma}{\sqrt{n}} = \frac{43}{\sqrt{4}} = 21.5 \text{ mV}$$

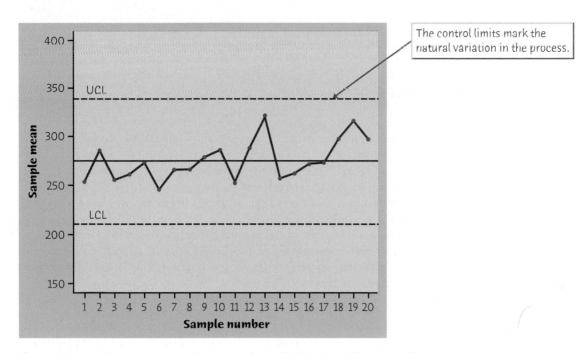

Figure 10.6 $\bar{x}$ chart for the mesh tension data of Table 10.1. The control limits are labeled UCL for upper control limit and LCL for lower control limit. No points lie outside the control limits.

As long as the mean remains at its target value $\mu = 275$ mV, the 99.7 part of the 68–95–99.7 rule says that almost all values of $\bar{x}$ will lie between

$$\mu - 3\frac{\sigma}{\sqrt{n}} = 275 - (3)(21.5) = 210.5$$

$$\mu + 3\frac{\sigma}{\sqrt{n}} = 275 + (3)(21.5) = 339.5$$

We therefore draw dashed *control limits* at these two levels on the plot. The control limits show the extent of the natural variation of $\bar{x}$-values when the process is in control. We now have an $\bar{x}$ *control chart*.

$\bar{x}$ CONTROL CHART

To evaluate the control of a process with given standards μ and σ, make an $\bar{x}$ **control chart** as follows:

- Plot the means $\bar{x}$ of regular samples of size n against time.
- Draw a horizontal **center line** at μ.
- Draw horizontal **control limits** at $\mu \pm 3\sigma/\sqrt{n}$.

Any $\bar{x}$ that does not fall between the control limits is evidence that the process is out of control.

EXAMPLE 10.9   Interpreting $\bar{x}$ charts

Figure 10.6 is a typical $\bar{x}$ chart for a process in control. The means of the 20 samples do vary, but all lie within the range of variation marked out by the control limits. We are seeing the natural variation of a stable process.

Figures 10.7 and 10.8 illustrate two ways in which the process can go out of control. In Figure 10.7, the process was disturbed sometime between sample 12 and sample 13. As a result, the mean tension for sample 13 falls above the upper control limit. It is common practice to mark all out-of-control points with an "x" to call attention to them. A search for the cause begins as soon as we see a point out of control. Investigation finds that the mounting of the tension-measuring device has slipped, resulting in readings that are too high. When the problem is corrected, samples 14 to 20 are again in control.

Figure 10.8 shows the effect of a steady upward drift in the process center, starting at sample 11. You see that some time elapses before the $\bar{x}$ for sample 18 is out of control. The one-point-out signal works better for detecting sudden large disturbances than for detecting slow drifts in a process.

$\bar{x}$ chart   An $\bar{x}$ control chart is often called simply an **$\bar{x}$ chart.** Because a control chart is a warning device, it is not necessary that our probability calculations be exactly correct. Approximate Normality is good enough. In that same spirit, control charts use the approximate Normal probabilities given by the 68–95–99.7 rule rather than more exact calculations using Table A.

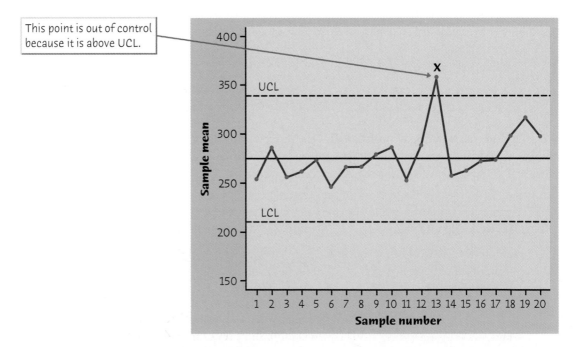

Figure 10.7 This $\bar{x}$ chart is identical to that in Figure 10.6, except that a disturbance has driven $\bar{x}$ for sample 13 above the upper control limit. The out-of-control point is marked with an x.

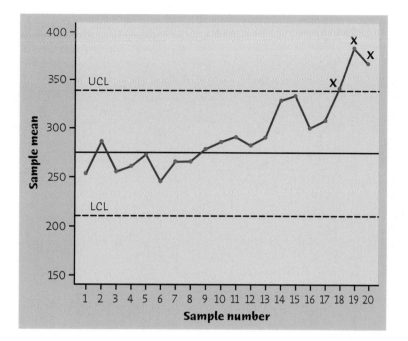

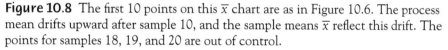

Figure 10.8 The first 10 points on this $\bar{x}$ chart are as in Figure 10.6. The process mean drifts upward after sample 10, and the sample means $\bar{x}$ reflect this drift. The points for samples 18, 19, and 20 are out of control.

APPLY YOUR KNOWLEDGE

10.13 **Auto thermostats.** A maker of auto air conditioners checks a sample of 4 thermostatic controls from each hour's production. The thermostats are set at 75° F and then placed in a chamber where the temperature is raised gradually. The temperature at which the thermostat turns on the air conditioner is recorded. The process mean should be $\mu = 75°$. Past experience indicates that the response temperature of properly adjusted thermostats varies with $\sigma = 0.5°$. The mean response temperature x for each hour's sample is plotted on an $\bar{x}$ control chart. Calculate the center line and control limits for this chart.

10.14 **Tablet hardness.** A pharmaceutical manufacturer forms tablets by compressing a granular material that contains the active ingredient and various fillers. The hardness of a sample from each lot of tablets is measured in order to control the compression process. The process has been operating in control with mean at the target value $\mu = 11.5$ and estimated standard deviation $\sigma = 0.2$. Table 10.2 gives three sets of data, each representing $\bar{x}$ for 20 successive samples of $n = 4$ tablets. One set remains in control at the target value. In a second set, the process mean μ shifts suddenly to a new value. In a third, the process mean drifts gradually.

TABLE 10.2 Three sets of $\bar{x}$'s from 20 samples of size 4

Sample	Data set A	Data set B	Data set C
1	11.602	11.627	11.495
2	11.547	11.613	11.475
3	11.312	11.493	11.465
4	11.449	11.602	11.497
5	11.401	11.360	11.573
6	11.608	11.374	11.563
7	11.471	11.592	11.321
8	11.453	11.458	11.533
9	11.446	11.552	11.486
10	11.522	11.463	11.502
11	11.664	11.383	11.534
12	11.823	11.715	11.624
13	11.629	11.485	11.629
14	11.602	11.509	11.575
15	11.756	11.429	11.730
16	11.707	11.477	11.680
17	11.612	11.570	11.729
18	11.628	11.623	11.704
19	11.603	11.472	12.052
20	11.816	11.531	11.905

(a) What are the center line and control limits for an $\bar{x}$ chart for this process?

(b) Draw a separate $\bar{x}$ chart for each of the three data sets. Mark any points that are beyond the control limits.

(c) Based on your work in (b) and the appearance of the control charts, which set of data comes from a process that is in control? In which case does the process mean shift suddenly and at about which sample do you think that the mean changed? Finally, in which case does the mean drift gradually?

Thinking about process control*

The purpose of a control chart is not to ensure good quality by inspecting most of the items produced. **Control charts focus on the process itself rather than on the individual products.** By checking the process at regular intervals, we can detect disturbances and correct them quickly. Statistical process control achieves high quality at a lower cost than inspecting all of the products. Small samples of 4 or 5 items are usually adequate for process control.

A process that is in control is stable over time, but stability alone does not guarantee good quality. The natural variation in the process may be so large that many of the products are unsatisfactory. Nonetheless, establishing control brings a number of advantages.

- In order to assess whether the process quality is satisfactory, we must observe the process operating in control free of breakdowns and other disturbances.

- A process in control is predictable. We can predict both the quantity and the quality of items produced.

- When a process is in control, we can easily see the effects of attempts to improve the process, which are not hidden by the unpredictable variation that characterizes lack of statistical control.

A process in control is doing as well as it can in its present state. If the process is not capable of producing adequate quality even when undisturbed, we must make some major change in the process, such as installing new machines or retraining the operators.

If the process is kept in control, we know what to expect in the finished product. The process mean μ and standard deviation σ remain stable over time, so (assuming Normal variation) the 99.7 part of the 68–95–99.7 rule tells us that almost all measurements on individual products will lie in the range $\mu \pm 3\sigma$. These are sometimes called the **natural tolerances** for the product. Be careful to distinguish $\mu \pm 3\sigma$, the range we expect for *individual measurements*, from the $\bar{x}$ chart control limits $\mu \pm 3\sigma/\sqrt{n}$, which mark off the expected range of *sample means*.

natural tolerances

EXAMPLE 10.10   Natural tolerances for mesh tension

The process of setting the mesh tension on computer monitors has been operating in control. The $\bar{x}$ chart is based on $\mu = 275$ mV and $\sigma = 43$ mV.

We are therefore confident that almost all individual monitors will have mesh tension between

$$\mu \pm 3\sigma = 275 \pm (3)(43) = 275 \pm 129$$

We expect mesh tension measurements to vary between 146 mV and 404 mV. You see that the spread of individual measurements is wider than the spread of sample means used for the control limits of the $\bar{x}$ chart.

APPLY YOUR KNOWLEDGE

10.15 Auto thermostats. Exercise 10.13 describes a process that produces auto thermostats. The temperature that turns on the thermostats has remained in control with mean $\mu = 75°$ F and standard deviation $\sigma = 0.5°$. What are the natural tolerances for this temperature? What range covers the middle 95% of response temperatures?

Chapter 10 SUMMARY

When we want information about the **population mean** μ for some variable, we often take an SRS and use the **sample mean** $\bar{x}$ to estimate the unknown parameter μ.

The **law of large numbers** states that the actually observed mean outcome $\bar{x}$ must approach the mean μ of the population as the number of observations increases.

The **sampling distribution** of $\bar{x}$ describes how the statistic $\bar{x}$ varies in all possible samples of the same size from the same population.

The **mean** of the sampling distribution is μ, so that $\bar{x}$ is an **unbiased estimator** of μ.

The **standard deviation** of the sampling distribution of $\bar{x}$ is $\sigma/\sqrt{n}$ for an SRS of size n if the population has standard deviation σ. That is, averages are less variable than individual observations.

If the population has a Normal distribution, so does $\bar{x}$.

The **central limit theorem** states that for large n the sampling distribution of $\bar{x}$ is approximately Normal for any population with finite standard deviation σ. That is, averages are more Normal than individual observations. We can use the $N(\mu, \sigma/\sqrt{n})$ distribution to calculate approximate probabilities for events involving $\bar{x}$.

All processes have variation. If the pattern of variation is stable over time, the process is **in statistical control. Control charts** are statistical plots intended to warn when a process is **out of control.**

An $\bar{x}$ **control chart** plots the means $\bar{x}$ of samples from a process against the time order in which the samples were taken. If the process has been in control with mean μ and standard deviation σ, **control limits** at $\mu \pm 3\sigma/\sqrt{n}$ mark off the range of variation we expect to see in the $\bar{x}$-values. Values outside the control limits suggest that the process has been disturbed.

Chapter 10 EXERCISES

10.16 Personal income. The government's Current Population Survey interviewed more than 131,000 people aged 25 or older in March 2002. The median income of the people with at least a bachelor's degree was **$44,776.** The median income of people with just a high school diploma was **$25,303.** Is each of the bold numbers a parameter or a statistic?

10.17 Women's heights. A random sample of female college students has a mean height of **65** inches, which is greater than the **64**-inch mean height of all young women. Is each of the bold numbers a parameter or a statistic?

10.18 Playing the numbers. The numbers racket is a well-entrenched illegal gambling operation in most large cities. One version works as follows: you choose one of the 1000 three-digit numbers 000 to 999 and pay your local numbers runner a dollar to enter your bet. Each day, one three-digit number is chosen at random and pays off $600. The mean payoff for the population of thousands of bets is $\mu = 60$ cents. Joe makes one bet every day for many years. Explain what the law of large numbers says about Joe's results as he keeps on betting.

10.19 Roulette. A roulette wheel has 38 slots, of which 18 are black, 18 are red, and 2 are green. When the wheel is spun, the ball is equally likely to come to rest in any of the slots. One of the simplest wagers chooses red or black. A bet of $1 on red returns $2 if the ball lands in a red slot. Otherwise, the player loses his dollar. When gamblers bet on red or black, the two green slots belong to the house. Because the probability of winning $2 is 18/38, the mean payoff from a $1 bet is twice 18/38, or 94.7 cents. Explain what the law of large numbers tells us about what will happen if a gambler makes very many bets on red.

10.20 Samples of incomes. In March 2000 the Bureau of Labor Statistics recorded the incomes of 55,899 people between the ages of 25 and 65 who had worked but whose main work was not in agriculture. We will treat these 55,899 people as a population. As is usually the case, the distribution of incomes in this population is strongly skewed to the right. To estimate the mean income in this population, we can select an SRS and use the sample mean $\overline{x}$ to estimate the unknown population mean. How will the sample mean behave when we take many samples?

    We used software to choose 1000 SRSs of size 25 and another 1000 SRSs of size 100. Figure 10.9 shows histograms of the two sets of 1000 sample means, using the same classes and drawn to the same scale for easy comparison.

(a) Which distribution is closer in shape to the bell curve of a Normal distribution? What important fact about sampling distributions does this comparison illustrate?

(b) About what is the range (from smallest to largest) of the sample

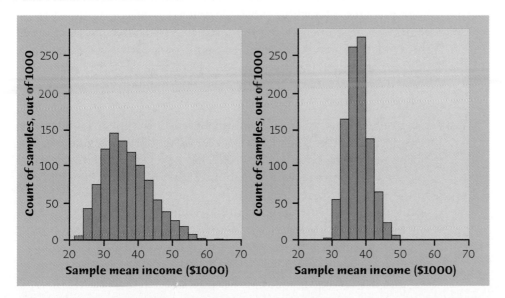

Figure 10.9 The distribution of sample means $\overline{x}$ for 1000 SRSs of size 25 (*left*) and of size 100 (*right*) from the same population, for Exercise 10.20.

means for samples of size 25? For samples of size 100? What important fact about sampling distributions does this comparison illustrate?

(c) Based on the sample means for samples of size 100, about what is the value of the mean income for this entire population?

10.21 **The cost of Internet access.** The amount that households pay service providers for access to the Internet varies quite a bit, but the mean monthly fee is $28 and the standard deviation is $10. The distribution is not Normal: many households pay about $10 for limited dial-up access or about $25 for unlimited dial-up access, but some pay much more for faster connections. A sample survey asks an SRS of 500 households with Internet access how much they pay. What is the probability that the average fee paid by the sample households exceeds $29?

10.22 **Dust in coal mines.** A laboratory weighs filters from a coal mine to measure the amount of dust in the mine atmosphere. Repeated measurements of the weight of dust on the same filter vary Normally with standard deviation $\sigma = 0.08$ milligram (mg) because the weighing is not perfectly precise. The dust on a particular filter actually weighs 123 mg. Repeated weighings will then have the Normal distribution with mean 123 mg and standard deviation 0.08 mg.

(Peter Turnley/CORBIS)

(a) The laboratory reports the mean of 3 weighings. What is the distribution of this mean?

(b) What is the probability that the laboratory reports a weight of 124 mg or higher for this filter?

10.23 **Making auto parts.** An automatic grinding machine in an auto parts plant prepares axles with a target diameter $\mu = 40.125$ millimeters (mm). The machine has some variability, so the standard deviation of the diameters is $\sigma = 0.002$ mm. A sample of 4 axles is inspected each hour for process control purposes, and records are kept of the sample mean diameter. What will be the mean and standard deviation of the numbers recorded?

10.24 **Glucose testing.** Shelia's doctor is concerned that she may suffer from gestational diabetes (high blood glucose levels during pregnancy). There is variation both in the actual glucose level and in the blood test that measures the level. A patient is classified as having gestational diabetes if the glucose level is above 140 milligrams per deciliter one hour after a sugary drink is ingested. Shelia's measured glucose level one hour after ingesting the sugary drink varies according to the Normal distribution with $\mu = 125$ mg/dl and $\sigma = 10$ mg/dl.

(a) If a single glucose measurement is made, what is the probability that Shelia is diagnosed as having gestational diabetes?

(b) If measurements are made instead on 4 separate days and the mean result is compared with the criterion 140 mg/dl, what is the probability that Shelia is diagnosed as having gestational diabetes?

10.25 Glucose testing, continued. Shelia's measured glucose level one hour after ingesting the sugary drink varies according to the Normal distribution with $\mu = 125$ mg/dl and $\sigma = 10$ mg/dl. Find the level L such that there is probability only 0.05 that the mean glucose level of 4 test results falls above L for Shelia's glucose level distribution. What is the value of L? (Hint: This requires a backward Normal calculation. See page 72 in Chapter 3 if you need to review.)

10.26 Pollutants in auto exhausts. The level of nitrogen oxides (NOX) in the exhaust of a particular car model varies with mean 0.9 grams per mile (g/mi) and standard deviation 0.15 g/mi. A company has 125 cars of this model in its fleet.

(a) What is the approximate distribution of the mean NOX emission level $\bar{x}$ for these cars?

(b) What is the level L such that the probability that $\bar{x}$ is greater than L is only 0.01? (Hint: This requires a backward Normal calculation. See page 72 in Chapter 3 if you need to review.)

10.27 Auto accidents. The number of accidents per week at a hazardous intersection varies with mean 2.2 and standard deviation 1.4. This distribution takes only whole-number values, so it is certainly not Normal.

(a) Let $\bar{x}$ be the mean number of accidents per week at the intersection during a year (52 weeks). What is the approximate distribution of $\bar{x}$ according to the central limit theorem?

(b) What is the approximate probability that $\bar{x}$ is less than 2?

(c) What is the approximate probability that there are fewer than 100 accidents at the intersection in a year? (Hint: Restate this event in terms of $\bar{x}$.)

10.28 How many people in a car? A study of rush-hour traffic in San Francisco counts the number of people in each car entering a freeway at a suburban interchange. Suppose that this count has mean 1.5 and standard deviation 0.75 in the population of all cars that enter at this interchange during rush hours.

(a) Could the exact distribution of the count be Normal? Why or why not?

(Kyle Krause/Inndex Stock Imagery/PictureQuest)

(b) Traffic engineers estimate that the capacity of the interchange is 700 cars per hour. According to the central limit theorem, what is the approximate distribution of the mean number of persons $\bar{x}$ in 700 randomly selected cars at this interchange?

(c) What is the probability that 700 cars will carry more than 1075 people? (Hint: Restate this event in terms of the mean number of people $\bar{x}$ per car.)

10.29 Returns on stocks. The distribution of annual returns on common stocks is roughly symmetric, but extreme observations are more frequent than in a Normal distribution. Because the distribution is not strongly non-Normal, the mean return over even a moderate number of years is close to Normal. In the long run, annual returns (not adjusted for inflation) on common stocks have varied with mean about 13% and standard deviation about 17%. Andrew plans to retire in 40 years and is considering investing in stocks. What is the probability (assuming that the past pattern of variation continues) that the mean annual return on common stocks over the next 45 years will exceed 15%? What is the probability that the mean return will be less than 10%?

10.30 Weights of eggs. The weight of the eggs produced by a certain breed of hen is Normally distributed with mean 65 grams (g) and standard deviation 5 g. Think of cartons of such eggs as SRSs of size 12 from the population of all eggs. What is the probability that the weight of a carton falls between 750 g and 825 g?

10.31 Generating a sampling distribution. Exercise 2.23 (page 51) gives the survival times of 72 guinea pigs in a medical experiment. Consider these 72 animals to be the population of interest.

(a) Make a histogram of the 72 survival times. This is the population distribution. It is strongly skewed to the right.

(b) Find the mean of the 72 survival times. This is the population mean μ. Mark μ on the x axis of your histogram.

(c) Label the members of the population 01 to 72 and use Table B to choose an SRS of size $n = 12$. What is the mean survival time $\bar{x}$ for your sample? Mark the value of $\bar{x}$ with a point on the axis of your histogram from (a).

(d) Choose four more SRSs of size 12, using different parts of Table B. Find $\bar{x}$ for each sample and mark the values on the axis of your histogram from (a). Would you be surprised if all five $\bar{x}$'s fell on the same side of μ? Why?

(e) If you chose a large number of SRSs of size 12 from this population and made a histogram of the $\bar{x}$-values, where would you expect the center of this sampling distribution to lie?

10.32 Generating a sampling distribution. We want to know what percent of American adults approve of legal gambling. This population proportion p is a parameter. To estimate p, take an SRS and find the proportion $\hat{p}$ in the sample who approve of gambling. If we take many SRSs of the same size, the proportion $\hat{p}$ will vary from sample to sample. The distribution of its values in all SRSs is the sampling distribution of this statistic.

Figure 10.10 is a small population. Each circle represents an adult. The colored circles are people who disapprove of legal gambling, and

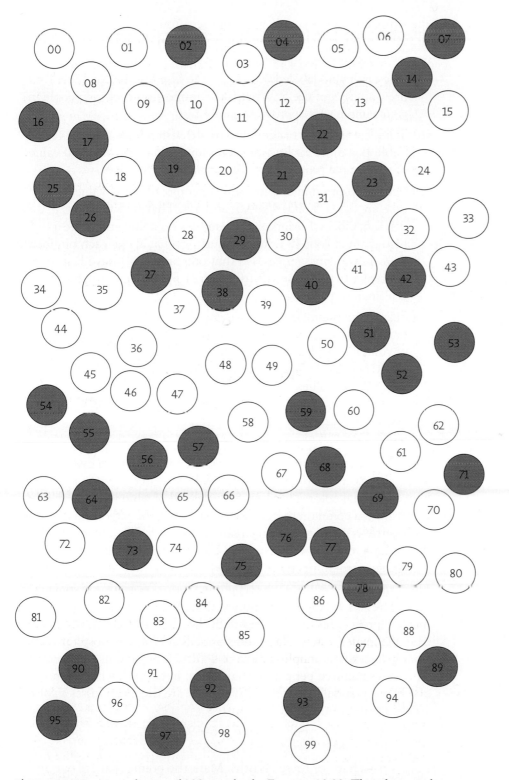

Figure 10.10 A population of 100 people, for Exercise 10.32. The white circles represent people who approve of legal gambling. The colored circles represent people who oppose gambling.

the white circles are people who approve. You can check that 60 of the 100 circles are white, so in this population the proportion who approve of gambling is $p = 60/100 = 0.6$.

(a) The circles are labeled 00, 01, ..., 99. Use line 101 of Table B to draw an SRS of size 5. What is the proportion $\hat{p}$ of the people in your sample who approve of gambling?

(b) Take 9 more SRSs of size 5 (10 in all), using lines 102 to 110 of Table B, a different line for each sample. You now have 10 values of the sample proportion $\hat{p}$. What are they?

(c) Because your samples have only 5 people, the only values $\hat{p}$ can take are 0/5, 1/5, 2/5, 3/5, 4/5, and 5/5. That is, $\hat{p}$ is always 0, 0.2, 0.4, 0.6, 0.8, or 1. Mark these numbers on a line and make a histogram of your 10 results by putting a bar above each number to show how many samples had that outcome. (You have begun to construct the sampling distribution of $\hat{p}$, although just 10 samples is a small start.)

(d) Taking samples of size 5 from a population of size 100 is not a practical setting, but let's look at your results anyway. How many of your 10 samples estimated the population proportion $p = 0.6$ exactly correctly? Is the true value 0.6 roughly in the center of your sample values?

The following exercises concern the optional material on statistical process control.

10.33 Dyeing yarn. The unique colors of the cashmere sweaters your firm makes result from heating undyed yarn in a kettle with a dye liquor. The pH (acidity) of the liquor is critical for regulating dye uptake and hence the final color. There are 5 kettles, all of which receive dye liquor from a common source. Twice each day, the pH of the liquor in each kettle is measured, giving a sample of size 5. The process has been operating in control with $\mu = 4.22$ and $\sigma = 0.127$. Give the center line and control limits for the $\bar{x}$ chart.

10.34 Hospital losses. A hospital struggling to contain costs investigates procedures on which it loses money. Government standards place medical procedures into Diagnostic Related Groups (DRGs). For example, major joint replacements are DRG 209. The hospital takes from its records a random sample of 8 DRG 209 patients each month. The losses incurred per patient have been in control, with mean $6400 and standard deviation $700. Here are the mean losses $\bar{x}$ for the samples taken in the next 15 months:

6244	6534	6080	6476	6469	6544	6415	6697
6497	6912	6638	6857	6659	7509	7374	

Make an $\bar{x}$ chart for these months. Mark the points that are out of control. What does the pattern on your chart suggest about the hospital's losses on major joint replacements?

10.35 **Dyeing yarn, continued.** What are the natural tolerances for the pH of an individual dye kettle in the setting of Exercise 10.33?

10.36 **Milling.** The width of a slot cut by a milling machine is important to the proper functioning of a hydraulic system for large tractors. The manufacturer checks the control of the milling process by measuring a sample of 5 consecutive items during each hour's production. The target width for the slot is $\mu = 0.8750$ inch. The process has been operating in control with center close to the target and $\sigma = 0.0012$ inch. What center line and control limits should be drawn on the $\bar{x}$ chart?

10.37 **Is the quality OK?** Statistical control means that a process is stable. It doesn't mean that this stable process produces high-quality items. Return to the mesh-tensioning process described in Examples 10.8 and 10.10. This process is in control with mean $\mu = 275$ mV and standard deviation $\sigma = 43$ mV.

(a) The current specifications set by customers for mesh tension are 100 to 400 mV. What percent of monitors meet these specifications?

(b) The customers now set tighter specifications, 150 to 350 mV. What percent meet the new specifications? The process has not changed, but product quality, measured by percent meeting the specifications, is no longer good.

10.38 **Improving the process.** The center of the new specifications for mesh tension in the previous exercise is 250 mV, but the center of our process is 275 mV. We can improve the process by adjusting the center to 250 mV. This is an easy adjustment that does not change the process variation. What percent of monitors now meet the new specifications?

Chapter 10 MEDIA EXERCISES

10.39 **The law of large numbers.** Suppose that you roll two balanced dice and look at the spots on the up-faces. There are 36 possible outcomes, displayed in Figure 9.2 (page 229). Because the dice are balanced, all 36 outcomes are equally likely. Add the spots on the up-faces. The average of the 36 totals is 7. This is the population mean μ for the idealized population that contains the results of rolling two dice forever. The law of large numbers says that the average $\bar{x}$ from a finite number of rolls gets closer and closer to 7 as we do more and more rolls.

(a) Click "More dice" in the *Law of Large Numbers* applet once to get two dice. Click "Show mean" to see the mean 7 on the graph. Leaving the number of rolls at 1, click "Roll dice" three times. Note the count of spots for each roll (what were they?) and the average for the three rolls. You see that the graph displays at each point the average number of spots for all rolls up to the last one. Now you understand the display.

(b) Set the number of rolls to 100 and click "Roll dice." The applet rolls the two dice 100 times. The graph shows how the average count of spots changes as we make more rolls. That is, the graph shows $\bar{x}$ as we continue to roll the dice. Make a rough sketch of the final graph.

(c) Repeat your work from (b). Click "Reset" to start over, then roll two dice 100 times. Make a sketch of the final graph of the mean $\bar{x}$ against the number of rolls. Your two graphs will often look very different. What they have in common is that the average eventually gets close to the population mean $\mu = 7$. The law of large numbers says that this will *always* happen if you keep on rolling the dice.

10.40 What's the mean? Suppose that you roll three balanced dice. We wonder what the mean number of spots on the up-faces of the three dice is. The law of large numbers says that we can find out by experience: roll three dice many times, and the average number of spots will eventually approach the true mean. Set up the *Law of Large Numbers* applet to roll three dice. Don't click "Show mean" yet. Roll the dice until you are confident you know the mean quite closely, then click "Show mean" to verify your discovery. What is the mean? Make a rough sketch of the path the averages $\bar{x}$ followed as you kept adding more rolls.

10.41 A sampling distribution. We can use the *Simple Random Sample* applet to help grasp the idea of a sampling distribution. Form a population labeled 1 to 100. We will choose an SRS of 10 of these numbers. That is, in this exercise, the numbers themselves are the population, not just labels for 100 individuals. The mean of the whole numbers 1 to 100 is $\mu = 50.5$. This is the population mean.

(a) Use the applet to choose an SRS of size 10. Which 10 numbers were chosen? What is their mean? This is the sample mean $\bar{x}$.

(b) Although the population and its mean $\mu = 50.5$ remain fixed, the sample mean changes as we take more samples. Take another SRS of size 10. (Use the "Reset" button to return to the original population before taking the second sample.) What are the 10 numbers in your sample? What is their mean? This is another value of $\bar{x}$.

(c) Take 8 more SRSs from this same population and record their means. You now have 10 values of the sample mean $\bar{x}$ from 10 SRSs of the same size from the same population. Make a histogram of the 10 values and mark the population mean $\mu = 50.5$ on the horizontal axis. Are your 10 sample values roughly centered at the population value μ? (If you kept going forever, your $\bar{x}$-values would form the sampling distribution of the sample mean; the population mean μ would indeed be the center of this distribution.)

10.42 Another sampling distribution. You can use the *Probability* applet to speed up and improve Exercise 10.32. You have a population in which 60% of the individuals approve of legal gambling. You want to take many small samples from this population to observe how the sample proportion who approve of gambling varies from sample to sample. Set the "Probability of heads" in the applet to 0.6 and the number of tosses to 5. This simulates an SRS of size 5 from a very large population, not just the 100 individuals in Figure 10.10. By alternating between "Toss" and "Reset" you can take many samples quickly.

(a) Take 50 samples, recording the number of heads (that is, the number in the sample who approve of gambling) in each sample. Make a histogram of the 50 sample proportions.

(b) Another population contains only 20% who approve of legal gambling. Take 50 samples of size 5 from this population, record the number in each sample who approve, and make a histogram of the 50 sample proportions. How do the centers of your two histograms reflect the differing truths about the two populations?

(Creasource/Series/PictureQuest)

General Rules
of Probability*

The mathematics of probability can provide models to describe the flow of traffic through a highway system, a telephone interchange, or a computer processor; the genetic makeup of populations; the energy states of subatomic particles; the spread of epidemics or rumors; and the rate of return on risky investments. Although we are interested in probability because of its usefulness in statistics, the mathematics of chance is important in many fields of study. This chapter presents a bit more of the theory of probability.

   Our study of probability in Chapter 9 concentrated on basic ideas and facts. Now we look at some details. With more probability at our command, we can model more complex random phenomena. We have already met and used four rules.

---

*This more advanced chapter gives more detail about probability. It is not needed to read the rest of the book.

RULES OF PROBABILITY

Rule 1. $0 \leq P(A) \leq 1$ for any event A

Rule 2. $P(S) = 1$

Rule 3. For any event A,

$$P(A \text{ does not occur}) = 1 - P(A)$$

Rule 4. **Addition rule:** If A and B are **disjoint** events, then

$$P(A \text{ or } B) = P(A) + P(B)$$

Independence and the multiplication rule

Rule 4, the addition rule for disjoint events, describes the probability that *one or the other* of two events A and B occurs in the special situation when A and B cannot occur together. Now we will describe the probability that *both* events A and B occur, again only in a special situation.

You may find it helpful to draw a picture to display relations among several events. A picture like Figure 11.1 that shows the sample space S as a rectangular area and events as areas within S is called a **Venn diagram.** The events A and B in Figure 11.1 are disjoint because they do not overlap. The Venn diagram in Figure 11.2 illustrates two events that are not disjoint. The event $\{A \text{ and } B\}$ appears as the overlapping area that is common to both A and B.

Venn diagram

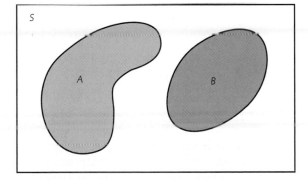

Figure 11.1 Venn diagram showing disjoint events A and B.

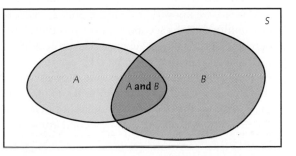

Figure 11.2 Venn diagram showing events A and B that are not disjoint. The event $\{A \text{ and } B\}$ consists of outcomes common to A and B.

Suppose that you toss a balanced coin twice. You are counting heads, so two events of interest are

$$A = \text{first toss is a head}$$
$$B = \text{second toss is a head}$$

The events A and B are not disjoint. They occur together whenever both tosses give heads. We want to find the probability of the event $\{A \text{ and } B\}$ that *both* tosses are heads.

The coin tossing of Buffon, Pearson, and Kerrich described at the beginning of Chapter 9 makes us willing to assign probability 1/2 to a head when we toss a coin. So

$$P(A) = 0.5$$
$$P(B) = 0.5$$

What is $P(A \text{ and } B)$? Our common sense says that it is 1/4. The first coin will give a head half the time and then the second will give a head on half of those trials, so both coins will give heads on $1/2 \times 1/2 = 1/4$ of all trials in the long run. This reasoning assumes that the second coin still has probability 1/2 of a head after the first has given a head. This is true—we can verify it by tossing two coins many times and observing the proportion of heads on the second toss after the first toss has produced a head. We say that the events "head on the first toss" and "head on the second toss" are **independent.** Independence means that the outcome of the first toss cannot influence the outcome of the second toss.

independence

EXAMPLE 11.1 Independent or not?

Because a coin has no memory and most coin tossers cannot influence the fall of the coin, it is safe to assume that successive coin tosses are independent. For a balanced coin this means that after we see the outcome of the first toss, we still assign probability 1/2 to heads on the second toss. On the other hand, the colors of successive cards dealt from the same deck are not independent. A standard 52-card deck contains 26 red and 26 black cards. For the first card dealt from a shuffled deck, the probability of a red card is $26/52 = 0.50$ (equally likely outcomes). Once we see that the first card is red, we know that there are only 25 reds among the remaining 51 cards. The probability that the second card is red is therefore only $25/51 = 0.49$. Knowing the outcome of the first deal changes the probabilities for the second.

If a doctor measures your blood pressure twice, it is reasonable to assume that the two results are independent because the first result does not influence the instrument that makes the second reading. But if you take an IQ test or other mental test twice in succession, the two test scores are not independent. The learning that occurs on the first attempt influences your second attempt.

MULTIPLICATION RULE FOR INDEPENDENT EVENTS

Two events A and B are **independent** if knowing that one occurs does not change the probability that the other occurs. If A and B are independent,

$$P(A \text{ and } B) = P(A)P(B)$$

EXAMPLE 11.2 Surviving?

During World War II, the British found that the probability that a bomber is lost through enemy action on a mission over occupied Europe was 0.05. The probability that the bomber returns safely from a mission was therefore 0.95. It is reasonable to assume that missions are independent. Take A_i to be the event that a bomber survives its ith mission. The probability of surviving 2 missions is

$$P(A_1 \text{ and } A_2) = P(A_1)P(A_2)$$
$$= (0.95)(0.95) = 0.9025$$

The multiplication rule applies to more than two events, provided that all are independent. So the probability of surviving 3 missions is

$$P(A_1 \text{ and } A_2 \text{ and } A_3) = P(A_1)P(A_2)P(A_3)$$
$$= (0.95)(0.95)(0.95) = 0.8574$$

The probability of surviving 20 missions is only

$$P(A_1 \text{ and } A_2 \text{ and } \ldots \text{ and } A_{20}) = P(A_1)P(A_2)\cdots P(A_{20})$$
$$= (0.95)(0.95)\cdots(0.95)$$
$$= (0.95)^{20} = 0.3585$$

The tour of duty for an airman was 30 missions.

The multiplication rule $P(A \text{ and } B) = P(A)P(B)$ holds if A and B are independent but not otherwise. The addition rule $P(A \text{ or } B) = P(A) + P(B)$ holds if A and B are disjoint but not otherwise. Resist the temptation to use these simple rules when the circumstances that justify them are not present. You must also be certain not to confuse disjointness and independence. If A and B are disjoint, then the fact that A occurs tells us that B cannot occur—look again at Figure 11.1. So disjoint events are not independent. Unlike disjointness, we cannot picture independence in a Venn diagram, because it involves the probabilities of the events rather than just the outcomes that make up the events.

APPLY YOUR KNOWLEDGE

11.1 Car colors. Exercise 1.3 (page 7) gives the distribution of colors for motor vehicles made in North America. This distribution will be approximately true for all vehicles sold in the past few years. You stand by a highway and choose two recent-model vehicles at random.

(a) It is reasonable to assume that the colors of the vehicles you choose are independent. Explain why.

(b) What is the probability that the two vehicles you choose are both black?

(c) What is the probability that the first vehicle is black and the second vehicle is white?

11.2 **Albinism.** The gene for albinism in humans is recessive. That is, carriers of this gene have probability 1/2 of passing it to a child, and the child is albino only if both parents pass the albinism gene. Parents pass their genes independently of each other. If both parents carry the albinism gene, what is the probability that their first child is albino? If they have two children (who inherit independently of each other), what is the probability that both are albino? That neither is albino?

11.3 **College-educated laborers?** Government data show that 27% of employed people have at least 4 years of college and that 14% of employed people work as laborers or operators of machines or vehicles. Nonetheless, we can't conclude that because $(0.27)(0.14) = 0.038$ about 3.8% of employed people are college-educated laborers or operators. Why not?

Applying the multiplication rule

If two events A and B are independent, the event that A does not occur is also independent of B, and so on. Suppose, for example, that 75% of all registered voters in a rural district are Republicans. If an opinion poll interviews two voters chosen independently, the probability that the first is a Republican and the second is not a Republican is $(0.75)(0.25) = 0.1875$. The multiplication rule also extends to collections of more than two events, provided that all are independent. Independence of events A, B, and C means that no information about any one or any two can change the probability of the remaining events. Independence is often assumed in setting up a probability model when the events we are describing seem to have no connection. We can then use the multiplication rule freely. Here is another example.

We want a boy

Misunderstanding independence can be disastrous. "Dear Abby" once published a letter from a mother of eight girls. She and her husband had planned a family of four children. When all four were girls, they kept trying. After seven girls, her doctor assured her that "the law of averages was in our favor 100 to 1." Unfortunately, having children is like tossing coins. Eight girls in a row is highly unlikely, but once seven girls have been born, it is not at all unlikely that the next child will be a girl—and it was.

EXAMPLE 11.3   Undersea cables

The first successful transatlantic telegraph cable was laid in 1866. The first telephone cable across the Atlantic did not appear until 1956—the barrier was designing "repeaters," amplifiers needed to boost the signal, that could operate for years on the sea bottom. This first cable had 52 repeaters. The last copper cable, laid in 1983 and retired in 1994, had 662 repeaters. The first fiber-optic cable was laid in 1988 and has 109 repeaters. There are now more than 400,000 miles of undersea cable, with more being laid every year to handle the flood of Internet traffic.

Repeaters in undersea cables must be very reliable. To see why, suppose that each repeater has probability 0.999 of functioning without failure for 25 years. Repeaters fail independently of each other. (This assumption means that there are no "common causes" such as earthquakes that would affect several repeaters at once.) Denote by A_i the event that the ith repeater operates successfully for 25 years.

The probability that 2 repeaters both last 25 years is

$$P(A_1 \text{ and } A_2) = P(A_1)P(A_2)$$
$$= 0.999 \times 0.999 = 0.998$$

For a cable with 10 repeaters the probability of no failures in 25 years is

$$P(A_1 \text{ and } A_2 \text{ and } \ldots \text{ and } A_{10}) = 0.999^{10} = 0.990$$

Cables with 2 or 10 repeaters would be quite reliable. Unfortunately, the last copper transatlantic cable had 662 repeaters. The probability that all 662 work for 25 years is

$$P(A_1 \text{ and } A_2 \text{ and } \ldots \text{ and } A_{662}) = 0.999^{662} = 0.516$$

This cable will fail to reach its 25-year design life about half the time if each repeater is 99.9% reliable over that period. The multiplication rule for probabilities shows that repeaters must be much more than 99.9% reliable.

By combining the rules we have learned, we can compute probabilities for rather complex events. Here is an example.

EXAMPLE 11.4   False positives in HIV testing

Screening large numbers of blood samples for HIV, the virus that causes AIDS, uses an enzyme immunoassay (EIA) test that detects antibodies to the virus. Applied to people who have no HIV antibodies, EIA has probability about 0.006 of producing a false positive (that is, a positive test result for a sample that has no antibodies). If the 140 employees of a medical clinic are tested and all 140 are free of HIV antibodies, what is the probability that at least 1 false positive will occur?

It is reasonable to assume as part of the probability model that the test results for different individuals are independent. The probability that the test is positive for a single person is 0.006, so the probability of a negative result is $1 - 0.006 = 0.994$. The probability of at least 1 false positive among the 140 people tested is therefore

$$P(\text{at least one positive}) = 1 - P(\text{no positives})$$
$$= 1 - P(140 \text{ negatives})$$
$$= 1 - 0.994^{140}$$
$$= 1 - 0.431 = 0.569$$

The probability is greater than 1/2 that at least 1 of the 140 people will test positive for HIV, even though no one has the virus. That is why samples that test positive are retested using a more accurate "Western blot" test.

APPLY YOUR KNOWLEDGE

11.4   Telemarketing. Telephone marketers and opinion polls use random-digit dialing equipment to call residential telephone numbers at random. The telephone polling firm Zogby International reports that the probability that a call reaches a live person is 0.2.[1] Calls are independent.

(a) A telemarketer places 5 calls. What is the probability that none of them reaches a person?

(b) When calls are made to New York City, the probability of reaching a person is only 0.08. What is the probability that none of 5 calls made to New York City reaches a person?

11.5 Bright lights? A string of holiday lights contains 20 lights. The lights are wired in series, so that if any light fails, the whole string will go dark. Each light has probability 0.02 of failing during the holiday season. The lights fail independently of each other. What is the probability that the string of lights will remain bright?

11.6 Playing the slots. Slot machines are now video games, with winning determined by electronic random number generators. In the old days, slot machines were like this: you pull the lever to spin three wheels; each wheel has 20 symbols, all equally likely to show when the wheel stops spinning; the three wheels are independent of each other. Suppose that the middle wheel has 9 bells among its 20 symbols, and the left and right wheels have 1 bell each.

(a) You win the jackpot if all three wheels show bells. What is the probability of winning the jackpot?

(b) There are three ways that the three wheels can show two bells and one symbol other than a bell. Find the probability of each of these ways. What is the probability that the wheels stop with exactly two bells showing among them?

The general addition rule

We know that if A and B are disjoint events, then $P(A \text{ or } B) = P(A) + P(B)$. This addition rule extends to more than two events that are disjoint in the sense that no two have any outcomes in common. The Venn diagram in Figure 11.3 shows three disjoint events A, B, and C. The probability that one of these events occurs is $P(A) + P(B) + P(C)$.

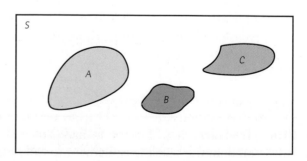

Figure 11.3 The addition rule for disjoint events: $P(A \text{ or } B \text{ or } C) = P(A) + P(B) + P(C)$ when events A, B, and C are disjoint.

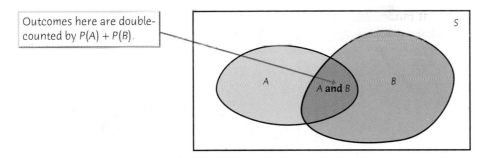

Outcomes here are double-counted by $P(A) + P(B)$.

Figure 11.4 The general addition rule: $P(A \text{ or } B) = P(A) + P(B) - P(A \text{ and } B)$ for any events A and B.

If events A and B are *not* disjoint, they can occur simultaneously. The probability that one or the other occurs is then *less* than the sum of their probabilities. As Figure 11.4 suggests, the outcomes common to both are counted twice when we add probabilities, so we must subtract this probability once. Here is the addition rule for any two events, disjoint or not.

GENERAL ADDITION RULE FOR ANY TWO EVENTS

For any two events A and B,

$$P(A \text{ or } B) = P(A) + P(B) - P(A \text{ and } B)$$

If A and B are disjoint, the event $\{A \text{ and } B\}$ that both occur contains no outcomes and therefore has probability 0. So the general addition rule includes Rule 4, the addition rule for disjoint events.

EXAMPLE 11.5   Making partner

Deborah and Matthew are anxiously awaiting word on whether they have been made partners of their law firm. Deborah guesses that her probability of making partner is 0.7 and that Matthew's is 0.5. (These are personal probabilities reflecting Deborah's assessment of chance.) This assignment of probabilities does not give us enough information to compute the probability that at least one of the two is promoted. In particular, adding the individual probabilities of promotion gives the impossible result 1.2. If Deborah also guesses that the probability that *both* she and Matthew are made partners is 0.3, then by the general addition rule

$$P(\text{at least one is promoted}) = 0.7 + 0.5 - 0.3 = 0.9$$

The probability that *neither* is promoted is then 0.1 by Rule 3.

Venn diagrams are a great help in finding probabilities because you can just think of adding and subtracting areas. Figure 11.5 shows some events and their probabilities for Example 11.5. What is the probability that Deborah is promoted and Matthew is not? The Venn diagram shows that this is the

Figure 11.5 Venn diagram and probabilities for Example 11.5.

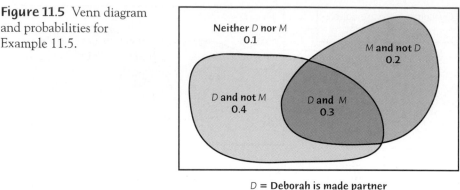

D = Deborah is made partner
M = Matthew is made partner

probability that Deborah is promoted minus the probability that both are promoted, $0.7 - 0.3 = 0.4$. Similarly, the probability that Matthew is promoted and Deborah is not is $0.5 - 0.3 = 0.2$. The four probabilities that appear in the figure add to 1 because they refer to four disjoint events that make up the entire sample space.

APPLY YOUR KNOWLEDGE

(Mark C, Burnett/Stock, Boston)

11.7 **Tastes in music.** Musical styles other than rock and pop are becoming more popular. A survey of college students finds that 40% like country music, 30% like gospel music, and 10% like both.

(a) Make a Venn diagram with these results.

(b) What percent of college students like country but not gospel?

(c) What percent like neither country nor gospel?

11.8 **Prosperity and education.** Call a household prosperous if its income exceeds $100,000. Call the household educated if the householder completed college. Select an American household at random, and let A be the event that the selected household is prosperous and B the event that it is educated. According to the Current Population Survey, in 2001 $P(A) = 0.138$, $P(B) = 0.261$, and the probability that a household is both prosperous and educated is $P(A \text{ and } B) = 0.082$.[2]

(a) Draw a Venn diagram that shows the relation between the events A and B. What is the probability $P(A \text{ or } B)$ that the household selected is either prosperous or educated?

(b) In your diagram, shade the event that the household is educated but not prosperous. What is the probability of this event?

Conditional probability

The probability we assign to an event can change if we know that some other event has occurred. This idea is the key to many applications of probability.

EXAMPLE 11.6   Race and ethnicity

The 2000 census allowed people to choose what race they consider themselves to be, and separately whether or not they consider themselves to be "Hispanic/Latino." Here are the probabilities for a randomly chosen American:

	Hispanic	Not Hispanic	Total
Asian	0.000	0.036	0.036
Black	0.003	0.121	0.124
White	0.060	0.691	0.751
Other	0.062	0.027	0.089
Total	0.125	0.875	

The "Total" row and column are obtained from the probabilities in the body of the table by the addition rule. For example, the probability that a randomly chosen American is black is

$$P(\text{black}) = P(\text{black and Hispanic}) + P(\text{black and not Hispanic})$$
$$= 0.003 + 0.121 = 0.124$$

Now we are told that the person chosen is Hispanic. That is, he or she is one of the 12.5% in the "Hispanic" column of the table. The probability that a person is black, *given the information that he or she is Hispanic*, is the proportion of blacks in the "Hispanic" column:

$$P(\text{black} \mid \text{Hispanic}) = \frac{0.003}{0.125} = 0.024$$

This is a **conditional probability.** You can read the bar | as "given the information that."

Although 12.4% of the entire population is black, only 2.4% of Hispanics are black. It's common sense that knowing that one event (the person is Hispanic) occurs often changes the probability of another event (the person is black). The example also shows how we should define conditional probability. The idea of a conditional probability $P(B \mid A)$ of one event B given that another event A occurs is the proportion *of all occurrences of A for which B also occurs.*

CONDITIONAL PROBABILITY

When $P(A) > 0$, the **conditional probability** of B given A is

$$P(B \mid A) = \frac{P(A \text{ and } B)}{P(A)}$$

Be sure to keep in mind the distinct roles of the events A and B in $P(B \mid A)$. Event A represents the information we are given, and B is the event whose

Politically correct

In 1950, the Russian mathematician B. V. Gnedenko (1912–1995) wrote *The Theory of Probability*, a text that was popular around the world. The introduction contains a mystifying paragraph that begins, "We note that the entire development of probability theory shows evidence of how its concepts and ideas were crystallized in a severe struggle between materialistic and idealistic conceptions." It turns out that "materialistic" is jargon for "Marxist-Leninist." It was good for the health of Russian scientists in the Stalin era to add such statements to their books.

conditional probability

probability we are calculating. The conditional probability $P(B \mid A)$ makes no sense if the event A can never occur, so we require that $P(A) > 0$ whenever we talk about $P(B \mid A)$.

EXAMPLE 11.7 More on race and ethnicity

What is the conditional probability that a randomly chosen person lists his or her race as "other," given that he or she is Hispanic? Using the definition,

$$P(\text{other} \mid \text{Hispanic}) = \frac{P(\text{other and Hispanic})}{P(\text{Hispanic})}$$

$$= \frac{0.062}{0.125} = 0.496$$

Almost half of all Hispanics did not choose any of the listed races. Although race and Hispanic origin are separate in the minds of the government, many Hispanics identify themselves simply as Hispanic and not as members of a race.

APPLY YOUR KNOWLEDGE

11.9 **College degrees.** Here are the counts (in thousands) of earned degrees in the United States in the 2001–2002 academic year, classified by level and by the sex of the degree recipient:[3]

	Bachelor's	Master's	Professional	Doctorate	Total
Female	645	227	32	18	922
Male	505	161	40	26	732
Total	1150	388	72	44	1654

(a) If you choose a degree recipient at random, what is the probability that the person you choose is a woman?

(b) What is the conditional probability that you choose a woman, given that the person chosen received a professional degree?

11.10 **Computer games.** Here is the distribution of computer games sold in 2001 by type of game:[4]

Game type	Probability
Strategy	0.254
Children's	0.142
Family	0.115
Action	0.101
Role playing	0.088
Sports	0.081
Other	0.219

What is the conditional probability that a computer game is an action game, given that it is not a children's game or a family game?

11.11 **Race and ethnicity again.** Choose an American at random and consider the events

$$A = \text{the person chosen is white}$$
$$B = \text{the person chosen is Hispanic}$$

It is easy to confuse $P(A \mid B)$ and $P(B \mid A)$. These are very different probabilities.

(a) What is the conditional probability notation for the proportion of whites who are Hispanic? Find this conditional probability from the table in Example 11.6.

(b) What is the conditional probability notation for the proportion of Hispanics who are white? Find this conditional probability.

The general multiplication rule

The definition of conditional probability reminds us that in principle all probabilities, including conditional probabilities, can be found from the assignment of probabilities to events that describes a random phenomenon. More often, however, conditional probabilities are part of the information given to us in a probability model. The definition of conditional probability then turns into a rule for finding the probability that both of two events occur.

GENERAL MULTIPLICATION RULE FOR ANY TWO EVENTS

The probability that both of two events A and B happen together can be found by

$$P(A \text{ and } B) = P(A)P(B \mid A)$$

Here $P(B \mid A)$ is the conditional probability that B occurs given the information that A occurs.

In words, this rule says that for both of two events to occur, first one must occur and then, given that the first event has occurred, the second must occur.

EXAMPLE 11.8   Slim wants diamonds

Slim is a professional poker player. At the moment, he wants very much to draw two diamonds in a row. As he sits at the table looking at his hand and at the upturned cards on the table, Slim sees 11 cards. Of these, 4 are diamonds. The full deck contains 13 diamonds among its 52 cards, so 9 of the 41 unseen cards are diamonds. Because the deck was carefully shuffled, each card that Slim draws is equally likely to be any of the cards that he has not seen.

Winning the lottery twice

To find Slim's probability of drawing 2 diamonds, first calculate

$$P(\text{first card diamond}) = \frac{9}{41}$$

$$P(\text{second card diamond} \mid \text{first card diamond}) = \frac{8}{40}$$

Slim finds both probabilities by counting cards. The probability that the first card drawn is a diamond is 9/41 because 9 of the 41 unseen cards are diamonds. If the first card is a diamond, that leaves 8 diamonds among the 40 remaining cards. So the *conditional* probability of another diamond is 8/40. The multiplication rule now says that

$$P(\text{both cards diamonds}) = \frac{9}{41} \times \frac{8}{40} = 0.044$$

Slim will need luck to draw his diamonds.

The multiplication rule extends to the probability that all of several events occur. The key is to condition each event on the occurrence of *all* of the preceding events. For example, we have for three events A, B, and C that

$$P(A \text{ and } B \text{ and } C) = P(A)P(B \mid A)P(C \mid A \text{ and } B)$$

EXAMPLE 11.9   Fundraising by telephone

A charity raises funds by calling a list of prospective donors to ask for pledges. It is able to talk with 40% of the names on its list. Of those the charity reaches, 30% make a pledge. But only half of those who pledge send a check. What percent of the donor list actually contribute?

Define these events:

$$A = \{\text{charity reaches a prospect}\}$$
$$B = \{\text{the prospect makes a pledge}\}$$
$$C = \{\text{the prospect makes a contribution}\}$$

The information we are given is

$$P(A) = 0.4$$
$$P(B \mid A) = 0.3$$
$$P(C \mid A \text{ and } B) = 0.5$$

The probability we want is therefore

$$P(A \text{ and } B \text{ and } C) = P(A)P(B \mid A)P(C \mid A \text{ and } B)$$
$$= 0.4 \times 0.3 \times 0.5 = 0.06$$

Only 6% of the prospective donors make a contribution.

Independence

The conditional probability $P(B \mid A)$ is generally not equal to the unconditional probability $P(B)$. That is because the occurrence of event A generally gives us some additional information about whether or not event B occurs. If knowing that A occurs gives no additional information about B, then A and B are independent events. The precise definition of independence is expressed in terms of conditional probability.

INDEPENDENT EVENTS

Two events A and B that both have positive probability are **independent** if

$$P(B \mid A) = P(B)$$

We now see that the multiplication rule for independent events, $P(A \text{ and } B) = P(A)P(B)$, is a special case of the general multiplication rule, $P(A \text{ and } B) = P(A)P(B \mid A)$, just as the addition rule for disjoint events is a special case of the general addition rule. We will rarely use the definition of independence, because most often independence is part of the information given to us in a probability model.

APPLY YOUR KNOWLEDGE

11.12 At the gym. Many conditional probability calculations are just common sense made automatic. For example, 10% of adults belong to health clubs, and 40% of these health club members go to the club at least twice a week. What percent of all adults go to a health club at least twice a week? Write the information given in terms of probabilities and use the general multiplication rule.

11.13 Buying from Japan. Functional Robotics Corporation buys electrical controllers from a Japanese supplier. The company's treasurer thinks that there is probability 0.4 that the dollar will fall in value against the Japanese yen in the next month. The treasurer also believes that *if* the dollar falls there is probability 0.8 that the supplier will demand renegotiation of the contract. What personal probability has the treasurer assigned to the event that the dollar falls and the supplier demands renegotiation?

11.14 The probability of a flush. A poker player holds a flush when all five cards in the hand belong to the same suit. We will find the probability of a flush when five cards are dealt. Remember that a deck contains 52 cards, 13 of each suit, and that when the deck is well shuffled, each card dealt is equally likely to be any of those that remain in the deck.

  (a) Concentrate on spades. What is the probability that the first card dealt is a spade? What is the conditional probability that the second card is a spade, given that the first is a spade? Continue to count the remaining cards to find the conditional probabilities of a spade on the third, the fourth, and the fifth card, given in each case that all previous cards are spades.

(b) The probability of being dealt five spades is the product of the five probabilities you have found. Why? What is this probability?

(c) The probability of being dealt five hearts or five diamonds or five clubs is the same as the probability of being dealt five spades. What is the probability of being dealt a flush?

Tree diagrams

Probability problems often require us to combine several of the basic rules into a more elaborate calculation. Here is an example that illustrates how to solve problems that have several stages.

tree diagram

EXAMPLE 11.10 **Fundraising by telephone**

A charity raises funds by calling prospective donors. Its list contains three classes of prospects. Half are active donors, who gave within the past year. Thirty percent of the list are past donors, who gave within the past three years but not in the past year. The remaining 20% are potential donors who have given to similar causes. The probabilities of obtaining a pledge and of receiving a check once a pledge is made are different for these classes. What is the probability that a randomly chosen prospect makes a contribution?

The **tree diagram** in Figure 11.6 organizes the information the charity has from its past fundraising efforts. Each segment in the tree is one stage of the problem. Each complete branch shows a path that contact with a prospect can take. The probability written on each segment is the conditional probability for that stage, given that the contact reached the point from which it branches. Starting at the left, each prospect is in one of the three classes of donors, with probabilities 0.5, 0.3, and 0.2. These probabilities mark the first-stage segments.

Looking at active donors, we see that the *conditional* probability of a pledge is 0.4. Because each prospect either does or does not pledge, the conditional probability of refusing to pledge is $1 - 0.4$, or 0.6. These two conditional probabilities appear on the segments going out from the "active donor" branch point. Continue along the top of the tree: once a pledge is made by an active donor, the conditional probability of an actual contribution is 0.8. The full tree shows the probabilities for all three classes of prospects.[5]

Now use the multiplication rule. The probability that a call goes to an active donor who pledges and then sends a check is

$$(0.5)(0.4)(0.8) = 0.16$$

This probability appears at the end of the topmost branch. You see that the probability of any complete branch in the tree is the product of the probabilities of all the segments in that branch.

There are three disjoint paths to a contribution, starting with the three classes of prospects. Because these paths are disjoint, the probability of a contribution is the

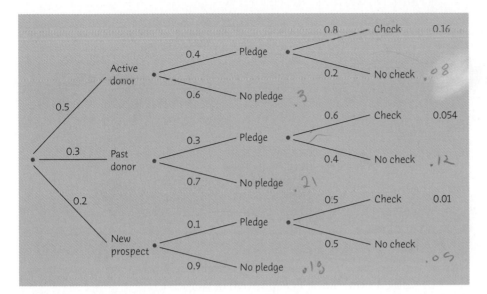

Figure 11.6 Tree diagram for Example 11.10. Tree diagrams organize probability models that have several stages.

sum of their probabilities:

$$P(\text{contribution}) = (0.5)(0.4)(0.8) + (0.3)(0.3)(0.6) + (0.2)(0.1)(0.5)$$
$$= 0.16 + 0.054 + 0.01 = 0.224$$

In all, 22.4% of prospects on the list make a contribution.

Tree diagrams combine the addition and multiplication rules. The multiplication rule says that the probability of reaching the end of any complete branch is the product of the probabilities written on its segments. The probability of any outcome, such as the event that a prospect contributes, is then found by adding the probabilities of all branches that are part of that event.

APPLY YOUR KNOWLEDGE

11.15 Fundraising by telephone. Use the tree diagram in Figure 11.6 to find the probability that a randomly chosen prospect makes a pledge and then fails to send a check.

11.16 Buying from Japan. In the setting of Exercise 11.13, the treasurer also thinks that if the dollar does not fall, there is probability 0.2 that the supplier will demand that the contract be renegotiated. What is the probability that the supplier will demand renegotiation? (Use a tree diagram to organize the information given.)

11.17 Testing for HIV. Enzyme immunoassay (EIA) tests are used to screen blood specimens for the presence of antibodies to HIV, the virus that causes AIDS. Antibodies indicate the presence of the virus. The test is quite accurate but is not always correct. Here are approximate probabilities of positive and negative EIA outcomes when the blood tested does and does not actually contain antibodies to HIV:[6]

	Test result	
	+	−
Antibodies present	0.9985	0.0015
Antibodies absent	0.0060	0.9940

Suppose that 1% of a large population carries antibodies to HIV in their blood.

(a) Draw a tree diagram for selecting a person from this population (outcomes: antibodies present or absent) and for testing his or her blood (outcomes: EIA positive or negative).

(b) What is the probability that the EIA test is positive for a randomly chosen person from this population?

11.18 **False HIV positives.** Use the information in the previous exercise and the definition of conditional probability to find the probability that a person has the antibody, given that the EIA test is positive. (This exercise illustrates a fact that is important when considering proposals for widespread testing for HIV, illegal drugs, or agents of biological warfare: if the condition being tested is uncommon in the population, many positives will be false positives.)

Chapter 11 SUMMARY

Events A and B are **disjoint** if they have no outcomes in common. In that case, $P(A \text{ or } B) = P(A) + P(B)$.

The **conditional probability** $P(B \mid A)$ of an event B given an event A is defined by

$$P(B \mid A) = \frac{P(A \text{ and } B)}{P(A)}$$

when $P(A) > 0$. In practice, we most often find conditional probabilities from directly available information rather than from the definition.

Events A and B are **independent** if knowing that one event occurs does not change the probability we would assign to the other event; that is, $P(B \mid A) = P(B)$. In that case, $P(A \text{ and } B) = P(A)P(B)$.

Any assignment of probability obeys these general rules:

Addition rule: If events $A, B, C, \ldots$ are all **disjoint** in pairs, then

$$P(\text{at least one of these events occurs}) = P(A) + P(B) + P(C) + \cdots$$

Multiplication rule: If events $A, B, C, \ldots$ are **independent,** then

$$P(\text{all of the events occur}) = P(A)P(B)P(C)\cdots$$

General addition rule: For any two events A and B,

$$P(A \text{ or } B) = P(A) + P(B) - P(A \text{ and } B)$$

General multiplication rule: For any two events A and B,

$$P(A \text{ and } B) = P(A)P(B \mid A)$$

Chapter 11 EXERCISES

11.19 Foreign language study. Choose a student in grades 9 to 12 at random and ask if he or she is studying a language other than English. Here is the distribution of results:

Language	Spanish	French	German	All others	None
Probability	0.26	0.09	0.03	0.03	0.59

What is the conditional probability that a student is studying Spanish, given that he or she is studying some language other than English?

11.20 Playing the lottery. An instant lottery game gives you probability 0.02 of winning on any one play. Plays are independent of each other. If you play 5 times, what is the probability that you win at least once?

11.21 Nonconforming chips. An automobile manufacturer buys computer chips from a supplier. The supplier sends a shipment of which 5% fail to conform to performance specifications. Each chip chosen from this shipment has probability 0.05 of being nonconforming, and each automobile uses 12 chips selected independently. What is the probability that all 12 chips in a car will work properly?

11.22 A random walk on Wall Street? The "random walk" theory of securities prices holds that price movements in disjoint time periods are independent of each other. Suppose that we record only whether the price is up or down each year, and that the probability that our portfolio rises in price in any one year is 0.65. (This probability is approximately correct for a portfolio containing equal dollar amounts of all common stocks listed on the New York Stock Exchange.)

(a) What is the probability that our portfolio goes up for three consecutive years?

(b) If you know that the portfolio has risen in price two years in a row, what probability do you assign to the event that it will go down next year?

(c) What is the probability that the portfolio's value moves in the same direction in both of the next two years?

11.23 College degrees. Exercise 11.9 gives the counts (in thousands) of earned degrees in the United States in 2001–2002. Use these data to answer the following questions.

(a) What is the probability that a randomly chosen degree recipient is a man?

(b) What is the conditional probability that the person chosen received a bachelor's degree, given that he is a man?

(c) Use the multiplication rule to find the probability of choosing a male bachelor's degree recipient. Check your result by finding this probability directly from the table of counts.

11.24 Getting into college. Ramon has applied to both Princeton and Stanford. He thinks the probability that Princeton will admit him is 0.4, the probability that Stanford will admit him is 0.5, and the probability that both will admit him is 0.2.

(a) Make a Venn diagram with the probabilities given marked.

(b) What is the probability that neither university admits Ramon?

(c) What is the probability that he gets into Stanford but not Princeton?

11.25 Geometric probability. Choose a point at random in the square with sides $0 \leq x \leq 1$ and $0 \leq y \leq 1$. This means that the probability that the point falls in any region within the square is equal to the area of that region. Let X be the x coordinate and Y the y coordinate of the point chosen. Find the conditional probability $P(Y < 1/2 \mid Y > X)$. (Hint: Draw a diagram of the square and the events $Y < 1/2$ and $Y > X$.)

11.26 The probability of a royal flush. A royal flush is the highest hand possible in poker. It consists of the ace, king, queen, jack, and ten of the same suit. Modify the outline given in Exercise 11.14 to find the probability of being dealt a royal flush in a five-card hand.

(Luc Beziat/Stone/Getty Images)

11.27 Income tax returns. Here is the distribution of the adjusted gross income (in thousands of dollars) reported on individual federal income tax returns in 2000:

Income	<15	15–29	30–49	50–99	100–199	≥200
Probability	0.30	0.23	0.19	0.20	0.06	0.02

(a) What is the probability that a randomly chosen return shows an adjusted gross income of $50,000 or more?

(b) Given that a return shows an income of at least $50,000, what is the conditional probability that the income is at least $100,000?

11.28 Prosperity and education. Call a household prosperous if its income exceeds $100,000. Call the household educated if the householder completed college. Select an American household at random, and let A be the event that the selected household is prosperous and B the event that it is educated. According to the Current Population Survey, $P(A) = 0.134$, $P(B) = 0.254$, and the probability that a household is both prosperous and educated is $P(A \text{ and } B) = 0.080$.

(a) Find the conditional probability that a household is educated, given that it is prosperous.

(b) Find the conditional probability that a household is prosperous, given that it is educated.

(c) Are events A and B independent? How do you know?

11.29 Inspecting switches. A shipment contains 10,000 switches. Of these, 1000 are bad. An inspector draws switches at random, so that each switch has the same chance to be drawn.

(a) Draw one switch. What is the probability that the switch you draw is bad? What is the probability that it is not bad?

(b) Suppose the first switch drawn is bad. How many switches remain? How many of them are bad? Draw a second switch at random. What is the conditional probability that this switch is bad?

(c) Answer the questions in (b) again, but now suppose that the first switch drawn is not bad.

Comment: Knowing the result of the first trial changes the conditional probability for the second trial, so the trials are not independent. But because the shipment is large, the probabilities change very little. The trials are almost independent.

11.30 Screening job applicants. A company retains a psychologist to assess whether job applicants are suited for assembly-line work. The psychologist classifies applicants as A (well suited), B (marginal), or C (not suited). The company is concerned about event D: an employee leaves the company within a year of being hired. Data on all people hired in the past five years give these probabilities:

$$P(A) = 0.4 \qquad P(B) = 0.3 \qquad P(C) = 0.3$$
$$P(A \text{ and } D) = 0.1 \qquad P(B \text{ and } D) = 0.1 \qquad P(C \text{ and } D) = 0.2$$

Sketch a Venn diagram of the events A, B, C, and D and mark on your diagram the probabilities of all combinations of psychological assessment and leaving (or not) within a year. What is $P(D)$, the probability that an employee leaves within a year?

(David Stoecklein/CORBIS)

11.31 Tastes in music. Musical styles other than rock and pop are becoming more popular. A survey of college students finds that 40% like country music, 30% like gospel music, and 10% like both.

(a) What is the conditional probability that a student likes gospel music if we know that he or she likes country music?

(b) What is the conditional probability that a student who does not like country music likes gospel music? (A Venn diagram may help you.)

11.32 The geometric distributions. You are tossing a balanced die that has probability 1/6 of coming up 1 on each toss. Tosses are independent. We are interested in how long we must wait to get the first 1.

(a) The probability of a 1 on the first toss is 1/6. What is the probability that the first toss is not a 1 and the second toss is a 1?

(b) What is the probability that the first two tosses are not 1s and the third toss is a 1? This is the probability that the first 1 occurs on the third toss.

(c) Now you see the pattern. What is the probability that the first 1 occurs on the fourth toss? On the fifth toss? Give the general result: what is the probability that the first 1 occurs on the kth toss?

geometric distribution

Comment: The distribution of the number of trials to the first success is called a **geometric distribution.** In this problem you have found geometric distribution probabilities when the probability of a success on each trial is $p = 1/6$. The same idea works for any p.

11.33 Income and savings. A sample survey chooses a sample of households and measures their annual income and their savings. Some events of interest are

$A =$ the household chosen has income at least $100,000
$C =$ the household chosen has at least $50,000 in savings

Based on this sample survey, we estimate that $P(A) = 0.07$ and $P(C) = 0.2$.

(a) We want to find the probability that a household has either income at least $100,000 *or* savings at least $50,000. Explain why we do not have enough information to find this probability. What additional information is needed?

(b) We want to find the probability that a household has income at least $100,000 *and* savings at least $50,000. Explain why we do not have enough information to find this probability. What additional information is needed?

11.34 Tendon surgery. You have torn a tendon and are facing surgery to repair it. The surgeon explains the risks to you: infection occurs in 3% of such operations, the repair fails in 14%, and both infection and failure occur together in 1%. What percent of these operations succeed and are free from infection?

11.35 Urban voters. The voters in a large city are 40% white, 40% black, and 20% Hispanic. (Hispanics may be of any race in official statistics, but here we are speaking of political blocks.) A black mayoral candidate anticipates attracting 30% of the white vote, 90% of the black vote, and 50% of the Hispanic vote. Draw a tree diagram with probabilities for the race (white, black, or Hispanic) and vote (for or against the candidate) of a randomly chosen voter. What percent of the overall vote does the candidate expect to get?

11.36 Preparing for the GMAT. A company that offers courses to prepare would-be MBA students for the Graduate Management Admission Test (GMAT) finds that 40% of its customers are currently undergraduate students and 60% are college graduates. After completing the course, 50% of the undergraduates and 70% of the graduates achieve scores of at least 600 on the GMAT. Draw a tree diagram that organizes this information. What percent of all customers score at least 600 on the GMAT?

11.37 Where do the votes come from? In the election described in Exercise 11.35, what percent of the candidate's votes come from black voters? (Write this as a conditional probability and use the definition of conditional probability.)

Working. In the language of government statistics, you are "in the labor force" if you are available for work and either working or actively seeking work. The unemployment rate is the proportion of the labor force (not of the entire population) who are unemployed. Here are data from the Current Population Survey for the civilian population aged 25 years and over. The table entries are counts in thousands of people. Exercises 11.38 to 11.40 concern these data.

Highest education	Total population	In labor force	Employed
Did not finish high school	27,325	12,073	11,139
High school but no college	57,221	36,855	35,137
Less than bachelor's degree	45,471	33,331	31,975
College graduate	47,371	37,281	36,259

11.38 Find the unemployment rate for people with each level of education. (This is the conditional probability of being unemployed, given an education level.) How does the unemployment rate change with education? Explain carefully why your results show that level of education and being employed are not independent.

11.39 (a) What is the probability that a randomly chosen person 25 years of age or older is in the labor force?

(b) If you know that the person chosen is a college graduate, what is the conditional probability that he or she is in the labor force?

(c) Are the events "in the labor force" and "college graduate" independent? How do you know?

11.40 (a) You know that a person is employed. What is the conditional probability that he or she is a college graduate?

(b) You know that a second person is a college graduate. What is the conditional probability that he or she is employed?

Blood types. *All human blood can be "ABO-typed" as one of O, A, B, or AB, but the distribution of the types varies a bit among groups of people. Here is the distribution of blood types for a randomly chosen person in the United States:*

Blood type	O	A	B	AB
U.S. probability	0.45	0.40	0.11	0.04

Exercises 11.41 to 11.44 use this information.

11.41 Maria has type B blood. She can safely receive transfusions only from persons with type B or type O blood. What is the probability that Maria's husband is an acceptable blood donor for her? (It is reasonable to assume that the blood types of husband and wife are independent.)

11.42 What is the probability that a wife and husband share the same blood type? (Assume that their blood types are independent.)

11.43 (a) What is the probability that the wife has type A blood and the husband has type B?

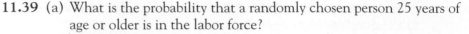

(b) What is the probability that one of the couple has type A blood and the other has type B?

11.44 Human blood is also typed as Rh-positive or Rh-negative. ABO type and Rh-factor are independent because they are governed by different genes. In the American population, 84% of people are Rh-positive. Give the probability distribution of blood type (ABO and Rh together) for a randomly chosen person.

11.45 Universal donors. People with type O-negative blood are universal donors. That is, any patient can receive a transfusion of O-negative blood. Only 7.2% of the American population have O-negative blood. If 10 people appear at random to give blood, what is the probability that at least one of them is a universal donor?

Mendelian inheritance. *Some traits of plants and animals depend on inheritance of a single gene. This is called Mendelian inheritance, after*

Gregor Mendel (1822–1884). Exercises 11.46 to 11.49 are based on the following information about Mendelian inheritance of blood type.

Each of us has an ABO blood type, which describes whether two characteristics called A and B are present. Every human being has two blood type alleles (gene forms), one inherited from our mother and one from our father. Each of these alleles can be A, B, or O. Which two we inherit determines our blood type. Here is a table that shows what our blood type is for each combination of two alleles:

Alleles inherited	Blood type
A and A	A
A and B	AB
A and O	A
B and B	B
B and O	B
O and O	O

We inherit each of a parent's two alleles with probability 0.5. We inherit independently from our mother and father.

11.46 Hannah and Jacob both have alleles A and B.

(a) What blood types can their children have?

(b) What is the probability that their next child has each of these blood types?

11.47 Nancy and David both have alleles B and O.

(a) What blood types can their children have?

(b) What is the probability that their next child has each of these blood types?

11.48 Jennifer has alleles A and O. José has alleles A and B. They have two children.

(a) What is the probability that both children have blood type A?

(b) What is the probability that both children have the same blood type?

11.49 Jasmine has alleles A and O. Joshua has alleles B and O.

(a) What is the probability that a child of these parents has blood type O?

(b) If Jasmine and Joshua have three children, what is the probability that all three have blood type O?

(c) What is the probability that the first child has blood type O and the next two do not?

(AP/Wide World Photos)

Binomial Distributions*

A basketball player shoots 5 free throws. How many does she make? A new treatment for pancreatic cancer is tried on 250 patients. How many survive for five years? You plant 10 dogwood trees. How many live through the winter? In all these situations, we want a probability model for a *count* of successful outcomes.

The binomial setting and binomial distributions

The distribution of a count depends on how the data are produced. Here is a common situation.

THE BINOMIAL SETTING

1. There are a fixed number n of observations.

2. The n observations are all **independent.** That is, knowing the result of one observation tells you nothing about the other observations.

3. Each observation falls into one of just two categories, which for convenience we call "success" and "failure."

4. The probability of a success, call it p, is the same for each observation.

---

*This more advanced chapter concerns a special topic in probability. It is not needed to read the rest of the book.

Think of tossing a coin n times as an example of the binomial setting. Each toss gives either heads or tails. Knowing the outcome of one toss doesn't tell us anything about other tosses, so the n tosses are independent. If we call heads a success, then p is the probability of a head and remains the same as long as we toss the same coin. The number of heads we count is a random variable X. The distribution of X is called a *binomial distribution*.

> ## BINOMIAL DISTRIBUTION
>
> The distribution of the count X of successes in the binomial setting is the **binomial distribution** with parameters n and p. The parameter n is the number of observations, and p is the probability of a success on any one observation. The possible values of X are the whole numbers from 0 to n.

The binomial distributions are an important class of probability distributions. Pay attention to the binomial setting, because not all counts have binomial distributions.

EXAMPLE 12.1   Blood types

Genetics says that children receive genes from their parents independently. Each child of a particular pair of parents has probability 0.25 of having type O blood. If these parents have 5 children, the number who have type O blood is the count X of successes in 5 independent trials with probability 0.25 of a success on each trial. So X has the binomial distribution with $n = 5$ and $p = 0.25$.

EXAMPLE 12.2   Dealing cards

Deal 10 cards from a shuffled deck and count the number X of red cards. There are 10 observations, and each gives either a red or a black card. A "success" is a red card. But the observations are *not* independent. If the first card is black, the second is more likely to be red because there are more red cards than black cards left in the deck. The count X does *not* have a binomial distribution.

Binomial distributions in statistical sampling

The binomial distributions are important in statistics when we wish to make inferences about the proportion p of "successes" in a population. Here is a typical example.

EXAMPLE 12.3   Choosing an SRS

An engineer chooses an SRS of 10 switches from a shipment of 10,000 switches. Suppose that (unknown to the engineer) 10% of the switches in the shipment are bad. The engineer counts the number X of bad switches in the sample.

Was he good or was he lucky?

When a baseball player hits .300, everyone applauds. A .300 hitter gets a hit in 30% of times at bat. Could a .300 year just be luck? Typical major leaguers bat about 500 times a season and hit about .260. A hitter's successive tries seem to be independent, so we have a binomial setting. From this model, we can calculate or simulate the probability of hitting .300. It is about 0.025. Out of 100 run-of-the-mill major league hitters, two or three each year will bat .300 because they were lucky.

This is not quite a binomial setting. Just as removing one card in Example 12.2 changes the makeup of the deck, removing one switch changes the proportion of bad switches remaining in the shipment. So the state of the second switch chosen is not independent of the first. But removing one switch from a shipment of 10,000 changes the makeup of the remaining 9999 switches very little. In practice, the distribution of X is very close to the binomial distribution with $n = 10$ and $p = 0.1$.

Example 12.3 shows how we can use the binomial distributions in the statistical setting of selecting an SRS. When the population is much larger than the sample, a count of successes in an SRS of size n has approximately the binomial distribution with n equal to the sample size and p equal to the proportion of successes in the population.

SAMPLING DISTRIBUTION OF A COUNT

Choose an SRS of size n from a population with proportion p of successes. When the population is much larger than the sample, the count X of successes in the sample has approximately the binomial distribution with parameters n and p.

APPLY YOUR KNOWLEDGE

In each of Exercises 12.1 to 12.3, X is a count. Does X have a binomial distribution? Give your reasons in each case.

12.1 M&Ms. Forty percent of all milk chocolate M&M candies are either red or yellow. X is the count of red or yellow candies in a package of 40.

12.2 More M&Ms. You choose M&M candies from a package until you get the first red or yellow. X is the number you choose before you stop.

12.3 Computer instruction. A student studies binomial distributions using computer-assisted instruction. After the lesson, the computer presents 10 problems. The student solves each problem and enters her answer. The computer gives additional instruction between problems if the answer is wrong. The count X is the number of problems that the student gets right.

12.4 I can't relax. Opinion polls find that 14% of Americans "never have time to relax."[1] If you take an SRS of 500 adults, what is the approximate distribution of the number in your sample who say they never have time to relax?

Binomial probabilities

We can find a formula for the probability that a binomial random variable takes any value by adding probabilities for the different ways of getting exactly that many successes in n observations. Here is the example we will use to show the idea.

EXAMPLE 12.4   Inheriting blood type

Each child born to a particular set of parents has probability 0.25 of having blood type O. If these parents have 5 children, what is the probability that exactly 2 of them have type O blood?

The count of children with type O blood is a binomial random variable X with $n = 5$ tries and probability $p = 0.25$ of a success on each try. We want $P(X = 2)$.

Because the method doesn't depend on the specific example, let's use "S" for success and "F" for failure for short. Do the work in two steps.

Step 1. Find the probability that a specific 2 of the 5 tries, say the first and the third, give successes. This is the outcome SFSFF. Because tries are independent, the multiplication rule for independent events applies. The probability we want is

$$P(\text{SFSFF}) = P(S)P(F)P(S)P(F)P(F)$$
$$= (0.25)(0.75)(0.25)(0.75)(0.75)$$
$$= (0.25)^2(0.75)^3$$

Step 2. Observe that the probability of *any one* arrangement of 2 S's and 3 F's has this same probability. This is true because we multiply together 0.25 twice and 0.75 three times whenever we have 2 S's and 3 F's. The probability that $X = 2$ is the probability of getting 2 S's and 3 F's in any arrangement whatsoever. Here are all the possible arrangements:

SSFFF	SFSFF	SFFSF	SFFFS	FSSFF
FSFSF	FSFFS	FFSSF	FFSFS	FFFSS

There are 10 of them, all with the same probability. The overall probability of 2 successes is therefore

$$P(X = 2) = 10(0.25)^2(0.75)^3 = 0.2637$$

The pattern of this calculation works for any binomial probability. To use it, we must count the number of arrangements of k successes in n observations. We use the following fact to do the counting without actually listing all the arrangements.

What looks random?

Toss a coin six times and record heads (H) or tails (T) on each toss. Which of these outcomes is more probable: HTHTTT or TTTHHH? Almost everyone says that HTHTTT is more probable, because TTTHHH does not "look random." In fact, both are equally probable. That heads has probability 0.5 says that about half of a very long sequence of tosses will be heads. It doesn't say that heads and tails must come close to alternating in the short run. The coin doesn't know what past outcomes were, and it can't try to create a balanced sequence.

BINOMIAL COEFFICIENT

The number of ways of arranging k successes among n observations is given by the **binomial coefficient**

$$\binom{n}{k} = \frac{n!}{k!\,(n-k)!}$$

for $k = 0, 1, 2, \ldots, n$.

factorial

The formula for binomial coefficients uses the **factorial** notation. For any positive whole number n, its factorial $n!$ is

$$n! = n \times (n-1) \times (n-2) \times \cdots \times 3 \times 2 \times 1$$

Also, $0! = 1$.

The larger of the two factorials in the denominator of a binomial coefficient will cancel much of the $n!$ in the numerator. For example, the binomial coefficient we need for Example 12.4 is

$$\binom{5}{2} = \frac{5!}{2!\,3!}$$

$$= \frac{(5)(4)(3)(2)(1)}{(2)(1) \times (3)(2)(1)}$$

$$= \frac{(5)(4)}{(2)(1)} = \frac{20}{2} = 10$$

The notation $\binom{n}{k}$ is *not* related to the fraction $\frac{n}{k}$. A helpful way to remember its meaning is to read it as "binomial coefficient n choose k." Binomial coefficients have many uses in mathematics, but we are interested in them only as an aid to finding binomial probabilities. The binomial coefficient $\binom{n}{k}$ counts the number of different ways in which k successes can be arranged among n observations. The binomial probability $P(X = k)$ is this count multiplied by the probability of any specific arrangement of the k successes. Here is the result we seek.

BINOMIAL PROBABILITY

If X has the binomial distribution with n observations and probability p of success on each observation, the possible values of X are $0, 1, 2, \ldots, n$. If k is any one of these values,

$$P(X = k) = \binom{n}{k} p^k (1-p)^{n-k}$$

EXAMPLE 12.5 Inspecting switches

The number X of switches that fail inspection in Example 12.3 has approximately the binomial distribution with $n = 10$ and $p = 0.1$.

The probability that no more than 1 switch fails is

$$P(X \leq 1) = P(X = 1) + P(X = 0)$$

$$= \binom{10}{1}(0.1)^1(0.9)^9 + \binom{10}{0}(0.1)^0(0.9)^{10}$$

$$= \frac{10!}{1!\,9!}(0.1)(0.3874) + \frac{10!}{0!\,10!}(1)(0.3487)$$

$$= (10)(0.1)(0.3874) + (1)(1)(0.3487)$$

$$= 0.3874 + 0.3487 = 0.7361$$

This calculation uses the facts that $0! = 1$ and that $a^0 = 1$ for any number a other than 0. We see that about 74% of all samples will contain no more than 1 bad switch. In fact, 35% of the samples will contain no bad switches. A sample of size 10 cannot be trusted to alert the engineer to the presence of unacceptable items in the shipment.

Using technology

The binomial probability formula is awkward to use, particularly for the probabilities of events that contain many outcomes. You can find tables of binomial probabilities $P(X = k)$ and cumulative probabilities $P(X \leq k)$ for selected values of n and p.

The most efficient way to do binomial calculations is to use technology. Figure 12.1 shows output for the calculation in Example 12.5 from a statistical software program, a spreadsheet, and a graphing calculator. We asked all three to give cumulative probabilities. Minitab and the TI-83 Plus have menu

Minitab

```
Session                                    _ □ ×

Cumulative Distribution Function

Binomial with n = 10 and p = 0.100000

        x       P( X <= x )
     0.00         0.3487
     1.00         0.7361
     2.00         0.9298
     3.00         0.9072
```

Microsoft Excel

```
Microsoft Excel - Book1                              _ □ ×
   A1              =    =BINOMDIST(1,10,0.1,TRUE)
            A                    B       C      D      E
  1              0.7361
  2
  Sheet1  Sheet2  Sheet3
```

Texas Instruments TI-83 Plus

```
binomcdf(10,0.1,
1)
              .7361
```

Figure 12.1 The binomial probability $P(X \leq 1)$ for Example 12.5. Output from statistical software, a spreadsheet, and a graphing calculator.

entries for binomial cumulative probabilities. Excel has no menu entry, but the worksheet function BINOMDIST is available. All three outputs agree with the result 0.7361 of Example 12.5.

APPLY YOUR KNOWLEDGE

12.5 Inheriting blood type. If the parents in Example 12.4 have 5 children, the number who have type O blood is a random variable X that has the binomial distribution with $n = 5$ and $p = 0.25$.

    (a) What are the possible values of X?

    (b) Find the probability of each value of X. Draw a histogram to display this distribution. (Because probabilities are long-run proportions, a histogram with the probabilities as the heights of the bars shows what the distribution of X would be in very many repetitions.)

12.6 Random-digit dialing. When an opinion poll or telemarketer calls residential telephone numbers at random, 20% of the calls reach a live person. You watch the random dialing machine make 15 calls. The number that reach a person has the binomial distribution with $n = 15$ and $p = 0.2$.

    (a) What is the probability that exactly 3 calls reach a person?

    (b) What is the probability that 3 or fewer calls reach a person?

12.7 Tax returns. The Internal Revenue Service reports that 8% of individual tax returns in 2000 showed an adjusted gross income of $100,000 or more. A random audit chooses 20 tax returns for careful study. What is the probability that more than 1 return shows an income of $100,000 or more? (Hint: It is easier to first find the probability that only 0 or 1 of the returns chosen shows an income this high.)

Binomial mean and standard deviation

If a count X has the binomial distribution based on n observations with probability p of success, what is its mean μ? That is, in very many repetitions of the binomial setting, what will be the average count of successes? We can guess the answer. If a basketball player makes 80% of her free throws, the mean number made in 10 tries should be 80% of 10, or 8. In general, the mean of a binomial distribution should be $\mu = np$. Here are the facts.

BINOMIAL MEAN AND STANDARD DEVIATION

If a count X has the binomial distribution with number of observations n and probability of success p, the **mean** and **standard deviation** of X are

$$\mu = np$$
$$\sigma = \sqrt{np(1 - p)}$$

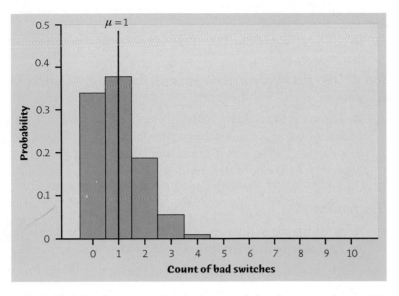

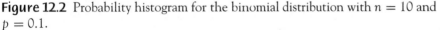

Figure 12.2 Probability histogram for the binomial distribution with $n = 10$ and $p = 0.1$.

Remember that these short formulas are good only for binomial distributions. They can't be used for other distributions.

EXAMPLE 12.6 Inspecting switches

Continuing Example 12.5, the count X of bad switches is binomial with $n = 10$ and $p = 0.1$. The histogram in Figure 12.2 displays this probability distribution. (Because probabilities are long-run proportions, using probabilities as the heights of the bars shows what the distribution of X would be in very many repetitions.) The distribution is strongly skewed. Although X can take any whole-number value from 0 to 10, the probabilities of values larger than 5 are so small that they do not appear in the histogram.

The mean and standard deviation of the binomial distribution in Figure 12.2 are

$$\mu = np = (10)(0.1) = 1$$
$$\sigma = \sqrt{np(1 - p)}$$
$$= \sqrt{(10)(0.1)(0.9)} = \sqrt{0.9} = 0.9487$$

The mean is marked on the probability histogram in Figure 12.2.

APPLY YOUR KNOWLEDGE

12.8 **Inheriting blood type.** What are the mean and standard deviation of the number of children with type O blood in Exercise 12.5? Mark the location of the mean on the probability histogram you made in that exercise.

12.9   Random-digit dialing

(a) What is the mean number of calls that reach a person in Exercise 12.6?

(b) What is the standard deviation σ of the count of calls that reach a person?

(c) If calls are made to New York City rather than nationally, the probability that a call reaches a person is only $p = 0.08$. What is σ for this p? What is σ if $p = 0.01$? What does your work show about the behavior of the standard deviation of a binomial distribution as the probability of a success gets closer to 0?

12.10   Tax returns

(a) What is the mean number of returns showing at least $100,000 of income among the 20 returns chosen in Exercise 12.7?

(b) What is the standard deviation σ of the number of returns with income at least $100,000?

(c) The probability that a return shows income less than $50,000 is 0.72. What is σ for the number of such returns in a sample of 20? The probability of income less than $200,000 is 0.98. What is σ for the count of these returns? What does your work show about the behavior of the standard deviation of a binomial distribution as the probability p of success gets closer to 1?

The Normal approximation to binomial distributions

The formula for binomial probabilities becomes awkward as the number of trials n increases. You can use software or a statistical calculator to handle some problems for which the formula is not practical. Here is another alternative: *as the number of trials n gets larger, the binomial distribution gets close to a Normal distribution.* When n is large, we can use Normal probability calculations to approximate hard-to-calculate binomial probabilities.

(AP/Wide World Photos)

EXAMPLE 12.7   Attitudes toward shopping

Are attitudes toward shopping changing? Sample surveys show that fewer people enjoy shopping than in the past. A survey asked a nationwide random sample of 2500 adults if they agreed or disagreed that "I like buying new clothes, but shopping is often frustrating and time-consuming."[2] The population that the poll wants to draw conclusions about is all U.S. residents aged 18 and over. Suppose that in fact 60% of all adult U.S. residents would say "Agree" if asked the same question. What is the probability that 1520 or more of the sample agree?

Because there are more than 210 million adults, we can take the responses of 2500 randomly chosen adults to be independent. So the number in our sample who agree that shopping is frustrating is a random variable X having the binomial distribution with $n = 2500$ and $p = 0.6$. To find the probability that

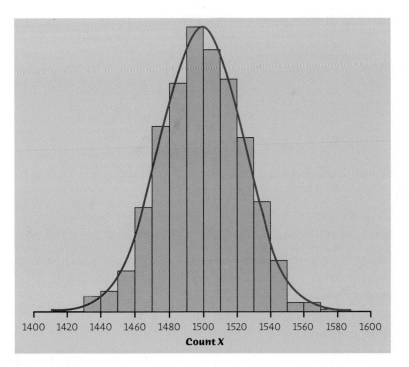

Figure 12.3 Histogram of 1000 binomial counts ($n = 2500$, $p = 0.6$) and the Normal density curve that approximates this binomial distribution.

at least 1520 of the people in the sample find shopping frustrating, we must add the binomial probabilities of all outcomes from $X = 1520$ to $X = 2500$. This isn't practical. Here are three ways to do this problem.

1. Use technology, as in Figure 12.1. The result is

$$P(X \geq 1520) = 0.2131$$

2. We can simulate a large number of repetitions of the sample. Figure 12.3 displays a histogram of the counts X from 1000 samples of size 2500 when the truth about the population is $p = 0.6$. Because 221 of these 1000 samples have X at least 1520, the probability estimated from the simulation is

$$P(X \geq 1520) = \frac{221}{1000} = 0.221$$

3. Both of the previous methods require software. Instead, look at the Normal curve in Figure 12.3. This is the density curve of the Normal distribution with the same mean and standard deviation as the binomial variable X:

$$\mu = np = (2500)(0.6) = 1500$$
$$\sigma = \sqrt{np(1-p)} = \sqrt{(2500)(0.6)(0.4)} = 24.49$$

As the figure shows, this Normal distribution approximates the binomial distribution quite well. So we can do a Normal calculation.

EXAMPLE 12.8   Normal calculation of a binomial probability

If we act as though the count X has the $N(1500, 24.49)$ distribution, here is the probability we want, using Table A:

$$P(X \geq 1520) = P\left(\frac{X - 1500}{24.49} \geq \frac{1520 - 1500}{24.49}\right)$$

$$= P(Z \geq 0.82)$$

$$= 1 - 0.7939 = 0.2061$$

The Normal approximation 0.2061 differs from the software result 0.2131 by only 0.007.

NORMAL APPROXIMATION FOR BINOMIAL DISTRIBUTIONS

Suppose that a count X has the binomial distribution with n trials and success probability p. When n is large, the distribution of X is approximately Normal, $N\left(np, \sqrt{np(1-p)}\right)$.

As a rule of thumb, we will use the Normal approximation when n and p satisfy $np \geq 10$ and $n(1-p) \geq 10$.

The Normal approximation is easy to remember because it says that X is Normal with its binomial mean and standard deviation. The accuracy of the Normal approximation improves as the sample size n increases. It is most accurate for any fixed n when p is close to 1/2 and least accurate when p is near 0 or 1. Whether or not you use the Normal approximation should depend on how accurate your calculations need to be. For most statistical purposes great accuracy is not required. Our "rule of thumb" for use of the Normal approximation reflects this judgment.

APPLY YOUR KNOWLEDGE

12.11 Using Benford's law. According to Benford's law (Example 9.5, page 231) the probability that the first digit of the amount of a randomly chosen invoice is a 1 or a 2 is 0.477. You examine 90 invoices from a vendor and find that 29 have first digits 1 or 2. If Benford's law holds, the count of 1s and 2s will have the binomial distribution with $n = 90$ and $p = 0.477$. Too few 1s and 2s suggests fraud. What is the approximate probability of 29 or fewer if the invoices follow Benford's law? Do you suspect that the invoice amounts are not genuine?

12.12 Mark McGwire's home runs. In 1998, Mark McGwire of the St. Louis Cardinals hit 70 home runs, a new major league record. Was this feat as surprising as most of us thought? In the three seasons before 1998, McGwire hit a home run in 11.6% of his times at bat. He went to bat 509 times in 1998. McGwire's home run count in 509 times at bat has approximately the binomial distribution with $n = 509$ and $p = 0.116$. What is the mean number of home runs he will hit in 509 times at bat? What is the probability of 70 or more home runs? (Use the Normal approximation.)

12.13 Checking for survey errors. One way of checking the effect of undercoverage, nonresponse, and other sources of error in a sample survey is to compare the sample with known facts about the population. About 12% of American adults are black. The number X of blacks in a random sample of 1500 adults should therefore vary with the binomial ($n = 1500$, $p = 0.12$) distribution.

(a) What are the mean and standard deviation of X?

(b) Use the Normal approximation to find the probability that the sample will contain 170 or fewer blacks. Be sure to check that you can safely use the approximation.

Chapter 12 SUMMARY

A count X of successes has a **binomial distribution** in the **binomial setting:** there are n observations; the observations are independent of each other; each observation results in a success or a failure; and each observation has the same probability p of a success.

The binomial distribution with n observations and probability p of success gives a good approximation to the sampling distribution of the count of successes in an SRS of size n from a large population containing proportion p of successes.

If X has the binomial distribution with parameters n and p, the possible values of X are the whole numbers $0, 1, 2, \ldots, n$. The **binomial probability** that X takes any value is

$$P(X = k) = \binom{n}{k} p^k (1 - p)^{n-k}$$

Binomial probabilities in practice are best found using software.

The **binomial coefficient**

$$\binom{n}{k} = \frac{n!}{k!\,(n-k)!}$$

counts the number of ways k successes can be arranged among n observations. Here the **factorial $n!$** is

$$n! = n \times (n - 1) \times (n - 2) \times \cdots \times 3 \times 2 \times 1$$

for positive whole numbers n, and $0! = 1$.

The **mean** and **standard deviation** of a binomial count X are

$$\mu = np$$

$$\sigma = \sqrt{np(1 - p)}$$

The **Normal approximation** to the binomial distribution says that if X is a count having the binomial distribution with parameters n and p, then when n is large, X is approximately $N\left(np, \sqrt{np(1-p)}\right)$. We will use this approximation when $np \geq 10$ and $n(1-p) \geq 10$.

Chapter 12 EXERCISES

12.14 Binomial setting? In each situation below, is it reasonable to use a binomial distribution for the random variable X? Give reasons for your answer in each case.

(AP/Wide World Photos)

(a) An auto manufacturer chooses one car from each hour's production for a detailed quality inspection. One variable recorded is the count X of finish defects (dimples, ripples, etc.) in the car's paint.

(b) The pool of potential jurors for a murder case contains 100 persons chosen at random from the adult residents of a large city. Each person in the pool is asked whether he or she opposes the death penalty; X is the number who say "Yes."

(c) Joe buys a ticket in his state's "Pick 3" lottery game every week; X is the number of times in a year that he wins a prize.

12.15 Binomial setting? In each of the following cases, decide whether or not a binomial distribution is an appropriate model, and give your reasons.

(a) Fifty students are taught about binomial distributions by a television program. After completing their study, all students take the same examination. The number of students who pass is counted.

(LWA-Stephen Welstead/CORBIS)

(b) A student studies binomial distributions using computer-assisted instruction. After the initial instruction is completed, the computer presents 10 problems. The student solves each problem and enters the answer; the computer gives additional instruction between problems if the student's answer is wrong. The number of problems that the student solves correctly is counted.

(c) A chemist repeats a solubility test 10 times on the same substance. Each test is conducted at a temperature 10° higher than the previous test. She counts the number of times that the substance dissolves completely.

12.16 Random digits. Each entry in a table of random digits like Table B has probability 0.1 of being a 0, and digits are independent of each other.

(a) What is the probability that a group of five digits from the table will contain at least one 0?

(b) What is the mean number of 0s in lines 40 digits long?

12.17 **Universal donors.** People with type O-negative blood are universal donors whose blood can be safely given to anyone. Only 7.2% of the population have O-negative blood. A blood center is visited by 20 donors in an afternoon. What is the probability that there are at least 2 universal donors among them?

12.18 **Testing ESP.** In a test for ESP (extrasensory perception), a subject is told that cards the experimenter can see but he cannot contain either a star, a circle, a wave, or a square. As the experimenter looks at each of 20 cards in turn, the subject names the shape on the card. A subject who is just guessing has probability 0.25 of guessing correctly on each card.

    (a) The count of correct guesses in 20 cards has a binomial distribution. What are n and p?

    (b) What is the mean number of correct guesses in many repetitions?

    (c) What is the probability of exactly 5 correct guesses?

12.19 **Random stock prices.** A believer in the "random walk" theory of stock markets thinks that an index of stock prices has probability 0.65 of increasing in any year. Moreover, the change in the index in any given year is not influenced by whether it rose or fell in earlier years. Let X be the number of years among the next 5 years in which the index rises.

    (a) X has a binomial distribution. What are n and p?

    (b) What are the possible values that X can take?

    (c) Find the probability of each value of X. Draw a probability histogram for the distribution of X.

    (d) What are the mean and standard deviation of this distribution? Mark the location of the mean on your histogram.

12.20 **How many cars?** Twenty percent of American households own three or more motor vehicles. You choose 12 households at random.

    (a) What is the probability that none of the chosen households owns three or more vehicles? What is the probability that at least one household owns three or more vehicles?

    (b) What are the mean and standard deviation of the number of households in your sample that own three or more vehicles?

    (c) What is the probability that your sample count is greater than the mean?

12.21 **False positives in testing for HIV.** The common test for the presence in the blood of antibodies to HIV, the virus that causes AIDS, gives a positive result with probability about 0.006 when a person who is free of HIV antibodies is tested. A clinic tests 1000 people who are all free of HIV antibodies.

    (a) What is the mean number of positive tests?

(b) What is the distribution of the number of positive tests?

(c) You cannot safely use the Normal approximation for this distribution. Explain why.

12.22 Multiple-choice tests. Here is a simple probability model for multiple-choice tests. Suppose that each student has probability p of correctly answering a question chosen at random from a universe of possible questions. (A strong student has a higher p than a weak student.) Answers to different questions are independent. Jodi is a good student for whom $p = 0.75$.

(a) Use the Normal approximation to find the probability that Jodi scores 70% or lower on a 100-question test.

(b) If the test contains 250 questions, what is the probability that Jodi will score 70% or lower?

12.23 Planning a survey. You are planning a sample survey of small businesses in your area. You will choose an SRS of businesses listed in the telephone book's Yellow Pages. Experience shows that only about half the businesses you contact will respond.

(a) If you contact 150 businesses, it is reasonable to use the binomial distribution with $n = 150$ and $p = 0.5$ for the number X who respond. Explain why.

(b) What is the mean number who respond to surveys like yours?

(c) What is the probability that 70 or fewer will respond? (Use the Normal approximation.)

(d) How large a sample must you take to increase the mean number of respondents to 100?

12.24 Survey demographics. According to the Census Bureau, 9.96% of American adults (age 18 and over) are Hispanics. An opinion poll plans to contact an SRS of 1200 adults.

(a) What is the mean number of Hispanics in such samples?

(b) What is the probability that the sample will contain fewer than 100 Hispanics? (Use the Normal approximation.)

12.25 Leaking gas tanks. Leakage from underground gasoline tanks at service stations can damage the environment. It is estimated that 25% of these tanks leak. You examine 15 tanks chosen at random, independently of each other.

(a) What is the mean number of leaking tanks in such samples of 15?

(b) What is the probability that 10 or more of the 15 tanks leak?

(c) Now you do a larger study, examining a random sample of 1000 tanks nationally. What is the probability that at least 275 of these tanks are leaking?

12.26 Language study. Of American high school students, 41% are studying a language other than English. An opinion poll plans to ask high school students about foreign affairs. Perhaps language study will influence attitudes. If the poll interviews an SRS of 500 students, what is the probability that between 35% and 50% of the sample are studying a foreign language? (Hint: First translate these percents into counts of the 500 students in the sample.)

12.27 Reaching dropouts. High school dropouts make up 13% of all Americans aged 18 to 24. A vocational school that wants to attract dropouts mails an advertising flyer to 25,000 persons between the ages of 18 and 24.

(a) If the mailing list can be considered a random sample of the population, what is the mean number of high school dropouts who will receive the flyer?

(b) What is the probability that at least 3500 dropouts will receive the flyer?

12.28 Is this coin balanced? While he was a prisoner of the Germans during World War II, John Kerrich tossed a coin 10,000 times. He got 5067 heads. Take Kerrich's tosses to be an SRS from the population of all possible tosses of his coin. If the coin is perfectly balanced, $p = 0.5$. Is there reason to think that Kerrich's coin gave too many heads to be balanced? To answer this question, find the probability that a balanced coin would give 5067 or more heads in 10,000 tosses. What do you conclude?

Chapter 12 MEDIA EXERCISES

12.29 Inspecting switches. Example 12.5 concerns the count of bad switches in inspection samples of size 10. The count has the binomial distribution with $n = 10$ and $p = 0.1$. Set these values for the number of tosses and probability of heads in the *Probability* applet. The example calculates that the probability of getting a sample with exactly 1 bad switch is 0.3874. Of course, when we inspect only a few lots, the proportion of samples with exactly 1 bad switch will differ from this probability. Click "Toss" and "Reset" repeatedly to simulate inspecting 20 lots. Record the number of bad switches (the count of heads) in each of the 20 samples. What proportion of the 20 lots had exactly 1 bad switch? Remember that probability tells us only what happens in the long run.

(David Lees/CORBIS)

Confidence Intervals: The Basics

After we have selected a sample, we know the responses of the individuals in the sample. Often we are not content with information about the sample. We want to *infer* from the sample data some conclusion about the wider population that the sample represents.

STATISTICAL INFERENCE

Statistical inference provides methods for drawing conclusions about a population from sample data.

We cannot be certain that our conclusions are correct—a different sample might lead to different conclusions. Statistical inference uses the language of probability to say how trustworthy its conclusions are.

This chapter and the following chapter introduce the two most common types of statistical inference. This chapter concerns *confidence intervals* for estimating the value of a population parameter. Chapter 14 introduces *tests of significance*, which assess the evidence for a claim about a population. Both types of inference are based on the sampling distributions of statistics. That is, both report probabilities that state what would happen if we used the inference method many times.

In these chapters, we concentrate on the reasoning of inference. To see the reasoning clearly, we act as if the world were simple, more simple than is realistic. Starting in Chapter 15, we will see how inference works in the real world of statistical practice. Here is the setting for our work in Chapters 13 and 14.

INFERENCE ABOUT A MEAN: SIMPLE CONDITIONS

1. We have an SRS from the population of interest.
2. The variable we measure has a perfectly Normal distribution $N(\mu, \sigma)$ in the population.
3. We don't know the population mean μ. Our task is to infer something about μ from the sample data. But we do know the population standard deviation σ.

In this chapter and the next, we will often say "Suppose we know" that the standard deviation σ has a particular value. Don't worry about how we might know this—later chapters will remove this unrealistic requirement.

Estimating with confidence

Young people have a better chance of good jobs and good wages if they are good with numbers. How strong are the quantitative skills of young Americans of working age? One source of data is the National Assessment of Educational Progress (NAEP) Young Adult Literacy Assessment Survey, which is based on a nationwide probability sample of households.

Beautiful theory, ugly facts

"Science is organized common sense where many a beautiful theory was killed by an ugly fact." So said Thomas Huxley. The job of statistical inference is to draw conclusions based on ugly facts. Educational theorists, for example, long pushed the "whole-language" approach to teaching reading and talked down the need for breaking words into basic sounds, called "phonics." In 2000, a national panel reviewed ugly facts from 52 randomized studies. Conclusion: no matter what theory says, phonics is essential in teaching reading.

EXAMPLE 13.1   NAEP quantitative scores

The NAEP survey includes a short test of quantitative skills, covering mainly basic arithmetic and the ability to apply it to realistic problems. Scores on the test range from 0 to 500. A person who scores 233 can add the amounts of two checks appearing on a bank deposit slip; someone scoring 325 can determine the price of a meal from a menu; a person scoring 375 can transform a price in cents per ounce into dollars per pound.

In a recent year, 840 men 21 to 25 years of age were in the NAEP sample. Their mean quantitative score was $\bar{x} = 272$. These 840 men are an SRS from the population of all young men. On the basis of this sample, we want to estimate the mean score μ in the population of all 9.5 million young men of these ages.[1]

Here is the reasoning of statistical estimation in a nutshell.

1. To estimate the unknown population mean μ, use the mean $\bar{x} = 272$ of a random sample.

2. We know that $\bar{x}$ is an unbiased estimate of μ. The law of large numbers says that $\bar{x}$ will always get close to μ if we take a large enough sample. Nonetheless, $\bar{x}$ will rarely be exactly equal to μ, so our estimate has some error.

3. We know the sampling distribution of $\bar{x}$. In repeated samples, $\bar{x}$ has the Normal distribution with mean μ and standard deviation $\sigma/\sqrt{n}$. Suppose we know that NAEP scores have standard deviation $\sigma = 60$. Then the average score $\bar{x}$ of 840 young men has standard deviation

$$\frac{\sigma}{\sqrt{n}} = \frac{60}{\sqrt{840}} = 2.1 \text{ (rounded off)}$$

4. The 95 part of the 68–95–99.7 rule for Normal distributions says that $\bar{x}$ and its mean μ are within 4.2 (that's two standard deviations) of each other in 95% of all samples. So if we estimate that μ lies somewhere in the interval from

$$\bar{x} - 4.2 = 272 - 4.2 = 267.8$$

to

$$\bar{x} + 4.2 = 272 + 4.2 = 276.2$$

we'll be right 95% of the time.

The big idea is that the sampling distribution of $\bar{x}$ tells us how big the error is likely to be when we use $\bar{x}$ to estimate μ. When we know the population standard deviation σ, this distribution is Normal, has known standard deviation, and is centered at the unknown population mean μ. (In practice, we rarely know σ. We'll deal with that real-world complication later.)

EXAMPLE 13.2   Statistical estimation in pictures

Figures 13.1 to 13.3 illustrate the reasoning of estimation in graphical form. Begin with Figure 13.1, and follow the flow from left to right. Starting with the population, imagine taking many SRSs of 840 young men. The first sample has mean NAEP score $\bar{x} = 272$, the second has mean $\bar{x} = 268$, the third has mean $\bar{x} = 273$, and so on. If we collect all these sample means and display their distribution, we get the Normal distribution with mean equal to the unknown μ and standard deviation 2.1.

Next, look at Figure 13.2. This is the same Normal curve as in Figure 13.1. The standard deviation of this curve is 2.1. The 68–95–99.7 rule says that in 95% of all samples, the sample mean $\bar{x}$ lies within two standard deviations of the population mean μ. That's the same as saying that the interval $\bar{x} \pm 4.2$ captures μ in 95% of all samples.

Finally, Figure 13.3 puts it all together. Starting with the population, imagine taking many SRSs of 840 young men. The recipe $\bar{x} \pm 4.2$ gives an interval based on each sample; 95% of these intervals capture the unknown population mean μ.

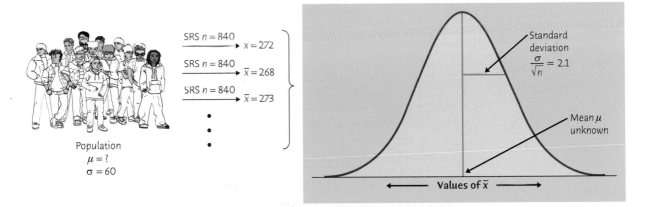

Figure 13.1 The sampling distribution of the mean score $\overline{x}$ of an SRS of 840 young men on the NAEP quantitative test.

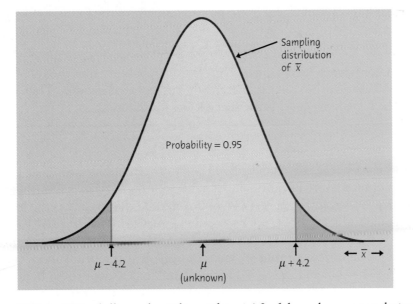

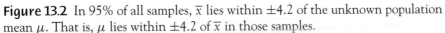

Figure 13.2 In 95% of all samples, $\overline{x}$ lies within ± 4.2 of the unknown population mean μ. That is, μ lies within ± 4.2 of $\overline{x}$ in those samples.

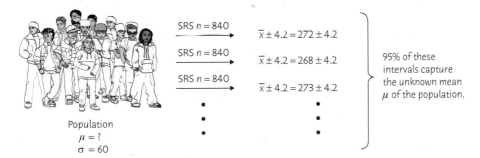

Figure 13.3 To say that $\overline{x} \pm 4.2$ is a 95% confidence interval for the population mean μ is to say that in repeated samples, 95% of these intervals capture μ.

The interval of numbers between the values $\bar{x} \pm 4.2$ is called a 95% *confidence interval* for μ. Like most confidence intervals we will meet, this one has the form

$$\text{estimate} \pm \text{margin of error}$$

margin of error

The estimate ($\bar{x}$ in this case) is our guess for the value of the unknown parameter. The **margin of error** ± 4.2 shows how accurate we believe our guess is, based on the variability of the estimate. This is a 95% confidence interval because it catches the unknown μ in 95% of all possible samples.

CONFIDENCE INTERVAL

A **level C confidence interval** for a parameter has two parts:

- An interval calculated from the data, usually of the form

$$\text{estimate} \pm \text{margin of error}$$

- A **confidence level** C, which gives the probability that the interval will capture the true parameter value in repeated samples. That is, the confidence level is the success rate for the method.

Users can choose the confidence level, most often 90% or higher because we most often want to be quite sure of our conclusions. Always remember that a confidence level is the success rate of the method that produces the interval. Our sample in Example 13.1 gives the first interval in Figure 13.3. We don't know whether this interval is one of the 95% that catch μ, or one of the unlucky 5%. The statement that we are 95% confident that the unknown μ lies between $272 - 4.2$ and $272 + 4.2$ is shorthand for "We got these numbers by a method that gives correct results 95% of the time."

Figure 13.3 is one way to picture the idea of a 95% confidence interval. Figure 13.4 illustrates the idea in a different form. Study these figures carefully. If you understand what they say, you have mastered one of the big ideas of statistics. Figure 13.4 shows the result of drawing many SRSs from the same population and calculating a 95% confidence interval from each sample. The center of each interval is at $\bar{x}$ and therefore varies from sample to sample. The sampling distribution of $\bar{x}$ appears at the top of the figure to show the long-term pattern of this variation. The 95% confidence intervals from 25 SRSs appear below. The center $\bar{x}$ of each interval is marked by a dot. The arrows on either side of the dot span the confidence interval. All except one of these 25 intervals cover the true value of μ. In a very large number of samples, 95% of the confidence intervals would contain μ. The *Confidence Interval* applet animates Figure 13.4. You can use the applet to watch confidence intervals from one sample after another capture or fail to capture the true parameter.

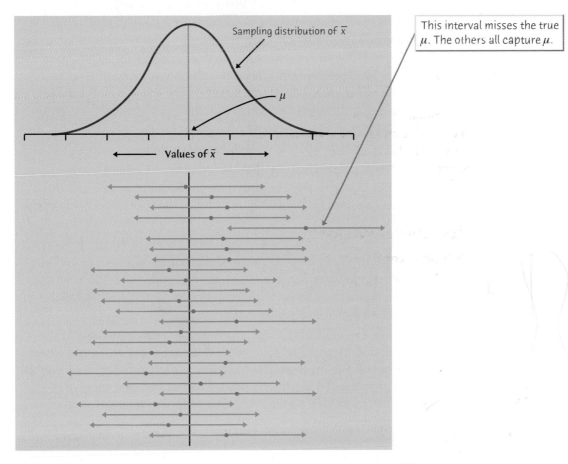

Figure 13.4 Twenty-five samples from the same population gave these 95% confidence intervals. In the long run, 95% of all samples give an interval that contains the population mean μ.

APPLY YOUR KNOWLEDGE

13.1 Losing weight. A Gallup Poll in November 2002 found that 51% of the people in its sample said "Yes" when asked, "Would you like to lose weight?" Gallup announced: "For results based on the total sample of national adults, one can say with 95% confidence that the margin of sampling error is ±3 percentage points."[2]

(a) What is the 95% confidence interval for the percent of all adults who want to lose weight?

(b) What does it mean to say that we have "95% confidence" in this interval?

13.2 Explaining confidence. A student reads that a 95% confidence interval for the mean NAEP quantitative score for men of ages 21 to 25 is 267.8 to 276.2. Asked to explain the meaning of this interval, the student says, "95% of all young men have scores between 267.8 and 276.2." Is the student right? Justify your answer.

13.3 **More on NAEP test scores.** Suppose that you give the NAEP test to an SRS of 1000 people from a large population in which the scores have mean 280 and standard deviation $\sigma = 60$. The mean $\bar{x}$ of the 1000 scores will vary if you take repeated samples.

(a) The sampling distribution of $\bar{x}$ is approximately Normal. It has mean $\mu = 280$. What is its standard deviation?

(b) Sketch the Normal curve that describes how $\bar{x}$ varies in many samples from this population. Mark the mean $\mu = 280$ and the values one, two, and three standard deviations on either side of the mean.

(c) According to the 68–95–99.7 rule, about 95% of all the values of $\bar{x}$ fall within _____ of the mean of this curve. What is the missing number? Call it m for "margin of error." Shade the region from the mean minus m to the mean plus m on the axis of your sketch, as in Figure 13.2.

(d) Whenever $\bar{x}$ falls in the region you shaded, the true value of the population mean, $\mu = 280$, lies in the confidence interval between $\bar{x} - m$ and $\bar{x} + m$. Draw the confidence interval below your sketch for one value of $\bar{x}$ inside the shaded region and one value of $\bar{x}$ outside the shaded region. (Use Figure 13.4 as a model for the drawing.)

(e) In what percent of all samples will the true mean $\mu = 280$ be covered by the confidence interval $\bar{x} \pm m$?

Confidence intervals for the mean μ

Our construction of a 95% confidence interval for the mean NAEP score of young men began by noting that any Normal distribution has probability about 0.95 within ±2 standard deviations of its mean. To construct a level C confidence interval, we first catch the central C area under a Normal curve. Because all Normal distributions are the same in the standard scale, we can obtain everything we need from the standard Normal curve.

Figure 13.5 shows the relationship between the central area C under a standard Normal curve and the points z^* and $-z^*$ that mark off this area. Numbers like z^* that mark off specified areas are called **critical values** of the standard Normal distribution. Values of z^* for many choices of C appear in the bottom row of Table C in the back of the book. This row is labeled z^*. Here are the entries for the most common confidence levels:

critical value

Confidence level C	90%	95%	99%
Critical value z^*	1.645	1.960	2.576

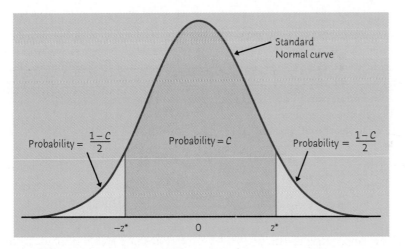

Figure 13.5 The critical value z^* is the number that catches central probability C under a standard Normal curve between $-z^*$ and z^*.

Notice that for C = 95% the table gives $z^* = 1.960$. This is slightly more precise than the value $z^* = 2$ based on the 68–95–99.7 rule.

As Figure 13.5 reminds us, there is area C under the standard Normal curve between $-z^*$ and z^*. If we start at the sample mean $\overline{x}$ and go out z^* standard deviations, we get an interval that contains the population mean μ in a proportion C of all samples. This interval is

$$\text{from } \overline{x} - z^* \frac{\sigma}{\sqrt{n}} \text{ to } \overline{x} + z^* \frac{\sigma}{\sqrt{n}}$$

or

$$\overline{x} \pm z^* \frac{\sigma}{\sqrt{n}}$$

It is a level C confidence interval for μ.

CONFIDENCE INTERVAL FOR THE MEAN OF A NORMAL POPULATION

Draw an SRS of size n from a Normal population having unknown mean μ and known standard deviation σ. A level C confidence interval for μ is

$$\overline{x} \pm z^* \frac{\sigma}{\sqrt{n}}$$

The critical value z^* is illustrated in Figure 13.5 and found in Table C.

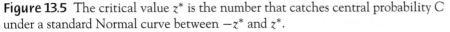

EXAMPLE 13.3 Analyzing pharmaceuticals

A manufacturer of pharmaceutical products analyzes a specimen from each batch of a product to verify the concentration of the active ingredient. The chemical analysis is not perfectly precise. Repeated measurements on the same specimen give slightly different results. Suppose we know that the results of repeated measurements follow a Normal distribution with mean μ equal to the true concentration and standard deviation $\sigma = 0.0068$ grams per liter. (That the mean of the population of all measurements is the true concentration says that the measurement process has no bias. The standard deviation describes the precision of the measurement.)

The laboratory analyzes each specimen three times and reports the mean result. Three analyses of one specimen give concentrations

$$0.8403 \quad 0.8363 \quad 0.8447$$

The lab reports the mean

$$\bar{x} = \frac{0.8403 + 0.8363 + 0.8447}{3} = 0.8404$$

We want a 99% confidence interval for the true concentration μ.

For 99% confidence, we see from Table C that $z^* = 2.576$. A 99% confidence interval for μ is therefore

$$\bar{x} \pm z^* \frac{\sigma}{\sqrt{n}} = 0.8404 \pm 2.576 \frac{0.0068}{\sqrt{3}}$$
$$= 0.8404 \pm 0.0101$$
$$= 0.8303 \text{ to } 0.8505$$

We are 99% confident that the true concentration lies between 0.8303 and 0.8505 grams per liter.

Suppose that a single measurement gave $x = 0.8404$, the same value that the sample mean took in Example 13.3. Repeating the calculation with $n = 1$ shows that the 99% confidence interval based on a single measurement is

$$\bar{x} \pm z^* \frac{\sigma}{\sqrt{1}} = 0.8404 \pm (2.576)(0.0068)$$
$$= 0.8404 \pm 0.0175$$
$$= 0.8229 \text{ to } 0.8579$$

The mean of three measurements gives a smaller margin of error and therefore a shorter interval than a single measurement. Figure 13.6 illustrates the gain from using three observations.

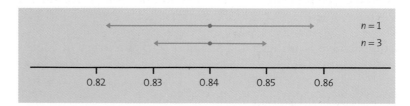

Figure 13.6 Confidence intervals for $n = 3$ and $n = 1$ for Example 13.3. Larger samples give shorter intervals.

APPLY YOUR KNOWLEDGE

13.4 Find a critical value. The critical value z^* for confidence level 97.5% is not in Table C. Use Table A of standard Normal probabilities to find z^*. Start by making a copy of Figure 13.5 with $C = 0.975$ that shows how much area is left in each tail when the central area is 0.975.

13.5 The tale of the scale. A laboratory scale is known to have a standard deviation of $\sigma = 0.001$ gram in repeated weighings. Scale readings in repeated weighings are Normally distributed, with mean equal to the true weight of the specimen. Three weighings of a specimen give (in grams)

$$3.412 \quad 3.414 \quad 3.415$$

Give a 95% confidence interval for the true weight of the specimen. What are the estimate and the margin of error in this interval?

13.6 IQ test scores. Here are the IQ test scores of 31 seventh-grade girls in a Midwest school district:[3]

114	100	104	89	102	91	114	114	103	105
108	130	120	132	111	128	118	119	86	72
111	103	74	112	107	103	98	96	112	112	93

(a) We expect the distribution of IQ scores to be close to Normal. Make a stemplot of the distribution of these 31 scores (split the stems) to verify that there are no major departures from Normality.

(b) Treat the 31 girls as an SRS of all seventh-grade girls in the school district. Suppose that the standard deviation of IQ scores in this population is known to be $\sigma = 15$. Give a 99% confidence interval for the mean score in the population.

How confidence intervals behave

The confidence interval $\bar{x} \pm z^* \sigma / \sqrt{n}$ for the mean of a Normal population illustrates several important properties that are shared by all confidence intervals in common use. The user chooses the confidence level, and the margin of error follows from this choice. We would like high confidence and also a small margin of error. High confidence says that our method almost always gives correct answers. A small margin of error says that we have pinned down the parameter quite precisely. The margin of error is

$$\text{margin of error} = z^* \frac{\sigma}{\sqrt{n}}$$

This expression has z^* and σ in the numerator and $\sqrt{n}$ in the denominator. So the margin of error gets smaller when

- z^* gets smaller. Smaller z^* is the same as smaller confidence level C (look at Figure 13.5 again). There is a trade-off between the confidence level and the margin of error. To obtain a smaller margin of error from the same data, you must be willing to accept lower confidence.

- σ gets smaller. The standard deviation σ measures the variation in the population. You can think of the variation among individuals in the population as noise that obscures the average value μ. It is easier to pin down μ when σ is small.

- n gets larger. Increasing the sample size n reduces the margin of error for any fixed confidence level. Because n appears under a square root sign, we must take four times as many observations in order to cut the margin of error in half.

EXAMPLE 13.4   Changing the margin of error

Suppose that the pharmaceutical manufacturer in Example 13.3 is content with 90% confidence rather than 99%. Table C gives the critical value for 90% confidence as $z^* = 1.645$. The 90% confidence interval for μ based on three repeated measurements with mean $\bar{x} = 0.8404$ is

$$\bar{x} \pm z^* \frac{\sigma}{\sqrt{n}} = 0.8404 \pm 1.645 \, \frac{0.0068}{\sqrt{3}}$$

$$= 0.8404 \pm 0.0065$$

$$= 0.8339 \text{ to } 0.8469$$

Settling for 90%, rather than 99%, confidence has reduced the margin of error from ± 0.0101 to ± 0.0065. Figure 13.7 compares these two intervals.

Increasing the number of measurements from 3 to 12 will also reduce the width of the 99% confidence interval in Example 13.3. Check that replacing $\sqrt{3}$ by $\sqrt{12}$ cuts the ± 0.0101 margin of error in half, because we now have four times as many observations.

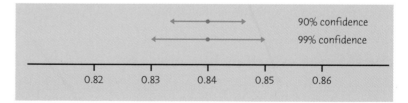

Figure 13.7 90% and 99% confidence intervals for Example 13.4. Higher confidence requires a wider interval.

APPLY YOUR KNOWLEDGE

13.7 **Confidence level and interval length.** Examples 13.3 and 13.4 give confidence intervals for the concentration μ based on 3 measurements with $\bar{x} = 0.8404$ and $\sigma = 0.0068$. The 99% confidence interval is 0.8303 to 0.8505 and the 90% confidence interval is 0.8339 to 0.8469.

   (a) Find the 80% confidence interval for μ.

   (b) Find the 99.9% confidence interval for μ.

(c) Make a sketch like Figure 13.7 to compare all four intervals. How does increasing the confidence level affect the length of the confidence interval?

13.8 Confidence level and margin of error. High school students who take the SAT mathematics exam a second time generally score higher than on their first try. The change in score has a Normal distribution with standard deviation $\sigma = 50$. A random sample of 1000 students gains an average of $\bar{x} = 22$ points on their second try.[4]

(a) Give a 95% confidence interval for the mean score gain μ in the population of all students.

(b) Give the 90% and 99% confidence intervals for μ.

(c) What are the margins of error for 90%, 95%, and 99% confidence? How does increasing the confidence level affect the margin of error of a confidence interval?

13.9 Sample size and margin of error. A sample of 1000 high school students gained an average of $\bar{x} = 22$ points in their second attempt at the SAT mathematics exam. The change in score has a Normal distribution with standard deviation $\sigma = 50$.

(a) Give a 95% confidence interval for the mean score gain μ in the population of all students.

(b) Suppose that the same result, $\bar{x} = 22$, had come from a sample of 250 students. Give the 95% confidence interval for the population mean μ in this case.

(c) Then suppose that a sample of 4000 students had produced the sample mean $\bar{x} = 22$. Again give the 95% confidence interval for μ.

(d) What are the margins of error for samples of size 250, 1000, and 4000? How does increasing the sample size affect the margin of error of a confidence interval?

Choosing the sample size

A wise user of statistics never plans data collection without at the same time planning the inference. You can arrange to have both high confidence and a small margin of error by taking enough observations. The margin of error of the confidence interval for the mean of a Normally distributed population is $m = z^* \sigma / \sqrt{n}$. To obtain a desired margin of error m, put in the value of z^* for your desired confidence level, and solve for the sample size n. Here is the result.

SAMPLE SIZE FOR DESIRED MARGIN OF ERROR

The confidence interval for the mean of a Normal population will have a specified margin of error m when the sample size is

$$n = \left(\frac{z^* \sigma}{m} \right)^2$$

This formula is not the proverbial free lunch. Taking observations costs time and money. The required sample size may be impossibly expensive. Notice that it is the size of the *sample* that determines the margin of error. The size of the *population* does not influence the sample size we need. (This is true as long as the population is much larger than the sample.)

EXAMPLE 13.5 **How many observations?**

Management asks the laboratory of Example 13.3 to produce results accurate to within ±0.005 with 95% confidence. How many measurements must be averaged to comply with this request?

The desired margin of error is $m = 0.005$. For 95% confidence, Table C gives $z^* = 1.960$. We know that $\sigma = 0.0068$. Therefore,

$$n = \left(\frac{z^*\sigma}{m}\right)^2 = \left(\frac{1.96 \times 0.0068}{0.005}\right)^2 = 7.1$$

Because 7 measurements will give a slightly larger margin of error than desired, and 8 measurements a slightly smaller margin of error, the lab must take 8 measurements on each specimen to meet management's demand. Always round *up* to the next higher whole number when finding n. On learning the cost of this many measurements, management may reconsider its request.

APPLY YOUR KNOWLEDGE

13.10 Calibrating a scale. To assess the accuracy of a laboratory scale, a standard weight known to weigh 10 grams is weighed repeatedly. The scale readings are Normally distributed with unknown mean (this mean is 10 grams if the scale has no bias). The standard deviation of the scale readings is known to be 0.0002 gram.

    (a) The weight is weighed five times. The mean result is 10.0023 grams. Give a 98% confidence interval for the mean of repeated measurements of the weight.

    (b) How many measurements must be averaged to get a margin of error of ±0.0001 with 98% confidence?

13.11 Estimating mean IQ. How large a sample of schoolgirls in Exercise 13.6 would be needed to estimate the mean IQ score μ within ±5 points with 99% confidence?

13.12 Improving SAT scores. How large a sample of high school students in Exercise 13.9 would be needed to estimate the mean change in SAT score μ to within ±2 points with 95% confidence?

Chapter 13 SUMMARY

A **confidence interval** uses sample data to estimate an unknown population parameter with an indication of how accurate the estimate is and of how confident we are that the result is correct.

Any confidence interval has two parts: an interval calculated from the data and a confidence level C. The **interval** often has the form

$$\text{estimate} \pm \text{margin of error}$$

The **confidence level** is the success rate of the method that produces the interval. That is, C is the probability that the method will give a correct answer. If you use 95% confidence intervals often, in the long run 95% of your intervals will contain the true parameter value. You do not know whether a 95% confidence interval calculated from a particular set of data contains the true parameter value.

A level C **confidence interval for the mean** μ of a Normal population with known standard deviation σ, based on an SRS of size n, is given by

$$\bar{x} \pm z^* \frac{\sigma}{\sqrt{n}}$$

The **critical value** z^* is chosen so that the standard Normal curve has area C between $-z^*$ and z^*.

Other things being equal, the **margin of error** of a confidence interval gets smaller as

- the confidence level C decreases,
- the population standard deviation σ decreases, and
- the sample size n increases.

The sample size required to obtain a confidence interval with specified margin of error m for a Normal mean is

$$n = \left(\frac{z^* \sigma}{m}\right)^2$$

where z^* is the critical value for the desired level of confidence. Always round n up when you use this formula.

Chapter 13 EXERCISES

13.13 Bone loss by nursing mothers. Breast-feeding mothers secrete calcium into their milk. Some of the calcium may come from their bones, so mothers may lose bone mineral. Researchers measured the percent change in mineral content of the spines of 47 mothers during three months of breast-feeding.[5] Here are the data:

−4.7	−2.5	−4.9	−2.7	−0.8	−5.3	−8.3	−2.1	−6.8	−4.3
2.2	−7.8	−3.1	−1.0	−6.5	−1.8	−5.2	−5.7	−7.0	−2.2
−6.5	−1.0	−3.0	−3.6	−5.2	−2.0	−2.1	−5.6	−4.4	−3.3
−4.0	−4.9	−4.7	−3.8	−5.9	−2.5	−0.3	−6.2	−6.8	1.7
0.3	−2.3	0.4	−5.3	0.2	−2.2	−5.1			

(SW Productions/Photo Disc/Getty Images)

(a) Make a stemplot of the data. (Don't forget that you need both a 0 and a −0 stem because there are both positive and negative values.) The data appear to follow a Normal distribution quite closely.

(b) Suppose that the percent change in the population of all nursing mothers has standard deviation $\sigma = 2.5\%$. Give a 99% confidence interval for the mean percent change in the population.

13.14 **Healing of skin wounds.** Biologists studying the healing of skin wounds measured the rate at which new cells closed a razor cut made in the skin of an anesthetized newt. Here are data from 18 newts, measured in micrometers (millionths of a meter) per hour:[6]

| 29 | 27 | 34 | 40 | 22 | 28 | 14 | 35 | 26 |
| 35 | 12 | 30 | 23 | 18 | 11 | 22 | 23 | 33 |

(a) Make a stemplot of the healing rates (split the stems). It is difficult to assess Normality from 18 observations, but look for outliers or extreme skewness. What do you find?

(b) Scientists usually assume that animal subjects are SRSs from their species or genetic type. Treat these newts as an SRS and suppose you know that the standard deviation of healing rates for this species of newt is 8 micrometers per hour. Give a 90% confidence interval for the mean healing rate for the species.

(c) A friend who knows almost no statistics uses the formula $\bar{x} \pm 1.96\sigma/\sqrt{n}$ in a biology lab manual to get a 95% confidence interval for the mean. Is her interval wider or narrower than yours? Explain to her why it makes sense that higher confidence changes the length of the interval.

13.15 **Pulling wood apart.** How heavy a load (pounds) is needed to pull apart pieces of Douglas fir 4 inches long and 1.5 inches square? Here are data from students doing a laboratory exercise:

33,190	31,860	32,590	26,520	33,280
32,320	33,020	32,030	30,460	32,700
23,040	30,930	32,720	33,650	32,340
24,050	30,170	31,300	28,730	31,920

(a) Suppose that the strength of pieces of wood like these follows a Normal distribution with standard deviation 3000 pounds. Give a 90% confidence interval for the mean load required to pull the wood apart.

(b) We are willing to regard the wood pieces prepared for the lab session as an SRS of all similar pieces of Douglas fir. Engineers also commonly assume that characteristics of materials vary Normally.

(David Austen/Stock, Boston)

Make a graph to show the shape of the distribution for these data. Does the Normality assumption appear safe?

13.16 This wine stinks. Sulfur compounds cause "off-odors" in wine, so winemakers want to know the odor threshold, the lowest concentration of a compound that the human nose can detect. The odor threshold for dimethyl sulfide (DMS) in trained wine tasters is about 25 micrograms per liter of wine (μg/l). The untrained noses of consumers may be less sensitive, however. Here are the DMS odor thresholds for 10 untrained students:

$$31 \quad 31 \quad 43 \quad 36 \quad 23 \quad 34 \quad 32 \quad 30 \quad 20 \quad 24$$

Assume that the standard deviation of the odor threshold for untrained noses is known to be $\sigma = 7 \ \mu$g/l.

(a) Make a stemplot to verify that the distribution is roughly symmetric with no outliers. (More data confirm that there are no systematic departures from Normality.)

(b) Give a 95% confidence interval for the mean DMS odor threshold among all students.

13.17 Engine crankshafts. Here are measurements (in millimeters) of a critical dimension on a sample of auto engine crankshafts:

$$
\begin{array}{cccccc}
224.120 & 224.001 & 224.017 & 223.982 & 223.989 & 223.961 \\
223.960 & 224.089 & 223.987 & 223.976 & 223.902 & 223.980 \\
224.098 & 224.057 & 223.913 & 223.999
\end{array}
$$

The data come from a production process that is known to have standard deviation $\sigma = 0.060$ mm. The process mean is supposed to be $\mu = 224$ mm but can drift away from this target during production.

(a) We expect the distribution of the dimension to be close to Normal. Make a stemplot or histogram of these data and describe the shape of the distribution.

(b) Give a 95% confidence interval for the process mean at the time these crankshafts were produced.

13.18 Student study times. A class survey in a large class for first-year college students asked, "About how many minutes do you study on a typical weeknight?" The mean response of the 269 students was $\bar{x} = 137$ minutes. Suppose that we know that the study time follows a Normal distribution with standard deviation $\sigma = 65$ minutes in the population of all first-year students at this university.

(a) Use the survey result to give a 99% confidence interval for the mean study time of all first-year students.

(b) What condition not yet mentioned is needed for your confidence interval to be valid?

13.19 A big toe deformity. Table I.1 (page 158) gives data on 38 consecutive patients who came to a medical center for treatment of hallux abducto valgus (HAV), a deformation of the big toe. It is reasonable to consider these patients as an SRS of people suffering from HAV. The seriousness of the deformity is measured by the angle (in degrees) of deformity.

(a) The data contain one high outlier. What is the angle for this outlier? The presence of the outlier violates the conditions for our confidence interval. Suppose that there is a good medical reason for removing the outlier.

(b) The remaining 37 observations follow a Normal distribution closely. Assume that angle has a Normal distribution with standard deviation $\sigma = 6.3$ degrees. Give a 95% confidence interval for the mean angle of deformity in the population.

13.20 An outlier strikes. There were actually 270 responses to the class survey in Exercise 13.18. One student claimed to study 30,000 minutes per night. We know he's joking, so we left out this value. If we did a calculation without looking at the data, we would get $\bar{x} = 248$ minutes for all 270 students. Now what is the 99% confidence interval for the population mean? (Continue to use $\sigma = 65$.) Compare the new interval with that in Exercise 13.18. The message is clear: always look at your data, because outliers can greatly change your result.

13.21 Healing of skin wounds, continued. How large a sample would enable you to estimate the mean healing rate of skin wounds in newts (see Exercise 13.14) within a margin of error of 1 micrometer per hour with 90% confidence?

13.22 Pulling wood apart, continued. You want to estimate the mean load needed to pull apart the pieces of wood in Exercise 13.15 to within ± 1000 pounds with 95% confidence. How large a sample is needed?

13.23 Crime. A Gallup Poll of 1002 adults in October 2002 found that 25% of the respondents said that their household had experienced a crime in the past year. Among respondents aged 18 to 29 years, 43% had been victims of a crime. Gallup says, "For results based on the total sample of national adults, one can say with 95% confidence that the margin of sampling error is ± 3 percentage points." Is the margin of error for adults aged 18 to 29 smaller or larger than ± 3 percentage points? Why?

13.24 A newspaper poll. A *New York Times* poll on women's issues interviewed 1025 women and 472 men randomly selected from the United States, excluding Alaska and Hawaii. The poll announced a margin of error of ± 3 percentage points for 95% confidence in conclusions about women. The margin of error for results concerning men was ± 4 percentage points. Why is this larger than the margin of error for women?

Exercises 13.25 to 13.27 are based on the following situation. A Gallup Poll in 2001 asked 1060 randomly selected adults, "How would you rate the overall quality of the environment in this country today—as excellent, good, only fair, or poor?" In all, 46% of the sample rated the environment as good or excellent. Gallup said that "one can say with 95% confidence that the margin of sampling error is ±3 percentage points."

(Getty Images)

13.25 Changing confidence. Would a 90% confidence interval based on the poll results have a margin of error less than ±3 percentage points, equal to ±3 percentage points, or greater than ±3 percentage points? Explain your answer.

13.26 A larger sample. If the poll had interviewed 1500 persons rather than 1060 (and still found 46% rating the environment as good or excellent), would the margin of error for 95% confidence be less than ±3 percentage points, equal to ±3 percentage points, or greater than ±3 percentage points? Explain your answer.

13.27 A different population. Suppose that the poll had obtained the outcome 46% by a similar random sampling method from all adults in New York State (population 19 million) instead of from all adults in the United States (population 281 million). Would the margin of error for 95% confidence be less than ±3 percentage points, equal to ±3 percentage points, or greater than ±3 percentage points? Explain your answer.

13.28 Would women govern better? A Gallup Poll in December 2000 asked, "Do you think this country would be governed better or governed worse if more women were in political office?" Of the 1026 adults in the sample, 57% said "better." Gallup added, "For results based on the total sample of National Adults, one can say with 95% confidence that the margin of sampling error is ±3 percentage points." Explain to someone who knows no statistics what the phrase "95% confidence" means here.

13.29 Explaining statistical confidence. Here is an explanation from the Associated Press concerning one of its opinion polls. Explain briefly but clearly in what way this explanation is incorrect.

For a poll of 1600 adults, the variation due to sampling error is no more than three percentage points either way. The error margin is said to be valid at the 95 percent confidence level. This means that, if the same questions were repeated in 20 polls, the results of at least 19 surveys would be within three percentage points of the results of this survey.

Chapter 13 MEDIA EXERCISES

13.30 80% confidence intervals. The idea of an 80% confidence interval is that the interval captures the true parameter value in 80% of all samples. That's not high enough confidence for practical use, but 80%

hits and 20% misses make it easy to see how a confidence interval behaves in repeated samples from the same population. Go to the *Confidence Interval* applet.

(a) Set the confidence level to 80%. Click "Sample" to choose an SRS and calculate the confidence interval. Do this 10 times to simulate 10 SRSs with their 10 confidence intervals. How many of the 10 intervals captured the true mean μ? How many missed?

(b) You see that we can't predict whether the next sample will hit or miss. The confidence level, however, tells us what percent will hit in the long run. Reset the applet and click "Sample 50" to get the confidence intervals from 50 SRSs. How many hit? Keep clicking "Sample 50" and record the percent of hits among 100, 200, 300, 400, 500, 600, 700, 800, and 1000 SRSs. Even 1000 samples is not truly "the long run," but we expect the percent of hits in 1000 samples to be fairly close to the confidence level, 80%.

13.31 What confidence means. Confidence tells us how often our method will produce an interval that captures the true population parameter if we use the method a very large number of times. The *Confidence Interval* applet allows us to actually use the method many times.

(a) Set the confidence level to 90%. Click "Sample 50" to choose 50 SRSs and calculate the confidence intervals. How many captured the true population mean μ? Keep clicking "Sample 50" until you have 1000 samples. What percent of the 1000 confidence intervals captured the true μ?

(b) Now choose 95% confidence. Look carefully when you first click "Sample 50." Are these intervals longer or shorter than the 90% confidence intervals? Again take 1000 samples. What percent of the intervals captured the true μ?

(c) Do the same thing for 99% confidence. What percent of 1000 samples gave confidence intervals that caught the true mean? Did the behavior of many intervals for the three confidence levels closely reflect the choice of confidence level?

13.32 An interactive table of critical values. The bottom row of Table C shows critical values for the standard Normal distribution. Use the *Normal Curve* applet to verify that the critical value needed for 95% confidence is $z = 1.96$. To do this, move the flags in the applet until they mark off the central area 0.95 under the curve. (Try dragging one flag across the other so that the applet tells you the area between them.) Make a sketch of the curve from the applet marked with the values of p and z.

13.33 Confidence level 92.5%. What standard normal critical value z^* is required for a 92.5% confidence interval for a population mean? The value isn't in Table C, but the *Normal Curve* applet allows you to find it by moving the limits in the applet until you have marked off the central 92.5% of a standard Normal curve.

Use your result to give the 92.5% confidence interval for the population mean in the following setting. The yield (bushels per acre) of a variety of corn has standard deviation $\sigma = 10$ bushels per acre. Fifteen plots have these yields:

| 138.0 | 139.1 | 113.0 | 132.5 | 140.7 | 109.7 | 118.9 | 134.8 |
| 109.6 | 127.3 | 115.6 | 130.4 | 130.2 | 111.7 | 105.5 | |

(Roger Ressmeyer/CORBIS)

Tests of Significance: The Basics

Confidence intervals are one of the two most common types of statistical inference. Use a confidence interval when your goal is to estimate a population parameter. The second common type of inference, called *tests of significance*, has a different goal: to assess the evidence provided by data about some claim concerning a population. Here is the reasoning of statistical tests in a nutshell.

EXAMPLE 14.1 I'm a great free-throw shooter

I claim that I make 80% of my basketball free throws. To test my claim, you ask me to shoot 20 free throws. I make only 8 of the 20. "Aha!" you say. "Someone who makes 80% of his free throws would almost never make only 8 out of 20. So I don't believe your claim."

Your reasoning is based on asking what would happen if my claim were true and we repeated the sample of 20 free throws many times—I would almost never make as few as 8. This outcome is so unlikely that it gives strong evidence that my claim is not true.

You can say how strong the evidence against my claim is by giving the probability that I would make as few as 8 out of 20 free throws if I really make 80% in the long run. This probability is 0.0001. I would make as few as 8 of 20 only once in 10,000 tries in the long run if my claim to make 80% is true. The small probability convinces you that my claim is false.

Significance tests use an elaborate vocabulary, but the basic idea is simple: an outcome that would rarely happen if a claim were true is good evidence that the claim is not true.

The reasoning of tests of significance

The reasoning of statistical tests, like that of confidence intervals, is based on asking what would happen if we repeated the sample or experiment many times. We will again start with the simple conditions listed on page 321: an SRS from an exactly Normal population with standard deviation σ known to us. Here is an example we will explore.

EXAMPLE 14.2   Sweetening colas

Diet colas use artificial sweeteners to avoid sugar. These sweeteners gradually lose their sweetness over time. Manufacturers therefore test new colas for loss of sweetness before marketing them. Trained tasters sip the cola along with drinks of standard sweetness and score the cola on a "sweetness score" of 1 to 10. The cola is then stored for a month at high temperature to imitate the effect of four months' storage at room temperature. Each taster scores the cola again after storage. This is a matched pairs experiment. Our data are the differences (score before storage minus score after storage) in the tasters' scores. The bigger these differences, the bigger the loss of sweetness.

Suppose we know that for any cola, the sweetness loss scores vary from taster to taster according to a Normal distribution with standard deviation $\sigma = 1$. The mean μ for all tasters measures loss of sweetness, and is different for different colas.

The following are the sweetness losses for a new cola, as measured by 10 trained tasters:

$$2.0 \quad 0.4 \quad 0.7 \quad 2.0 \quad -0.4 \quad 2.2 \quad -1.3 \quad 1.2 \quad 1.1 \quad 2.3$$

Most are positive. That is, most tasters found a loss of sweetness. But the losses are small, and two tasters (the negative scores) thought the cola gained sweetness. The average sweetness loss is given by the sample mean:

$$\bar{x} = \frac{2.0 + 0.4 + \cdots + 2.3}{10} = 1.02$$

Are these data good evidence that the cola lost sweetness in storage?

The reasoning is the same as in Example 14.1. We make a claim and ask if the data give evidence *against* it. We seek evidence that there *is* a sweetness loss, so the claim we test is that there *is not* a loss. In that case, the mean loss for the population of all trained testers would be $\mu = 0$.

• If the claim that $\mu = 0$ is true, the sampling distribution of $\bar{x}$ from 10 tasters is Normal with mean $\mu = 0$ and standard deviation

$$\frac{\sigma}{\sqrt{n}} = \frac{1}{\sqrt{10}} = 0.316$$

Figure 14.1 If the cola does not lose sweetness in storage, the mean score $\bar{x}$ for 10 tasters will have this sampling distribution. The actual result for one cola was $\bar{x} = 0.3$. That could easily happen just by chance. Another cola had $\bar{x} = 1.02$. That's so far out on the Normal curve that it is good evidence that this cola did lose sweetness.

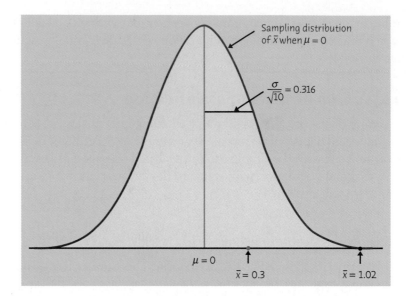

Figure 14.1 shows this sampling distribution. We can judge whether any observed $\bar{x}$ is surprising by locating it on this distribution.

- Suppose that the 10 tasters had mean loss $\bar{x} = 0.3$. It is clear from Figure 14.1 that an $\bar{x}$ this large could easily occur just by chance when the population mean is $\mu = 0$. That 10 tasters find $\bar{x} = 0.3$ is not evidence of a sweetness loss.

- In fact, the taste test produced $\bar{x} = 1.02$. That's way out on the Normal curve in Figure 14.1—so far out that an observed value this large would rarely occur just by chance if the true μ were 0. This observed value is good evidence that in fact the true μ is greater than 0, that is, that the cola lost sweetness. The manufacturer must reformulate the cola and try again.

APPLY YOUR KNOWLEDGE

14.1 **Anemia.** Hemoglobin is a protein in red blood cells that carries oxygen from the lungs to body tissues. People with less than 12 grams of hemoglobin per deciliter of blood (g/dl) are anemic. A public health official in Jordan suspects that the mean μ for all children in Jordan is less than 12. He measures a sample of 50 children. Suppose we know that hemoglobin level for all children this age follows a Normal distribution with standard deviation $\sigma = 1.6$ g/dl.

(a) We seek evidence *against* the claim that $\mu = 12$. What is the sampling distribution of $\bar{x}$ in many samples of size 50 if in fact $\mu = 12$? Make a sketch of the Normal curve for this distribution. (Sketch a Normal curve, then mark the axis using what you know about locating the mean and standard deviation on a Normal curve.)

(b) The sample mean was $\bar{x} = 11.3$. Mark this outcome on the sampling distribution. Also mark the outcome $\bar{x} = 11.8$ g/dl of a different study of 50 children. Explain carefully from your sketch why one of these outcomes is good evidence that μ is lower than 12, and also why the other outcome is not good evidence for this conclusion.

14.2   **Student attitudes.** The Survey of Study Habits and Attitudes (SSHA) is a psychological test that measures students' study habits and attitude toward school. Scores range from 0 to 200. The mean score for college students is about 115, and the standard deviation is about 30. A teacher suspects that the mean μ for older students is higher than 115. She gives the SSHA to an SRS of 25 students who are at least 30 years old. Suppose we know that scores in the population of older students are Normally distributed with standard deviation $\sigma = 30$.

(Will & Deni Mcintyre/Photo Researchers)

(a) We seek evidence *against* the claim that $\mu = 115$. What is the sampling distribution of the mean score $\overline{x}$ of a sample of 25 students if the claim is true? Sketch the density curve of this distribution. (Sketch a Normal curve, then mark the axis using what you know about locating the mean and standard deviation on a Normal curve.)

(b) Suppose that the sample data give $\overline{x} = 118.6$. Mark this point on the axis of your sketch. In fact, the result was $\overline{x} = 125.8$. Mark this point on your sketch. Using your sketch, explain in simple language why one result is good evidence that the mean score of all older students is greater than 115 and why the other outcome is not.

Stating hypotheses

A statistical test starts with a careful statement of the claims we want to compare. In Example 14.2, we asked whether the taste test data are plausible if, in fact, there is no loss of sweetness. Because the reasoning of tests looks for evidence *against* a claim, we start with the claim we seek evidence against, such as "no loss of sweetness."

NULL HYPOTHESIS H_0

The statement being tested in a statistical test is called the **null hypothesis.** The test is designed to assess the strength of the evidence against the null hypothesis. Usually the null hypothesis is a statement of "no effect" or "no difference."

The claim about the population that we are trying to find evidence *for* is the **alternative hypothesis,** written H_a. In Example 14.2, we are seeking evidence of a loss in sweetness. The null hypothesis says "no loss" on the average in a large population of tasters. The alternative hypothesis says "there is a loss." So the hypotheses are

alternative hypothesis

$$H_0: \mu = 0$$
$$H_a: \mu > 0$$

Hypotheses always refer to some population or model, not to a particular outcome. Be sure to state H_0 and H_a in terms of population parameters. Because H_a expresses the effect that we hope to find evidence *for*, it is often easier to begin by stating H_a and then set up H_0 as the statement that the hoped-for effect is not present. Stating H_a is not always straightforward. It is not always clear, in particular, whether H_a should be one-sided or two-sided.

one-sided alternative

The alternative H_a: $\mu > 0$ is **one-sided** because we are interested only in whether the cola lost sweetness. Here is an example in which the alternative hypothesis is two-sided.

EXAMPLE 14.3   Studying job satisfaction

Does the job satisfaction of assembly workers differ when their work is machine-paced rather than self-paced? Assign workers either to an assembly line moving at a fixed pace or to a self-paced setting. All subjects work in both settings, in random order. This is a matched pairs design. After two weeks in a work setting, the workers take a test of job satisfaction. The response variable is the difference in satisfaction scores, self-paced minus machine-paced.

The parameter of interest is the mean μ of the differences in scores in the population of all assembly workers. The null hypothesis says that there is no difference between self-paced and machine-paced work; that is,

$$H_0: \mu = 0$$

The authors of the study wanted to know if the two work conditions have different levels of job satisfaction. They did not specify the direction of the difference. The alternative hypothesis is therefore **two-sided**:

$$H_a: \mu \neq 0$$

two-sided alternative

The alternative hypothesis should express the hopes or suspicions we bring to the data. It is cheating to first look at the data and then frame H_a to fit what the data show. Thus, the fact that the workers in the study of Example 14.3 were more satisfied with self-paced work should not influence our choice of H_a. If you do not have a specific direction firmly in mind in advance, use a two-sided alternative.

APPLY YOUR KNOWLEDGE

14.3   **Anemia.** State the null and alternative hypotheses for the anemia study described in Exercise 14.1.

14.4   **Student attitudes.** State the null and alternative hypotheses for the study of older students' attitudes described in Exercise 14.2.

14.5   **Gas mileage.** Larry's car averages 26 miles per gallon on the highway. He switches to a new motor oil that is advertised as increasing gas mileage. After driving 3000 highway miles with the new oil, he wants to determine if his average gas mileage has increased. What are the null and alternative hypotheses? Explain briefly what the parameter μ in your hypotheses represents.

14.6 **Diameter of a part.** The diameter of a spindle in a small motor is supposed to be 5 millimeters. If the spindle is either too small or too large, the motor will not work properly. The manufacturer measures the diameter in a sample of motors to determine whether the mean diameter has moved away from the target. What are the null and alternative hypotheses? Explain briefly the distinction between the mean μ in your hypotheses and the mean $\bar{x}$ of the spindles the manufacturer measures.

Test statistics

A significance test uses data in the form of a **test statistic.** Here are some prin- *test statistic*
ciples that apply to most tests:

- The test is based on a statistic that compares the value of the parameter stated by the null hypothesis with an estimate of the parameter from the sample data. The estimate is usually the same one used in a confidence interval for the parameter.

- Large values of the test statistic indicate that the estimate is far from the parameter value specified by H_0. These values give evidence against H_0. The alternative hypothesis determines which directions count against H_0.

EXAMPLE 14.4   Sweetening colas: the test statistic

In Example 14.2, the null hypothesis is $H_0: \mu = 0$ and the estimate of μ is $\bar{x} = 1.02$. The test statistic for hypotheses about the mean μ of a Normal distribution is the standardized version of $\bar{x}$:

$$z = \frac{\bar{x} - \mu}{\sigma/\sqrt{n}}$$

The statistic z says how far $\bar{x}$ is from μ in standard deviation units. For Example 14.2,

$$z = \frac{1.02 - 0}{1/\sqrt{10}} = 3.23$$

Because the sample result is more than 3 standard deviations above the hypothesized mean 0, it gives good evidence that the mean sweetness loss is not 0, but positive.

APPLY YOUR KNOWLEDGE

14.7 **Sweetening colas.** Figure 14.1 compares two possible results for the taste test of Example 14.2. Mean $\bar{x} = 1.02$ is far out on the Normal curve and so is good evidence against $H_0: \mu = 0$. Mean $\bar{x} = 0.3$ is not far enough out to convince us that the population mean is greater than 0. Example 14.4 shows that the test statistic is $z = 3.23$ for $\bar{x} = 1.02$. What is z for $\bar{x} = 0.3$? The standard scale makes it easier to compare the two results.

14.8 **Anemia.** What are the values of the test statistic z for the two outcomes in the anemia study of Exercise 14.1?

14.9   **Student attitudes.** What are the values of the test statistic z for the two outcomes for mean SSHA of older students in Exercise 14.2?

P-values

The null hypothesis H_0 states the claim we are seeking evidence against. The test statistic measures how far the sample data diverge from the null hypothesis. If the test statistic is large and is in the direction suggested by the alternative hypothesis H_a, we have data that would be unlikely if H_0 were true. We make "unlikely" precise by calculating a probability.

P-VALUE

The probability, computed assuming that H_0 is true, that the test statistic would take a value as extreme or more extreme than that actually observed is called the **P-value** of the test. The smaller the P-value, the stronger the evidence against H_0 provided by the data.

Small P-values are evidence against H_0, because they say that the observed result is unlikely to occur when H_0 is true. Large P-values fail to give evidence against H_0.

EXAMPLE 14.5   Sweetening colas: the P-value

The 10 tasters in Example 14.2 found mean sweetness loss $\overline{x} = 1.02$. This is far from the value $H_0: \mu = 0$. The test statistic says just how far, in the standard scale,

$$z = \frac{1.02 - 0}{1/\sqrt{10}} = 3.23$$

The alternative H_a says that $\mu > 0$, so positive values of z favor H_a over H_0.

When H_0 is true, $\mu = 0$. The test statistic z is then the standardized version of the sample mean $\overline{x}$. Because $\overline{x}$ has a Normal distribution, z has the standard Normal distribution.

Figure 14.2 shows the P-value on the standard Normal curve that displays the distribution of z. The P-value is the probability that a standard Normal variable is 3.23 or larger. Using Table A,

$$P = P(Z > 3.23) = 1 - 0.9994 = 0.0006$$

We would very rarely observe a sample sweetness loss as large as 1.02 if H_0 were true. The small P-value provides strong evidence against H_0 and in favor of the alternative $H_a: \mu > 0$.

The P-value in Example 14.5 is the probability of getting a z *as large or larger* than the observed $z = 3.23$. The alternative hypothesis sets the direction that counts as evidence against H_0. If the alternative is two-sided, both directions count. Here is an example of the P-value for a two-sided test.

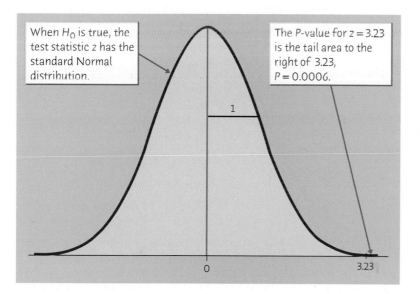

Figure 14.2 The *P*-value for the value $z = 3.23$ of the test statistic in Example 14.5. The *P*-value is the probability (when H_0 is true) that z takes a value as large or larger than the actually observed value.

EXAMPLE 14.6   Job satisfaction: the *P*-value

Suppose we know that differences in job satisfaction scores in Example 14.3 follow a Normal distribution with standard deviation $\sigma = 60$. If there is no difference in job satisfaction between the two work environments, the mean is $\mu = 0$. This is H_0. The alternative hypothesis says simply "there is a difference," $H_a: \mu \neq 0$.

Data from 18 workers gave $\overline{x} = 17$. That is, these workers preferred the self-paced environment on the average. The test statistic is

$$z = \frac{\overline{x} - 0}{\sigma/\sqrt{n}}$$

$$= \frac{17 - 0}{60/\sqrt{18}} = 1.20$$

Because the alternative is two-sided, the *P*-value is the probability of getting a z *at least as far from 0 in either direction* as the observed $z = 1.20$. As always, calculate the *P*-value taking H_0 to be true. When H_0 is true, $\mu = 0$ and z has the standard Normal distribution. Figure 14.3 shows the *P*-value as an area under the standard Normal curve. It is

$$P = P(Z < -1.20 \text{ or } Z > 1.20) = 2P(Z < -1.20)$$

$$= (2)(0.1151) = 0.2302$$

Values as far from 0 as $\overline{x} = 17$ would happen 23% of the time when the true population mean is $\mu = 0$. An outcome that would occur so often when H_0 is true is not good evidence against H_0.

The conclusion of Example 14.6 is *not* that H_0 is true. The study looked for evidence against $H_0: \mu = 0$ and failed to find strong evidence. That is all we

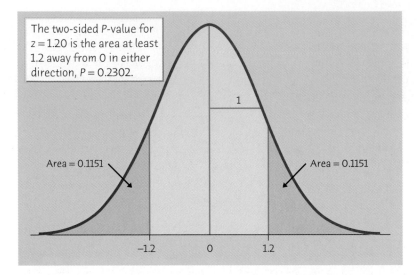

The two-sided P-value for $z = 1.20$ is the area at least 1.2 away from 0 in either direction, $P = 0.2302$.

Area = 0.1151

Area = 0.1151

1

−1.2   0   1.2

Figure 14.3 The P-value for the two-sided test in Example 14.6. The observed value of the test statistic is $z = 1.20$.

can say. No doubt the mean μ for the population of all assembly workers is not exactly equal to 0. A large enough sample would give evidence of the difference, even if it is very small. Tests of significance assess the evidence *against* H_0. If the evidence is strong, we can confidently reject H_0 in favor of the alternative. Failing to find evidence against H_0 means only that the data are consistent with H_0, not that we have clear evidence that H_0 is true.

APPLY YOUR KNOWLEDGE

14.10 Sweetening colas. Figure 14.1 shows that the outcome $\bar{x} = 0.3$ from the cola taste test is not good evidence that the mean sweetness loss is greater than 0. What is the P-value for this outcome? This P-value says, "A sample outcome this large or larger would often occur just by chance when the true mean is really 0."

14.11 Anemia. What are the P-values for the two outcomes of the anemia study in Exercise 14.1? Explain briefly why these values tell us that one outcome is strong evidence against the null hypothesis and that the other outcome is not.

14.12 Student attitudes. What are the P-values for the two outcomes of the study of SSHA scores of older students in Exercise 14.2? Explain briefly why these values tell us that one outcome is strong evidence against the null hypothesis and that the other outcome is not.

14.13 Job satisfaction with a larger sample. Suppose that the job satisfaction study had produced exactly the same outcome $\bar{x} = 17$ as in Example 14.6, but from a sample of 75 workers rather than just 18 workers. Find the test statistic z and its two-sided P-value. Do the data give good evidence that the population mean is not zero?

Statistical significance

We sometimes take one final step to assess the evidence against H_0. We can compare the P-value with a fixed value that we regard as decisive. This amounts to announcing in advance how much evidence against H_0 we will insist on. The decisive value of P is called the **significance level.** We write it as α, the Greek letter alpha. If we choose $\alpha = 0.05$, we are requiring that the data give evidence against H_0 so strong that it would happen no more than 5% of the time (1 time in 20 samples in the long run) when H_0 is true. If we choose $\alpha = 0.01$, we are insisting on stronger evidence against H_0, evidence so strong that it would appear only 1% of the time (1 time in 100 samples) if H_0 is in fact true.

significance level

STATISTICAL SIGNIFICANCE

If the P-value is as small or smaller than α, we say that the data are **statistically significant at level α.**

"Significant" in the statistical sense does not mean "important." It means simply "not likely to happen just by chance." The significance level α makes "not likely" more exact. Significance at level 0.01 is often expressed by the statement "The results were significant ($P < 0.01$)." Here P stands for the P-value. The actual P-value is more informative than a statement of significance, because it allows us to assess significance at any level we choose. For example, a result with $P = 0.03$ is significant at the $\alpha = 0.05$ level but is not significant at the $\alpha = 0.01$ level.

Down with driver ed!

Who could object to driver-training courses in schools? The killjoy who looks at data, that's who. Careful studies show no significant effect of driver training on the behavior of teenage drivers. Because many states allow those who take driver ed to get a license at a younger age, the programs may actually increase accidents and road deaths by increasing the number of young and risky drivers.

APPLY YOUR KNOWLEDGE

14.14 **Anemia.** In Exercises 14.8 and 14.11, you found the z test statistic and the P-value for the outcome $\bar{x} = 11.8$ in the anemia study of Exercise 14.1. Is this outcome statistically significant at the $\alpha = 0.05$ level? At the $\alpha = 0.01$ level?

14.15 **Student attitudes.** In Exercises 14.9 and 14.12, you found the z test statistic and the P-value for the outcome $\bar{x} = 125.8$ in the attitudes study of Exercise 14.2. Is this outcome statistically significant at the $\alpha = 0.05$ level? At the $\alpha = 0.01$ level?

14.16 **Protecting ultramarathon runners.** Exercise 8.14 (page 212) describes an experiment designed to learn whether taking vitamin C reduces the incidence of respiratory infections among ultramarathon runners. The report of the study said:

> *Sixty-eight percent of the runners in the placebo group reported the development of symptoms of upper respiratory tract infection after the race; this was significantly more ($P < 0.01$) than that reported by the vitamin C-supplemented group (33%).*

(a) Explain to someone who knows no statistics why "significantly more" means there is good reason to think that vitamin C works.

(b) Now explain more exactly: What does $P < 0.01$ mean?

Tests for a population mean

There are four steps in carrying out a significance test:

1.   State the hypotheses.
2.   Calculate the test statistic.
3.   Find the P-value.
4.   State your conclusion in the context of your specific setting.

Once you have stated your hypotheses and identified the proper test, you or your computer can do Steps 2 and 3 by following a recipe. Here is the recipe for the test we have used in our examples.

---

z TEST FOR A POPULATION MEAN

Draw an SRS of size n from a Normal population that has unknown mean μ and known standard deviation σ. To test the null hypothesis that μ has a specified value,

$$H_0: \mu = \mu_0$$

calculate the **one-sample z statistic**

$$z = \frac{\overline{x} - \mu_0}{\sigma/\sqrt{n}}$$

In terms of a variable Z having the standard Normal distribution, the P-value for a test of H_0 against

$H_a: \mu > \mu_0$ is $P(Z \geq z)$

$H_a: \mu < \mu_0$ is $P(Z \leq z)$

$H_a: \mu \neq \mu_0$ is $2P(Z \geq |z|)$

EXAMPLE 14.7   Executives' blood pressures

The National Center for Health Statistics reports that the systolic blood pressure for males 35 to 44 years of age has mean 128 and standard deviation 15. The medical director of a large company looks at the medical records of 72 executives in this age group and finds that the mean systolic blood pressure in this sample is $\bar{x} = 126.07$. Is this evidence that the company's executives have a different mean blood pressure from the general population?

Suppose we know that executives' blood pressures follow a Normal distribution with standard deviation $\sigma = 15$.

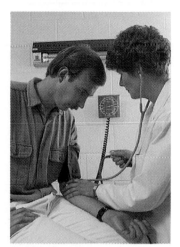
(W&D Mcintyre/Photo Researchers)

Step 1. Hypotheses. The null hypothesis is "no difference" from the national mean $\mu_0 = 128$. The alternative is two-sided, because the medical director did not have a particular direction in mind before examining the data. So the hypotheses about the unknown mean μ of the executive population are

$$H_0: \mu = 128$$
$$H_a: \mu \neq 128$$

Step 2. Test statistic. The one-sample z statistic is

$$z = \frac{\bar{x} - \mu_0}{\sigma/\sqrt{n}} = \frac{126.07 - 128}{15/\sqrt{72}}$$
$$= -1.09$$

Step 3. P-value. To help find a P-value, sketch the standard Normal curve and mark on it the observed value of z. Figure 14.4 shows that the P-value is the probability that a standard Normal variable Z takes a value at least 1.09 away from zero. From Table A we find that this probability is

$$P = 2P(Z \geq 1.09) = 2(1 - 0.8621) = 0.2758$$

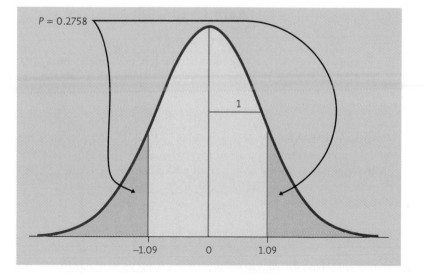

Figure 14.4 The P-value for the two-sided test in Example 14.7. The observed value of the test statistic is $z = -1.09$.

Conclusion. More than 27% of the time, an SRS of size 72 from the general male population would have a mean blood pressure at least as far from 128 as that of the executive sample. The observed $\bar{x} = 126.07$ is therefore not good evidence that executives differ from other men.

The z test requires that the 72 executives in the sample are an SRS from the population of all middle-aged male executives in the company. We should check this requirement by asking how the data were produced. If medical records are available only for executives with recent medical problems, for example, the data are of little value for our purpose. It turns out that all executives are given a free annual medical exam, and that the medical director selected 72 exam results at random.

EXAMPLE 14.8 Can you balance your checkbook?

In a discussion of the education level of the American workforce, someone says, "The average young person can't even balance a checkbook." The National Assessment of Educational Progress says that a score of 275 or higher on its quantitative test (see Example 13.1 on page 321) reflects the skill needed to balance a checkbook. The NAEP random sample of 840 young men had a mean score of $\bar{x} = 272$, a bit below the checkbook-balancing level. Is this sample result good evidence that the mean for *all* young men is less than 275? As in Example 13.1, suppose we know that $\sigma = 60$.

Step 1. **Hypotheses.** The hypotheses are

$$H_0: \mu = 275$$
$$H_a: \mu < 275$$

Step 2. **Test statistic.** The z statistic is

$$z = \frac{\bar{x} - \mu_0}{\sigma/\sqrt{n}} = \frac{272 - 275}{60/\sqrt{840}}$$
$$= -1.45$$

Step 3. **P-value.** Because H_a is one-sided on the low side, small values of z count against H_0. Figure 14.5 illustrates the P-value. Using Table A, we find that

$$P = P(Z \leq -1.45) = 0.0735$$

Conclusion. A mean score as low as 272 would occur about 7 times in 100 samples if the population mean were 275. This is modest evidence that the mean NAEP score for all young men is less than 275. It is significant at the $\alpha = 0.10$ level but not at the $\alpha = 0.05$ level.

APPLY YOUR KNOWLEDGE

14.17 Water quality. An environmentalist group collects a liter of water from each of 45 random locations along a stream and measures the amount of dissolved oxygen in each specimen. The mean is

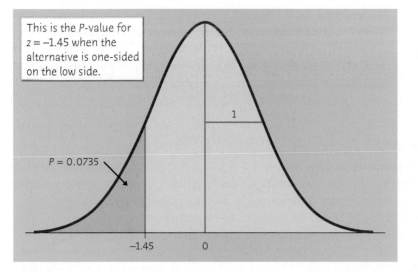

This is the P-value for $z = -1.45$ when the alternative is one-sided on the low side.

$P = 0.0735$

1

−1.45   0

Figure 14.5 The P-value for the one-sided test in Example 14.8. The observed value of the test statistic is $z = -1.45$.

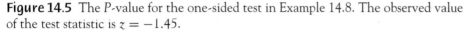

4.62 milligrams (mg). Is this strong evidence that the stream has a mean oxygen content of less than 5 mg per liter? (Suppose we know that dissolved oxygen varies among locations according to a Normal distribution with $\sigma = 0.92$ mg.)

14.18 Improving your SAT score. We suspect that on the average students will score higher on their second attempt at the SAT mathematics exam than on their first attempt. Suppose we know that the changes in score (second try minus first try) follow a Normal distribution with standard deviation $\sigma = 50$. Here are the results for 46 randomly chosen high school students:

```
−30    24    47   70  −62    55  −41  −32   128  −11
−43   122  −10   56    32  −30  −28  −19     1   17
 57   −14  −58   77    27  −33   51   17   −67   29
 94   −11    2   12  −53  −49   49    8   −24   96
120     2  −33  −2  −39    99
```

Do these data give good evidence that the mean change in the population is greater than zero? State hypotheses, calculate a test statistic and its P-value, and state your conclusion.

14.19 Engine crankshafts. Here are measurements (in millimeters) of a critical dimension on a sample of automobile engine crankshafts:

```
224.120  224.001  224.017  223.982  223.989  223.961
223.960  224.089  223.987  223.976  223.902  223.980
224.098  224.057  223.913  223.999
```

The manufacturing process is known to vary Normally with standard deviation $\sigma = 0.060$ mm. The process mean is supposed to be 224 mm. Do these data give evidence that the process mean is not equal to the target value 224 mm? State hypotheses and calculate a test statistic and its P-value. Are you convinced that the process mean is not 224 mm?

P-values and significance levels

Sometimes we demand a specific degree of evidence in order to reject the null hypothesis. A level of significance α says how much evidence we require. In terms of the P-value, the outcome of a test is significant at level α if $P \leq \alpha$. Significance at any level is easy to assess once you have the P-value. When you do not use software, the P-value can be difficult to calculate. Fortunately, you can decide whether a result is statistically significant by using a table of critical values, the same table we use for confidence intervals. The table also allows you to approximate the P-value without calculation. Here is an example.

EXAMPLE 14.9 Is it significant?

In Example 14.8, we examined whether the mean NAEP quantitative score of young men is less than 275. The hypotheses are

$$H_0: \mu = 275$$
$$H_a: \mu < 275$$

The z statistic takes the value $z = -1.45$. How significant is the evidence against H_0?

To determine significance, compare the observed $z = -1.45$ with the critical values z^* in the last row of Table C. The tail area for each z^* appears at the top of the table. The value $z = -1.45$ (ignoring its sign) falls between the critical values 1.282 and 1.645. Because z is farther from 0 than 1.282, the critical value for tail area 0.10, the test *is* significant at level $\alpha = 0.10$. Because $z = 1.45$ is *not* farther from 0 than the critical value 1.645 for tail area 0.05, the test is *not* significant at level $\alpha = 0.05$.

Figure 14.6 locates $z = -1.45$ between the two tabled critical values, with minus signs added because the alternative is one-sided on the low side. The figure also shows how the critical value $z^* = -1.645$ separates values of z that are significant at the $\alpha = 0.05$ level from values that are not significant.

The P-value for $z = -1.45$ is the area under the curve in Figure 14.6 to the left of -1.45. This area is greater than the area 0.05 to the left of $z^* = -1.645$ and less than the area 0.10 to the left of $z^* = -1.282$. We can say without any calculations that $0.05 < P < 0.10$.

EXAMPLE 14.10 Is the concentration OK?

The analytical laboratory of Example 13.4 (page 330) is asked to evaluate the claim that the concentration of the active ingredient in a specimen is 0.86%. The lab

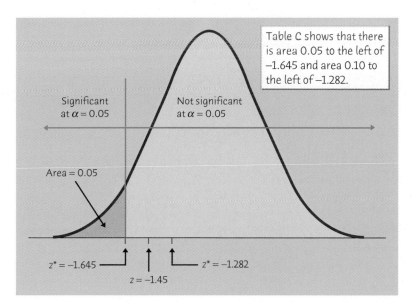

Table C shows that there is area 0.05 to the left of −1.645 and area 0.10 to the left of −1.282.

Significant at $\alpha = 0.05$

Not significant at $\alpha = 0.05$

Area = 0.05

$z^* = -1.645$ $z^* = -1.282$

$z = -1.45$

Figure 14.6 Deciding whether a z statistic is significant at the $\alpha = 0.05$ level in the one-sided test of Example 14.9. The observed value $z = -1.45$ of the test statistic is not significant because it is not in the extreme 5% of the standard Normal distribution.

makes 3 repeated analyses of the specimen. The mean result is $\bar{x} = 0.8404$. The true concentration is the mean μ of the population of all analyses of the specimen. The standard deviation of the analysis process is known to be $\sigma = 0.0068$. Is there significant evidence at the 1% level that $\mu \neq 0.86$?

Step 1. Hypotheses. The hypotheses are

$$H_0: \mu = 0.86$$
$$H_a: \mu \neq 0.86$$

Step 2. Test statistic. The z statistic is

$$z = \frac{0.8404 - 0.86}{0.0068/\sqrt{3}} = -4.99$$

Step 3. Significance. Because the alternative is two-sided, the P-value is the area under the standard Normal curve below −4.99 and above 4.99. This area is double the area in either tail alone. For significance at level $\alpha = 0.01$, z must be in the extreme 0.005 ($\alpha/2$) in either tail.

Compare $z = -4.99$ (ignoring its sign) with the critical value for tail area 0.005 from Table C. This critical value is $z^* = 2.576$. Figure 14.7 locates $z = -4.99$ and the critical values on the standard Normal curve. Because z is farther from 0 than the critical values, we have significant evidence ($P < 0.01$) that the concentration is not as claimed.

In fact, $z = -4.99$ lies beyond all the critical values in Table C. The largest critical value is 3.291, for tail area 0.0005. So we can say that the two-sided

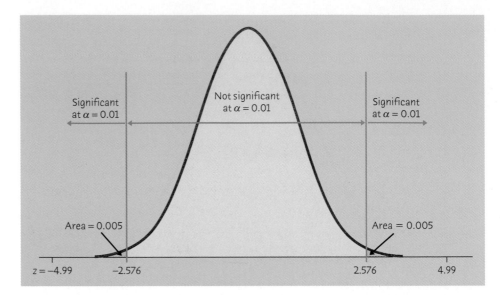

Figure 14.7 Deciding whether a z statistic is significant at the $\alpha = 0.01$ level in the two-sided test of Example 14.10. The observed value $z = -4.99$ is significant because it is in the extreme 1% of the standard Normal distribution.

test is significant at the 0.001 level, not just at the 0.01 level. Software gives the exact P-value as

$$P = 2P(Z \geq 4.99) = 0.0000006$$

No wonder Figure 14.7 places $z = -4.99$ so far out that the Normal curve is not visibly above its axis.

Because the practice of statistics almost always employs software that calculates P-values automatically, tables of critical values are becoming outdated. Tables of critical values such as Table C appear in this book for learning purposes and to rescue students without good computing facilities.

APPLY YOUR KNOWLEDGE

14.20 Significance. You are testing $H_0: \mu = 0$ against $H_a: \mu \neq 0$ based on an SRS of 20 observations from a Normal population. What values of the z statistic are statistically significant at the $\alpha = 0.005$ level?

14.21 Significance. You are testing $H_0: \mu = 0$ against $H_a: \mu > 0$ based on an SRS of 20 observations from a Normal population. What values of the z statistic are statistically significant at the $\alpha = 0.005$ level?

14.22 Testing a random number generator. A random number generator is supposed to produce random numbers that are uniformly distributed on the interval from 0 to 1. If this is true, the numbers generated come from a population with $\mu = 0.5$ and $\sigma = 0.2887$. A command to generate 100 random numbers gives outcomes with mean $\bar{x} = 0.4365$. Assume that the population σ remains fixed. We want to test

$$H_0: \mu = 0.5$$
$$H_a: \mu \neq 0.5$$

(a) Calculate the value of the z test statistic.

(b) Is the result significant at the 5% level ($\alpha = 0.05$)?

(c) Is the result significant at the 1% level ($\alpha = 0.01$)?

(d) Between which two Normal critical values in the bottom row of Table C does z lie? Between what two numbers does the P-value lie? What do you conclude?

14.23 Bone loss by nursing mothers. Exercise 13.13 (page 333) gives the percent change in the mineral content of the spine for 47 mothers during three months of nursing a baby. As in that exercise, suppose that the percent change in the population of all nursing mothers has a Normal distribution with standard deviation $\sigma = 2.5\%$. Do these data give good evidence that on the average nursing mothers lose bone mineral? State hypotheses, calculate the z test statistic, and use Table C to assess significance. What do you conclude?

Tests from confidence intervals

The calculation in Example 14.10 for a 1% significance test is very similar to that in Example 13.3 for a 99% confidence interval. In fact, a two-sided test at significance level α can be carried out directly from a confidence interval with confidence level $C = 1 - \alpha$.

CONFIDENCE INTERVALS AND TWO-SIDED TESTS

A level α two-sided significance test rejects a hypothesis $H_0: \mu = \mu_0$ exactly when the value μ_0 falls outside a level $1 - \alpha$ confidence interval for μ.

EXAMPLE 14.11   Tests from a confidence interval

The 99% confidence interval for μ in Example 13.3 is

$$\bar{x} \pm z^* \frac{\sigma}{\sqrt{n}} = 0.8404 \pm 0.0101$$

$$= 0.8303 \text{ to } 0.8505$$

The hypothesized value $\mu_0 = 0.86$ in Example 14.10 falls outside this confidence interval, so we reject

$$H_0: \mu = 0.86$$

Figure 14.8 Values of μ falling outside a 99% confidence interval can be rejected at the 1% significance level. Values falling inside the interval cannot be rejected.

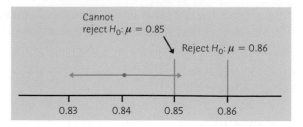

at the 1% significance level. On the other hand, we cannot reject

$$H_0: \mu = 0.85$$

at the 1% level in favor of the two-sided alternative $H_a: \mu \neq 0.85$, because 0.85 lies inside the 99% confidence interval for μ. Figure 14.8 illustrates both cases.

APPLY YOUR KNOWLEDGE

14.24 Test and confidence interval. The P-value for a two-sided test of the null hypothesis $H_0: \mu = 10$ is 0.06.

(a) Does the 95% confidence interval include the value 10? Why?

(b) Does the 90% confidence interval include the value 10? Why?

14.25 Confidence interval and test. A 95% confidence interval for a population mean is 31.5 ± 3.5.

(a) Can you reject the null hypothesis that $\mu = 34$ at the 5% significance level? Why?

(b) Can you reject the null hypothesis that $\mu = 36$ at the 5% significance level? Why?

Chapter 14 SUMMARY

A **test of significance** assesses the evidence provided by data against a **null hypothesis** H_0 in favor of an **alternative hypothesis** H_a.

Hypotheses are stated in terms of population parameters. Usually H_0 is a statement that no effect is present, and H_a says that a parameter differs from its null value in a specific direction (**one-sided alternative**) or in either direction (**two-sided alternative**).

The essential reasoning of a significance test is as follows. Suppose for the sake of argument that the null hypothesis is true. If we repeated our data production many times, would we often get data as inconsistent with H_0 as the data we actually have? If the data are unlikely when H_0 is true, they provide evidence against H_0.

A test is based on a **test statistic.** The **P-value** is the probability, computed supposing H_0 to be true, that the test statistic will take a value at least as extreme as that actually observed. Small *P*-values indicate strong evidence against H_0. Calculating *P*-values requires knowledge of the sampling distribution of the test statistic when H_0 is true.

If the *P*-value is as small or smaller than a specified value α, the data are **statistically significant** at significance level α.

Significance tests for the hypothesis H_0: $\mu = \mu_0$ concerning the unknown mean μ of a population are based on the **one-sample z statistic**

$$z = \frac{\bar{x} - \mu_0}{\sigma/\sqrt{n}}$$

The z test assumes an SRS of size n, known population standard deviation σ, and either a Normal population or a large sample. *P*-values are computed from the Normal distribution (Table A). Fixed α tests use the table of standard Normal **critical values** (bottom row of Table C).

Chapter 14 EXERCISES

14.26 Student study times. A student group claims that first-year students at a university must study 2.5 hours per night during the school week. A skeptic suspects that they study less than that on the average. A class survey finds that the average study time claimed by 269 students is $\bar{x} = 137$ minutes. Regard these students as a random sample of all first-year students and suppose we know that study times follow a Normal distribution with standard deviation 65 minutes. Carry out a test of H_0: $\mu = 150$ against H_a: $\mu < 150$. What do you conclude?

14.27 IQ test scores. Exercise 13.6 (page 329) gives the IQ test scores of 31 seventh-grade girls in a Midwest school district. IQ scores follow a Normal distribution with standard deviation $\sigma = 15$. Treat these 31 girls as an SRS of all seventh-grade girls in this district. IQ scores in a broad population are supposed to have mean $\mu = 100$. Is there evidence that the mean in this district differs from 100? State hypotheses, find the test statistic and its *P*-value, and state your conclusion.

14.28 This wine stinks. Sulfur compounds cause "off-odors" in wine, so winemakers want to know the odor threshold, the lowest concentration of a compound that the human nose can detect. The odor threshold for dimethyl sulfide (DMS) in trained wine tasters is about 25 micrograms per liter of wine (μg/l). The untrained noses of consumers may be less sensitive, however. Here are the DMS odor thresholds for 10 untrained students:

31   31   43   36   23   34   32   30   20   24

Assume that the odor threshold for untrained noses is Normal with $\sigma = 7$ μg/l. Is there evidence that the mean threshold for untrained tasters is greater than 25 μg/l?

14.29 Healing of skin wounds. Exercise 13.14 (page 334) gives data and information about the rate at which skin wounds heal in newts. A newt expert says that 25 micrometers per hour is the usual rate. Do the data give evidence against this claim?

14.30 P without pain. You can approximate P for any z by comparing z with the critical values in the bottom row of Table C. Between what values from Table C does the P-value for your z in Exercise 14.26 lie?

14.31 P without pain. Use Table C to approximate the P-value for the z you obtained in Exercise 14.27 without a probability calculation. That is, give two values from the table between which P must lie.

14.32 Tracking the placebo effect. The placebo effect (see Chapter 8, page 202) is particularly strong in patients with Parkinson's disease. To understand the workings of the placebo effect, scientists made chemical measurements at a key point in the brain when patients received a placebo that they thought was an active drug and also when no treatment was given.[1] They hoped to find that the placebo reduced the mean response. State H_0 and H_a for the significance test.

14.33 Tracking the placebo effect, continued. The report of the study described in the previous exercise says that the placebo reduced the chemical response by an average of 17% and that $P < 0.005$. What can you conclude?

14.34 Fortified breakfast cereals. The Food and Drug Administration recommends that breakfast cereals be fortified with folic acid. In a matched pairs study, volunteers ate either fortified or unfortified cereal for some time, then switched to the other cereal. The response variable is the difference in blood folic acid, fortified minus unfortified. Does eating fortified cereal raise the level of folic acid in the blood? State H_0 and H_a for a test to answer this question. State carefully what the parameter μ in your hypotheses is.

14.35 What's the P-value? A test of the null hypothesis H_0: $\mu = 0$ gives test statistic $z = 1.8$.

(a) What is the P-value if the alternative is H_a: $\mu > 0$?

(b) What is the P-value if the alternative is H_a: $\mu < 0$?

(c) What is the P-value if the alternative is H_a: $\mu \neq 0$?

14.36 P and significance. The P-value for a significance test is 0.078.

(a) Do you reject the null hypothesis at level $\alpha = 0.05$? Explain your answer.

(b) Do you reject the null hypothesis at level $\alpha = 0.01$? Explain your answer.

14.37 P and significance. The P-value for a significance test is 0.033.

(a) Do you reject the null hypothesis at level $\alpha = 0.05$? Explain your answer.

(Image Source/elektraVision/ PictureQuest)

(b) Do you reject the null hypothesis at level $\alpha = 0.01$? Explain your answer.

14.38 The Supreme Court speaks. Court cases in such areas as employment discrimination often involve statistical evidence. The Supreme Court has said that z-scores beyond $z^* = 2$ or 3 are generally convincing statistical evidence. For a two-sided test, what significance level α corresponds to $z^* = 2$? To $z^* = 3$?

14.39 Diet and diabetes. Does eating more fiber reduce the blood cholesterol level of patients with diabetes? A randomized clinical trial compared normal and high-fiber diets. Here is part of the researchers' conclusion:

> The high-fiber diet reduced plasma total cholesterol concentrations by 6.7 percent ($P = 0.02$), triglyceride concentrations by 10.2 percent ($P = 0.02$), and very-low-density lipoprotein cholesterol concentrations by 12.5 percent ($P = 0.01$).[2]

A doctor who knows no statistics says that a drop of 6.7% in cholesterol isn't a lot—maybe it's just an accident due to the chance assignment of patients to the two diets. Explain in simple language how "$P = 0.02$" answers this objection.

14.40 Diet and bowel cancer. It has long been thought that eating a healthier diet reduces the risk of bowel cancer. A large study cast doubt on this advice. The subjects were 2079 people who had polyps removed from their bowels in the past six months. Such polyps may lead to cancer. The subjects were randomly assigned to a low-fat, high-fiber diet or to a control group in which subjects ate their usual diets. All subjects were checked for polyps over the next four years.[3]

(a) Outline the design of this experiment.

(b) Surprisingly, the occurrence of new polyps "did not differ significantly between the two groups." Explain clearly what this finding means.

14.41 How to show that you are rich. Every society has its own marks of wealth and prestige. In ancient China, it appears that owning pigs was such a mark. Evidence comes from examining burial sites. The skulls of sacrificed pigs tend to appear along with expensive ornaments, which suggests that the pigs, like the ornaments, signal the wealth and prestige of the person buried. A study of burials from around 3500 B.C. concluded that "there are striking differences in grave goods between burials with pig skulls and burials without them.... A test indicates that the two samples of total artifacts are significantly different at the 0.01 level."[4] Explain clearly why "significantly different at the 0.01 level" gives good reason to think that there really is a systematic difference between burials that contain pig skulls and those that lack them.

(Carl and Ann Purcell/CORBIS)

14.42 **Forests and windstorms.** Does the destruction of large trees in a windstorm change forests in any important way? Here is the conclusion of a study that found that the answer is no:

> We found surprisingly little divergence between treefall areas and adjacent control areas in the richness of woody plants ($P = 0.62$), in total stem densities ($P = 0.98$), or in population size or structure for any individual shrub or tree species.[5]

The two P-values refer to null hypotheses that say "no change" in measurements between treefall and control areas. Explain clearly why these values provide no evidence of change.

14.43 **Reporting P.** The report of a study of seat belt use by drivers says, "Hispanic drivers were not significantly more likely than White/non-Hispanic drivers to overreport safety belt use (27.4 vs. 21.1%, respectively; $z = 1.33$, $P > 1.0$."[6] How do you know that the P-value given is incorrect? What is the correct one-sided P-value for test statistic $z = 1.33$?

14.44 **5% versus 1%.** Make a sketch that shows why a value of the z test statistic that is significant at the 1% level must always be significant at the 5% level. If z is significant at the 5% level, what can you say about its significance at the 1% level?

14.45 **Is this what P means?** When asked to explain the meaning of "the P-value was $P = 0.03$," a student says, "This means there is only probability 0.03 that the null hypothesis is true." Is this an essentially correct explanation? Explain your answer.

14.46 **Is this what significance means?** Another student, when asked why statistical significance appears so often in research reports, says, "Because saying that results are significant tells us that they cannot easily be explained by chance variation alone." Do you think that this statement is essentially correct? Explain your answer.

14.47 **Workers' earnings.** The Bureau of Labor Statistics generally uses 90% confidence in its reports. One report gives a 90% confidence interval for the mean hourly earnings of American workers in 2000 as $15.49 to $16.11. This result was calculated from the National Compensation Survey, a multistage probability sample of businesses.

(a) Would a 95% confidence interval be wider or narrower?

(b) Would the null hypothesis that the 2000 mean hourly earnings of all workers was $16 be rejected at the 10% significance level in favor of the two-sided alternative? What about the null hypothesis that the mean was $15?

14.48 **Pulling wood apart.** In Exercise 13.15 (page 334), you found a 90% confidence interval for the mean load required to pull apart pieces of

Douglas fir. Use this interval (or calculate it anew here) to answer these questions:

(a) Is there significant evidence at the $\alpha = 0.90$ level against the hypothesis that the mean is 32,000 pounds for the two-sided alternative?

(b) Is there significant evidence at the $\alpha = 0.90$ level against the hypothesis that the mean is 31,500 pounds for the two-sided alternative?

Chapter 14 MEDIA EXERCISES

14.49 I'm a great free-throw shooter. The *Test of Significance* applet animates Example 14.1. That example asks if a basketball player's actual performance gives evidence against the claim that he or she makes 80% of free throws. The parameter in question is the percent p of free throws that the player will make if he or she shoots free throws forever. The population is all free throws the player will ever shoot. The null hypothesis is always the same, that the player makes 80% of shots taken:

$$H_0: p = 80\%$$

The applet does not do a formal statistical test. Instead, it allows you to ask the player to shoot until you are reasonably confident that the true percent of hits is or is not very close to 80%.

I claim that I make 80% of my free throws. To test my claim, we go to the gym and I shoot 20 free throws. Set the applet to take 20 shots. Check "Show null hypothesis" so that my claim is visible in the graph.

(a) Click "Shoot." How many of the 20 shots did I make? Are you convinced that I really make less than 80%?

(b) If you are not convinced, click "Shoot" again for 20 more shots. Keep going until *either* you are convinced that I don't make 80% of my shots *or* it appears that my true percent made is pretty close to 80%. How many shots did you watch me shoot? How many did I make? What did you conclude? Then click "Show true %" to reveal the truth. Was your conclusion correct?

Comment: You see why statistical tests say how strong the evidence is *against* some claim. If I make only 10 of 40 shots, you are pretty sure I can't make 80% in the long run. But even if I make exactly 80 of 100, my true long-term percent might be 78% or 81% instead of 80%. It's hard to be convinced that I make exactly 80%.

(David Madison/The Image Bank/Getty Images)

14.50 Significance at the 0.0125 level. The *Normal Curve* applet allows you to find critical values of the standard Normal distribution and to visualize the values of the z statistic that are significant at any level.

Max is interested in whether a one-sided z test is statistically significant at the $\alpha = 0.0125$ level. Use the *Normal Curve* applet to tell Max what values of z are significant. Sketch the standard Normal curve marked with the values that led to your result.

14.51 Significance at the 0.011 level. Jaran wants to know if a two-sided z test statistic is significant at the $\alpha = 0.011$ level. Use the *Normal Curve* applet to say what values of z are significant at this level. Sketch the standard Normal curve marked with the values that led to your result.

Inference in Practice

To this point, we have met just two procedures for statistical inference. Both concern inference about the mean μ of a Normal population based on data from an SRS. If we know the standard deviation σ of the population, a confidence interval for the mean μ is

$$\overline{x} \pm z^* \frac{\sigma}{\sqrt{n}}$$

To test a hypothesis $H_0: \mu = \mu_0$ we use the one-sample z statistic:

$$z = \frac{\overline{x} - \mu_0}{\sigma/\sqrt{n}}$$

We call these **z procedures** because they both start with the one-sample z statistic and use the standard Normal distribution.

z procedures

In later chapters we will modify these procedures for inference about a population mean to make them more useful in practice. We will also introduce procedures for confidence intervals and tests in most of the settings we met in learning to explore data. There are libraries—both of books and of software—full of more elaborate statistical techniques. The reasoning of confidence intervals and tests is the same, no matter how elaborate the details of the procedure are.

There is a saying among statisticians that "mathematical theorems are true; statistical methods are effective when used with judgment." That the one-sample z statistic has the standard Normal distribution when the null hypothesis is true is a mathematical theorem. Effective use of statistical methods requires more than knowing such facts. It requires even more than understanding the underlying reasoning.

This chapter begins the process of helping you develop the judgment needed to use statistics in practice. That process will continue in examples and exercises through the rest of this book.

Where did the data come from?

The most important requirement for any inference procedure is that the data come from a process to which the laws of probability apply. Inference is most reliable when the data come from a probability sample or a randomized comparative experiment. Probability samples use chance to choose respondents. Randomized comparative experiments use chance to assign subjects to treatments. The deliberate use of chance ensures that the laws of probability apply to the outcomes, and this in turn ensures that statistical inference makes sense.

WHERE THE DATA COME FROM MATTERS

When you use statistical inference, you are acting as if your data are a probability sample or come from a randomized experiment.

Statistical confidence intervals and tests cannot remedy basic flaws in producing the data, such as voluntary response samples or uncontrolled experiments.

If your data don't come from a probability sample or a randomized comparative experiment, your conclusions may be open to challenge. To answer the challenge, you will most often rely on subject-matter knowledge, not on statistics. It is common to apply statistics to data that are not produced by random selection. When you see such a study, ask whether the data can be trusted as a basis for the conclusions of the study.

Don't touch the plants

We know that confounding can distort inference. We don't always recognize how easy it is to confound data. Consider the innocent scientist who visits plants in the field once a week to measure their size. A study of six plant species found that one touch a week significantly increased leaf damage by insects in two species and significantly decreased damage in another species.

EXAMPLE 15.1   The psychologist and the sociologist

A psychologist is interested in how our visual perception can be fooled by optical illusions. Her subjects are students in Psychology 101 at her university. Most psychologists would agree that it's safe to treat the students as an SRS of all people with normal vision. There is nothing special about being a student that changes visual perception.

A sociologist at the same university uses students in Sociology 101 to examine attitudes toward poor people and antipoverty programs. Students as a group are younger than the adult population as a whole. Even among young people, students as a group come from more prosperous and better-educated homes. Even among students, this university isn't typical of all campuses. Even on this campus, students in a sociology course may have opinions that are quite different from those of engineering students. The sociologist can't reasonably act as if these students are a random sample from any interesting population.

EXAMPLE 15.2   Mammary artery ligation

Angina is the severe pain caused by inadequate blood supply to the heart. Perhaps we can relieve angina by tying off the mammary arteries to force the body to develop other routes to supply blood to the heart. Surgeons tried this procedure, called "mammary artery ligation." Patients reported a statistically significant reduction in angina pain.

Statistical significance says that something other than chance is at work, but it does not say what that something is. The mammary artery ligation experiment was uncontrolled, so that the reduction in pain might be nothing more than the placebo effect. Sure enough, a randomized comparative experiment showed that ligation was no more effective than a placebo. Surgeons abandoned the operation at once.[1]

APPLY YOUR KNOWLEDGE

15.1 A TV station takes a poll. A local television station announces a question for a call-in opinion poll on the six o'clock news and then gives the response on the eleven o'clock news. Today's question is "What yearly pay do you think members of the City Council should get? Call us with your number." In all, 958 people call. The mean pay they suggest is $\bar{x} = \$8740$ per year, and the standard deviation of the responses is $s = \$1125$. For a large sample such as this, s is very close to the unknown population σ, so take $\sigma = \$1125$. The station calculates the 95% confidence interval for the mean pay μ that all citizens would propose for council members to be $8669 to $8811.

(a) Is the station's calculation correct?

(b) Does their conclusion describe the population of all the city's citizens? Explain your answer.

Cautions about the z procedures

Any confidence interval or significance test can be used only under specific conditions. It's up to you to understand these conditions and judge whether they fit your problem. If statistical procedures carried warning labels like those on drugs, most inference methods would have long labels indeed. With that in mind, let's look back at the "simple conditions" of Chapters 13 and 14 for the z confidence interval and test.

• *The data must be an SRS from the population.* We are completely safe if we actually carried out the random selection of an SRS. We are not in great danger if the data can plausibly be thought of as observations taken at random from a population. We are willing to argue that the 10 cola taste

Dropping out

An experiment found that weight loss is significantly more effective than exercise for reducing high cholesterol and high blood pressure. The 170 subjects were randomly assigned to a weight-loss program, an exercise program, or a control group. Only 111 of the 170 subjects completed their assigned treatment, and the analysis used data from these 111. Did the dropouts create bias? Always ask about details of the data before trusting inference.

testers of Example 14.2 (page 341) are a plausible SRS from the population of all trained tasters—there is nothing that would make them unrepresentative. In the chemical analysis of Example 13.3 (page 327), we are willing to think of the three measurements as a random sample from the population resulting from a very large number of repeated analyses of the same specimen.

- *Different methods are needed for different designs.* The z procedures aren't correct for probability samples more complex than an SRS. Later chapters give methods for some other designs, but we won't discuss inference for complex settings such as multistage samples and experiments with several factors. Always be sure that you (or your statistical consultant) know how to carry out the inference your design calls for.

- *Outliers can distort the result.* Because $\bar{x}$ is strongly influenced by a few extreme observations, outliers can have a large effect on the z confidence interval and test. Always explore your data before doing inference. In particular, you should search for outliers and try to correct them or justify their removal before performing the z procedures. If the outliers cannot be removed, ask your statistical consultant about procedures that are not sensitive to outliers.

- *The shape of the population distribution matters.* Our "simple conditions" state that the population distribution is Normal. Outliers or extreme skewness make the z procedures untrustworthy unless the sample is large. Other violations of Normality are often not critical in practice. The z procedures use Normality of the sample mean $\bar{x}$, not Normality of individual observations. The central limit theorem tells us that $\bar{x}$ is more Normal than the individual observations. In practice, the z procedures are reasonably accurate for any reasonably symmetric distribution for samples of even moderate size. If the sample is large, $\bar{x}$ will be close to Normal even if individual measurements are strongly skewed. Chapter 16 gives practical guidelines.

- *You must know the standard deviation σ of the population.* This condition is rarely satisfied in practice. Because of it, the z procedures are of little use. We will see in Chapter 16 that simple changes give very useful procedures that don't require that σ be known. When the sample is large, the sample standard deviation s will be close to σ, so in effect we do know σ if we have a large sample. Even in this situation, it is better to use the procedures of Chapter 16.

Every inference procedure that we will meet has its own list of warnings. Because many of the warnings are similar to those above, we will not print the full warning label each time. It is easy to state (from the mathematics of probability) conditions under which a method of inference is exactly correct. These conditions are never fully met in practice. For example, no population is exactly Normal. Deciding when a statistical procedure should be used often requires judgment assisted by exploratory analysis of the data.

15.2 Running red lights. A survey of licensed drivers inquired about running red lights. One question asked, "Of every ten motorists who run a red light, about how many do you think will be caught?" The mean result for 880 respondents was $\bar{x} = 1.92$ and the standard deviation was $s = 1.83$.[2] For this large sample, s will be close to the population standard deviation σ, so suppose we know that $\sigma = 1.83$.

(a) Give a 95% confidence interval for the mean opinion in the population of all licensed drivers.

(b) The distribution of responses is skewed to the right rather than Normal. This will not strongly affect the z confidence interval for this sample. Why not?

(c) The 880 respondents are an SRS from completed calls among 45,956 calls to randomly chosen residential telephone numbers listed in telephone directories. Only 5029 of the calls were completed. This information gives two reasons to suspect that the sample may not represent all licensed drivers. What are these reasons?

Cautions about confidence intervals

The most important caution about confidence intervals in general is a consequence of the use of a sampling distribution. A sampling distribution shows how a statistic such as $\bar{x}$ varies in repeated sampling. This variation causes "random sampling error" because the statistic misses the true parameter by a random amount. No other source of variation or bias in the sample data influences the sampling distribution. So the margin of error in a confidence interval also ignores everything except the sample-to-sample variation due to choosing the sample randomly.

THE MARGIN OF ERROR DOESN'T COVER ALL ERRORS

The margin of error in a confidence interval covers only random sampling errors.

Practical difficulties such as undercoverage and nonresponse are often more serious than random sampling error. The margin of error does not take such difficulties into account.

Remember this unpleasant fact when reading the results of an opinion poll or other sample survey. The practical conduct of the survey influences the trustworthiness of its results in ways that are not included in the announced margin of error.

15.3 Rating the environment. A Gallup Poll asked the question "How would you rate the overall quality of the environment in this country today—as excellent, good, only fair, or poor?" In all, 46% of the sample rated the environment as good or excellent. Gallup announced the poll's margin of error for 95% confidence as ±3 percentage points. Which of the following sources of error are included in the margin of error?

(a) The poll dialed telephone numbers at random and so missed all people without phones.

(b) Nonresponse—some people whose numbers were chosen never answered the phone in several calls or answered but refused to participate in the poll.

(c) There is chance variation in the random selection of telephone numbers.

15.4 Holiday spending. "How much do you plan to spend for gifts this holiday season?" An interviewer asks this question of 250 customers at a large shopping mall. The sample mean and standard deviation of the responses are $\overline{x} = \$237$ and $s = \$65$.

(a) The distribution of spending is skewed, but we can act as though $\overline{x}$ is Normal. Why?

(b) For this large sample, we can act as if $\sigma = \$65$ because the sample s will be close to the population σ. Use the sample result to give a 99% confidence interval for the mean gift spending of all adults.

(c) This confidence interval can't be trusted because the sample responses may be badly biased. Suggest some reasons why the responses may be biased.

Cautions about significance tests

Significance tests are widely used in reporting the results of research in many fields of applied science and in industry. New pharmaceutical products require significant evidence of effectiveness and safety. Courts inquire about statistical significance in hearing class action discrimination cases. Marketers want to know whether a new ad campaign significantly outperforms the old one, and medical researchers want to know whether a new therapy performs significantly better. In all these uses, statistical significance is valued because it points to an effect that is unlikely to occur simply by chance.

The reasoning of tests is less straightforward than the reasoning of confidence intervals, and the cautions needed are more elaborate. Here are some points to keep in mind when using or interpreting significance tests.

How small a *P* is convincing? The purpose of a test of significance is to describe the degree of evidence provided by the sample against the null hypothesis. The *P*-value does this. But how small a *P*-value is convincing evidence against the null hypothesis? This depends mainly on two circumstances:

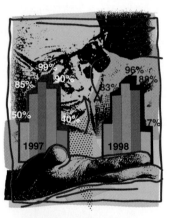

Should tests be banned?

Significance tests don't tell us how large or how important an effect is. Research psychologists have emphasized tests, so much so that some think their weaknesses should ban them from use. The American Psychological Association asked a group of experts. They said: use anything that sheds light on your study. Use more data analysis and confidence intervals. But: "The task force does not support any action that could be interpreted as banning the use of null hypothesis significance testing or *P*-values in psychological research and publication."

- *How plausible is* H_0? If H_0 represents an assumption that the people you must convince have believed for years, strong evidence (small P) will be needed to persuade them.

- *What are the consequences of rejecting* H_0? If rejecting H_0 in favor of H_a means making an expensive changeover from one type of product packaging to another, you need strong evidence that the new packaging will boost sales.

These criteria are a bit subjective. Different people will often insist on different levels of significance. Giving the P-value allows each of us to decide individually if the evidence is sufficiently strong.

Users of statistics have often emphasized standard levels of significance such as 10%, 5%, and 1%. For example, courts have tended to accept 5% as a standard in discrimination cases.[3] This emphasis reflects the time when tables of critical values rather than software dominated statistical practice. The 5% level ($\alpha = 0.05$) is particularly common. **There is no sharp border between "significant" and "insignificant," only increasingly strong evidence as the P-value decreases.** There is no practical distinction between the P-values 0.049 and 0.051. It makes no sense to treat $P \leq 0.05$ as a universal rule for what is significant.

APPLY YOUR KNOWLEDGE

15.5 Is it significant? Suppose that in the absence of special preparation SAT mathematics (SATM) scores vary Normally with mean $\mu = 475$ and $\sigma = 100$. One hundred students go through a rigorous training program designed to raise their SATM scores by improving their mathematics skills. Carry out a test of

$$H_0: \mu = 475$$
$$H_a: \mu > 475$$

In each of the following situations:

(a) The students' average score is $\bar{x} = 491.4$. Is this result significant at the 5% level?

(b) The average score is $\bar{x} = 491.5$. Is this result significant at the 5% level?

The difference between the two outcomes in (a) and (b) is of no importance. Beware attempts to treat $\alpha = 0.05$ as sacred.

Statistical significance and practical significance When a null hypothesis ("no effect" or "no difference") can be rejected at the usual levels, $\alpha = 0.05$ or $\alpha = 0.01$, there is good evidence that an effect is present. But that effect may be very small. When large samples are available, even tiny deviations from the null hypothesis will be significant.

EXAMPLE 15.3   It's significant. So what?

We are testing the hypothesis of no correlation between two variables. With 1000 observations, an observed correlation of only $r = 0.08$ is significant evidence at the $\alpha = 0.01$ level that the correlation in the population is not zero but positive. The low significance level does not mean there is a strong association, only that there is strong evidence of some association. The true population correlation is probably quite close to the observed sample value, $r = 0.08$. We might well conclude that for practical purposes we can ignore the association between these variables, even though we are confident (at the 1% level) that the correlation is positive.

Exercise 15.6 demonstrates in detail how increasing the sample size drives down P. Remember the wise saying: **Statistical significance is not the same thing as practical significance.**

The remedy for attaching too much importance to statistical significance is to pay attention to the actual data as well as to the P-value. Plot your data and examine them carefully. Are there outliers or other deviations from a consistent pattern? A few outlying observations can produce highly significant results if you blindly apply common tests of significance. Outliers can also destroy the significance of otherwise convincing data. The foolish user of statistics who feeds the data to a computer without exploratory analysis will often be embarrassed. Is the effect you are seeking visible in your plots? If not, ask yourself if the effect is large enough to be practically important. Give a confidence interval for the parameter in which you are interested. A confidence interval actually estimates the size of an effect rather than simply asking if it is too large to reasonably occur by chance alone. Confidence intervals are not used as often as they should be, while tests of significance are perhaps overused.

APPLY YOUR KNOWLEDGE

15.6 **Coaching and the SAT.**   Suppose that SAT mathematics scores in the absence of coaching vary Normally with mean $\mu = 475$ and $\sigma = 100$. Suppose also that coaching may change μ but does not change σ. An increase in an SAT score from 475 to 478 is of no importance in seeking admission to college, but this unimportant change can be statistically very significant. To see this, calculate the P-value for the test of

$$H_0: \mu = 475$$
$$H_a: \mu > 475$$

in each of the following situations:

(a) A coaching service coaches 100 students. Their SATM scores average $\overline{x} = 478$.

(b) By the next year, the service has coached 1000 students. Their SATM scores average $\overline{x} = 478$.

(c) An advertising campaign brings the number of students coached to 10,000. Their average score is still $\overline{x} = 478$.

15.7 Confidence intervals help. Give a 99% confidence interval for the mean SATM score μ after coaching in each part of the previous exercise. For large samples, the confidence interval tells us, "Yes, the mean score is higher than 475 after coaching, but only by a small amount."

15.8 How far do rich parents take us? How much education children get is strongly associated with the wealth and social status of their parents. In social science jargon, this is "socioeconomic status," or SES. But the SES of parents has little influence on whether children who have graduated from college go on to yet more education. One study looked at whether college graduates took the graduate admissions tests for business, law, and other graduate programs. The effects of the parents' SES on taking the LSAT test for law school were "both statistically insignificant and small."

(a) What does "statistically insignificant" mean?

(b) Why is it important that the effects were small in size as well as insignificant?

Beware of multiple analyses Statistical significance ought to mean that you have found an effect that you were looking for. The reasoning behind statistical significance works well if you decide what effect you are seeking, design a study to search for it, and use a test of significance to weigh the evidence you get. In other settings, significance may have little meaning.

EXAMPLE 15.4 Cell phones and brain cancer

Might the radiation from cell phones be harmful to users? Many studies have found little or no connection between using cell phones and various illnesses. Here is part of a news account of one study:

> A hospital study that compared brain cancer patients and a similar group without brain cancer found no statistically significant association between cell phone use and a group of brain cancers known as gliomas. But when 20 types of glioma were considered separately an association was found between phone use and one rare form. Puzzlingly, however, this risk appeared to decrease rather than increase with greater mobile phone use.[4]

Think for a moment: Suppose that the 20 null hypotheses for these 20 significance tests are all true. Then each test has a 5% chance of being significant at the 5% level. That's what $\alpha = 0.05$ means: results this extreme occur only 5% of the time just by chance when the null hypothesis is true. Because 5% is 1/20, we expect about 1 of 20 tests to give a significant result just by chance. Running one test and reaching the $\alpha = 0.05$ level is reasonably good evidence that you have found something; running 20 tests and reaching that level only once is not.

The caution about multiple analyses applies to confidence intervals as well. A single 95% confidence interval has probability 0.95 of capturing the true parameter each time you use it. The probability that all of 20 confidence intervals will capture their parameters is much less than 95%.

Honestly significant?

You may recall (page 89) that Chinese and Japanese, for whom the number 4 is unlucky, die more often on the fourth day of the month than on other days. Deaths on that day are "significantly more frequent" than on other days, say the authors of a study. Can we trust this? Not if the authors looked at all days, picked the ones with the most deaths, then searched for an explanation. A critic raised that issue, and the authors replied: No, we had day 4 in mind in advance, so our test was legitimate and the extra deaths on that day are honestly significant.

Searching data for suggestive patterns is certainly legitimate. Exploratory data analysis is an important part of statistics. But the reasoning of formal inference does not apply when your search for a striking effect in the data is successful. The remedy is clear. Once you have a hypothesis, design a study to search specifically for the effect you now think is there. If the result of this study is statistically significant, you have real evidence.

APPLY YOUR KNOWLEDGE

15.9 Searching for ESP. A researcher looking for evidence of extrasensory perception (ESP) tests 500 subjects. Four of these subjects do significantly better ($P < 0.01$) than random guessing.

  (a) Is it proper to conclude that these four people have ESP? Explain your answer.

  (b) What should the researcher now do to test whether any of these four subjects have ESP?

The power of a test*

We assess the performance of a confidence interval in two ways: by its confidence level and by its margin of error. The confidence level says how often the method succeeds in capturing the true parameter. The margin of error says how closely the interval pins down the value of the parameter.

We should also assess the performance of a statistical test in two ways. A good test will rarely reject the null hypothesis H_0 when it is really true. We may insist on significance at levels such as 0.05 and 0.01 to guarantee that we will reject a true hypothesis only 1 time in 20 tests or 1 time in 100 tests on the average. A good test has another property: it usually rejects H_0 when the alternative hypothesis is true.

We use the z test for hypotheses such as

$$H_0: \mu = 0$$
$$H_a: \mu > 0$$

Suppose we choose a fixed significance level α as a criterion for the level of evidence we require to reject H_0. Then the probability of wrongly rejecting H_0 when in fact $\mu = 0$ is α. This probability is one way to assess the performance of the test. What about the second way, "usually rejects H_0 when the alternative hypothesis is true"? The probability of correctly rejecting H_0 when H_a is true is different for different values of μ. When μ is very close to 0, the test will find it hard to distinguish μ from 0, and the probability of rejecting will be small. When μ is far from 0, it should be easy to see from the data that H_0 isn't true, and the probability of rejecting should be large. We have to specify the *specific alternative* μ we have in mind before we can ask if the test usually rejects H_0.

---

*The remainder of this chapter introduces additional ideas used to describe statistical tests. This more advanced material is not needed to read the rest of the book.

The probability that a test rejects the null hypothesis when an alternative is true is called the *power* of the test for that alternative. The higher the power, the more sensitive the test.

POWER

The probability that a fixed level α significance test will reject H_0 when a particular alternative value of the parameter is true is called the **power** of the test against that alternative.

EXAMPLE 15.5 Sweetening colas: power

The cola maker of Example 14.2 (page 341) determines that a sweetness loss is too large to accept if the mean response for all tasters is $\mu = 1.1$. Will a 5% significance test of the hypotheses

$$H_0: \mu = 0$$
$$H_a: \mu > 0$$

based on a sample of 10 tasters usually detect a change this great?

We want the power of the test against the alternative $\mu = 1.1$. This is the probability that the test rejects H_0 when $\mu = 1.1$ is true.

Step 1. *Write the rule for rejecting H_0 in terms of $\bar{x}$.* We know that $\sigma = 1$, so the z test rejects H_0 at the $\alpha = 0.05$ level when

$$z = \frac{\bar{x} - 0}{1/\sqrt{10}} \geq 1.645$$

This is the same as

$$\bar{x} \geq 0 + 1.645 \, \frac{1}{\sqrt{10}}$$

or, doing the arithmetic,

$$\text{Reject } H_0 \text{ when } \bar{x} \geq 0.520$$

This step just restates the rule for the test. It pays no attention to the specific alternative we have in mind.

Step 2. *The power is the probability of this event under the condition that the alternative $\mu = 1.1$ is true.* To calculate this probability, standardize $\bar{x}$ using $\mu = 1.1$.

$$\text{power} = P(\bar{x} \geq 0.520 \text{ when } \mu = 1.1)$$
$$= P\left(\frac{\bar{x} - 1.1}{1/\sqrt{10}} \geq \frac{0.520 - 1.1}{1/\sqrt{10}}\right)$$
$$= P(Z \geq -1.83)$$
$$= 1 - 0.0336 = 0.9664 \text{ using Table A}$$

The cola maker is satisfied. The test will declare that the cola loses sweetness only 5% of the time when it actually does not ($\alpha = 0.05$) and 97% of the time when the true mean sweetness loss is 1.1 (power = 0.97).

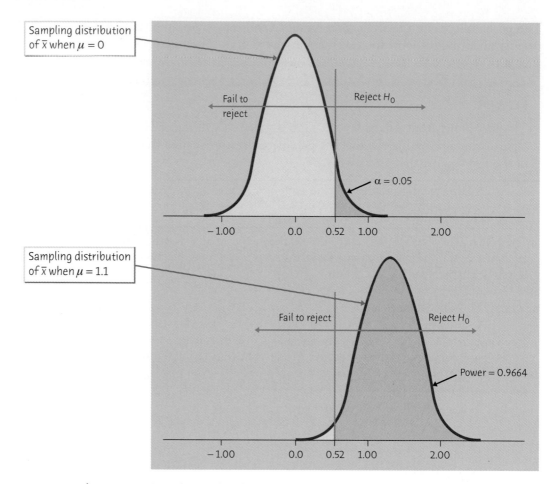

Sampling distribution of $\bar{x}$ when $\mu = 0$

Sampling distribution of $\bar{x}$ when $\mu = 1.1$

Figure 15.1 Significance level and power for the test of Example 15.5. The test rejects H_0 when $\bar{x} \geq 0.52$. The level α is the probability of this event when the null hypothesis is true. The power is the probability of the same event when the alternative hypothesis is true.

The idea behind the calculation in Example 15.5 is that the z test statistic is standardized taking the null hypothesis to be true. If an alternative is true, this is no longer the correct standardization. So we go back to $\bar{x}$ and standardize again taking the alternative to be true. Figure 15.1 illustrates the rule for the test and the two sampling distributions, one under the null hypothesis $\mu = 0$ and the other under the alternative $\mu = 1.1$. The level α is the probability of rejecting H_0 when H_0 is true, so it is an area under the top curve. The power is the probability of rejecting H_0 when H_a is true, so it is an area under the bottom curve.

APPLY YOUR KNOWLEDGE

15.10 NAEP scores: power. You have the NAEP quantitative scores for an SRS of 840 young men. You plan to test hypotheses about the population mean score,

$$H_0: \mu = 275$$
$$H_a: \mu < 275$$

at the 1% level of significance. The population standard deviation is known to be $\sigma = 60$. The z test statistic is

$$z = \frac{\bar{x} - 275}{60/\sqrt{840}}$$

(a) What is the rule for rejecting H_0 in terms of z?

(b) What is the rule for rejecting H_0 restated in terms of $\bar{x}$?

(c) You want to know whether this test will usually reject H_0 when the true population mean is 270, 5 points lower than the null hypothesis claims. Answer this question by calculating the power when $\mu = 270$.

15.11 Cola bottles: power. Bottles of a popular cola are supposed to contain 300 milliliters (ml) of cola. There is some variation from bottle to bottle because the filling machinery is not perfectly precise. The distribution of the contents is Normal with standard deviation $\sigma = 3$ ml. Will inspecting 6 bottles discover underfilling? The hypotheses are

$$H_0: \mu = 300$$
$$H_a: \mu < 300$$

A 5% significance test rejects H_0 if $z \le -1.645$, where the test statistic z is

$$z = \frac{\bar{x} - 300}{3/\sqrt{6}}$$

Power calculations help us see how large a shortfall in the bottle contents the test can be expected to detect.

(a) Find the power of this test against the alternative $\mu = 299$.

(b) Find the power against the alternative $\mu = 295$.

(c) Is the power against $\mu = 290$ higher or lower than the value you found in (b)? (Don't actually calculate that power.) Explain your answer.

15.12 Sample size and power. Increasing the sample size increases the power of a test when the level α is unchanged. In the previous exercise, $n = 6$. Suppose that a sample of n bottles had been measured. The 5% significance test still rejects H_0 when $z \le -1.645$, but the z statistic is now

$$z = \frac{\bar{x} - 300}{3/\sqrt{n}}$$

(a) Find the power of this test against the alternative $\mu = 299$ when $n = 25$.

(b) Find the power against $\mu = 299$ when $n = 100$.

Type I and Type II errors*

Fish, fishermen, and Type II errors

Are the stocks of cod in the ocean off eastern Canada declining? Studies over many years failed to find significant evidence of a decline. These studies had low power—that is, they might fail to find a decline even if one was present. When it became clear that the cod were vanishing, quotas on fishing ravaged the economy in parts of Canada. It appears that the earlier studies had made Type II errors. If they had seen the decline, early action might have reduced the economic and environmental costs.

We can assess the performance of a test by giving two probabilities: the level α and the power for an alternative that we want to be able to detect. In practice, part of planning a study is to calculate power against a range of alternatives to learn which alternatives the test is likely to detect and which it is likely to miss. If the test does not have high enough power against alternatives that we want to detect, the remedy is to increase the size of the sample. That can be expensive, so the planning process must balance good statistical properties against cost.

The level of a test is the probability of making the *wrong* decision when the null hypothesis is true. The power for a specific alternative is the probability of making the *right* decision when that alternative is true. We can just as well describe the test by giving the probability of a *wrong* decision under both conditions.

> **TYPE I AND TYPE II ERRORS**
>
> If we reject H_0 when in fact H_0 is true, this is a **Type I error.**
>
> If we fail to reject H_0 when in fact H_a is true, this is a **Type II error.**
>
> The **significance level** α of any fixed level test is the probability of a Type I error.
>
> The **power** of a test against any alternative is 1 minus the probability of a Type II error for that alternative.

The possibilities are summed up in Figure 15.2. If H_0 is true, our decision is correct if we accept H_0 and is a Type I error if we reject H_0. If H_a is true, our decision is either correct or a Type II error. Only one error is possible at one time.

		Truth about the population	
		H_0 true	H_a true
Decision based on sample	Reject H_0	Type I error	Correct decision
	Accept H_0	Correct decision	Type II error

Figure 15.2 The two types of error in testing hypotheses.

EXAMPLE 15.6   Sweetening colas: error probabilities

When we select the significance level α of a test, we are setting the probability of a Type I error. Calculating the probability of a Type II error is just like calculating the power, except that we find the probability of the wrong decision (failing to reject H_0) rather than the probability of the right decision (rejecting).

The cola maker of Example 15.5 wants to test

$$H_0: \mu = 0$$
$$H_a: \mu > 0$$

The example shows that the z test rejects the null hypothesis at level $\alpha = 0.05$ when the mean sweetness loss assigned by 10 tasters satisfies $\bar{x} \geq 0.520$. The two error probabilities are

$$P(\text{Type I error}) = P(\text{reject } H_0 \text{ when } \mu = 0)$$
$$= P(\bar{x} \geq 0.520 \text{ when } \mu = 0) = 0.05$$
$$P(\text{Type II error}) = P(\text{fail to reject } H_0 \text{ when } \mu = 1.1)$$
$$= P(\bar{x} < 0.520 \text{ when } \mu = 1.1) = 0.0336$$

We did the calculation for the probability of a Type II error in Example 15.5 on the way to finding the power.

Calculations of power (or of error probabilities) are useful for planning studies because we can make these calculations before we have any data. Once we actually have data, it is more common to report a *P*-value rather than a reject-or-not decision at a fixed significance level α. The *P*-value measures the strength of the evidence provided by the data against H_0 and in favor of H_a. It leaves any action or decision based on that evidence up to each individual. Different people may require different strengths of evidence.

APPLY YOUR KNOWLEDGE

15.13 **Error probabilities.**   Your company markets a computerized medical diagnostic program. The program scans the results of routine medical tests (pulse rate, blood tests, etc.) and either clears the patient or refers the case to a doctor. The program is used to screen thousands of people who do not have specific medical complaints. The program makes a decision about each person.

(a) What are the two hypotheses and the two types of error that the program can make? Describe the two types of error in terms of "false positive" and "false negative" test results.

(b) The program can be adjusted to decrease one error probability, at the cost of an increase in the other error probability. Which error probability would you choose to make smaller, and why? (This is a matter of judgment. There is no single correct answer.)

15.14 **NAEP scores: error probabilities.**   You have the NAEP quantitative scores for an SRS of 840 young men. You plan to test hypotheses about the population mean score,

$$H_0: \mu = 275$$
$$H_a: \mu < 275$$

at the 1% level of significance. The population standard deviation is known to be $\sigma = 60$. The z test statistic is

$$z = \frac{\overline{x} - 275}{60/\sqrt{840}}$$

(a) What is the rule for rejecting H_0 in terms of z?

(b) What is the probability of a Type I error?

(c) You want to know whether this test will usually reject H_0 when the true population mean is 270, 5 points lower than the null hypothesis claims. Answer this question by calculating the probability of a Type II error when $\mu = 270$.

15.15 **Find the error probabilities.**   You have an SRS of size $n = 9$ from a Normal distribution with $\sigma = 1$. You wish to test

$$H_0: \mu = 0$$
$$H_a: \mu > 0$$

You decide to reject H_0 if $\overline{x} > 0$ and to accept H_0 otherwise.

(a) Find the probability of a Type I error. That is, find the probability that the test rejects H_0 when in fact $\mu = 0$.

(b) Find the probability of a Type II error when $\mu = 0.3$. This is the probability that the test accepts H_0 when in fact $\mu = 0.3$.

(c) Find the probability of a Type II error when $\mu = 1$.

Chapter 15 SUMMARY

A specific recipe for a confidence interval or test is correct only under specific conditions. The most important conditions concern the method used to produce the data. Other factors such as the shape of the population distribution may also be important.

Whenever you use statistical inference, you are acting as if your data are a probability sample or come from a randomized comparative experiment.

Always do data analysis before inference to detect outliers or other problems that would make inference untrustworthy.

The margin of error in a confidence interval accounts for only the chance variation due to random sampling. In practice, errors due to nonresponse or undercoverage are often more serious.

There is no universal rule for how small a P-value is convincing. Beware of placing too much weight on traditional significance levels such as $\alpha = 0.05$.

Very small effects can be highly significant (small P) when a test is based on a large sample. A statistically significant **effect** need not be practically important. Plot the data to display the effect you are seeking, and use confidence intervals to estimate the actual values of parameters.

On the other hand, lack of significance does not imply that H_0 is true. Even a large effect can fail to be significant when a test is based on a small sample.

Many tests run at once will probably produce some significant results by chance alone, even if all the null hypotheses are true.

The **power** of a significance test measures its ability to detect an alternative hypothesis. The power against a specific alternative is the probability that the test will reject H_0 when the alternative is true.

We can describe the performance of a test at fixed level α by giving the probabilities of two types of error. A **Type I error** occurs if we reject H_0 when it is in fact true. A **Type II error** occurs if we fail to reject H_0 when in fact H_a is true.

In a fixed level α significance test, the significance level α is the probability of a Type I error, and the power against a specific alternative is 1 minus the probability of a Type II error for that alternative.

Increasing the size of the sample increases the power (reduces the probability of a Type II error) when the significance level remains fixed.

Chapter 15 EXERCISES

15.16 Don't do inference. Give an example of a set of data for which statistical inference is not valid.

15.17 When to use pacemakers. A medical panel prepared guidelines for when cardiac pacemakers should be implanted in patients with heart problems. The panel reviewed a large number of medical studies to judge the strength of the evidence supporting each recommendation. For each recommendation, they ranked the evidence as level A (strongest), B, or C (weakest). Here, in scrambled order, are the panel's descriptions of the three levels of evidence.[5] Which is A, which B, and which C? Explain your ranking.

Evidence was ranked as level ___ when data were derived from a limited number of trials involving comparatively small numbers of patients or from well-designed data analysis of nonrandomized studies or observational data registries.

Evidence was ranked as level ___ if the data were derived from multiple randomized clinical trials involving a large number of individuals.

Evidence was ranked as level ___ when consensus of expert opinion was the primary source of recommendation.

15.18 Why are larger samples better? Statisticians prefer large samples. Describe briefly the effect of increasing the size of a sample (or the number of subjects in an experiment) on each of the following:

(a) The margin of error of a 95% confidence interval.

(b) The *P*-value of a test, when H_0 is false and all facts about the population remain unchanged as n increases.

(c) (Optional) The power of a fixed level α test, when α, the alternative hypothesis, and all facts about the population remain unchanged.

15.19 What is significance good for? Which of the following questions does a test of significance answer?

(a) Is the sample or experiment properly designed?

(b) Is the observed effect due to chance?

(c) Is the observed effect important?

15.20 Sensitive questions. The National AIDS Behavioral Surveys found that 170 individuals in its random sample of 2673 adult heterosexuals said they had multiple sexual partners in the past year. That's 6.36% of the sample. Why is this estimate likely to be biased? Does the margin of error of a 95% confidence interval for the proportion of all adults with multiple partners allow for this bias?

15.21 Ages of presidents. Joe is writing a report on the backgrounds of American presidents. He looks up the ages of all 43 presidents when they entered office. Because Joe took a statistics course, he uses these 43 numbers to get a 95% confidence interval for the mean age of all men who have been president. This makes no sense. Why not?

15.22 Supermarket shoppers. A marketing consultant observes 50 consecutive shoppers at a supermarket. Here are the amounts (in dollars) spent in the store by these shoppers:

3.11	8.88	9.26	10.81	12.69	13.78	15.23	15.62	17.00	17.39
18.36	18.43	19.27	19.50	19.54	20.16	20.59	22.22	23.04	24.47
24.58	25.13	26.24	26.26	27.65	28.06	28.08	28.38	32.03	34.98
36.37	38.64	39.16	41.02	42.97	44.08	44.67	45.40	46.69	48.65
50.39	52.75	54.80	59.07	61.22	70.32	82.70	85.76	86.37	93.34

(Daly and Newton/The Image Bank/Getty Images)

(a) Why is it risky to regard these 50 shoppers as an SRS from the population of all shoppers at this store? Name some factors that might make 50 consecutive shoppers at a particular time unrepresentative of all shoppers.

(b) Make a stemplot of the data. The stemplot suggests caution in using the z procedures for these data. Why?

15.23 Predicting success of trainees. What distinguishes managerial trainees who eventually become executives from those who don't succeed and leave the company? We have abundant data on past trainees—data on their personalities and goals, their college preparation and performance, even their family backgrounds and their hobbies. Statistical software makes it easy to perform dozens of significance tests on these dozens of variables to see which ones best predict later success. We find that future executives are significantly more likely than washouts to have an urban or suburban upbringing and an undergraduate degree in a technical field.

Explain clearly why using these "significant" variables to select future trainees is not wise. Then design a follow-up study using this year's new trainees as subjects that should clarify the importance of the variables identified by the first study.

15.24 Color blindness in Africa. An anthropologist suspects that color blindness is less common in societies that live by hunting and gathering than in settled agricultural societies. He tests a number of adults in two populations in Africa, one of each type. The proportion of color-blind people is significantly lower ($P < 0.05$) in the hunter-gatherer population. What additional information would you want to help you decide whether you accept the claim about color blindness?

15.25 Internet users. A survey of users of the Internet found that males outnumbered females by nearly 2 to 1. This was a surprise, because earlier surveys had put the ratio of men to women closer to 9 to 1. Later in the article we find this information:

> Detailed surveys were sent to more than 13,000 organizations on the Internet; 1,468 usable responses were received. According to Mr. Quarterman, the margin of error is 2.8 percent, with a confidence level of 95 percent.[6]

(a) What was the *response rate* for this survey? (The response rate is the percent of the planned sample that responded.)

(b) Do you think that the small margin of error is a good measure of the accuracy of the survey's results? Explain your answer.

15.26 Prayer in the schools? A *New York Times*/CBS News poll asked the question "Do you favor an amendment to the Constitution that would permit organized prayer in public schools?" Sixty-six percent of the sample answered "Yes." The article describing the poll says that it "is based on telephone interviews conducted from Sept. 13 to Sept. 18 with 1,664 adults around the United States, excluding Alaska and Hawaii.... the telephone numbers were formed by random digits, thus permitting access to both listed and unlisted residential numbers." The article gives the margin of error as 3 percentage points. Opinion polls customarily announce margins of error for 95% confidence, so we are 95% confident that the percent of all adults who favor prayer in the schools lies in the interval 66% ± 3%.

The news article goes on to say: "The theoretical errors do not take into account a margin of additional error resulting from the various practical difficulties in taking any survey of public opinion." List some of the "practical difficulties" that may cause errors in addition to the

±3% margin of error. Pay particular attention to the news article's description of the sampling method.

15.27 Comparing package designs. A company compares two package designs for a laundry detergent by placing bottles with both designs on the shelves of several markets. Checkout scanner data on more than 5000 bottles bought show that more shoppers bought Design A than Design B. The difference is statistically significant ($P = 0.02$). Can we conclude that consumers strongly prefer Design A? Explain your answer.

15.28 Plagiarizing online. The Excite Poll can be found online at **poll.excite.com.** The question appears on the screen, and you click to choose a response. On October 14, 2002, the question was

> It is now possible for school students to log on to Internet sites and download homework. Everything from book reports to doctoral dissertations can be downloaded free or for a fee. Do you believe giving a student who is caught plagiarizing an "F" for their assignment is the right punishment?

Of the 20,125 people who responded, 14,793 clicked "Yes." That's 73.5% of the sample. Based on this sample, a 95% confidence interval for the percent of the population who would say "Yes" is 73.5% ± 0.61% (details in Chapter 18). Why is this confidence interval worthless?

15.29 What distinguishes schizophrenics? A group of psychologists once measured 77 variables on a sample of schizophrenic people and a sample of people who were not schizophrenic. They compared the two samples using 77 separate significance tests. Two of these tests were significant at the 5% level. Suppose that there is in fact no difference on any of the 77 variables between people who are and people who are not schizophrenic in the adult population. Then all 77 null hypotheses are true.

(a) What is the probability that one specific test shows a difference significant at the 5% level?

(b) Why is it not surprising that 2 of the 77 tests were significant at the 5% level?

15.30 Helping welfare mothers. A study compares two groups of mothers with young children who were on welfare two years ago. One group attended a voluntary training program that was offered free of charge at a local vocational school and was advertised in the local news media. The other group did not choose to attend the training program. The study finds a significant difference ($P < 0.01$) between the proportions of the mothers in the two groups who are still on welfare. The difference is not only significant but quite large. The report says that with 95% confidence the percent of the nonattending group still on welfare is 21% ± 4% higher than that of the group who attended the program. You are on the staff of a member of Congress who is interested in the plight of welfare mothers, and who asks you about the report.

(a) Explain in simple language what "a significant difference $(P < 0.01)$" means.

(b) Explain clearly and briefly what "95% confidence" means.

(c) Is this study good evidence that requiring job training of all welfare mothers would greatly reduce the percent who remain on welfare for several years?

The following exercises concern the optional material on error probabilities and power.

15.31 **Two-sided tests: power.** Power calculations for two-sided tests follow the same outline as for one-sided tests. Example 14.10 (page 354) presents a test of

$$H_0: \mu = 0.86$$
$$H_a: \mu \neq 0.86$$

at the 1% level of significance. The sample size is $n = 3$ and $\sigma = 0.0068$. We will find the power of this test against the alternative $\mu = 0.845$.

(a) The test in Example 14.10 rejects H_0 when $|z| \geq 2.576$. The test statistic z is

$$z = \frac{\bar{x} - 0.86}{0.0068/\sqrt{3}}$$

Write the rule for rejecting H_0 in terms of the values of $\bar{x}$. (Because the test is two-sided, it rejects when $\bar{x}$ is either too large or too small.)

(b) Now find the probability that $\bar{x}$ takes values that lead to rejecting H_0 if the true mean is $\mu = 0.845$. This probability is the power.

(c) What is the probability that this test makes a Type II error when $\mu = 0.845$?

15.32 **Find the power.** In Example 14.7 (page 351), a company medical director failed to find significant evidence that the mean blood pressure of a population of executives differed from the national mean $\mu = 128$. The medical director now wonders if the test used would detect an important difference if one were present. For the SRS of size 72 from a population with standard deviation $\sigma = 15$, the z statistic is

$$z = \frac{\bar{x} - 128}{15/\sqrt{72}}$$

The two-sided test rejects $H_0: \mu = 128$ at the 5% level of significance when $|z| \geq 1.96$.

(a) Find the power of the test against the alternative $\mu = 134$.

(b) Find the power of the test against $\mu = 122$. Can the test be relied on to detect a mean that differs from 128 by 6?

(c) If the alternative were farther from H_0, say $\mu = 136$, would the power be higher or lower than the values calculated in (a) and (b)?

15.33 Power and error probabilities. In Exercise 15.11 you found the power of a test against the alternative $\mu = 295$. Use the result of that exercise to find the probabilities of Type I and Type II errors for that test and that alternative.

15.34 Error probabilities and power. In Exercise 15.14 you found the probabilities of the two types of error for a test of H_0: $\mu = 275$, with the specific alternative that $\mu = 270$. Use the result of that exercise to give the power of the test against the alternative $\mu = 270$.

(Stephen Chernin/Getty Images)

15.35 Is the stock market efficient? You are reading an article in a business journal that discusses the "efficient market hypothesis" for the behavior of securities prices. The author admits that most tests of this hypothesis have failed to find significant evidence against it. But he says this failure is a result of the fact that the tests used have low power. "The widespread impression that there is strong evidence for market efficiency may be due just to a lack of appreciation of the low power of many statistical tests."[7]

Explain in simple language why tests having low power often fail to give evidence against a hypothesis even when the hypothesis is really false.

Chapter 15 MEDIA EXERCISES

15.36 Radar detectors and speeding. Researchers observed the speed of cars on a rural highway (speed limit 55 miles per hour) before and after police radar was directed at them. They compared the speed of cars that had radar detectors with the speed of cars without detectors. Here are the mean speeds (miles per hour) observed for 22 cars with radar detectors and 46 cars without:

	Radar Detector?	
	Yes	No
Before radar	70	68
After radar	59	67

The study report says: "Those vehicles with radar detectors were significantly faster ($P < 0.01$) than those without them before radar exposure...and significantly slower immediately after ($P < 0.0001$)."

(a) Explain in simple language why these P-values are good evidence that drivers with radar detectors do behave differently.

(b) Despite the fact that $P < 0.01$, there is little difference in the mean speeds of drivers with and without radar detectors before the radar is turned on. Explain in simple language how so small a difference can be statistically significant.

From Exploration to Inference Part II Review

Designs for producing data are essential parts of statistics in practice. Figures II.1 and II.2 display the big ideas visually. Random sampling and randomized comparative experiments are perhaps the most important statistical inventions of the twentieth century. Both were slow to gain acceptance, and you will still see many voluntary response samples and uncontrolled experiments. You should now understand good techniques for producing data and also why bad techniques often produce worthless data. The deliberate use of chance in producing data is a central idea in statistics. It not only reduces bias but allows use of the laws of probability to analyze data. Fortunately, we need only some basic facts about probability in order to understand statistical inference.

Statistical inference draws conclusions about a population on the basis of sample data and uses probability to indicate how reliable the conclusions are. A confidence interval estimates an unknown parameter. A significance test shows how strong the evidence is for some claim about a parameter.

The probabilities in both confidence intervals and tests tell us what would happen if we used the recipe for the interval or test very many times. A confidence level is the success rate of the recipe for a confidence interval, that is, the probability that the recipe actually produces an interval that contains the

Figure II.1 STATISTICS IN SUMMARY
Simple Random Sample

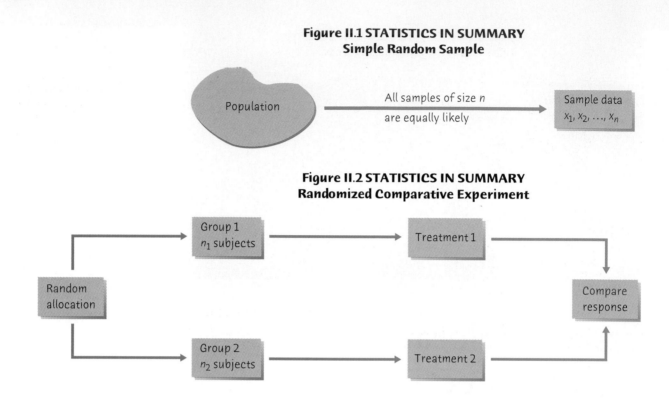

Figure II.2 STATISTICS IN SUMMARY
Randomized Comparative Experiment

unknown parameter. A 95% confidence interval gives a correct result 95% of the time when we use it repeatedly. A P-value tells us how surprising the observed outcome would be if the null hypothesis were true. That is, P is the probability that the test would produce a result at least as extreme as the observed result if the null hypothesis really were true. Very surprising outcomes (small P-values) are good evidence that the null hypothesis is not true.

Figures II.3 and II.4 use the z procedures introduced in Chapters 13 and 14 to present in picture form the big ideas of confidence intervals and significance

Figure II.3 STATISTICS IN SUMMARY
The Idea of a Confidence Interval

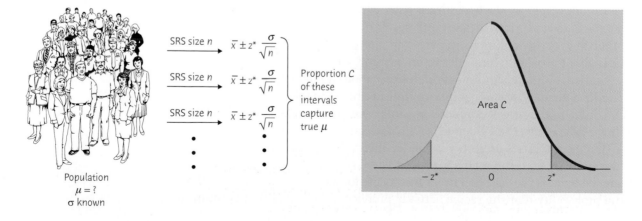

Figure II.4 STATISTICS IN SUMMARY
The Idea of a Significance Test

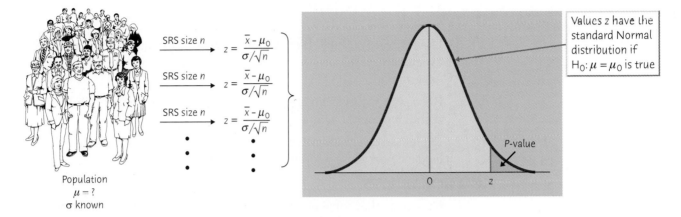

Population
$\mu = ?$
σ known

SRS size n → $z = \dfrac{\bar{x} - \mu_0}{\sigma/\sqrt{n}}$

SRS size n → $z = \dfrac{\bar{x} - \mu_0}{\sigma/\sqrt{n}}$

SRS size n → $z = \dfrac{\bar{x} - \mu_0}{\sigma/\sqrt{n}}$

Values z have the standard Normal distribution if $H_0: \mu = \mu_0$ is true

P-value

tests. These ideas are the foundation for the rest of this book. We will have much to say about many statistical methods and their use in practice. In every case, the basic reasoning of confidence intervals and significance tests remains the same.

Part II SUMMARY

Here are the most important skills you should have acquired from reading Chapters 7 to 15.

A. SAMPLING

1. Identify the population in a sampling situation.
2. Recognize bias due to voluntary response samples and other inferior sampling methods.
3. Use software or Table B of random digits to select a simple random sample (SRS) from a population.
4. Recognize the presence of undercoverage and nonresponse as sources of error in a sample survey. Recognize the effect of the wording of questions on the responses.
5. Use random digits to select a stratified random sample from a population when the strata are identified.

B. EXPERIMENTS

1. Recognize whether a study is an observational study or an experiment.
2. Recognize bias due to confounding of explanatory variables with lurking variables in either an observational study or an experiment.
3. Identify the factors (explanatory variables), treatments, response variables, and individuals or subjects in an experiment.

4. Outline the design of a completely randomized experiment using a diagram like that in Figure II.2. The diagram in a specific case should show the sizes of the groups, the specific treatments, and the response variable.

5. Use software or Table B of random digits to carry out the random assignment of subjects to groups in a completely randomized experiment.

6. Recognize the placebo effect. Recognize when the double-blind technique should be used.

7. Explain why a randomized comparative experiment can give good evidence for cause-and-effect relationships.

C. PROBABILITY

1. Recognize that some phenomena are random. Probability describes the long-run regularity of random phenomena.

2. Understand that the probability of an event is the proportion of times the event occurs in very many repetitions of a random phenomenon. Use the idea of probability as long-run proportion to think about probability.

3. Use basic probability facts to detect illegitimate assignments of probability: Any probability must be a number between 0 and 1, and the total probability assigned to all possible outcomes must be 1.

4. Use basic probability facts to find the probabilities of events that are formed from other events: The probability that an event does not occur is 1 minus its probability. If two events are disjoint, the probability that one or the other occurs is the sum of their individual probabilities.

5. Find probabilities of events in a finite sample space by adding the probabilities of their outcomes. Find probabilities of events as areas under a density curve.

6. Use the notation of random variables to make compact statements about random outcomes, such as $P(\bar{x} \leq 4) = 0.3$. Be able to interpret such statements.

D. SAMPLING DISTRIBUTIONS

1. Identify parameters and statistics in a sample or experiment.

2. Recognize the fact of sampling variability: a statistic will take different values when you repeat a sample or experiment.

3. Interpret a sampling distribution as describing the values taken by a statistic in all possible repetitions of a sample or experiment under the same conditions.

4. Interpret the sampling distribution of a statistic as describing the probabilities of its possible values.

E. THE SAMPLING DISTRIBUTION OF A SAMPLE MEAN

1. Recognize when a problem involves the mean $\bar{x}$ of a sample. Understand that $\bar{x}$ estimates the mean μ of the population from which the sample is drawn.

2. Use the law of large numbers to describe the behavior of $\bar{x}$ as the size of the sample increases.

3. Find the mean and standard deviation of a sample mean $\bar{x}$ from an SRS of size n when the mean μ and standard deviation σ of the population are known.

4. Understand that $\bar{x}$ is an unbiased estimator of μ and that the variability of $\bar{x}$ about its mean μ gets smaller as the sample size increases.

5. Understand that $\bar{x}$ has approximately a Normal distribution when the sample is large (central limit theorem). Use this Normal distribution to calculate probabilities that concern $\bar{x}$.

F. GENERAL RULES OF PROBABILITY (Optional)

1. Use Venn diagrams to picture relationships among several events.

2. Use the general addition rule to find probabilities that involve overlapping events.

3. Understand the idea of independence. Judge when it is reasonable to assume independence as part of a probability model.

4. Use the multiplication rule for independent events to find the probability that all of several independent events occur.

5. Use the multiplication rule for independent events in combination with other probability rules to find the probabilities of complex events.

6. Understand the idea of conditional probability. Find conditional probabilities for individuals chosen at random from a table of counts of possible outcomes.

7. Use the general multiplication rule to find $P(A \text{ and } B)$ from $P(A)$ and the conditional probability $P(B \mid A)$.

8. Use tree diagrams to organize several-stage probability models.

G. BINOMIAL DISTRIBUTIONS (Optional)

1. Recognize the binomial setting: a fixed number n of independent success-failure trials with the same probability P of success on each trial.

2. Recognize and use the binomial distribution of the count of successes in a binomial setting.

3. Use the binomial probability formula to find probabilities of events involving the count X of successes in a binomial setting for small values of n.

4. Find the mean and standard deviation of a binomial count X.

5. Recognize when you can use the Normal approximation to a binomial distribution. Use the Normal approximation to calculate probabilities that concern a binomial count X.

H. CONFIDENCE INTERVALS

1. State in nontechnical language what is meant by "95% confidence" or other statements of confidence in statistical reports.

2. Calculate a confidence interval for the mean μ of a Normal population with known standard deviation σ, using the recipe $\bar{x} \pm z^*\sigma/\sqrt{n}$.

3. Understand how the margin of error of a confidence interval changes with the sample size and the level of confidence C.

4. Find the sample size required to obtain a confidence interval of specified margin of error m when the confidence level and other information are given.

5. Identify sources of error in a study that are *not* included in the margin of error of a confidence interval, such as undercoverage or nonresponse.

I. SIGNIFICANCE TESTS

1. State the null and alternative hypotheses in a testing situation when the parameter in question is a population mean μ.

2. Explain in nontechnical language the meaning of the P-value when you are given the numerical value of P for a test.

3. Calculate the one-sample z statistic and the P-value for both one-sided and two-sided tests about the mean μ of a Normal population.

4. Assess statistical significance at standard levels α, either by comparing P to α or by comparing z to standard Normal critical values.

5. Recognize that significance testing does not measure the size or importance of an effect. Explain why a small effect can be significant in a large sample and why a large effect can fail to be significant in a small sample.

6. Recognize that any inference procedure acts as if the data were properly produced. The z confidence interval and test require that the data be an SRS from the population.

Review EXERCISES

Review exercises are short and straightforward exercises that help you solidify the basic ideas and skills in each part of this book.

II.1   **Tom Clancy's writing.** Different types of writing can sometimes be distinguished by the lengths of the words used. A student interested in this fact wants to study the lengths of words used by Tom Clancy in his novels. She opens a Clancy novel at random and records the lengths of each of the first 250 words on the page. What is the population in this study? What is the sample? What is the variable measured?

II.2   **Marijuana and driving.** A sample of young people in New Zealand found a positive association between use of marijuana (cannabis) and traffic accidents caused by the members of the sample. Both cannabis use and accidents were measured by interviewing the young people themselves. The study report says, "It is unlikely that self reports of cannabis use and accident rates will be perfectly accurate."[1] Is the response bias likely to make the reported association stronger or weaker than the true association? Why?

II.3   **Sample space.** A randomly chosen subject arrives for a study of exercise and fitness. Describe a sample space for each of the following. (In some cases, you may have some freedom in your choice of S.)

   (a) The subject is either female or male.

   (b) After 10 minutes on an exercise bicycle, you ask the subject to rate his or her effort on the Rate of Perceived Exertion (RPE) scale. RPE ranges in whole-number steps from 6 (no exertion at all) to 20 (maximal exertion).

   (c) You measure VO2, the maximum volume of oxygen consumed per minute during exercise. VO2 is generally between 2.5 liters per minute and 6 l/min.

   (d) You measure the maximum heart rate (beats per minute).

II.4   **Support the arts.** A local arts group surveys teenagers. It finds that teens who participate in performing arts programs (music, dance, theater) are less likely to use drugs, have criminal records, or drop out of school. The group launches a publicity campaign with the theme that arts prevent juvenile delinquency. Politely explain to the group why their survey doesn't prove this.

II.5   **Effects of day care.** The Carolina Abecedarian Project investigated the effect of high-quality preschool programs on children from poor families. Children were randomly assigned to two groups. One group participated in a year-round preschool program from the age of three months. The control group received social services but no preschool. At age 21, **35%** of the treatment group and **14%** of the control group were attending a four-year college or had already graduated from college. Is each of the boldface numbers a parameter or a statistic? Why?

II.6   **Estimating blood cholesterol.** The distribution of blood cholesterol level in the population of young men aged 20 to 34 years is close to

Normal with standard deviation $\sigma = 41$ milligrams per deciliter (mg/dl). You measure the blood cholesterol of 14 cross-country runners. The mean level is $\overline{x} = 172$ mg/dl. Assuming that σ is the same as in the general population, give a 90% confidence interval for the mean level μ among cross-country runners.

II.7 **Smaller margin of error.** How large a sample is needed to cut the margin of error in the previous problem in half? How large a sample is needed to cut the margin of error to ± 5 mg/dl?

II.8 **Testing blood cholesterol.** The mean blood cholesterol level for all men aged 20 to 34 years is $\mu = 188$ mg/dl. We suspect that the mean for cross-country runners is lower. State hypotheses, use the information in Exercise II.6 to find the test statistic, and give the P-value. Is the result significant at the $\alpha = 0.10$ level? At $\alpha = 0.05$? At $\alpha = 0.01$?

II.9 **More significant results.** In Exercise II.7 you increased the sample of cross-country runners enough to cut in half the margin of error of a confidence interval. Suppose that this larger sample gave the same mean level, $\overline{x} = 172$ mg/dl. Redo the test in the previous exercise. What is the P-value now? At which of the levels $\alpha = 0.10$, $\alpha = 0.05$, $\alpha = 0.01$ is the result significant? What general fact about significance tests does your result illustrate?

II.10 **Pesticides in whale blubber.** The level of pesticides found in the blubber of whales is a measure of pollution of the oceans by runoff from land areas and can also be used to identify different populations of whales. A sample of 8 male minke whales in the West Greenland area of the North Atlantic found the mean concentration of the insecticide dieldrin to be $\overline{x} = 357$ nanograms per gram of blubber (ng/g).[2] Suppose that the concentration in all such whales varies Normally with standard deviation $\sigma = 50$ ng/g. What can you say with 90% confidence about the mean level?

II.11 **Other confidence levels.** Use the information in the previous exercise to give an 80% confidence interval and a 95% confidence interval for the mean concentration of dieldrin in the whale population. What general fact about confidence intervals do the margins of error of your three intervals illustrate?

II.12 **Testing pesticide level.** The Food and Drug Administration regulates the amount of dieldrin in raw food. For some foods, no more than 100 ng/g is allowed. Using the information in Exercise II.10, is there good evidence that the mean concentration in whale blubber is above this level? State hypotheses, find the z statistic, and use Table C to assess significance.

II.13 **Low birth weight and IQ.** Infants weighing less than 1500 grams at birth are classed as "very low birth weight." Low birth weight carries many risks. One study followed 113 male infants with very low birth weight to adulthood. At age 20, the mean IQ score for these men was

$\bar{x} = 87.6.^3$ IQ scores vary Normally with standard deviation $\sigma = 15$. Give a 95% confidence interval for the mean IQ score at age 20 for all very-low-birth-weight males.

II.14 **Low birth rate and IQ, continued.** IQ tests are scaled so that the mean score in a large population should be $\mu = 100$. We suspect that the very-low-birth-weight population has mean score less than 100. Does the study described in the previous exercise give good evidence that this is true? State hypotheses, find the z statistic, and use Table C to assess significance.

II.15 **More on low birth weight and IQ.** Very-low-birth-weight babies are more likely to be born to unmarried mothers and to mothers who did not complete high school.

 (a) Explain why the study of Exercise II.13 was not an experiment.

 (b) Explain clearly why confounding prevents us from concluding that very low birth weight in itself reduces adult IQ.

II.16 **Support groups for breast cancer.** It is widely believed that participating in a support group can extend the lives of women with breast cancer. There is no good evidence for this claim, but it was hard to carry out randomized comparative experiments because breast cancer patients believe that support groups help and want to be in one. When the first such experiment was finally completed, it showed that support groups have no effect on survival time.[4] The experiment assigned 235 women with advanced breast cancer to two groups: 158 to "expressive group therapy" and 77 to a control group. Outline the design of this experiment. How would you label the list of subjects? Use Table B, starting at line 110, to choose the first 5 members of the therapy group.

II.17 **Support groups for breast cancer, continued.** Here are some of the results of the medical study described in the previous exercise. Women in the treatment group reported less pain ($P = 0.04$), but there was no significant difference between the groups in median survival time ($P = 0.72$). Explain carefully why $P = 0.04$ is evidence that the treatment *does* make a difference and why $P = 0.72$ means that there is no evidence that support groups prolong life.

II.18 **Fortified breakfast cereals.** The Food and Drug Administration (FDA) recommends that breakfast cereals be fortified with folic acid. This is supposed to reduce the level of a potentially harmful substance called homocysteine in the blood. A randomized comparative experiment compared the effects on homocysteine of eating breakfast cereals that provided several levels of folic acid: the FDA recommendation of 127 micrograms (μg) daily, 499 μg, 665 μg, and a placebo cereal with no folic acid supplement.[5] Outline the design of such an experiment.

II.19 **Fortified breakfast cereals, continued.** Here are some results of the study described in the previous exercise. The 127 μg treatment reduced mean homocysteine by only 3.7 percent ($P = 0.24$) relative to the placebo. The 499 and 665 μg treatments produced reductions of 11% ($P < 0.001$) and 14% ($P = 0.001$), respectively. Write a brief conclusion based on these results. What do you think the FDA should do?

II.20 TV programming. Your local television station wonders if its viewers would rather watch your college basketball team play or an NBA game scheduled at the same time. It schedules the NBA game and receives 89 calls asking that it show the local game instead. The 89 callers are a sample. What population does the station want information about? Does the sample give trustworthy information about the sample? Explain your answer.

II.21 Missile defense. The question of whether the United States should develop a system intended to protect against a missile attack is controversial. Opinion polls give quite different results depending on the wording of the question asked. For each of the following items of information, say whether including it in the question would *increase* or *decrease* the percent of a poll sample who support missile defense.[6]

(a) The system is intended to protect the nation against a nuclear attack.

(b) The system would cost $60 billion.

(c) Deploying a missile defense system would interfere with existing arms-control treaties.

(d) Many scientists say such a system is unlikely to work.

II.22 Nutrition of premature infants. Substances called LCPs are important to the development of premature infants. It isn't clear how best to provide LCPs in the diet. An experiment assigned premature infants to four treatments: breast milk (20 infants), formula without added LCPs (19 infants), formula with LCPs from eggs (19 infants), and formula with LCPs from single-cell organisms.[7] The response variables were the amounts of several LCPs absorbed in the infants' intestines. Outline the design of the experiment and use Table B or the *Simple Random Sample* applet to assign infants to treatments.

II.23 How much do students earn? A university's financial aid office wants to know how much it can expect students to earn from summer employment. This information will be used to set the level of financial aid. The population contains 3478 students who have completed at least one year of study but have not yet graduated. The university will send a questionnaire to an SRS of 100 of these students, drawn from an alphabetized list.

(a) Describe how you will label the students in order to select the sample.

(b) Use Table B, beginning at line 105, to select the *first 5* students in the sample.

II.24 Bottling cola. A bottling company uses a filling machine to fill plastic bottles with cola. The bottles are supposed to contain 300 milliliters (ml). In fact, the contents vary according to a Normal distribution with mean $\mu = 298$ ml and standard deviation $\sigma = 3$ ml.

(a) What is the probability that an individual bottle contains less than 295 ml?

(b) What is the probability that the mean contents of the bottles in a six-pack is less than 295 ml?

II.25 An IQ test. The Wechsler Adult Intelligence Scale (WAIS) is a common "IQ test" for adults. The distribution of WAIS scores for persons over 16 years of age is approximately Normal with mean 100 and standard deviation 15.

(a) What is the probability that a randomly chosen individual has a WAIS score of 105 or higher?

(b) What are the mean and standard deviation of the average WAIS score $\bar{x}$ for an SRS of 60 people?

(c) What is the probability that the average WAIS score of an SRS of 60 people is 105 or higher?

(d) Would your answers to any of (a), (b), or (c) be affected if the distribution of WAIS scores in the adult population were distinctly non-Normal?

II.26 Poker hands. Deal a five-card poker hand from a shuffled deck. The probabilities of several types of hand are approximately as follows:

Hand	Worthless	One pair	Two pairs	Better hands
Probability	0.50	0.42	0.05	?

What must be the probability of getting a hand better than two pairs? What is the probability of getting a hand that is not worthless?

II.27 A supermarket prize game. A supermarket gives its customers cards that may win them a prize when matched with other cards. The back of the card announces the following probabilities of winning various amounts if a customer visits the store 10 times:

Amount	$1000	$200	$50	$10
Probability	1/10,000	1/1000	1/100	1/20

What is the probability of winning nothing?

II.28 Elephants and bees. Elephants can do substantial damage to crops in Africa. It turns out that elephants dislike bees. They recognize beehives in areas where they are common and avoid them. Can this be

(Ralph A. Clevenger/CORBIS)

used to keep elephants away from trees? A group in Kenya placed active beehives in some trees, and empty beehives in others. Will elephant damage be less in trees with hives? Will even empty hives keep elephants away? Outline the design of an experiment using 72 acacia trees (be sure to include a control group).[8]

II.29 **Moving up.** A study of social mobility in England looked at the social class reached by the sons of lower-class fathers. Social classes are numbered from 1 (low) to 5 (high). Take the random variable X to be the class of a randomly chosen son of a father in Class 1. The study found that the distribution of X is

Son's class	1	2	3	4	5
Probability	0.48	0.38	0.08	0.05	0.01

(a) What percent of the sons of lower-class (Class 1) fathers reach the highest class, Class 5?

(b) Check that this distribution satisfies the two requirements for a legitimate assignment of probabilities to individual outcomes.

(c) What is $P(X \le 3)$? (Be careful: the event "$X \le 3$" includes the value 3.)

(d) What is $P(X < 3)$?

(e) Write the event "a son of a lower-class father reaches one of the two highest classes" in terms of values of X. What is the probability of this event?

II.30 **Size of apartments.** The mean area of the several thousand apartments in a new development is advertised to be 1250 square feet. A tenant group thinks that the apartments are smaller than advertised. They hire an engineer to measure a sample of apartments to test their suspicion. What are the null hypothesis H_0 and alternative hypothesis H_a?

II.31 **California brush fires.** We often see televised reports of brush fires threatening homes in California. Some people argue that the modern practice of quickly putting out small fires allows fuel to accumulate and so increases the damage done by large fires. A detailed study of historical data suggests that this is wrong—the damage has risen simply because there are more houses in risky areas.[9] As usual, the study report gives statistical information tersely. Here is the summary of a regression of number of fires on decade (9 data points, for the 1910s to the 1990s): "Collectively, since 1910, there has been a highly significant increase ($r^2 = 0.61$, $P < 0.01$) in the number of fires per decade." How would you explain this statement to someone who knows no statistics? Include an explanation of both the description given by r^2 and its statistical significance.

(CORBIS)

II.32 Ancient Egypt. Settlements in Egypt before the time of the pharaohs are dated by measuring the presence of forms of carbon that decay over time. The first datings of settlements in the Nagada region used hair that had been excavated 60 years earlier. Now researchers have used newer methods and more recently excavated material.[10] Do the dates differ? Here is the conclusion about one location: "There are two dates from Site KH6. Statistically, the two dates are not significantly different. They provide a weighted average corrected date of 3715 ± 90 B.C." Explain to someone interested in ancient Egypt but not interested in statistics what "not significantly different" means.

II.33 What's a gift worth? Do people value gifts from others more highly than they value the money it would take to buy the gift? We would like to think so, because we hope that "the thought counts." A survey of 209 adults asked them to list three recent gifts and then asked, "Aside from any sentimental value, if, without the giver ever knowing, you could receive an amount of money instead of the gift, what is the minimum amount of money that would make you equally happy?" It turned out that most people would need more money than the gift cost to be equally happy. The magic words "significant ($P < 0.01$)" appear in the report of this finding.[11]

(a) The sample consisted of students and staff in a graduate program and of "members of the general public at train stations and airports in Boston and Philadelphia." The report says this sample is "not ideal." What's wrong with the sample?

(b) In simple language, what does it mean to say that the sample thought their gifts were worth "significantly more" than their actual cost?

(c) Now be more specific: what does "significant ($P < 0.01$)" mean?

Supplementary EXERCISES

Supplementary exercises apply the skills you have learned in ways that require more thought or more elaborate use of technology.

II.34 Sampling students. You want to investigate the attitudes of students at your school about the school's policy on sexual harassment. You have a grant that will pay the costs of contacting about 500 students.

(a) Specify the exact population for your study. For example, will you include part-time students?

(b) Describe your sample design. Will you use a stratified sample?

(c) Briefly discuss the practical difficulties that you anticipate. For example, how will you contact the students in your sample?

II.35 The placebo effect. A survey of physicians found that some doctors give a placebo to a patient who complains of pain for which the physician can find no cause. If the patient's pain improves, these doctors conclude that it had no physical basis. The medical school researchers who conducted the survey claimed that these doctors do not understand the placebo effect. Why?

II.36 **Informed consent.** The requirement that human subjects give their informed consent to participate in an experiment can greatly reduce the number of available subjects. For example, a study of new teaching methods asks the consent of parents for their children to be taught by either a new method or the standard method. Many parents do not return the forms, so their children must continue to follow the standard curriculum. Why is it not correct to consider these children as part of the control group along with children who are randomly assigned to the standard method?

II.37 **Roulette.** A roulette wheel has 18 red slots among its 38 slots. You observe many spins and record the number of times that red occurs. Now you want to use these data to test whether the probability p of a red has the value that is correct for a fair roulette wheel. State the hypotheses H_0 and H_a that you will test. (We will describe the test for this situation in Chapter 18.)

II.38 **Making french fries.** Few people want to eat discolored french fries. Potatoes are kept refrigerated before being cut for french fries to prevent spoiling and preserve flavor. But immediate processing of cold potatoes causes discoloring due to complex chemical reactions. The potatoes must therefore be brought to room temperature before processing. Design an experiment in which tasters will rate the color and flavor of french fries prepared from several groups of potatoes. The potatoes will be fresh picked or stored for a month at room temperature or stored for a month refrigerated. They will then be sliced and cooked either immediately or after an hour at room temperature.

(a) What are the factors and their levels, the treatments, and the response variables?

(b) Describe and outline the design of this experiment.

(c) It is efficient to have each taster rate fries from all treatments. How will you use randomization in presenting fries to the tasters?

II.39 **A 14-sided die.** An ancient Korean drinking game involves a 14-sided die. The players roll the die in turn and must submit to whatever humiliation is written on the up-face: something like, "Keep still when tickled on face." Six of the 14 faces are squares. Let's call them A, B, C, D, E, and F for short. The other eight faces are triangles, which we will call 1, 2, 3, 4, 5, 6, 7, and 8. Each of the squares is equally likely. Each of the triangles is also equally likely, but the triangle probability differs from the square probability. The probability of getting a square is 0.72. Give the probability model for the 14 possible outcomes.

II.40 **Alcohol and mortality.** It appears that people who drink alcohol in moderation have lower death rates than either people who drink heavily or people who do not drink at all. The protection offered by moderate drinking is concentrated among people over 50 and on deaths from heart disease. The Nurses Health Study played an essential role in establishing these facts for women. This part of the study followed 85,709 female nurses for 12 years, during which time 2658 of the subjects died. The nurses completed a questionnaire that described their diet, including their use of alcohol. They were reexamined every two years. Conclusion: "As compared with nondrinkers and heavy drinkers, light-to-moderate drinkers had a significantly lower risk of death."[12]

(a) Was this study an experiment? Explain your answer.

(b) What does "significantly lower risk of death" mean in simple language?

(c) Suggest some lurking variables that might be confounded with how much a person drinks. The investigators used advanced statistical methods to adjust for many such variables before concluding that the moderate drinkers really have a lower risk of death.

II.41 **Alcohol and mortality, continued.** The Nurses Health Study (see the previous exercise) reported the *relative risk* of death for women who drink different amounts of alcohol. (A relative risk of 2 means people in this group were twice as likely as nondrinkers to die during the study period. The risks were adjusted to correct for other differences among the groups.) Here are the facts:

Group	Relative risk compared with nondrinkers	95% confidence interval for risk
Light	0.83	0.74 to 0.93
Moderate	0.88	0.80 to 0.98
Heavy	1.19	1.02 to 1.38

Roughly speaking, light drinkers report 1 to 3 drinks per week, moderate drinkers 4 to 20 drinks per week (without bingeing!), and heavy drinkers more than 20 drinks per week.

(a) Can you be quite confident that light drinkers have a lower risk of death than nondrinkers? Why?

(b) Can you be quite confident that light drinkers have a lower risk of death than moderate drinkers? Why?

(c) Can you be quite confident that heavy drinkers have a higher risk of death than nondrinkers? Why?

II.42 Distributions: means versus individuals. The z confidence interval and test are based on the sampling distribution of the sample mean $\bar{x}$. Suppose that the distribution of the scores of young men on the National Assessment of Educational Progress quantitative test is Normal with mean $\mu = 272$ and standard deviation $\sigma = 60$.

(a) You sample 100 young men. According to the 99.7 part of the 68–95–99.7 rule, what is the range of scores you expect to see in your sample?

(b) What is the range of sample mean scores $\bar{x}$ for almost all samples of this size? You see once again that averages are less variable than individuals.

II.43 Distributions: larger samples. In the setting of the previous exercise, how many men must you sample to cut the range of values of $\bar{x}$ in half? This will also cut the margin of error of a confidence interval for μ in half. Do you expect the range of individual scores in the new sample to also be much less than in a sample of size 100? Why?

Optional EXERCISES

These exercises concern the optional material in Chapters 11 and 12.

II.44 Is business success just chance? Investors like to think that some companies are consistently successful. Academic researchers looked at data for many companies to determine whether each firm's sales growth was above the median for all firms in each year. They found that a simple "just chance" model fit well: years are independent, and the probability of being above the median in any one year is 1/2.[13] If this model holds, what is the probability that a particular firm is above average for two consecutive years? For all of four years?

II.45 A hot stock. You purchase a hot stock for $1000. The stock gains 30% or loses 25% each day, each with probability 0.5, and its behaviors on consecutive days are independent of each other. You plan to sell the stock after two days. What are the possible values of the stock after two days, and what is the probability for each value? What is the probability that the stock is worth more after two days than the $1000 you paid for it? (Hint: Remember that the value is multiplied by 1.30 on a day of 30% increase and multiplied by 0.75 on a day of 25% loss.)

II.46 Life tables. The National Center for Health Statistics produces a "life table" for the American population. For each year of age, the table gives the probability that a randomly chosen U.S. resident will die during that year of life. These are *conditional* probabilities, given that the person lived to the birthday that marks the beginning of the year. Here is an excerpt from the table:

Year of life	Probability of death
51	0.00439
52	0.00473
53	0.00512
54	0.00557
55	0.00610

What is the probability that a person who lives to age 50 (the beginning of the 51st year) will live to age 55?

II.47 Cystic fibrosis. Cystic fibrosis is a lung disorder that often results in death. It is inherited but can be inherited only if both parents are carriers of an abnormal gene. In 1989, the CF gene that is abnormal in carriers of cystic fibrosis was identified. The probability that a randomly chosen person of European ancestry carries an abnormal CF gene is 1/25. (The probability is less in other ethnic groups.) The CF20m test detects most but not all harmful mutations of the CF gene. The test is positive for 90% of people who are carriers. It is (ignoring human error) never positive for people who are not carriers. What is the probability that a randomly chosen person of European ancestry tests positive?

II.48 Cystic fibrosis, continued. Jason tests positive on the CF20m test. What is the probability that he is a carrier of the abnormal CF gene?

II.49 Albinism. People with albinism have little pigment in their skin, hair, and eyes. The gene that governs albinism has two forms (called alleles), which we denote by a and A. Each person has a pair of these genes, one inherited from each parent. A child inherits one of each parent's two alleles, independently with probability 0.5. Albinism is a recessive trait, so a person is albino only if the inherited pair is aa.

(a) Beth's parents are not albino but she has an albino brother. This implies that both of Beth's parents have type Aa. Why?

(b) Which of the types aa, Aa, and AA could a child of Beth's parents have? What is the probability of each type?

II.50 Albinism, continued. Beth is not albino. What are the conditional probabilities for Beth's possible genetic types, given this fact? (Use the definition of conditional probability and results from the previous exercise.)

II.51 Teenage drivers. An insurance company has the following information about drivers aged 16 to 18 years: 20% are involved in accidents each year; 10% in this age group are A students; among those involved in an accident, 5% are A students.

(a) Let A be the event that a young driver is an A student and C the event that a young driver is involved in an accident this year. State the information given in terms of probabilities and conditional probabilities for the events A and C.

(b) What is the probability that a randomly chosen young driver is an A student and is involved in an accident?

II.52 Teenage drivers, continued. Use your work from the previous exercise to find the percent of A students who are involved in accidents. (Start by expressing this as a conditional probability.)

II.53 Bomber crews. During World War II, the British estimated that each of their bombers had probability 0.05 of being lost due to enemy action on a mission over occupied Europe. A tour of duty for a member of the crew was 30 missions.

(a) What is the probability that a bomber survived 30 missions?

(b) What is the probability that a bomber survived 5 or fewer missions?

II.54 Survival of trees. A study of trees with diameter 12 inches in the Ozark Highlands area of Missouri found that the probability that such a tree survives the next five years is 0.94. A lot contains 345 trees in this class.

(a) What is the mean number that survive for five years?

(b) What is the probability that between 325 and 335 of these trees survive for five years?

II.55 Do the rich stay that way? We like to think that anyone can rise to the top. That's possible, but it's easier if you start near the top. Divide families by the income of the parents into the top 20%, the bottom 20%, and the middle 60%. Here are the conditional probabilities that a child of each class of parents ends up in each income class as an adult.[14] For example, a child of parents in the top 20% has probability 0.42 of also being in the top 20%.

	Child's class		
Parent's class	Top 20%	Middle 60%	Bottom 20%
Top 20%	0.42	0.52	0.06
Middle 60%	0.15	0.68	0.17
Bottom 20%	0.07	0.56	0.37

Suppose that these probabilities stay the same for three generations. Draw a tree diagram to show the path of a child and grandchild of parents in the top 20% of incomes. For example, the child might drop to the middle and the grandchild might then rise back to the top. What is the probability that the grandchild of people in the top 20% is also in the top 20%?

II.56 Many tests. Long ago, a group of psychologists carried out 77 separate significance tests and found that 2 were significant at the 5% level.

Suppose that these tests are independent of each other. (In fact, they were not independent, because all involved the same subjects.) If all of the null hypotheses are true, each test has probability 0.05 of being significant at the 5% level. Use the binomial distribution to find the probability that 2 or more of the tests are significant.

Inference about Variables

With the principles in hand, we proceed to "inference in practice." In the remaining chapters of this book, you will meet many of the most commonly used statistical procedures. We have grouped these procedures into two classes, corresponding to our division of data analysis into exploring variables and distributions and exploring relationships. The four chapters of Part III concern inference about the distribution of a single variable and inference for comparing the distributions of two variables. In Chapters 16 and 17, we use data on quantitative variables. We begin with the familiar Normal distribution for a quantitative variable. Chapters 18 and 19 concern categorical variables, so that inference begins with counts and proportions of outcomes. Part IV deals with inference for relationships among variables.

In all of these chapters, the examples and exercises involve both carrying out inference and thinking about inference in practice. Any inference method is useful only under certain conditions. We must learn to judge these conditions, usually by asking where the data came from and by performing data analysis to examine the data before rushing to inference.

D. Fradon

Quantitative Response Variable

Categorical Response Variable

Inference about Variables Review

(David A. Northcott/CORBIS)

Inference about a Population Mean

This chapter describes confidence intervals and significance tests for the mean μ of a population. We used the z procedures in exactly this setting to introduce the ideas of confidence intervals and tests. Now we discard the unrealistic condition that we know the population standard deviation σ and present procedures for practical use. We also pay more attention to the real-data setting of our work.

Conditions for inference

Confidence intervals and tests of significance for the mean μ of a Normal population are based on the sample mean $\overline{x}$. The sampling distribution of $\overline{x}$ has μ as its mean. That is, $\overline{x}$ is an unbiased estimator of the unknown μ. The spread of $\overline{x}$ depends on the sample size and also on the population standard deviation σ. In practice, σ is unknown. We must estimate σ from the data even though we are primarily interested in μ. The need to estimate σ changes some details of tests and confidence intervals for μ, but not their interpretation.

Here are the conditions under which we now do inference about a population mean.

> **CONDITIONS FOR INFERENCE ABOUT A MEAN**
>
> • Our data are a **simple random sample** (SRS) of size n from the population. This condition is very important.
>
> • Observations from the population have a **Normal distribution** with mean μ and standard deviation σ. In practice, it is enough that the distribution be symmetric and single-peaked unless the sample is very small. Both μ and σ are unknown parameters.

In this setting, the sample mean $\overline{x}$ has the Normal distribution with mean μ and standard deviation $\sigma/\sqrt{n}$. Because we don't know σ, we estimate it by the sample standard deviation s. We then estimate the standard deviation of $\overline{x}$ by $s/\sqrt{n}$. This quantity is called the *standard error* of the sample mean $\overline{x}$.

> **STANDARD ERROR**
>
> When the standard deviation of a statistic is estimated from data, the result is called the **standard error** of the statistic. The standard error of the sample mean $\overline{x}$ is $s/\sqrt{n}$.

APPLY YOUR KNOWLEDGE

16.1 Measuring blood pressure. A medical study finds that $\overline{x} = 114.9$ and $s = 9.3$ for the seated systolic blood pressure of the 27 members of one treatment group. What is the standard error of the mean?

16.2 Shrimp embryos. Biologists studying the levels of several compounds in shrimp embryos reported their results in a table, with the note, "Values are means ± SEM for three independent samples." The table entry for the compound ATP was 0.84 ± 0.01. Readers are supposed to understand that $\overline{x} = 0.84$ based on $n = 3$ measurements and that the standard error of the mean (SEM) is 0.01. What was the sample standard deviation s for these measurements?

The t distributions

When we know the value of σ, we base confidence intervals and tests for μ on the one-sample z statistic:

$$z = \frac{\overline{x} - \mu}{\sigma/\sqrt{n}}$$

This z statistic has the standard Normal distribution $N(0, 1)$. When we do not know σ, we substitute the standard error $s/\sqrt{n}$ of $\overline{x}$ for its standard deviation $\sigma/\sqrt{n}$. The statistic that results does not have a Normal distribution. It has a distribution that is new to us, called a *t distribution*.

THE ONE-SAMPLE t STATISTIC AND THE t DISTRIBUTIONS

Draw an SRS of size n from a population that has the Normal distribution with mean μ and standard deviation σ. The **one-sample t statistic**

$$t = \frac{\overline{x} - \mu}{s/\sqrt{n}}$$

has the **t distribution** with $n - 1$ degrees of freedom.

The t statistic has the same interpretation as any standardized statistic: it says how far $\overline{x}$ is from its mean μ in standard deviation units. There is a different t distribution for each sample size. We specify a particular t distribution by giving its **degrees of freedom.** The degrees of freedom for the one-sample t statistic *degrees of freedom* come from the sample standard deviation s in the denominator of t. We saw in Chapter 2 (page 44) that s has $n - 1$ degrees of freedom. There are other t statistics with different degrees of freedom, some of which we will meet later. We will write the t distribution with k degrees of freedom as $t(k)$ for short.

Figure 16.1 compares the density curves of the standard Normal distribution and the t distributions with 2 and 9 degrees of freedom. The figure illustrates these facts about the t distributions:

- The density curves of the t distributions are similar in shape to the standard Normal curve. They are symmetric about 0, single-peaked, and bell-shaped.

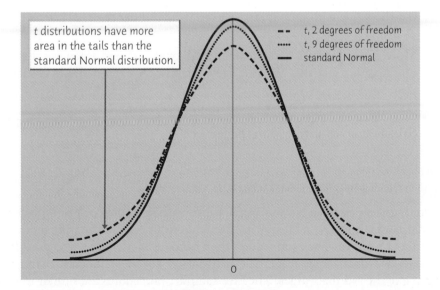

Figure 16.1 Density curves for the t distributions with 2 and 9 degrees of freedom and the standard Normal distribution. All are symmetric with center 0. The t distributions are somewhat more spread out.

- The spread of the t distributions is a bit greater than that of the standard Normal distribution. The t distributions in Figure 16.1 have more probability in the tails and less in the center than does the standard Normal. This is true because substituting the estimate s for the fixed parameter σ introduces more variation into the statistic.

- As the degrees of freedom k increase, the $t(k)$ density curve approaches the $N(0, 1)$ curve ever more closely. This happens because s estimates σ more accurately as the sample size increases. So using s in place of σ causes little extra variation when the sample is large.

Table C in the back of the book gives critical values for the t distributions. Each row in the table contains critical values for one of the t distributions; the degrees of freedom appear at the left of the row. For convenience, we label the table entries both by p, the upper-tail probability needed for significance tests, and by the confidence level C (in percent) required for confidence intervals. You have already used the standard Normal critical values z^* in the bottom row of Table C. By looking down any column, you can check that the t critical values approach the Normal values as the degrees of freedom increase. As in the case of the Normal table, statistical software often makes Table C unnecessary.

Better statistics, better beer

The t distribution and the t inference procedures were invented by William S. Gossett (1876–1937). Gossett worked for the Guinness brewery, and his goal in life was to make better beer. He used his new t procedures to find the best varieties of barley and hops. Gosset's statistical work helped him become head brewer, a more interesting title than professor of statistics.

APPLY YOUR KNOWLEDGE

16.3 Critical values. Use Table C to find

    (a) the value with probability 0.05 to its right under the $t(5)$ density curve.

    (b) the value with probability 0.99 to its left under the $t(21)$ density curve.

16.4 More critical values. You have an SRS of size 25 and calculate the one-sample t statistic. What is the critical value t^* from Table C such that

    (a) t has probability 0.025 to the right of t^*?

    (b) t has probability 0.75 to the left of t^*?

The one-sample t confidence interval

To analyze samples from Normal populations with unknown σ, just replace the standard deviation $\sigma/\sqrt{n}$ of $\overline{x}$ by its standard error $s/\sqrt{n}$ in the z procedures of Chapters 13 and 14. The z procedures then become *one-sample t procedures*. Use P-values or critical values from the t distribution with $n - 1$ degrees of freedom in place of the Normal values. The one-sample t procedures are similar in both reasoning and computational detail to the z procedures. So we will now pay more attention to questions about using these methods in practice.

THE ONE-SAMPLE *t* CONFIDENCE INTERVAL

Draw an SRS of size n from a population having unknown mean μ.
A level C confidence interval for μ is

$$\bar{x} \pm t^* \frac{s}{\sqrt{n}}$$

where t^* is the critical value for the $t(n-1)$ density curve with area C
between $-t^*$ and t^*. This interval is exact when the population
distribution is Normal and is approximately correct for large n in other
cases.

EXAMPLE 16.1   Healing of skin wounds

Biologists studying the healing of skin wounds measured the rate at which new cells
closed a razor cut made in the skin of an anesthetized newt. Here are data from 18
newts, measured in micrometers (millionths of a meter) per hour:[1]

29	27	34	40	22	28	14	35	26
35	12	30	23	18	11	22	23	33

We want to calculate the 95% *t* confidence interval for the mean healing rate μ in
the population of all newts.

First calculate that

$$\bar{x} = 25.67 \quad \text{and} \quad s = 8.324$$

The degrees of freedom are $n - 1 = 17$. From Table C we find that for 95% confidence $t^* = 2.110$. The confidence interval is

$$\bar{x} \pm t^* \frac{s}{\sqrt{n}} = 25.67 \pm 2.110 \frac{8.324}{\sqrt{18}}$$

$$= 25.67 \pm 4.14$$

$$= 21.53 \text{ to } 29.81 \text{ micrometers per hour}$$

The healing-rate data come from a larger study that compared healing rates of
wounds under different conditions, using newts as subjects. The data in this example
are for the control group of newts, left to heal naturally. Comparing this estimate of
the mean rate with those under other conditions gave new insight into the healing
of wounds, not only in newts but in animals in general.

The one-sample *t* confidence interval has the form

$$\text{estimate} \pm t^* SE_{\text{estimate}}$$

where "SE" stands for "standard error." We will meet a number of confidence
intervals that have this common form. In Example 16.1, the standard error of
the mean $\bar{x}$ is

$$SE_{\bar{x}} = \frac{s}{\sqrt{n}} = \frac{8.324}{\sqrt{18}} = 1.962$$

```
1 | 124
1 | 8
2 | 2233
2 | 6789
3 | 034
3 | 55
4 | 0
```

Figure 16.2 Stemplot of the healing rates in Example 16.1.

Can we trust the t confidence interval in Example 16.1? We must examine two conditions. **SRS:** Scientists are accustomed to treating laboratory animals, whether newts, mice, or rabbits, as random samples from their populations. In the case of mice, many standard varieties with known genetic makeup are sold for laboratory use. The newts were assigned at random to several conditions for healing. We are willing to regard these 18 newts as an SRS from all newts of this variety. **Normality:** Figure 16.2 is a stemplot of the 18 healing rates. It is a bit irregular, as is common in small samples, but shows no outliers, clusters, or strong skewness. As we will see, the t procedures are not sensitive to non-Normality when the distribution is roughly symmetric. The t confidence interval in Example 16.1 will be quite accurate.

APPLY YOUR KNOWLEDGE

16.5 Critical values. What critical value t^* from Table C would you use for a confidence interval for the mean of the population in each of the following situations?

(a) A 95% confidence interval based on $n = 10$ observations.

(b) A 99% confidence interval from an SRS of 20 observations.

(c) An 80% confidence interval from a sample of size 7.

16.6 Red wine is good for the heart. Observational studies suggest that moderate use of alcohol reduces heart attacks, and that red wine may have special benefits. One reason may be that red wine contains polyphenols, substances that do good things to cholesterol in the blood and so may reduce the risk of heart attacks. In an experiment, healthy men were assigned at random to several groups. One group of 9 men drank half a bottle of red wine each day for two weeks. The level of polyphenols in their blood was measured before and after the two-week period. Here are the percent changes in level:[2]

3.5 8.1 7.4 4.0 0.7 4.9 8.4 7.0 5.5

Make a stemplot of the data. It is difficult to assess Normality from just 9 observations. Give a 90% t confidence interval for the mean percent change in blood polyphenols among all healthy men if all drank this amount of red wine.

16.7 Ancient air. The composition of the earth's atmosphere may have changed over time. To try to discover the nature of the atmosphere long ago, we can examine the gas in bubbles inside ancient amber. Amber is tree resin that has hardened and been trapped in rocks. The gas in bubbles within amber should be a sample of the atmosphere at the time the amber was formed. Measurements on specimens of amber from the late Cretaceous era (75 to 95 million years ago) give these percents of nitrogen:[3]

63.4 65.0 64.4 63.3 54.8 64.5 60.8 49.1 51.0

Assume (this is not yet agreed on by experts) that these observations are an SRS from the late Cretaceous atmosphere.

(a) Graph the data, and comment on skewness and outliers. The t procedures will be only approximate for these data.

(b) Give a 95% t confidence interval for the mean percent of nitrogen in ancient air.

The one-sample t test

Like the confidence interval, the t test is close in form to the z test we met earlier. We will use the t test to do a realistic analysis of the cola-sweetening example from Chapter 14.

THE ONE-SAMPLE t TEST

Draw an SRS of size n from a population having unknown mean μ. To test the hypothesis $H_0: \mu = \mu_0$ based on an SRS of size n, compute the one-sample t statistic:

$$t = \frac{\overline{x} - \mu_0}{s / \sqrt{n}}$$

In terms of a variable T having the $t(n-1)$ distribution, the P-value for a test of H_0 against

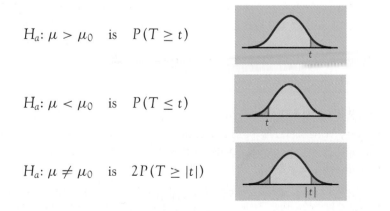

$$H_a: \mu > \mu_0 \quad \text{is} \quad P(T \geq t)$$

$$H_a: \mu < \mu_0 \quad \text{is} \quad P(T \leq t)$$

$$H_a: \mu \neq \mu_0 \quad \text{is} \quad 2P(T \geq |t|)$$

These P-values are exact if the population distribution is Normal and are approximately correct for large n in other cases.

EXAMPLE 16.2 Sweetening colas

Cola makers test new recipes for loss of sweetness during storage. Trained tasters rate the sweetness before and after storage. Here are the sweetness losses (sweetness before storage minus sweetness after storage) found by 10 tasters for one new cola recipe:

$$2.0 \quad 0.4 \quad 0.7 \quad 2.0 \quad -0.4 \quad 2.2 \quad -1.3 \quad 1.2 \quad 1.1 \quad 2.3$$

Are these data good evidence that the cola lost sweetness?

As before, we are willing to regard these 10 carefully trained tasters as an SRS from the population of all trained tasters. Figure 16.3 is a stemplot of the data. We can't judge Normality from just 10 observations, but there are no outliers, clusters, or extreme skewness. *P*-values for the *t* test will be reasonably accurate.

Step 1. Hypotheses. Tasters vary in their perception of sweetness loss. So we ask the question in terms of the mean loss μ for a large population of tasters. The null hypothesis is "no loss," and the alternative hypothesis says "there is a loss."

$$H_0: \mu = 0$$
$$H_a: \mu > 0$$

Step 2. Test statistic. The basic statistics are

$$\bar{x} = 1.02 \quad \text{and} \quad s = 1.196$$

The one-sample *t* test statistic is

$$t = \frac{\bar{x} - \mu_0}{s/\sqrt{n}} = \frac{1.02 - 0}{1.196/\sqrt{10}} = 2.70$$

Step 3. P-value. The *P*-value for $t = 2.70$ is the area to the right of 2.70 under the *t* distribution curve with degrees of freedom $n - 1 = 9$. Figure 16.4 shows this area. We can't find the exact value of *P* without software. But we can pin *P* between two values by using Table C. Search the $df = 9$ row of Table C for entries that bracket $t = 2.70$. Because the observed *t* lies between the critical values for 0.02 and 0.01, the *P*-value lies between 0.01 and 0.02. There is quite strong evidence for a loss of sweetness.

```
-1 | 3
-0 | 4
 0 | 47
 1 | 12
 2 | 0023
```

Figure 16.3 Stemplot of the sweetness losses in Example 16.2.

df = 9

p	.02	.01
t^*	2.398	2.821

APPLY YOUR KNOWLEDGE

16.8 **Is it significant?** The one-sample *t* statistic for testing

$$H_0: \mu = 0$$
$$H_a: \mu > 0$$

from a sample of $n = 15$ observations has the value $t = 1.82$.

(a) What are the degrees of freedom for this statistic?

(b) Give the two critical values t^* from Table C that bracket t. What are the right-tail probabilities p for these two entries?

(c) Between what two values does the *P*-value of the test fall?

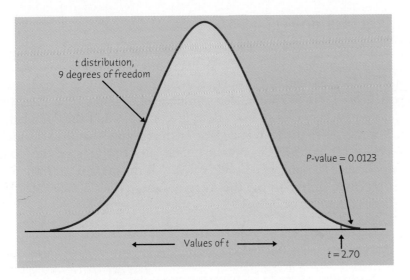

Figure 16.4 The P-value for the one-sided t test in Example 16.2.

(d) Is the value $t = 1.82$ significant at the 5% level? Is it significant at the 1% level?

16.9 Is it significant? The one-sample t statistic from a sample of $n = 25$ observations for the two-sided test of

$$H_0: \mu = 64$$
$$H_a: \mu \neq 64$$

has the value $t = 1.12$.

(a) What are the degrees of freedom for t?

(b) Locate the two critical values t^* from Table C that bracket t. What are the right-tail probabilities p for these two values?

(c) Between what two values does the P-value of the test fall? (Note that H_a is two-sided.)

(d) Is the value $t = 1.12$ statistically significant at the 10% level? At the 5% level?

16.10 Ancient air, continued. The data of Exercise 16.7 suggest that the percent of nitrogen in the air during the Cretaceous era was quite different from the present 78.1%. Carry out a test of

$$H_0: \mu = 78.1$$
$$H_a: \mu \neq 78.1$$

to assess the significance of the difference. (Give the test statistic, its P-value, and your conclusion.)

Using technology

Any technology suitable for statistics will implement the one-sample t procedures. As usual, you can read and use almost any output now that you know what to look for. Figure 16.5 displays output from Minitab, Excel, and the TI-83 calculator for the 95% confidence interval of Example 16.1. The Minitab and TI-83 outputs are easy to read. Both give n, $\bar{x}$, s, and the confidence interval. Minitab gives the standard error of the mean as well. Excel includes the pieces of the confidence interval in its "descriptive statistics" output. The entry labeled "Confidence Level (95.0%)" is the margin of error. You can use this together with $\bar{x}$ to get the interval using either a calculator or the spreadsheet's formula capability.

Minitab

```
Session                                                    _ □ X

One-Sample T: Rate

Variable        N      Mean     StDev    SE Mean        95.0% CI
Rate            18     25.67    8.32     1.96     (   21.53,   29.81)
```

Excel

```
Microsoft Excel - Book1                                    _ □ X
            B                    C        D      E      F
 1                 Rate
 2
 3   Mean                        25.6667
 4   Standard Error              1.9621
 5   Median                      26.5
 6   Standard Deviation          8.3243
 7   Sample Variance             69.2941
 8   Range                       29
 9   Minimum                     11
10   Maximum                     40
11   Count                       18
12   Confidence Level(95.0%)     4.1396
13
  Sheet1  Sheet2  Sheet3
```

This is the estimate $\bar{x}$.

This is the margin of error $\pm t^*$ SE.

Texas Instruments TI-83 Plus

```
TInterval
 (21.527,29.806)
 x̄=25.6667
 Sx=8.3243
 n=18.0000
```

Figure 16.5 Output for the t confidence interval of Example 16.1 from Minitab, the Excel spreadsheet program, and the TI-83 graphing calculator.

Figure 16.6 displays output for the t test in Example 16.2. From either Minitab or the TI-83, we see that the P-value is 0.012. The ability to use the t distribution to calculate P-values accurately is the most valuable feature of software for the t procedures. Excel lacks a one-sample t test menu selection but does have a function named TDIST for tail areas under t density curves. The Excel output shows functions for the t statistic and its P-value to the right of the "descriptive statistics" display, along with their values $t = 0.269669$ and $P = 0.01226$.

Matched pairs t procedures

The study of healing in Example 16.1 estimated the mean healing rate for newts under natural conditions, but the researchers then compared results under several conditions. The taste test in Example 16.2 was a matched pairs study in which the same 10 tasters rated before-and-after sweetness. Comparative studies are more convincing than single-sample investigations. For that reason, one-sample inference is less common than comparative inference. One common design to compare two treatments makes use of one-sample procedures. In a **matched pairs design,** subjects are matched in pairs and each treatment is given to one subject in each pair. The experimenter can toss a coin to assign two treatments to the two subjects in each pair. Another situation calling for matched pairs is before-and-after observations on the same subjects, as in the taste test of Example 16.2.

matched pairs design

MATCHED PAIRS t PROCEDURES

To compare the responses to the two treatments in a matched pairs design, apply the one-sample t procedures to the observed differences.

The parameter μ in a matched pairs t procedure is the mean difference in the responses to the two treatments within matched pairs of subjects in the entire population.

EXAMPLE 16.3   Floral scents and learning

We hear that listening to Mozart improves students' performance on tests. In the EESEE story "Floral Scents and Learning," investigators asked whether pleasant odors have a similar effect. Twenty-one subjects worked a paper-and-pencil maze while wearing a mask. The mask was either unscented or carried a floral scent. The response variable is their average time on three trials. Each subject worked the maze with both masks, in a random order. The randomization is important because subjects tend to improve their times as they work a maze repeatedly. Table 16.1 gives the subjects' average times with both masks.

To analyze these data, subtract the scented time from the unscented time for each subject. The 21 differences form a single sample. They appear in the "Difference" column in Table 16.1. The first subject, for example, was 7.37 seconds slower wearing

FESEE

Minitab

```
 Session                                          _ □ X

 One-Sample T: Loss

 Test of mu = 0 vs mu > 0

 Variable          N      Mean     StDev    SE Mean
 Loss             10     1.020     1.196      0.378

 Variable     95.0% Lower Bound        T        P
 Loss                       0.327     2.70    0.012
```

Excel

```
 Microsoft Excel                                  _ □ X

 Book1

        A                    B        C          D        E
 1          Sweetness loss
 2                                   (B3-0)/B4   2.69669  ◄──  This is the t statistic.
 3   Mean                   1.0200   TDIST (D3,9,1) 0.01226 ◄──
 4   Standard Error         0.3782                               This is the P-value.
 5   Median                 1.15
 6   Standard Deviation     1.1961
 7   Sample Variance        1.4307
 8   Range                  3.6
 9   Minimum                -1.3
 10  Maximum                2.3
 11  Count                  10
 12  Confidence Level(95.0%) 0.8556
 13
```

Texas Instruments TI-83 Plus

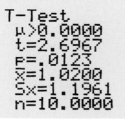

```
T-Test
 μ>0.0000
 t=2.6967
 P=.0123
 x̄=1.0200
 Sx=1.1961
 n=10.0000
```

Figure 16.6 Output for the *t* test of Example 16.2 from Minitab, the Excel spreadsheet program, and the TI-83 graphing calculator.

TABLE 16.1 Average time to complete a maze

Subject	Unscented (seconds)	Scented (seconds)	Difference	Subject	Unscented (seconds)	Scented (seconds)	Difference
1	30.60	37.97	−7.37	12	58.93	83.50	−24.57
2	48.43	51.57	−3.14	13	54.47	38.30	16.17
3	60.77	56.67	4.10	14	43.53	51.37	−7.84
4	36.07	40.47	−4.40	15	37.93	29.33	8.60
5	68.47	49.00	19.47	16	43.50	54.27	−10.77
6	32.43	43.23	−10.80	17	87.70	62.73	24.97
7	43.70	44.57	−0.87	18	53.53	58.00	−4.47
8	37.10	28.40	8.70	19	64.30	52.40	11.90
9	31.17	28.23	2.94	20	47.37	53.63	−6.26
10	51.23	68.47	−17.24	21	53.67	47.00	6.67
11	65.40	51.10	14.30				

the scented mask, so the difference is negative. Because shorter times represent better performance, positive differences show that the subject did better when wearing the scented mask. A stemplot of the differences shows that their distribution is symmetric and reasonably Normal in shape:

```
−2 | 5
−1 | 711
−0 | 8764431
 0 | 34799
 1 | 2469
 2 | 5
```

The data in this stemplot are rounded to the nearest whole second. Notice that the stem 0 must appear twice, to display differences between 0 and 9 and between 0 and −9.

Step 1. Hypotheses. To assess whether the floral scent significantly improved performance, we test

$$H_0: \mu = 0$$
$$H_a: \mu > 0$$

Here μ is the mean difference in the population from which the subjects were drawn. The null hypothesis says that no improvement occurs, and H_a says that unscented times are longer than scented times on the average.

Step 2. Test statistic. The 21 differences have

$$\bar{x} = 0.9567 \quad \text{and} \quad s = 12.5479$$

The one-sample *t* statistic is therefore

$$t = \frac{\bar{x} - 0}{s/\sqrt{n}} = \frac{0.9567 - 0}{12.5479/\sqrt{21}}$$
$$= 0.349$$

df = 20		
p	.25	.20
t*	0.687	0.860

Step 3.   P-value. Find the P-value from the $t(20)$ distribution. (Remember that the degrees of freedom are 1 less than the sample size.) Table C shows that 0.349 is less than the 0.25 critical value of the $t(20)$ distribution. The P-value is therefore greater than 0.25. Statistical software gives the value $P = 0.3652$.

Conclusion. The data do not support the claim that floral scents improve performance. The average improvement is small, just 0.96 seconds over the 50 seconds that the average subject took when wearing the unscented mask. This small improvement is not statistically significant at even the 25% level.

Example 16.3 illustrates how to turn matched pairs data into single-sample data by taking differences within each pair. We are making inferences about a single population, the population of all differences within matched pairs. It is incorrect to ignore the pairs and analyze the data as if we had two samples, one from subjects who wore unscented masks and a second from subjects who wore scented masks. Inference procedures for comparing two samples assume that the samples are selected independently of each other. This assumption does not hold when the same subjects are measured twice. The proper analysis depends on the design used to produce the data.

APPLY YOUR KNOWLEDGE

Many exercises from this point on ask you to give the P-value of a t test. If you have a suitable technology, give the exact P-value. Otherwise, use Table C to give two values between which P lies.

16.11 Right versus left. The design of controls and instruments affects how easily people can use them. A student project investigated this effect by asking 25 right-handed students to turn a knob (with their right hands) that moved an indicator by screw action. There were two identical instruments, one with a right-hand thread (the knob turns clockwise) and the other with a left-hand thread (the knob must be turned counterclockwise). Table 16.2 gives the times in seconds each subject took to move the indicator a fixed distance.[4]

(a) Each of the 25 students used both instruments. Discuss briefly how you would use randomization in arranging the experiment.

(b) The project hoped to show that right-handed people find right-hand threads easier to use. What is the parameter μ for a matched pairs t test? State H_0 and H_a in terms of μ.

(c) Carry out a test of your hypotheses. Give the P-value and report your conclusions.

16.12 Mutual-fund performance. Do "index funds" that simply buy and hold all the stocks in one of the stock market indexes, such as the Standard & Poor's 500 Index, perform better than actively managed mutual funds? Compare the percent total return (price change plus dividends) of a large actively managed fund with that of the Vanguard Index 500 fund for the 24 years from 1977 to 2000. Vanguard did

TABLE 16.2 Performance times (seconds) using right-hand and left-hand threads

Subject	Right thread	Left thread	Subject	Right thread	Left thread
1	113	137	14	107	87
2	105	105	15	118	166
3	130	133	16	103	146
4	101	108	17	111	123
5	138	115	18	104	135
6	118	170	19	111	112
7	87	103	20	89	93
8	116	145	21	78	76
9	75	78	22	100	116
10	96	107	23	89	78
11	122	84	24	85	101
12	103	148	25	88	123
13	116	147			

better by an average of 2.83% per year, and the standard deviation of the 24 annual differences was 11.65%. Is there convincing evidence that the index fund does better?

(a) Describe in words the parameter μ for this comparison.

(b) State the hypotheses H_0 and H_a.

(c) Find the matched pairs t statistic and its P-value. What do you conclude?

16.13 Right versus left, continued. Give a 90% confidence interval for the mean time advantage of right-hand over left-hand threads in the setting of Exercise 16.11. Do you think that the time saved would be of practical importance if the task were performed many times—for example, by an assembly-line worker? To help answer this question, find the mean time for right-hand threads as a percent of the mean time for left-hand threads.

Robustness of *t* procedures

The *t* confidence interval and test are exactly correct when the distribution of the population is exactly Normal. No real data are exactly Normal. The usefulness of the *t* procedures in practice therefore depends on how strongly they are affected by lack of Normality.

ROBUST PROCEDURES

A confidence interval or significance test is called **robust** if the confidence level or P-value does not change very much when the conditions for use of the procedure are violated.

Catching cheaters

A certification test for surgeons asks 277 multiple-choice questions. Smith and Jones have 193 common right answers and 53 identical wrong choices. The computer flags their 246 identical answers as evidence of possible cheating. They sue. The court wants to know how unlikely it is that exams this similar would occur just by chance. That is, the court wants a P-value. Statisticians offer several P-values based on different models for the exam-taking process. They all say that results this similar would almost never happen just by chance. Smith and Jones fail the exam.

The condition that the population be Normal rules out outliers, so the presence of outliers shows that this condition is not fulfilled. The t procedures are not robust against outliers, because $\bar{x}$ and s are not resistant to outliers.

Fortunately, the t procedures are quite robust against non-Normality of the population except when outliers or strong skewness are present. (Skewness is more serious than other kinds of non-Normality.) As the size of the sample increases, the central limit theorem ensures that the distribution of the sample mean $\bar{x}$ becomes more nearly Normal and that the t distribution becomes more accurate for critical values and P-values of the t procedures.

Always make a plot to check for skewness and outliers before you use the t procedures for small samples. For most purposes, you can safely use the one-sample t procedures when $n \geq 15$ unless an outlier or quite strong skewness is present. Here are practical guidelines for inference on a single mean.[5]

USING THE t PROCEDURES

- Except in the case of small samples, the assumption that the data are an SRS from the population of interest is more important than the assumption that the population distribution is Normal.

- *Sample size less than 15*: Use t procedures if the data appear close to Normal (symmetric, single peak, no outliers). If the data are skewed or if outliers are present, do not use t.

- *Sample size at least 15*: The t procedures can be used except in the presence of outliers or strong skewness.

- *Large samples*: The t procedures can be used even for clearly skewed distributions when the sample is large, roughly $n \geq 40$.

EXAMPLE 16.4   Can we use t?

Figure 16.7 shows plots of several data sets from Chapter 1. For which of these can we safely use the t procedures?

- Figure 16.7(a) is a histogram of the percent of each state's residents who are Hispanic. *We have data on the entire population of 50 states, so formal inference makes no sense.* We can calculate the exact mean for the population. There is no uncertainty due to having only a sample from the population, and no need for a confidence interval or test. *If these data were an SRS from a larger population, strong skewness and outliers would make use of the t procedures risky even for $n = 50$.*

- Figure 16.7(b) is a stemplot of the force required to pull apart 20 pieces of Douglas fir. *The data are strongly skewed to the left, so we cannot trust the t procedures for $n = 20$.*

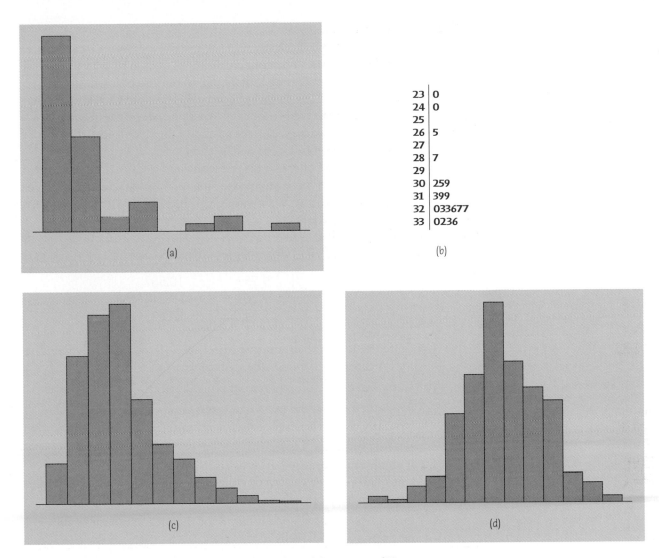

23 | 0
24 | 0
25 |
26 | 5
27 |
28 | 7
29 |
30 | 259
31 | 399
32 | 033677
33 | 0236

Figure 16.7 Can we use *t* procedures for these data? (a) Percent of Hispanic residents in the 50 states. *No*, this is an entire population, not a sample. (b) Force required to pull apart 20 pieces of Douglas fir. *No*, there are just 20 observations and strong skewness. (c) Word lengths in Shakespeare's plays. *Yes, if the sample is large enough* to overcome the right skewness. (d) Heights of college students. *Yes, for any size sample*, because the distribution is close to Normal.

- Figure 16.7(c) shows the distribution of word lengths in Shakespeare's plays. *The histogram is somewhat skewed to the right, but there are no outliers. We can use the t distributions for such data if we have at least 40 observations.*

- Figure 16.7(d) is a histogram of the heights of the students in a college class. *This distribution is quite symmetric and appears close to Normal. We can use the t procedures for any sample size.*

APPLY YOUR KNOWLEDGE

16.14 **Growing trees faster.** The concentration of carbon dioxide (CO_2) in the atmosphere is increasing rapidly due to our use of fossil fuels. Because plants use CO_2 to fuel photosynthesis, more CO_2 may cause trees and other plants to grow faster. An elaborate apparatus allows researchers to pipe extra CO_2 to a 30-meter circle of forest. They selected two nearby circles in each of three parts of a pine forest and randomly chose one of each pair to receive extra CO_2. The response variable is the mean increase in base area for 30 to 40 trees in a circle during a growing season. We measure this in percent increase per year. Here are one year's data:[6]

Pair	Control plot	Treated plot
1	9.752	10.587
2	7.263	9.244
3	5.742	8.675

(a) State the null and alternative hypotheses. Explain clearly why the investigators used a one-sided alternative.

(b) Carry out a test and report your conclusion in simple language.

(c) The investigators did use a t test, the same test that you used in (b). Any use of the t procedures with samples this size is somewhat risky. Why?

16.15 **How much oil?** Exercise I.10 (page 158) gives data on the total amount of oil recovered from 64 oil wells in the same area. Take these wells to be an SRS of wells in this area.[7]

(a) Give a 95% t confidence interval for the mean amount of oil recovered from all wells in this area.

(b) The data are very skewed, with several high outliers. A new computer-intensive method that gives accurate confidence intervals without assuming any specific shape for the distribution gives a 95% confidence interval of 40.28 to 60.32.[8] How does the t interval compare with this?

16.16 **Will they charge more?** A bank wonders whether omitting the annual credit card fee for customers who charge at least $2400 in a year will increase the amount charged on its credit cards. The bank makes this offer to an SRS of 200 of its credit card customers. It then compares how much these customers charge this year with the amount that they charged last year. The mean increase in the sample is $332, and the standard deviation is $108.

(a) Is there significant evidence at the 1% level that the mean amount charged increases under the no-fee offer? State H_0 and H_a and carry out a t test.

(b) Give a 99% confidence interval for the mean amount of the increase.

(c) The distribution of the amount charged is skewed to the right, but outliers are prevented by the credit limit that the bank enforces on each card. Use of the t procedures is justified in this case even though the population distribution is not Normal. Explain why.

(d) A critic points out that the customers would probably have charged more this year than last even without the new offer because the economy is more prosperous and interest rates are lower. Briefly describe the design of an experiment to study the effect of the no-fee offer that would avoid this criticism.

Chapter 16 SUMMARY

Tests and confidence intervals for the mean μ of a Normal population are based on the sample mean $\bar{x}$ of an SRS. Because of the central limit theorem, the resulting procedures are approximately correct for other population distributions when the sample is large.

The standardized sample mean is the **one-sample z statistic,**

$$z = \frac{\bar{x} - \mu}{\sigma/\sqrt{n}}$$

When we know σ, we use the z statistic and the standard Normal distribution. In practice, we do not know σ. Replace the standard deviation $\sigma/\sqrt{n}$ of $\bar{x}$ by the **standard error** $s/\sqrt{n}$ to get the **one-sample t statistic,**

$$t = \frac{\bar{x} - \mu}{s/\sqrt{n}}$$

The t statistic has the **t distribution** with $n - 1$ degrees of freedom.

There is a t distribution for every positive **degrees of freedom** k. All are symmetric distributions similar in shape to the standard Normal distribution. The $t(k)$ distribution approaches the $N(0, 1)$ distribution as k increases.

An exact level C **confidence interval** for the mean μ of a Normal population is

$$\bar{x} \pm t^* \frac{s}{\sqrt{n}}$$

The **critical value** t^* is chosen so that the t curve with $n - 1$ degrees of freedom has area C between $-t^*$ and t^*.

Significance tests for H_0: $\mu = \mu_0$ are based on the t statistic. Use P-values or fixed significance levels from the $t(n - 1)$ distribution.

Use these one-sample procedures to analyze **matched pairs** data by first taking the difference within each matched pair to produce a single sample.

The t procedures are relatively **robust** when the population is non-Normal, especially for larger sample sizes. The t procedures are useful for non-Normal data when $n \geq 15$ unless the data show outliers or strong skewness.

Chapter 16 EXERCISES

16.17 Testing. The one-sample t statistic for a test of

$$H_0: \mu = 10$$
$$H_a: \mu < 10$$

based on $n = 10$ observations has the value $t = -2.25$.

(a) What are the degrees of freedom for this statistic?

(b) Between what two probabilities p from Table C does the P-value of the test fall?

(Anna Clopet/CORBIS)

16.18 Worker absenteeism. A study of unexcused absenteeism among factory workers looked at a year's records for 668 workers in an English factory. The mean number of days absent was 9.88 and the standard deviation was 17.847 days.[9] Regard these workers in this year as a random sample of all workers in all years as long as this factory does not change work conditions or worker benefits. What can you say with 99% confidence about the mean number of unexcused absences for all workers?

16.19 Conditions for inference. The number of days absent for a worker in a year cannot be less than 0. The workers in the previous exercise had standard deviation much greater than the mean days absent. What does this say about the shape of the distribution of days absent? Why can we nonetheless use a t confidence interval for the population mean?

16.20 Market research. A manufacturer of small appliances employs a market research firm to estimate retail sales of its products by gathering information from a sample of retail stores. This month an SRS of 75 stores in the Midwest sales region finds that these stores sold an average of 24 of the manufacturer's hand mixers, with standard deviation 11.

(a) Give a 95% confidence interval for the mean number of mixers sold by all stores in the region.

(b) The distribution of sales is strongly right-skewed because there are many smaller stores and a few very large stores. The use of t in (a) is reasonably safe despite this violation of the Normality assumption. Why?

16.21 Auto crankshafts. The following are measurements (in millimeters) of a critical dimension for 16 auto engine crankshafts:

224.120   224.001   224.017   223.982   223.989   223.961
223.960   224.089   223.987   223.976   223.902   223.980
224.098   224.057   223.913   223.999

The dimension is supposed to be 224 mm and the variability of the manufacturing process is unknown. Is there evidence that the mean dimension is not 224 mm?

(a) Check the data graphically for outliers or strong skewness that might threaten the validity of the t procedures. What do you conclude?

(b) State H_0 and H_a and carry out a t test. Give the P-value. What do you conclude?

16.22 Cockroach metabolism. To study the metabolism of insects, researchers fed cockroaches measured amounts of a sugar solution. After 2, 5, and 10 hours, they dissected some of the cockroaches and measured the amount of sugar in various tissues.[10] Five roaches fed the sugar D-glucose and dissected after 10 hours had the following amounts (in micrograms) of D-glucose in their hindguts:

$$55.95 \quad 68.24 \quad 52.73 \quad 21.50 \quad 23.78$$

The researchers gave a 95% confidence interval for the mean amount of D-glucose in cockroach hindguts under these conditions. The insects are a random sample from a uniform population grown in the laboratory. We therefore expect responses to be Normal. What confidence interval did the researchers give?

16.23 The cost of Internet access. How much do users pay for Internet service? Here are the monthly fees (in dollars) paid by a random sample of 50 users of commercial Internet service providers in August 2000:[11]

20	40	22	22	21	21	20	10	20	20
20	13	18	50	20	18	15	8	22	25
22	10	20	22	22	21	15	23	30	12
9	20	40	22	29	19	15	20	20	20
20	15	19	21	14	22	21	35	20	22

(Ryan McVay/PhotoDisc/Getty Images)

(a) Make a stemplot of the data. The data are not Normal: there are stacks of observations taking the same values, and the distribution is more spread out in both directions and somewhat skewed to the right. The t procedures are nonetheless approximately correct because $n = 50$ and there are no extreme outliers.

(b) Give a 95% confidence interval for the mean monthly cost of Internet access in August 2000.

16.24 The cost of Internet access, continued. The data in Exercise 16.23 show that many people paid $20 per month for Internet access, presumably because major providers such as AOL charged this amount. Do the data give good reason to think that the mean cost for all Internet users differs from $20 per month?

16.25 Economic impact of the Internet. Exercise 16.23 gives the fees paid for Internet access by a national random sample of clients of Internet service providers in 2000. The Census Bureau estimates that 44 million households had Internet access in 2000. Use your confidence interval from Exercise 16.23 to give a 95% confidence interval for the total amount these households paid in Internet access fees. This is one aspect of the national economic impact of the Internet.

16.26 A big toe problem. Hallux abducto valgus (call it HAV) is a deformation of the big toe that is not common in youth and often requires surgery. Doctors used X-rays to measure the angle (in degrees) of deformity in 38 consecutive patients under the age of 21 who came to a medical center for surgery to correct HAV. The angle is a measure of the seriousness of the deformity. Here are the data:[12]

28	32	25	34	38	26	25	18	30	26	28	13	20
21	17	16	21	23	14	32	25	21	22	20	18	26
16	30	30	20	50	25	26	28	31	38	32	21	

It is reasonable to regard these patients as a random sample of young patients who require HAV surgery. Give a 95% confidence interval for the mean HAV angle in the population of all such patients.

16.27 A big toe problem, continued. The data in the previous problem follow a Normal distribution quite closely except for one patient with HAV angle 50 degrees, a high outlier.

(a) Find the 95% confidence interval for the population mean based on the 37 patients who remain after you drop the outlier.

(b) Compare your interval in (a) with your interval from the previous problem. What is the most important effect of removing the outlier?

16.28 Significance. You are testing H_0: $\mu = 0$ against H_a: $\mu \neq 0$ based on an SRS of 20 observations from a Normal population. What values of the t statistic are statistically significant at the $\alpha = 0.005$ level?

16.29 What critical value? You have an SRS of 15 observations from a Normally distributed population. What critical value would you use to obtain a 98% confidence interval for the mean μ of the population?

16.30 Mutual-fund performance. Exercise 4.21 gives data on the annual returns (percent) for the Vanguard International Growth Fund and its benchmark index, the Morgan Stanley EAFE index. Does the fund significantly outperform its benchmark?

(a) Explain clearly why the matched pairs t test is the proper choice to answer this question.

(b) Make a stemplot of the differences (fund − EAFE) for the 20 years. There is no reason to doubt the approximate Normality of the differences. (More detailed study shows that the differences follow a Normal distribution quite closely.)

(c) Carry out the test and state your conclusion about the fund's performance.

16.31 Sharks. Great white sharks are big and hungry. Here are the lengths in feet of 44 great whites:[13]

(David Fleetham/Taxi/Getty Images)

18.7	12.3	18.6	16.4	15.7	18.3	14.6	15.8	14.9	17.6	12.1
16.4	16.7	17.8	16.2	12.6	17.8	13.8	12.2	15.2	14.7	12.4
13.2	15.8	14.3	16.6	9.4	18.2	13.2	13.6	15.3	16.1	13.5
19.1	16.2	22.8	16.8	13.6	13.2	15.7	19.7	18.7	13.2	16.8

(a) Examine these data for shape, center, spread, and outliers. The distribution is reasonably Normal except for one outlier in each direction. Because these are not extreme and preserve the symmetry of the distribution, use of the t procedures is safe with 44 observations.

(b) Give a 95% confidence interval for the mean length of great white sharks. Based on this interval, is there significant evidence at the 5% level to reject the claim "Great white sharks average 20 feet in length"?

(c) It isn't clear exactly what parameter μ you estimated in (b). What information do you need to say what μ is?

16.32 Calcium and blood pressure. In a randomized comparative experiment on the effect of calcium in the diet on blood pressure, researchers divided 54 healthy white males at random into two groups. One group received calcium; the other, a placebo. At the beginning of the study, the researchers measured many variables on the subjects. The paper reporting the study gives $\bar{x} = 114.9$ and $s = 9.3$ for the seated systolic blood pressure of the 27 members of the placebo group.

(a) Give a 95% confidence interval for the mean blood pressure in the population from which the subjects were recruited.

(b) What assumptions about the population and the study design are required by the procedure you used in (a)? Which of these assumptions are important for the validity of the procedure in this case?

16.33 The placebo effect. The placebo effect (see Chapter 8, page 202) is particularly strong in patients with Parkinson's disease. To understand the workings of the placebo effect, scientists made chemical measurements at a key point in the brain when patients received a placebo that they thought was an active drug and also when no treatment was given.[14] The same six patients were measured both with and without the placebo, at different times.

(a) Explain why the proper procedure to compare the mean response to placebo with control (no treatment) is a matched pairs t test.

(b) The six differences (treatment minus control) had $\bar{x} = -0.326$ and $s = 0.181$. Is there significant evidence of a difference between treatment and control?

16.34 Calibrating an instrument. Gas chromatography is a sensitive technique used to measure small amounts of compounds. The response of a gas chromatograph is calibrated by repeatedly testing specimens containing a known amount of the compound to be measured. A calibration study for a specimen containing 1 nanogram (that's 10^{-9} gram) of a compound gave the following response readings:[15]

$$21.6 \quad 20.0 \quad 25.0 \quad 21.9$$

The response is known from experience to vary according to a Normal distribution unless an outlier indicates an error in the analysis. Estimate the mean response to 1 nanogram of this substance, and give the margin of error for your choice of confidence level. Then explain to a chemist who knows no statistics what your margin of error means.

(David A. Northcott/CORBIS)

16.35 Does nature heal best? Differences of electric potential occur naturally from point to point on a body's skin. Is the natural electric field strength best for helping wounds to heal? If so, changing the field will slow healing. The research subjects are anesthetized newts. Make a razor cut in both hind limbs. Let one heal naturally (the control). Use an electrode to change the electric field in the other to half its usual value. After two hours, measure the healing rate. Table 16.3 gives the healing rates (in micrometers per hour) for 14 newts.[16]

TABLE 16.3   Healing rates (micrometers per hour) for newts

Newt	Experimental limb	Control limb	Difference in healing
13	24	25	−1
14	23	13	10
15	47	44	3
16	42	45	−3
17	26	57	−31
18	46	42	4
19	38	50	−12
20	33	36	−3
21	28	35	−7
22	28	38	10
23	21	43	−22
24	27	31	−4
25	25	26	−1
26	45	48	−3

(a) As is usual, the paper did not report these raw data. Readers are expected to be able to interpret the summaries that the paper did report. The paper summarized the differences in the table as "-5.71 ± 2.82" and said, "All values are expressed as means $\pm$ standard error of the mean." Show carefully where the numbers -5.71 and 2.82 come from.

(b) The researchers want to know if changing the electric field reduces the mean healing rate for all newts. State hypotheses, carry out a test, and give your conclusion. Is the result statistically significant at the 5% level? At the 1% level? (The researchers compared several field strengths and concluded that the natural strength is about right for fastest healing.)

16.36 Does nature heal best, continued. Give a 90% confidence interval for the amount by which changing the field changes the rate of healing. Then explain in a sentence what it means to say that you are "90% confident" of your result.

16.37 Comparing two drugs. Makers of generic drugs must show that they do not differ significantly from the "reference" drugs that they imitate. One aspect in which drugs might differ is their extent of absorption in the blood. Table 16.4 gives data taken from 20 healthy nonsmoking male subjects for one pair of drugs.[17] This is a matched pairs design. Subjects 1 to 10 received the generic drug first, and Subjects 11 to 20 received the reference drug first. In all cases, a washout period separated the two drugs so that the first had disappeared from the blood before the subject took the second. The subject numbers in the table were assigned at random to decide the order of the drugs for each subject.

(a) Do a data analysis of the differences between the absorption measures for the generic and reference drugs. Is there any reason not to apply t procedures?

(b) Use a t test to answer the key question: do the drugs differ significantly in absorption?

Chapter 16 MEDIA EXERCISES

16.38 Is caffeine dependence real? The EESEE story "Is Caffeine Dependence Real?" contains data on 11 people diagnosed as being dependent on caffeine. Each subject was barred from coffee, colas, and other substances containing caffeine. Instead, they took capsules containing their usual caffeine intake. During a different time period, they took placebo capsules. The order in which subjects took caffeine and the placebo was randomized. Table 16.5 contains data on two of several tests given to the subjects. "Depression" is the score on the Beck Depression Inventory. Higher scores show more symptoms of

TABLE 16.4 Absorption extent for two versions of a drug

Subject	Reference drug	Generic drug
15	4108	1755
3	2526	1138
9	2779	1613
13	3852	2254
12	1833	1310
8	2463	2120
18	2059	1851
20	1709	1878
17	1829	1682
2	2594	2613
4	2344	2738
16	1864	2302
6	1022	1284
10	2256	3052
5	938	1287
7	1339	1930
14	1262	1964
11	1438	2549
1	1735	3340
19	1020	3050

TABLE 16.5 Results of a caffeine-deprivation study

Subject	Depression (caffeine)	Depression (placebo)	Beats (caffeine)	Beats (placebo)
1	5	16	281	201
2	5	23	284	262
3	4	5	300	283
4	3	7	421	290
5	8	14	240	259
6	5	24	294	291
7	0	6	377	354
8	0	3	345	346
9	2	15	303	283
10	11	12	340	391
11	1	0	408	411

depression. "Beats" is the beats per minute the subject achieved when asked to press a button 200 times as quickly as possible. We are interested in whether being deprived of caffeine affects these outcomes.

(a) The study was double-blind. What does this mean?

(b) Does this matched pairs study give evidence that being deprived of caffeine raises depression scores? Make a stemplot to check that the differences are not strikingly non-Normal, state hypotheses, carry out a test, and state your conclusion.

(c) Now make a stemplot of the differences in beats per minute with and without caffeine. You should hesitate to use the t procedures on these data. Why?

<table>
<tr><td>CHAPTER</td></tr>
<tr><td>17</td></tr>
</table>

(CORBIS)

Two-Sample Problems

Comparing two populations or two treatments is one of the most common situations encountered in statistical practice. We call such situations *two-sample problems*.

TWO-SAMPLE PROBLEMS

- The goal of inference is to compare the responses to two treatments or to compare the characteristics of two populations.

- We have a separate sample from each treatment or each population.

Two-sample problems

A two-sample problem can arise from a randomized comparative experiment that randomly divides subjects into two groups and exposes each group to a different treatment. Comparing random samples separately selected from two populations is also a two-sample problem. Unlike the matched pairs designs studied earlier, there is no matching of the units in the two samples, and the two samples can be of different sizes. Inference procedures for two-sample data differ from those for matched pairs. Here are some typical two-sample problems.

EXAMPLE 17.1   Two-sample problems

(a) A medical researcher is interested in the effect on blood pressure of added calcium in our diet. She conducts a randomized comparative experiment in which one group of subjects receives a calcium supplement and a control group gets a placebo.

(b) A psychologist develops a test that measures social insight. He compares the social insight of male college students with that of female college students by giving the test to a sample of students of each gender.

(c) A bank wants to know which of two incentive plans will most increase the use of its credit cards. It offers each incentive to a random sample of credit card customers and compares the amount charged during the following six months.

We may wish to compare either the *centers* or the *spreads* of the two groups in a two-sample setting. This chapter emphasizes the most common inference procedures, those for comparing two population means. We comment briefly on the issue of comparing spreads (standard deviations), where simple inference is much less satisfactory.

APPLY YOUR KNOWLEDGE

> *Which data design?* *Each situation described in Exercises 17.1 to 17.4 requires inference about a mean or means. Identify each as involving (1) a single sample, (2) matched pairs, or (3) two independent samples. The procedures of Chapter 16 apply to designs (1) and (2). We are about to learn procedures for (3).*

17.1 Looking back on love. Choose 40 romantically attached couples in their midtwenties. Interview the man and woman separately about a romantic attachment they had at age 15 or 16. Compare the attitudes of men and women.

17.2 Community service. Choose a random sample of college students. Use a questionnaire to discover which of the students have ever done volunteer work in the community and which have not. Compare the attitudes of the two groups toward people of other races.

17.3 Chemical analysis. To check a new analytical method, a chemist obtains a reference specimen of known concentration from the National Institute of Standards and Technology. She then makes 20 measurements of the concentration of this specimen with the new method and checks for bias by comparing the mean result with the known concentration.

(Creasource/Series/PictureQuest)

17.4 Chemical analysis, continued. Another chemist is checking the same new method. He has no reference specimen, but a familiar analytic method is available. He wants to know if the new and old methods

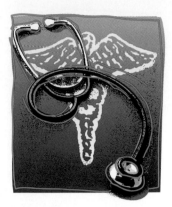

Sounds good—but no comparison

Most women have mammograms to check for breast cancer once they reach middle age. Could a fancier test do a better job of finding cancers early? PET scans are a fancier (and more expensive) test. Doctors used PET scans on 14 women with tumors and got the detailed diagnosis right in 12 cases. That's promising. But there were no controls, and 14 cases are not statistically significant. Medical standards require randomized comparative experiments and statistically significant results. Only then can we be confident that the fancy test really is better.

agree. He takes a specimen of unknown concentration and measures the concentration 10 times with the new method and 10 times with the old method.

Comparing two population means

We can examine two-sample data graphically by comparing stemplots (for small samples) or histograms or boxplots (for larger samples). Now we will apply the ideas of formal inference in this setting. When both population distributions are symmetric, and especially when they are at least approximately Normal, a comparison of the mean responses in the two populations is the most common goal of inference. Here are the conditions for inference.

CONDITIONS FOR COMPARING TWO MEANS

- We have **two SRSs,** from two distinct populations. The samples are **independent.** That is, one sample has no influence on the other. Matching violates independence, for example. We measure the same variable for both samples.

- Both populations are **Normally distributed.** The means and standard deviations of the populations are unknown. In practice, it is enough that the distributions have similar shapes and that the data have no strong outliers.

Call the variable we measure x_1 in the first population and x_2 in the second because the variable may have different distributions in the two populations. Here is the notation we will use to describe the two populations:

Population	Variable	Mean	Standard deviation
1	x_1	μ_1	σ_1
2	x_2	μ_2	σ_2

There are four unknown parameters, the two means and the two standard deviations. The subscripts remind us which population a parameter describes. We want to compare the two population means, either by giving a confidence interval for their difference $\mu_1 - \mu_2$ or by testing the hypothesis of no difference, $H_0: \mu_1 = \mu_2$.

We use the sample means and standard deviations to estimate the unknown parameters. Again, subscripts remind us which sample a statistic comes from. Here is the notation that describes the samples:

Population	Sample size	Sample mean	Sample standard deviation
1	n_1	$\overline{x}_1$	s_1
2	n_2	$\overline{x}_2$	s_2

To do inference about the difference $\mu_1 - \mu_2$ between the means of the two populations, we start from the difference $\overline{x}_1 - \overline{x}_2$ between the means of the two samples.

EXAMPLE 17.2   Does polyester decay?

How quickly do synthetic fabrics such as polyester decay in landfills? A researcher buried polyester strips in the soil for different lengths of time, then dug up the strips and measured the force required to break them. Breaking strength is easy to measure and is a good indicator of decay. Lower strength means the fabric has decayed.

Part of the study buried 10 polyester strips in well-drained soil in the summer. Five of the strips, chosen at random, were dug up after 2 weeks; the other 5 were dug up after 16 weeks. Here are the breaking strengths in pounds:[1]

2 weeks	118	126	126	120	129
16 weeks	124	98	110	140	110

From the data, calculate the summary statistics:

Group	Treatment	n	$\overline{x}$	s
1	2 weeks	5	123.80	4.60
2	16 weeks	5	116.40	16.09

The fabric that was buried longer has somewhat lower mean strength, along with more variation. The observed difference in mean strengths is

$$\overline{x}_1 - \overline{x}_2 = 123.80 - 116.40 = 7.40 \text{ pounds}$$

Is this good evidence that polyester decays more in 16 weeks than in 2 weeks?

Example 17.2 fits the two-sample setting. We write hypotheses in terms of the mean breaking strengths in the entire population of polyester fabric, μ_1 for fabric buried for 2 weeks and μ_2 for fabric buried for 16 weeks. The hypotheses are

$$H_0: \mu_1 = \mu_2$$
$$H_a: \mu_1 > \mu_2$$

We want to test these hypotheses and also estimate the amount of the decrease in mean breaking strength, $\mu_1 - \mu_2$.

Are the conditions for inference met? Because of the randomization, we are willing to regard the two groups of fabric strips as two independent SRSs.

Although the samples are small, we check for serious non-Normality by examining the data. Here is a back-to-back stemplot of the responses:

```
        2 weeks            16 weeks
                       9  | 8
                      10  |
               8      11  | 00
            9660      12  | 4
                      13  |
                      14  | 0
```

The 16-week group is much more spread out, as its larger standard deviation suggests. As far as we can tell from so few observations, there are no departures from Normality that prevent the use of t procedures.

Two-sample t procedures

To assess the significance of the observed difference between the means of our two samples, we follow a familiar path. Whether an observed difference between two samples is surprising depends on the spread of the observations as well as on the two means. Widely different means can arise just by chance if the individual observations vary a great deal. To take variation into account, we would like to standardize the observed difference $\overline{x}_1 - \overline{x}_2$ by dividing by its standard deviation. This standard deviation is

$$\sqrt{\frac{\sigma_1^2}{n_1} + \frac{\sigma_2^2}{n_2}}$$

This standard deviation gets larger as either population gets more variable, that is, as σ_1 or σ_2 increases. It gets smaller as the sample sizes n_1 and n_2 increase.

Because we don't know the population standard deviations, we estimate them by the sample standard deviations from our two samples. The result is *standard error* the **standard error,** or estimated standard deviation, of the difference in sample means:

$$\text{SE} = \sqrt{\frac{s_1^2}{n_1} + \frac{s_2^2}{n_2}}$$

When we standardize the estimate by dividing it by its standard error, the result *two-sample t statistic* is the **two-sample t statistic:**

$$t = \frac{\overline{x}_1 - \overline{x}_2}{\text{SE}}$$

The statistic t has the same interpretation as any z or t statistic: it says how far $\overline{x}_1 - \overline{x}_2$ is from 0 in standard deviation units.

The two-sample t statistic has approximately a t distribution. It does not have exactly a t distribution even if the populations are both exactly Normal. In practice, however, the approximation is very accurate. There is a catch: the degrees of freedom of the *t* distribution we want to use is calculated from the data by a somewhat messy formula; moreover, the degrees of freedom need not be a whole number. There are two practical options for using the two-sample *t* procedures:

Option 1. With software, use the statistic *t* with accurate critical values from the approximating *t* distribution.

Option 2. Without software, use the statistic *t* with critical values from the *t* distribution with degrees of freedom equal to the smaller of $n_1 - 1$ and $n_2 - 1$. These procedures are always conservative for any two Normal populations.

The two options are exactly the same except for the degrees of freedom used for *t* critical values and *P*-values. The Using Technology section (page 449) illustrates how software uses Option 1. Some details of Option 1 appear in the optional section on page 452. We recommend that you use Option 2 when doing calculations without software. Here is a description of the Option 2 procedures that includes a statement of just how they are "conservative."

THE TWO-SAMPLE *t* PROCEDURES

Draw an SRS of size n_1 from a Normal population with unknown mean μ_1, and draw an independent SRS of size n_2 from another Normal population with unknown mean μ_2. A **confidence interval for $\mu_1 - \mu_2$** is given by

$$(\overline{x}_1 - \overline{x}_2) \pm t^* \sqrt{\frac{s_1^2}{n_1} + \frac{s_2^2}{n_2}}$$

Here t^* is the critical value for the $t(k)$ density curve with area C between $-t^*$ and t^*. The degrees of freedom *k* are equal to the smaller of $n_1 - 1$ and $n_2 - 1$. This interval has confidence level *at least* C no matter what the population standard deviations may be.

To **test the hypothesis $H_0: \mu_1 = \mu_2$,** calculate the two-sample *t* statistic

$$t = \frac{\overline{x}_1 - \overline{x}_2}{\sqrt{\dfrac{s_1^2}{n_1} + \dfrac{s_2^2}{n_2}}}$$

and use *P*-values or critical values for the $t(k)$ distribution. The true *P*-value or fixed significance level will always be *equal to or less than* the value calculated from $t(k)$ no matter what values the unknown population standard deviations have.

The two-sample t confidence interval again has the form

$$\text{estimate} \pm t^{*}\, \text{SE}_{\text{estimate}}$$

These two-sample t procedures always err on the safe side, reporting *higher P-* values and *lower* confidence than are actually true. The gap between what is reported and the truth is quite small unless the sample sizes are both small and unequal. As the sample sizes increase, probability values based on t with degrees of freedom equal to the smaller of $n_1 - 1$ and $n_2 - 1$ become more accurate.[2]

APPLY YOUR KNOWLEDGE

17.5 Treating scrapie in hamsters. Scrapie is a degenerative disease of the nervous system. A study of the substance IDX as a treatment for scrapie used as subjects 20 infected hamsters. Ten, chosen at random, were injected with IDX. The other 10 were untreated. The researchers recorded how long each hamster lived. They reported, "Thus, although all infected control hamsters had died by 94 days after infection (mean ± SEM = 88.5 ± 1.9 days), IDX-treated hamsters lived up to 128 days (mean ± SEM = 116 ± 5.6 days)."[3] Readers are supposed to know that SEM stands for "standard error of the mean."

(a) Fill in the values in this summary table:

Group	Treatment	n	$\bar{x}$	s
1	IDX	?	?	?
2	Untreated	?	?	?

(b) What degrees of freedom would you use in the conservative two-sample t procedures to compare the two treatments?

17.6 Going bankrupt. A business school study compared a sample of Greek firms that went bankrupt with a sample of healthy Greek businesses. One measure of a firm's financial health is the ratio of current assets to current liabilities, called CA/CL. For the year before bankruptcy, the study found the mean CA/CL to be 1.72565 in the healthy group and 0.78640 in the group that failed. The paper reporting the study says that $t = 7.36$.[4]

(a) You can draw a conclusion from this t without using a table and even without knowing the sizes of the samples (as long as the samples are not tiny). What is your conclusion? Why don't you need the sample sizes and a table?

(b) In fact, the study looked at 33 firms that failed and 68 healthy firms. What degrees of freedom would you use for the t test if you follow the conservative approach recommended for use without software?

Examples of the two-sample t procedures

EXAMPLE 17.3   Does polyester decay?

We will use the two-sample t procedures to compare the fabric strengths in Example 17.2. The test statistic for the null hypothesis $H_0: \mu_1 = \mu_2$ is

$$t = \frac{\bar{x}_1 - \bar{x}_2}{\sqrt{\dfrac{s_1^2}{n_1} + \dfrac{s_2^2}{n_2}}}$$

$$= \frac{123.8 - 116.4}{\sqrt{\dfrac{4.60^2}{5} + \dfrac{16.09^2}{5}}}$$

$$= \frac{7.4}{7.484} = 0.9889$$

Without software, use the conservative Option 2. That is, use the t table with 4 degrees of freedom, since $n_1 - 1 = 4$ and $n_2 - 1 = 4$. Because H_a is one-sided on the high side, the P-value is the area to the right of $t = 0.9889$ under the $t(4)$ curve. Figure 17.1 illustrates this P-value. Table C shows that it lies between 0.15 and

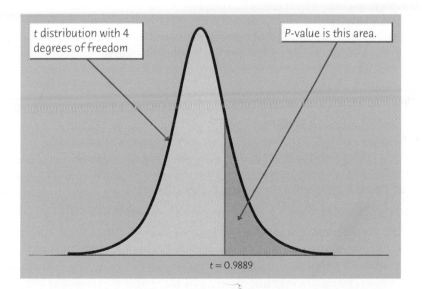

Figure 17.1 The P-value in Example 17.3. This example uses the conservative Option 2, which leads to the t distribution with 4 degrees of freedom.

df = 4		
p	.20	.15
t*	0.941	1.190

0.20. This is enough for a clear conclusion: the experiment did not find convincing evidence that polyester decays more in 16 weeks than in 2 weeks.

For a 90% confidence interval, Table C shows that the $t(4)$ critical value is $t^* = 2.132$. We are 90% confident that the mean strength change between 2 and 16 weeks, $\mu_1 - \mu_2$, lies in the interval

$$(\bar{x}_1 - \bar{x}_2) \pm t^* \sqrt{\frac{s_1^2}{n_1} + \frac{s_2^2}{n_2}}$$

$$= (123.8 - 116.4) \pm 2.132\sqrt{\frac{4.60^2}{5} + \frac{16.09^2}{5}}$$

$$= 7.40 \pm 15.96$$

$$= -8.56 \text{ to } 23.36$$

That the 90% confidence interval covers 0 tells us that we cannot reject $H_0: \mu_1 = \mu_2$ in favor of the two-sided alternative at the $\alpha = 0.10$ level of significance.

Sample size strongly influences the P-value of a test. An effect that fails to be significant at a level α in a small sample may be significant in a larger sample. In the light of the small samples in Example 17.3, we suspect that more data might show that longer burial time does significantly reduce strength. Even if significant, the reduction may be quite small. Our data suggest that buried polyester decays slowly, as the mean breaking strength dropped only from 123.8 pounds to 116.4 pounds between 2 and 16 weeks.

Meta-analysis

Small samples have large margins of error. Large samples are expensive. Often we can find several studies of the same issue; if we could combine their results, we would have a large sample with a small margin of error. That is the idea of "meta-analysis." Of course, we can't just lump the studies together, because of differences in design and quality. Statisticians have more sophisticated ways of combining the results. Meta-analysis has been applied to issues ranging from the effect of secondhand smoke to whether coaching improves SAT scores.

EXAMPLE 17.4 **Community service and attachment to friends**

Do college students who have volunteered for community service work differ from those who have not? A study obtained data from 57 students who had done service work and 17 who had not. One of the response variables was a measure of attachment to friends (roughly, secure relationships), measured by the Inventory of Parent and Peer Attachment. Here are the results:[5]

Group	Condition	n	$\bar{x}$	s
1	Service	57	105.32	14.68
2	No service	17	96.82	14.26

The paper reporting the study results applied a two-sample t test.

Step 1. Hypotheses. The investigator had no specific direction for the difference in mind before looking at the data, so the alternative is two-sided. The hypotheses are

$$H_0: \mu_1 = \mu_2$$
$$H_a: \mu_1 \neq \mu_2$$

Step 2. Test statistic. The two-sample t statistic is

$$t = \frac{\bar{x}_1 - \bar{x}_2}{\sqrt{\dfrac{s_1^2}{n_1} + \dfrac{s_2^2}{n_2}}}$$

$$= \frac{105.32 - 96.82}{\sqrt{\dfrac{14.68^2}{57} + \dfrac{14.26^2}{17}}}$$

$$= \frac{8.5}{3.9677} = 2.142$$

Step 3. P-value. Without software, use Option 2. There are 16 degrees of freedom, the smaller of

$$n_1 - 1 = 57 - 1 = 56 \quad \text{and} \quad n_2 - 1 = 17 - 1 = 16$$

Figure 17.2 illustrates the P-value. Find it by comparing 2.142 with critical values for the $t(16)$ distribution and then doubling p because the alternative is two-sided. Table C shows that $t = 2.142$ lies between the 0.025 and 0.02 critical values. The P-value is therefore between 0.05 and 0.04. The data give moderately strong evidence that students who have engaged in community service are on the average more attached to their friends.

df = 16		
p	.025	.02
t^*	2.120	2.235

Is the t test in Example 17.4 justified? The student subjects were "enrolled in a course on U.S. Diversity at a large mid-western university." Unless this course is required of all students, the subjects cannot be considered a random sample even from this campus. Students were placed in the two groups on the basis of a

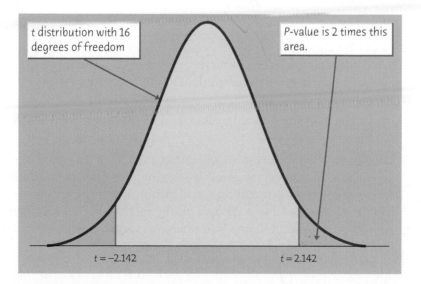

Figure 17.2 The P-value in Example 17.4. To find P, find the area above $t = 2.142$ and double it because the alternative is two-sided.

questionnaire: 39 in the "no service" group and 71 in the "service" group. The data were gathered from a follow-up survey two years later; 17 of the 39 "no service" students responded (44%), compared with 80% response (57 of 71) in the "service" group. Nonresponse is confounded with group: students who had done community service were much more likely to respond. Finally, 75% of the "service" respondents were women, compared with 47% of the "no service" respondents. Gender, which can strongly affect attachment, is badly confounded with the presence or absence of community service. The data are so far from meeting the SRS condition for inference that the t test is meaningless. Difficulties like these are common in social science research, where confounding variables have stronger effects than is usual when biological or physical variables are measured. This researcher honestly disclosed the weaknesses in data production but left it to readers to decide whether to trust her inferences.

APPLY YOUR KNOWLEDGE

17.7   **Is red wine better than white wine?** Observational studies suggest that moderate use of alcohol reduces heart attacks, and that red wine may have special benefits. One reason may be that red wine contains polyphenols, substances that do good things to cholesterol in the blood and so may reduce the risk of heart attacks. In an experiment, healthy men were assigned at random to drink half a bottle of either red or white wine each day for two weeks. The level of polyphenols in their blood was measured before and after the two-week period. Here are the percent changes in level for the subjects in both groups:[6]

Red wine	3.5	8.1	7.4	4.0	0.7	4.9	8.4	7.0	5.5
White wine	3.1	0.5	−3.8	4.1	−0.6	2.7	1.9	−5.9	0.1

   (a) Is there good evidence that red wine drinkers gain more polyphenols on the average than white wine drinkers?

   (b) Does this study give reason to think that it is drinking red wine, rather than some lurking variable, that causes the increase in blood polyphenols?

17.8   **Logging in the rainforest.** "Conservationists have despaired over destruction of tropical rainforest by logging, clearing, and burning." These words begin a report on a statistical study of the effects of logging in Borneo.[7] Here are data on the number of tree species in 12 unlogged forest plots and 9 similar plots logged 8 years earlier:

Unlogged	22	18	22	20	15	21	13	13	19	13	19	15
Logged	17	4	18	14	18	15	15	10	12			

(a) The study report says, "Loggers were unaware that the effects of logging would be assessed." Why is this important? The study report also explains why the plots can be considered to be randomly assigned.

(b) Does logging significantly reduce the mean number of species in a plot after 8 years? State the hypotheses and do a t test. Is the result significant at the 5% level? At the 1% level?

17.9 Red wine versus white wine, continued. Using the data in Exercise 17.7, give a 95% confidence interval for the difference (red minus white) in mean change in blood polyphenol levels.

17.10 Logging in the rainforest, continued. Use the data in Exercise 17.8 to give a 90% confidence interval for the difference in mean number of species between unlogged and logged plots.

Using technology

In Examples 17.3 and 17.4, we used the conservative Option 2 for the degrees of freedom. Software should use Option 1 to give more accurate confidence intervals and P-values. Unfortunately, there is variation in how well software implements the t distribution used in Option 1. Figure 17.3 displays output from Minitab, Excel, and the TI-83 calculator for the test and 90% confidence interval of Example 17.3. All three claim to use Option 1. The two-sample t statistic is exactly as in Example 17.3, $t = 0.9889$. You can find this in all three outputs (Minitab rounds to 0.99).

Only the TI-83 gets Option 1 completely right. The accurate approximation uses the t distribution with 4.65 degrees of freedom. The 90% confidence interval for $\mu_1 - \mu_2$ is -7.933 to 22.733. This is a bit narrower than the conservative interval in Example 17.3. The P-value for $t = 0.9889$ is $P = 0.1857$.

Minitab uses Option 1, but it truncates the exact degrees of freedom to the next smaller whole number to get critical values and P-values. In this example, the exact df $= 4.65$ is truncated to df $= 4$, so that Minitab's results agree with the conservative Option 2 except for rounding. Minitab reports the two-sided P-value as 0.379. The one-sided value is half that, $P = 0.1895$.

The output from Excel's menu command gives only the test, not the confidence interval. Excel rounds the exact degrees of freedom to the nearest whole number, so that df $= 4.65$ becomes df $= 5$. Excel's P-value $P = 0.1841$ is therefore slightly smaller than is correct. Because Excel suggests that the evidence against H_0 is stronger than is actually the case, its output is a bit misleading. Excel also gives the critical value 2.015 for a test at the $\alpha = 0.05$ level. You can find this value in Table C.

Excel's label for the test, "Two-Sample Assuming Unequal Variances," is seriously misleading. **The two-sample t procedures we have described work whether or not the two populations have the same variance.** There is an old-fashioned special procedure that only works when the two variances are equal.

Minitab

```
┌─ Session ──────────────────────────────────────────────── _ □ × ─┐
│                                                                   │
│  Two-sample T for Strength                                        │
│                                                                   │
│  C1          N       Mean      StDev    SE Mean                   │
│   2          5      123.80      4.60       2.1                    │
│  16          5      116.4      16.1        7.2                    │
│                                                                   │
│  Difference = mu ( 2) - mu (16)                                   │
│  Estimate for difference:  7.40                                   │
│  95% CI for difference: (-13.38, 28.18)                           │
│  T-Test of difference = 0 (vs not =): T-Value = 0.99  P-Value = 0.379   DF = 4  │
│                                                                   │
└───────────────────────────────────────────────────────────────────┘
```

Excel

	A	B	C
1	t-Test: Two-Sample Assuming Unequal Variances		
2			
3		2 weeks	16 weeks
4	Mean	123.8	116.4
5	Variance	21.2	258.8
6	Observations	5	5
7	Hypothesized Mean Difference	0	
8	df	5	
9	t Stat	0.9889	
10	P(T<=t) one-tail	0.1841	
11	t Critical one-tail	2.0150	
12	P(T<=t) Two-tail	0.3681	
13	t Critical two-tail	2.5706	
14			
15			

Texas Instruments TI-83 Plus

```
2-SampTInt              2-SampTTest
 (-7.933,22.733)         μ1>μ2
 df=4.6510               t=.9889
 x̄1=123.8000            P=.1857
 x̄2=116.4000            df=4.6510
 Sx1=4.6043              x̄1=123.8000
↓Sx2=16.0873           ↓x̄2=116.4000
```

Figure 17.3 The two-sample t procedures applied to the polyester decay data: output from Minitab, the Excel spreadsheet program, and the TI-83 graphing calculator.

We discuss this method in an optional section on page 454, but you should never use it.

The three technologies gave us three P-values: 0.1841, 0.1857, and 0.1895. The TI-83's $P = 0.1857$ from df $= 4.65$ is accurate; the other two differ slightly because they are based on whole-number degrees of freedom, df $= 4$ and df $= 5$. In practice, just trust whatever technology you use. The small differences in

Nothing After

P don't affect the conclusion. Even "between 0.15 and 0.20" from Table C is close enough for practical purposes.

Robustness again

The two-sample *t* procedures are more robust than the one-sample *t* methods, particularly when the distributions are not symmetric. When the sizes of the two samples are equal and the two populations being compared have distributions with similar shapes, probability values from the *t* table are quite accurate for a broad range of distributions when the sample sizes are as small as $n_1 = n_2 = 5$.[8] When the two population distributions have different shapes, larger samples are needed.

As a guide to practice, adapt the guidelines given on page 426 for the use of one-sample *t* procedures to two-sample procedures by replacing "sample size" with the "sum of the sample sizes," $n_1 + n_2$. These guidelines err on the side of safety, especially when the two samples are of equal size. In planning a two-sample study, you should usually choose equal sample sizes. The two-sample *t* procedures are most robust against non-Normality in this case, and the conservative probability values are most accurate.

APPLY YOUR KNOWLEDGE

17.11 Bone loss by nursing mothers. Exercise 2.19 (page 49) gives the percent change in the mineral content of the spine for 47 mothers during three months of nursing a baby and for a control group of 22 women of similar age who were neither pregnant nor lactating.

(a) What two populations did the investigators want to compare? We must be willing to regard the women recruited for this observational study as SRSs from these populations.

(b) Make graphs to check for clear violations of the Normality condition. Use of the *t* procedures is safe for these distributions.

(c) Do these data give good evidence that on the average nursing mothers lose bone mineral? State hypotheses, calculate the two-sample *t* test statistic, and give a *P*-value. What do you conclude?

17.12 Weeds among the corn. Exercise I.41 (page 169) gives these corn yields (bushels per acre) for experimental plots controlled to have 1 weed per meter of row and 3 weeds per meter of row:

1 weed/meter	166.2	157.3	166.7	161.1
3 weeds/meter	158.6	176.4	153.1	156.0

Explain carefully why a two-sample t confidence interval for the difference in mean yields may not be accurate.

Details of the t approximation*

The exact distribution of the two-sample t statistic is not a t distribution. Moreover, the distribution changes as the unknown population standard deviations σ_1 and σ_2 change. However, an excellent approximation is available. We called this Option 1 for t procedures.

APPROXIMATE DISTRIBUTION OF THE TWO-SAMPLE t STATISTIC

The distribution of the two-sample t statistic is very close to the t distribution with degrees of freedom df given by

$$\mathrm{df} = \frac{\left(\dfrac{s_1^2}{n_1} + \dfrac{s_2^2}{n_2}\right)^2}{\dfrac{1}{n_1-1}\left(\dfrac{s_1^2}{n_1}\right)^2 + \dfrac{1}{n_2-1}\left(\dfrac{s_2^2}{n_2}\right)^2}$$

This approximation is accurate when both sample sizes n_1 and n_2 are 5 or larger.

The t procedures remain exactly as before except that we use the t distribution with df degrees of freedom to give critical values and P-values.

EXAMPLE 17.5   Does polyester decay?

In the experiment of Examples 17.2 and 17.3, the data on buried polyester fabric gave

Group	Treatment	n	$\overline{x}$	s
1	2 weeks	5	123.80	4.60
2	16 weeks	5	116.40	16.09

The two-sample t test statistic calculated from these values is $t = 0.9889$.

  The one-sided P-value is the area to the right of 0.9889 under a t density curve, as in Figure 17.1. The conservative Option 2 uses the t distribution with 4 degrees of freedom. Option 1 finds a very accurate P-value by using the t distribution with degrees of freedom df given by

---

*This section can be omitted unless you are using software and wish to understand what the software does.

$$df = \frac{\left(\dfrac{4.60^2}{5} + \dfrac{16.09^2}{5}\right)^2}{\dfrac{1}{4}\left(\dfrac{4.60^2}{5}\right)^2 + \dfrac{1}{4}\left(\dfrac{16.09^2}{5}\right)^2}$$

$$= \frac{3137.08}{674.71} = 4.65$$

This is the degrees of freedom calculated by the software displayed in Figure 17.3. Two of the three then moved to a whole-number degrees of freedom before doing probability calculations.

The degrees of freedom df is generally not a whole number. It is always at least as large as the smaller of $n_1 - 1$ and $n_2 - 1$. The larger degrees of freedom that results from Option 1 give slightly shorter confidence intervals and slightly smaller P-values than the conservative Option 2 produces. There is a t distribution for any positive degrees of freedom, even though Table C contains entries only for whole-number degrees of freedom.

The difference between the t procedures using Options 1 and 2 is rarely of practical importance. That is why we recommend the simpler, conservative Option 2 for inference without a computer. With a computer, the more accurate Option 1 procedures are painless.

APPLY YOUR KNOWLEDGE

17.13 DDT poisoning. In a randomized comparative experiment, researchers compared 6 white rats poisoned with DDT with a control group of 6 unpoisoned rats. Electrical measurements of nerve activity are the main clue to the nature of DDT poisoning. When a nerve is stimulated, its electrical response shows a sharp spike followed by a much smaller second spike. The experiment found that the second spike is larger in rats fed DDT than in normal rats.[9]

The researchers measured the height of the second spike as a percent of the first spike when a nerve in the rat's leg was stimulated. Here are the results:

Poisoned	12.207	16.869	25.050	22.429	8.456	20.589
Unpoisoned	11.074	9.686	12.064	9.351	8.182	6.642

Figure 17.4 shows the TI-83 output (two screens) for the two-sample t test. Starting from the calculator's results for $\bar{x}_i$ and s_i, verify that the values for the test statistic and the degrees of freedom are approximately $t = 2.99$ and $df = 5.9$.

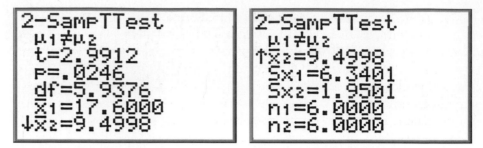

Figure 17.4 Two-sample *t* output from the TI-83 calculator, for Exercise 17.13.

17.14 Students' self-concept. A study of the self-concept of seventh-grade students asked if male and female students differ in mean score on the Piers-Harris Self Concept Scale. Software gave these summary results:[10]

Gender	n	Mean	Std dev	Std err	t	df	P
F	31	55.5161	12.6961	2.2803	−0.8276	62.8	0.4110
M	47	57.9149	12.2649	1.7890			

Starting from the sample means and standard deviations, verify each of these entries: the standard errors of the means; the degrees of freedom for two-sample *t*; the value of *t*.

17.15 DDT poisoning, continued. Do poisoned rats differ significantly from unpoisoned rats in the study of Exercise 17.13? Write a summary in a sentence or two, including *t*, df, *P*, and a conclusion.

17.16 Students' self-concept, continued. Write a sentence or two summarizing the comparison of female and male students in Exercise 17.14, as if you were preparing a report for publication.

Avoid the pooled two-sample *t* procedures*

Most software, including Minitab, Excel, and the TI-83, offers two two-sample *t* statistics. One is often labeled for "unequal" variances, the other for "equal" variances. The "unequal" variance procedure is our two-sample *t*. This test is valid whether or not the population variances are equal. The other choice is a special version of the two-sample *t* statistic that assumes that the two populations have the same variance. This procedure averages (the statistical term is "pools") the two sample variances to estimate the common population variance. The resulting statistic is called the *pooled two-sample t statistic*. It is equal to our *t* statistic if the two sample sizes are the same, but not otherwise. We could choose to use the pooled *t* for tests and confidence intervals.

---

*The remaining sections of this chapter concern optional special topics. They are needed only as background for Chapter 22.

The pooled t statistic has exactly the t distribution with $n_1 + n_2 - 2$ degrees of freedom *if* the two population variances really are equal and the population distributions are exactly Normal. The pooled t was in common use before software made it easy to use Option 1 for our two-sample t statistic. Of course, in the real world distributions are not exactly Normal and population variances are not exactly equal. In practice, the Option 1 t procedures are almost always more accurate than the pooled procedures. Our advice: **Never use the pooled t procedures if you have software that will implement Option 1.**

Avoid inference about standard deviations*

Two basic descriptive features of a distribution are its center and spread. In a Normal population, we measure center by the mean and spread by the standard deviation. We use the t procedures for inference about population means for Normal populations, and we know that t procedures are widely useful for non-Normal populations as well. It is natural to turn next to inference about the standard deviations of Normal populations. Our advice here is short and clear: Don't do it without expert advice.

There are methods for inference about the standard deviations of Normal populations. We will describe the most common such method: the F test for comparing the spread of two Normal populations. Unlike the t procedures for means, the F test and other procedures for standard deviations are extremely sensitive to non-Normal distributions. This lack of robustness does not improve in large samples. It is difficult in practice to tell whether a significant F-value is evidence of unequal population spreads or simply a sign that the populations are not Normal.

The deeper difficulty underlying the very poor robustness of Normal population procedures for inference about spread already appeared in our work on describing data. The standard deviation is a natural measure of spread for Normal distributions but not for distributions in general. In fact, because skewed distributions have unequally spread tails, no single numerical measure does a good job of describing the spread of a skewed distribution. In summary, the standard deviation is not always a useful parameter, and even when it is (for symmetric distributions), the results of inference are not trustworthy. Consequently, **we do not recommend trying to do inference about population standard deviations in basic statistical practice.**[11]

The F test for comparing two standard deviations*

Because of the limited usefulness of procedures for inference about the standard deviations of Normal distributions, we will describe only one such procedure. Suppose that we have independent SRSs from two Normal populations, a sample of size n_1 from $N(\mu_1, \sigma_1)$ and a sample of size n_2 from $N(\mu_2, \sigma_2)$. The population means and standard deviations are all unknown. The two-sample t test examines whether the means are equal in this setting. To test the hypothesis of equal spread,

$$H_0: \sigma_1 = \sigma_2$$
$$H_a: \sigma_1 \neq \sigma_2$$

we use the ratio of sample variances. This is the F *statistic*.

THE F STATISTIC AND F DISTRIBUTIONS

When s_1^2 and s_2^2 are sample variances from independent SRSs of sizes n_1 and n_2 drawn from Normal populations, the **F statistic**

$$F = \frac{s_1^2}{s_2^2}$$

has the **F distribution** with $n_1 - 1$ and $n_2 - 1$ degrees of freedom when $H_0: \sigma_1 = \sigma_2$ is true.

The F distributions are a family of distributions with two parameters. The parameters are the degrees of freedom of the sample variances in the numerator and denominator of the F statistic. The numerator degrees of freedom are always mentioned first. Interchanging the degrees of freedom changes the distribution, so the order is important. Our brief notation will be $F(j, k)$ for the F distribution with j degrees of freedom in the numerator and k in the denominator. The F distributions are right-skewed. The density curve in Figure 17.5 illustrates the shape. Because sample variances cannot be negative, the F statistic takes only positive values, and the F distribution has no probability below 0. The peak of the F density curve is near 1. When the two populations have the same standard deviation, we expect the two sample variances to be close in size, so that F takes a value near 1. Values of F far from 1 in either direction provide evidence against the hypothesis of equal standard deviations.

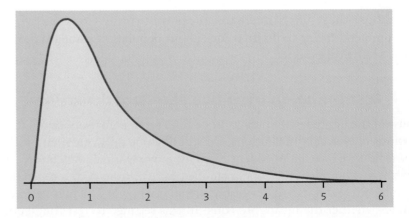

Figure 17.5 The density curve for the $F(9, 10)$ distribution. The F distributions are skewed to the right.

Tables of F critical points are awkward, because we need a separate table for every pair of degrees of freedom j and k. Table D in the back of the book gives upper p critical points of the F distributions for $p = 0.10, 0.05, 0.025, 0.01$, and 0.001. For example, these critical points for the $F(9, 10)$ distribution shown in Figure 17.5 are

p	.10	.05	.025	.01	.001
F^*	2.35	3.02	3.78	4.94	8.96

The skewness of the F distributions causes additional complications. In the symmetric Normal and t distributions, the point with probability 0.05 below it is just the negative of the point with probability 0.05 above it. This is not true for F distributions. We therefore need either tables of both the upper and lower tails or some way to eliminate the need for lower-tail critical values. Statistical software that does away with the need for tables is very convenient. If you do not use statistical software, arrange the two-sided F test as follows.

CARRYING OUT THE F TEST

Step 1. Take the test statistic to be

$$F = \frac{\text{larger } s^2}{\text{smaller } s^2}$$

This amounts to naming the populations so that Population 1 has the larger of the observed sample variances. The resulting F is always 1 or greater.

Step 2. Compare the value of F with critical values from Table D. Then *double* the significance levels from the table to obtain the significance level for the two-sided F test.

The idea is that we calculate the probability in the upper tail and double it to obtain the probability of all ratios on either side of 1 that are at least as improbable as that observed. Remember that the order of the degrees of freedom is important in using Table D.

EXAMPLE 17.6 Comparing variability

Example 17.2 describes an experiment to compare the mean breaking strengths of polyester fabric after being buried for 2 weeks and for 16 weeks. Here are the data summaries:

Group	Treatment	n	$\bar{x}$	s
2	2 weeks	5	123.80	4.60
1	16 weeks	5	116.40	16.09

We might also compare the standard deviations to see whether strength loss is more or less variable after 16 weeks. We want to test

$$H_0: \sigma_1 = \sigma_2$$
$$H_a: \sigma_1 \neq \sigma_2$$

Note that we relabeled the groups so that Group 1 (16 weeks) has the larger standard deviation. The F test statistic is

$$F = \frac{\text{larger } s^2}{\text{smaller } s^2} = \frac{16.09^2}{4.60^2} = 12.23$$

Compare the calculated value $F = 12.23$ with critical points for the $F(4, 4)$ distribution. Table D shows that 12.23 lies between the 0.025 and 0.01 critical values of the $F(4, 4)$ distribution. So the two-sided P-value lies between 0.05 and 0.02. The data show significantly unequal spreads at the 5% level. The P-value depends heavily on the assumption that both samples come from Normally distributed populations.

APPLY YOUR KNOWLEDGE

In all exercises calling for use of the F test, assume that both population distributions are very close to Normal. The actual data are not always sufficiently Normal to justify use of the F test.

17.17 F distributions. The F statistic $F = s_1^2/s_2^2$ is calculated from samples of sizes $n_1 = 10$ and $n_2 = 8$. (Remember that n_1 is the numerator sample size.)

(a) What is the upper 5% critical value for this F?

(b) In a test of equality of standard deviations against the two-sided alternative, this statistic has the value $F = 3.45$. Is this value significant at the 10% level? Is it significant at the 5% level?

17.18 F distributions. The F statistic for equality of standard deviations based on samples of sizes $n_1 = 21$ and $n_2 = 16$ takes the value $F = 2.78$.

(a) Is this significant evidence of unequal population standard deviations at the 5% level? At the 1% level?

(b) Between which two values obtained from Table D does the P-value of the test fall?

17.19 Treating scrapie. The report in Exercise 17.5 suggests that hamsters in the longer-lived group also had more variation in their length of life. It is common to see s increase along with $\bar{x}$ when we compare groups. Do the data give significant evidence of unequal standard deviations?

17.20 Red and white wine. Is there a statistically significant difference between the standard deviations of blood polyphenol level change in the red and white wine groups in Exercise 17.7?

17.21 Logging in the rainforest. Variation in species counts as well as mean counts may be of interest to ecologists. Do the data in Exercise 17.8

give evidence that logging affects the variation in species counts among plots?

17.22 DDT poisoning. The sample variance for the treatment group in the DDT experiment of Exercise 17.13 is more than 10 times as large as the sample variance for the control group. Calculate the F statistic. Can you reject the hypothesis of equal population standard deviations at the 5% significance level? At the 1% level?

Chapter 17 SUMMARY

The data in a **two-sample problem** are two independent SRSs, each drawn from a separate population.

Tests and confidence intervals for the difference between the means μ_1 and μ_2 of two Normal populations start from the difference $\overline{x}_1 - \overline{x}_2$ of the two sample means. Because of the central limit theorem, the resulting procedures are approximately correct for other population distributions when the sample sizes are large.

Draw independent SRSs of sizes n_1 and n_2 from two Normal populations with parameters μ_1, σ_1 and μ_2, σ_2. The **two-sample t statistic** is

$$t = \frac{(\overline{x}_1 - \overline{x}_2) - (\mu_1 - \mu_2)}{\sqrt{\dfrac{s_1^2}{n_1} + \dfrac{s_2^2}{n_2}}}$$

The statistic t has approximately a t distribution.

For conservative inference procedures to compare μ_1 and μ_2, use the two-sample t statistic with the $t(k)$ distribution. The degrees of freedom k is the smaller of $n_1 - 1$ and $n_2 - 1$. Software produces more accurate probability values from the $t(\text{df})$ distribution with degrees of freedom df calculated from the data.

The **confidence interval** for $\mu_1 - \mu_2$ is

$$(\overline{x}_1 - \overline{x}_2) \pm t^* \sqrt{\dfrac{s_1^2}{n_1} + \dfrac{s_2^2}{n_2}}$$

The confidence level is very close to C if t^* is the $t(\text{df})$ critical value. It is guaranteed to be at least C if t^* is the $t(k)$ critical value.

Significance tests for $H_0: \mu_1 = \mu_2$ are based on

$$t = \frac{\overline{x}_1 - \overline{x}_2}{\sqrt{\dfrac{s_1^2}{n_1} + \dfrac{s_2^2}{n_2}}}$$

P-values calculated from the $t(df)$ distribution are very accurate. A *P*-value calculated from $t(k)$ is slightly larger than the true *P*.

The guidelines for practical use of two-sample *t* procedures are similar to those for one-sample *t* procedures. Equal sample sizes are recommended.

Inference procedures for comparing the standard deviations of two Normal populations are based on the **F statistic,** which is the ratio of sample variances

$$F = \frac{s_1^2}{s_2^2}$$

If an SRS of size n_1 is drawn from Population 1 and an independent SRS of size n_2 is drawn from Population 2, the *F* statistic has the **F distribution** $F(n_1 - 1, n_2 - 1)$ if the two population standard deviations σ_1 and σ_2 are in fact equal.

The *F* test for $H_0: \sigma_1 = \sigma_2$ and other procedures for inference on the spread of one or more Normal distributions are so strongly affected by lack of Normality that we do not recommend them for regular use.

Chapter 17 EXERCISES

In exercises that call for two-sample t procedures, use Option 2 (degrees of freedom the smaller of $n_1 - 1$ and $n_2 - 1$) unless you have software that implements the more accurate Option 1. Many of these exercises ask you to think about issues of statistical practice as well as to carry out procedures.

17.23 Education and income. In March 2002 the Bureau of Labor Statistics contacted a large random sample of households for its Annual Demographic Supplement. Among the many questions asked were the education level and 2001 income of all persons aged 25 years or more in the household. The 24,042 whose education ended with a high school diploma had mean income $33,107.51 and standard deviation $30,698.36. There were 16,018 college graduates who had no higher degrees. Their mean income was $59,851.63 and the standard deviation was $57,001.32.

(a) It is clear without any calculations that the difference between the mean incomes for the two groups is significant at any practical level. Why?

(b) Give a 99% confidence interval for the difference between the mean incomes in the populations of all adults with these levels of education.

(c) Income distributions are usually strongly skewed to the right. How does comparing $\overline{x}$ and *s* for these samples show the skewness? Why does the strong skewness not rule out using the two-sample t confidence interval?

17.24 Road rage. "The phenomenon of road rage has been frequently discussed but infrequently examined." So begins a report based on interviews with randomly selected drivers.[12] The respondents' answers to interview questions produced scores on an "angry/threatening driving scale" with values between 0 and 19. Here are summaries of the scores:

(Anthony Redpath/CORBIS)

Group	n	$\bar{x}$	s
Male	596	1.78	2.79
Female	769	0.97	1.84

(a) We suspect that men are more susceptible to road rage than women. Carry out a test of that hypothesis. (State hypotheses, find the test statistic and P-value, and state your conclusion.)

(b) The subjects were selected using random-digit dialing. The large sample sizes make the Normality condition unnecessary. There is one aspect of the data production that might reduce the validity of the data. What is it?

17.25 Eating potato chips. Healthy women aged 18 to 40 participated in a study of eating habits. Subjects were given bags of potato chips and bottled water and invited to snack freely. Interviews showed that some women were trying to restrain their diet out of concern about their weight. How much effect did these good intentions have on their eating habits? Here are the data on grams of potato chips consumed (note that the study report gave the standard error of the mean rather than the standard deviation):[13]

Group	n	$\bar{x}$	SEM
Unrestrained	9	59	7
Restrained	11	32	10

Give a 90% confidence interval that describes the effect of restraint. Based on this interval, is there a significant difference between the two groups? At what significance level does the interval allow this conclusion?

17.26 Depressed teens. To study depression among adolescents, investigators administered the Children's Depression Inventory (CDI) to teenagers in rural Newfoundland, Canada. As is often the case in social science studies, there is some question about whether the subjects can be considered a random sample from an interesting population. We will ignore this issue. One finding was that "older adolescents scored significantly higher on the CDI." Higher scores indicate symptoms of depression. Here are summary data for two grades:[14]

Group	n	$\bar{x}$	s
Grade 9	84	6.94	6.03
Grade 11	70	8.98	7.08

Do an analysis to verify the quoted conclusion.

17.27 **Depressed teens, continued.** The study in the previous exercise also concluded that there were no sex differences in depression. Use these summary data to verify this finding:

Group	n	$\bar{x}$	s
Males	112	7.38	6.95
Females	104	7.15	6.31

17.28 **Each day I am getting better in math.** A "subliminal" message is below our threshold of awareness but may nonetheless influence us. Can subliminal messages help students learn math? A group of students who had failed the mathematics part of the City University of New York Skills Assessment Test agreed to participate in a study to find out.

TABLE 17.1 Mathematics skills scores before and after a subliminal message

Treatment Group		Control Group	
Before	After	Before	After
18	24	18	29
18	25	24	29
21	33	20	24
18	29	18	26
18	33	24	38
20	36	22	27
23	34	15	22
23	36	19	31
21	34		
17	27		

All received a daily subliminal message, flashed on a screen too rapidly to be consciously read. The treatment group of 10 students (chosen at random) was exposed to "Each day I am getting better in math." The control group of 8 students was exposed to a neutral message, "People are walking on the street." All students participated in a summer program designed to raise their math skills, and all took the assessment test again at the end of the program. Table 17.1 gives data on the subjects' scores before and after the program.[15] Is there good evidence that the treatment brought about a greater improvement in math scores than the neutral message? State hypotheses, carry out a test, and state your conclusion. Is your result significant at the 5% level? At the 10% level?

17.29 **Students' attitudes.** The Survey of Study Habits and Attitudes (SSHA) is a psychological test that measures the motivation, attitude toward school, and study habits of students. Scores range from 0 to 200. A selective private college gives the SSHA to an SRS of both male and female first-year students. The data for the women are as follows:

(Bob Llewellyn/ImageState-Pictor/ PictureQuest)

| 154 | 109 | 137 | 115 | 152 | 140 | 154 | 178 | 101 |
| 103 | 126 | 126 | 137 | 165 | 165 | 129 | 200 | 148 |

Here are the scores of the men:

| 108 | 140 | 114 | 91 | 180 | 115 | 126 | 92 | 169 | 146 |
| 109 | 132 | 75 | 88 | 113 | 151 | 70 | 115 | 187 | 104 |

(a) Examine each sample graphically, with special attention to outliers and skewness. Is use of a t procedure acceptable for these data?

(b) Most studies have found that the mean SSHA score for men is lower than the mean score in a comparable group of women. Is this true for first-year students at this college? Carry out a test and give your conclusions.

17.30 **Getting better at math, continued.** Using the data in Table 17.1, give a 90% confidence interval for the mean difference in gains between treatment and control.

17.31 **Students' attitudes, continued.** Use the data in Exercise 17.29 to give a 90% confidence interval for the mean difference between the SSHA scores of male and female first-year students at this college.

Do birds learn to time their breeding? Blue titmice eat caterpillars. The birds would like lots of caterpillars around when they have young to feed, but they must breed much earlier. Do the birds learn from one year's experience when they time breeding the next year? Researchers randomly assigned 7 pairs of birds to have the natural caterpillar supply supplemented while feeding their young and another 6 pairs to serve as a control group relying on natural food supply. The next year, they

(Hugh Clark/Frank Lane Picture Agency/ CORBIS)

measured how many days after the caterpillar peak the birds produced their nestlings.[16] Exercises 17.32 to 17.34 are based on this experiment.

17.32 Did the randomization produce similar groups? First, compare the two groups in the first year. The only difference should be the chance effect of the random assignment. The study report says: "In the experimental year, the degree of synchronization did not differ between food-supplemented and control females." For this comparison, the report gives $t = -1.05$. What type of t statistic (paired or two-sample) is this? Show that this t leads to the quoted conclusion.

17.33 Did the treatment have an effect? The investigators expected the control group to adjust their breeding date the next year, whereas the well-fed supplemented group had no reason to change. The report continues: "but in the following year food-supplemented females were more out of synchrony with the caterpillar peak than the controls." Here are the data (days behind the caterpillar peak):

Control	4.6	2.3	7.7	6.0	4.6	−1.2	
Supplemented	15.5	11.3	5.4	16.5	11.3	11.4	7.7

Carry out a t test and show that it leads to the quoted conclusion.

17.34 Year-to-year comparison. Rather than comparing the two groups in each year, we could compare the behavior of each group in the first and second years. The study report says: "Our main prediction was that females receiving additional food in the nestling period should not change laying date the next year, whereas controls, which (in our area) breed too late in their first year, were expected to advance their laying date in the second year."

Comparing days behind the caterpillar peak in Years 1 and 2 gave $t = 0.63$ for the control group and $t = -2.63$ for the supplemented group. What type of t statistic (paired or two-sample) are these? What are the degrees of freedom for each t? Show that these t-values do *not* agree with the prediction.

17.35 Treating scrapie. Exercise 17.5 reports the results of a study to determine whether IDX is an effective treatment for scrapie.

(a) Is there good evidence that hamsters treated with IDX live longer on the average?

(b) Give a 95% confidence interval for the mean amount by which IDX prolongs life.

17.36 Extraterrestrial handedness? Molecules often have "left-handed" and "right-handed" versions. Some classes of molecules found in life on earth are almost entirely left-handed. Did this left-handedness precede the origin of life? To find out, scientists analyzed meteorites from space. To correct for bias in the sensitive analysis, they also analyzed standard compounds known to be even-handed. Here are the results for the percents of left-handed forms of one molecule in two analyses:[17]

Analysis	Meteorite			Standard		
	n	$\overline{x}$	s	n	$\overline{x}$	s
1	5	52.6	0.5	14	48.8	1.9
2	10	51.7	0.4	13	49.0	1.3

The researchers used the t test to see if the meteorite had a significantly higher percent than the standard. Carry out the tests and report the results. The researchers concluded: "The observations suggest that organic matter of extraterrestrial origin could have played an essential role in the origin of terrestrial life."

17.37 **Active versus passive learning.** A study of computer-assisted learning examined the learning of "Blissymbols" by children. Blissymbols are pictographs (think of Egyptian hieroglyphs) that are sometimes used to help learning-impaired children communicate. The researcher designed two computer lessons that taught the same content using the same examples. One lesson required the children to interact with the material, while in the other the children controlled only the pace of the lesson. Call these two styles "Active" and "Passive." After the lesson, the computer presented a quiz that asked the children to identify 56 Blissymbols. Here are the numbers of correct identifications by the 24 children in the Active group:[18]

```
29  28  24  31  15  24  27  23  20  22  23  21
24  35  21  24  44  28  17  21  21  20  28  16
```

The 24 children in the Passive group had these counts of correct identifications:

```
16  14  17  15  26  17  12  25  21  20  18  21
20  16  18  15  26  15  13  17  21  19  15  12
```

(a) Is there good evidence that active learning is superior to passive learning? State hypotheses, give a test and its P-value, and state your conclusion.

(b) What conditions does your test require? Which of these conditions can you use the data to check? Examine the data and report your results.

17.38 IQ scores for boys and girls. Here are the IQ test scores of 31 seventh-grade girls in a Midwest school district:[19]

114	100	104	89	102	91	114	114	103	105
108	130	120	132	111	128	118	119	86	72
111	103	74	112	107	103	98	96	112	112	93

The IQ test scores of 47 seventh-grade boys in the same district are

111	107	100	107	115	111	97	112	104	106	113
109	113	128	128	118	113	124	127	136	106	123
124	126	116	127	119	97	102	110	120	103	115
93	123	79	119	110	110	107	105	105	110	77
90	114	106								

(a) Find the mean scores for girls and for boys. It is common for boys to have somewhat higher scores on standardized tests. Is that true here?

(b) Make stemplots or histograms of both sets of data. Because the distributions are reasonably symmetric with no extreme outliers, the t procedures will work well.

(c) Treat these data as SRSs from all seventh-grade students in the district. Is there good evidence that girls and boys differ in their mean IQ scores?

(d) What other information would you ask for before accepting the results as describing all seventh-graders in the school district?

17.39 Active versus passive learning, continued.

(a) Use the data in Exercise 17.37 to give a 90% confidence interval for the difference in mean number of Blissymbols identified correctly by children after active and passive learning.

(b) Give a 90% confidence interval for the mean number of Blissymbols identified correctly by children after the Active computer lesson.

17.40 IQ scores for boys and girls, continued. Use the data in Exercise 17.38 to give a 90% confidence interval for the difference between the mean IQ scores of all boys and girls in the district.

17.41 Coaching and SAT scores. Coaching companies claim that their courses can raise the SAT scores of high school students. Of course, students who retake the SAT without paying for coaching generally raise their scores. A random sample of students who took the SAT twice found 427 who were coached and 2733 who were uncoached.[20] Starting with their verbal scores on the first and second tries, we have these summary statistics:

(Stewart Cohen/Index Stock Imagery/ PictureQuest)

	Try 1		Try 2		Gain	
	Mean	Std. Dev.	Mean	Std. Dev.	Mean	Std. Dev.
Coached	500	92	529	97	29	59
Uncoached	506	101	527	101	21	52

Let's first ask if students who are coached significantly increased their scores.

(a) You could use the information given to carry out either a two-sample t test comparing Try 1 with Try 2 for coached students or a matched pairs t test using Gain. Which is the correct test? Why?

(b) Carry out the proper test. What do you conclude?

(c) Give a 99% confidence interval for the mean gain of all students who are coached.

17.42 Coaching and SAT scores, continued. What we really want to know is whether coached students improve more than uncoached students, and whether any advantage is large enough to be worth paying for. Use the information in the previous exercise to answer these questions.

(a) Is there good evidence that coached students gained more on the average than uncoached students?

(b) How much more do coached students gain on the average? Give a 99% confidence interval.

(c) Based on your work, what is your opinion: Do you think coaching courses are worth paying for?

17.43 Coaching and SAT scores: critique. The data you used in the previous two problems came from a random sample of students who took the SAT twice. The response rate was 63%, which is pretty good for nongovernment surveys, so let's accept that the respondents do represent all students who took the exam twice. Nonetheless, we can't be sure that coaching actually *caused* the coached students to gain more than the uncoached students. Explain briefly but clearly why this is so.

Chapter 17 MEDIA EXERCISES

17.44 Learning to solve a maze. Table 16.1 (page 423) contains the times required to complete a maze for 21 subjects wearing scented and unscented masks. Example 17.3 used the matched pairs t test to show that the scent makes no significant difference in the time. Now we ask whether there is a learning effect, so that subjects complete the maze faster on their second trial. All of the odd-numbered subjects in Table 16.1 first worked the maze wearing the unscented mask. Even-numbered subjects wore the scented mask first. The numbers were assigned at random.

(a) We will compare the unscented times for "unscented first" subjects with the unscented times for the "scented first" subjects. Explain why this comparison requires two-sample procedures.

(b) We suspect that on the average subjects are slower when the unscented time is their first trial. Make a back-to-back stemplot of unscented times for "scented first" and "unscented first" subjects. Find the mean unscented times for these two groups. Do the data appear to support our suspicion? Do the data have features that prevent use of the t procedures?

(c) Do the data give statistically significant support to our suspicion? State hypotheses, carry out a test, and report your conclusion.

17.45 **Stepping up your heart rate.** A student project asked subjects to step up and down for three minutes and measured their heart rates before and after the exercise. Here are data for five subjects and two treatments: stepping at a low rate (14 steps per minute) and at a medium rate (21 steps per minute). For each subject, we give the resting heart rate (beats per minutes) and the heart rate at the end of the exercise. (This is a partial and simplified version of the EESEE story "Stepping Up Your Heart Rate.")

	Low Rate		Medium Rate	
Subject	Resting	Final	Resting	Final
1	60	75	63	84
2	90	99	69	93
3	87	93	81	96
4	78	87	75	90
5	84	84	90	108

(a) Does exercise at the low rate raise heart rate significantly? By how much? (Use a 90% confidence interval to answer the second question.)

(b) Does exercise at the medium rate raise heart rate significantly? By how much?

(c) Does medium exercise raise heart rate more than low exercise?

(Tony Arruza/CORBIS)

Inference about a Population Proportion

Our discussion of statistical inference to this point has concerned making inferences about population *means*. Now we turn to questions about the *proportion* of some outcome in a population. Here are some examples that call for inference about population proportions.

EXAMPLE 18.1   Risky behavior in the age of AIDS

How common is behavior that puts people at risk of AIDS? The National AIDS Behavioral Surveys interviewed a random sample of 2673 adult heterosexuals. Of these, 170 had more than one sexual partner in the past year. That's 6.36% of the sample.[1] Based on these data, what can we say about the percent of all adult heterosexuals who have multiple partners? We want to *estimate a single population proportion*. This chapter concerns inference about one proportion.

EXAMPLE 18.2   Does preschool make a difference?

Do preschool programs for poor children make a difference in later life? A study looked at 62 children who were enrolled in a Michigan preschool in the late 1960s and at a control group of 61 similar children who were not enrolled. At 27 years of age, 61% of the preschool group and 80% of the control group had required the help of a social service agency (mainly welfare) in the previous ten years.[2] Is this significant evidence that preschool for poor children reduces later use of social services? We want to *compare two population proportions*. This is the topic of Chapter 19.

To do inference about a population mean μ, we use the mean $\bar{x}$ of a random sample from the population. The reasoning of inference starts with the sampling distribution of $\bar{x}$. Now we follow the same pattern, replacing means by proportions.

The sample proportion $\hat{p}$

We are interested in the unknown proportion p of a population that has some outcome. For convenience, call the outcome we are looking for a "success." In Example 18.1, the population is adult heterosexuals, and the parameter p is the proportion who have had more than one sexual partner in the past year. To estimate p, the National AIDS Behavioral Surveys used random dialing of telephone numbers to contact a sample of 2673 people. Of these, 170 said they had multiple sexual partners. The statistic that estimates the parameter p is the **sample proportion**

sample proportion

$$\hat{p} = \frac{\text{count of successes in the sample}}{\text{count of observations in the sample}}$$

$$= \frac{170}{2673} = 0.0636$$

Read the sample proportion $\hat{p}$ as "p-hat."

APPLY YOUR KNOWLEDGE

In each of the following settings:

(a) Describe the population and explain in words what the parameter p is.

(b) Give the numerical value of the statistic $\hat{p}$ that estimates p.

18.1 **Do college students pray?** A study of religious practices among college students interviewed a sample of 127 students; 107 of the students said that they prayed at least once in a while.

18.2 **Information online.** A random sample of 1318 Internet users was asked where they will go for information the next time they need information about health or medicine; 606 said that they would use the Internet.

The sampling distribution of $\hat{p}$

How good is the statistic $\hat{p}$ as an estimate of the parameter p? To find out, we ask, "What would happen if we took many samples?" The sampling distribution of $\hat{p}$ answers this question. Here are the facts.

> ### SAMPLING DISTRIBUTION OF A SAMPLE PROPORTION
>
> Choose an SRS of size n from a large population that contains population proportion p of "successes." Let $\hat{p}$ be the **sample proportion** of successes,
>
> $$\hat{p} = \frac{\text{count of successes in the sample}}{n}$$
>
> Then:
>
> - As the sample size increases, the sampling distribution of $\hat{p}$ becomes **approximately Normal.**
> - The **mean** of the sampling distribution is p.
> - The **standard deviation** of the sampling distribution is
>
> $$\sqrt{\frac{p(1-p)}{n}}$$

Figure 18.1 summarizes these facts in a form that helps you recall the big idea of a sampling distribution. The behavior of sample proportions $\hat{p}$ is similar to the behavior of sample means $\bar{x}$. When the sample size n is large, the sampling distribution is approximately Normal. The larger the sample, the more nearly Normal the distribution is. The mean of the sampling distribution is the true value of the population proportion p. That is, $\hat{p}$ is an unbiased estimator of p. The standard deviation of $\hat{p}$ gets smaller as the sample size n gets larger, so that estimation is likely to be more accurate when the sample is larger. But the standard deviation gets smaller only at the rate $\sqrt{n}$. We need four times as many observations to cut the standard deviation in half.

You should not use the Normal approximation to the distribution of $\hat{p}$ when the sample size n is small. What is more, the formula given for the standard

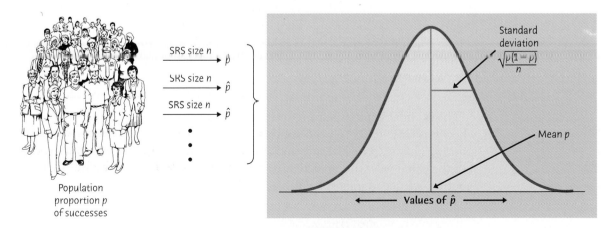

Figure 18.1 Select a large SRS from a population of which the proportion p are successes. The sampling distribution of the proportion $\hat{p}$ of successes in the sample is approximately Normal. The mean is p and the standard deviation is $\sqrt{p(1-p)/n}$.

deviation of $\hat{p}$ is not accurate unless the population is much larger than the sample—say, at least 10 times larger.[3] We will give guidelines to help you decide when methods for inference based on this sampling distribution are trustworthy.

EXAMPLE 18.3   Asking about risky behavior

Suppose that in fact 6% of all adult heterosexuals had more than one sexual partner in the past year (and would admit it when asked). The National AIDS Behavioral Surveys interviewed a random sample of 2673 people from this population. What is the probability that at least 5% of such a sample admit to having more than one partner?

If the sample size is $n = 2673$ and the population proportion is $p = 0.06$, the sample proportion $\hat{p}$ has mean 0.06 and standard deviation

$$\sqrt{\frac{p(1-p)}{n}} = \sqrt{\frac{(0.06)(0.94)}{2673}}$$

$$= \sqrt{0.0000211} = 0.00459$$

We want the probability that $\hat{p}$ is 0.05 or greater.

Standardize $\hat{p}$ by subtracting the mean 0.06 and dividing by the standard deviation 0.00459. This produces a new statistic that has approximately the standard Normal distribution. As usual, we call this statistic z:

$$z = \frac{\hat{p} - 0.06}{0.00459}$$

Figure 18.2 shows the probability we want as an area under the standard Normal curve.

$$P(\hat{p} \geq 0.05) = P\left(\frac{\hat{p} - 0.06}{0.00459} \geq \frac{0.05 - 0.06}{0.00459}\right)$$

$$= P(Z \geq -2.18)$$

$$= 1 - 0.0146 = 0.9854$$

If we repeat the National AIDS Behavioral Surveys many times, more than 98% of all the samples will contain at least 5% of respondents who admit to more than one sexual partner.

The Normal approximation for the sampling distribution of $\hat{p}$ works poorly when p is close to 0. You can see that if $p = 0$, any sample must contain only failures. That is, $\hat{p} = 0$ always and there is no Normal distribution in sight. In just the same way, the approximation works poorly when p is close to 1. In practice, we need larger n for values of p near 0 or 1.

APPLY YOUR KNOWLEDGE

18.3   Student drinking. The College Alcohol Study interviewed an SRS of 14,941 college students about their drinking habits. Suppose that

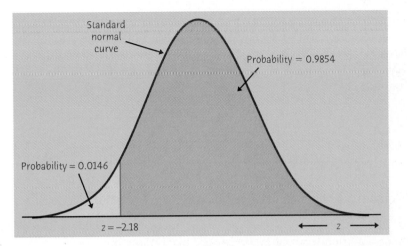

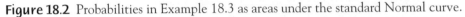

Figure 18.2 Probabilities in Example 18.3 as areas under the standard Normal curve.

half of all college students "drink to get drunk" at least once in a while. That is, $p = 0.5$.

(a) What are the mean and standard deviation of the proportion $\hat{p}$ of the sample who drink to get drunk?

(b) Use the Normal approximation to find the probability that $\hat{p}$ is between 0.49 and 0.51.

18.4   Harley motorcycles. Harley-Davidson motorcycles make up 14% of all the motorcycles registered in the United States. You plan to interview an SRS of 500 motorcycle owners.

(a) What is the approximate distribution of the proportion of your sample who own Harleys?

(b) Is your sample likely to contain 20% or more who own Harleys? Is it likely to contain at least 15% Harley owners? Do Normal probability calculations to answer these questions.

18.5   Student drinking, continued. Suppose that half of all college students drink to get drunk at least once in a while. Exercise 18.3 asks the probability that the sample proportion $\hat{p}$ estimates $p = 0.5$ within ± 1 percentage point. Find this probability for SRSs of sizes 1000, 4000, and 16,000. What general fact do your results illustrate?

Conditions for inference

Inference about a population proportion p is based on the sampling distribution of the sample proportion $\hat{p}$. More specifically, we use the **z statistic** that results from standardizing $\hat{p}$:

z statistic

$$z = \frac{\hat{p} - p}{\sqrt{\dfrac{p(1-p)}{n}}}$$

The statistic z has approximately the standard Normal distribution $N(0, 1)$ if the sample is not too small and the sample is not a large part of the entire population.

To test the hypothesis H_0: $p = p_0$, use the test statistic

$$z = \frac{\hat{p} - p_0}{\sqrt{\dfrac{p_0(1 - p_0)}{n}}}$$

This statistic has approximately the standard Normal distribution when H_0 is true.

To obtain a level C confidence interval for p, we would like to use

$$\hat{p} \pm z^* \sqrt{\frac{p(1 - p)}{n}}$$

with the critical value z^* chosen to cover the central area C under the standard Normal curve. Figure 18.3 shows why. Because we don't know the value of p, we replace the standard deviation by the **standard error of $\hat{p}$**

standard error of $\hat{p}$

$$\mathrm{SE} = \sqrt{\frac{\hat{p}(1 - \hat{p})}{n}}$$

to get a confidence interval of the familiar form

$$\text{estimate} \pm z^* \mathrm{SE}_{\text{estimate}}$$

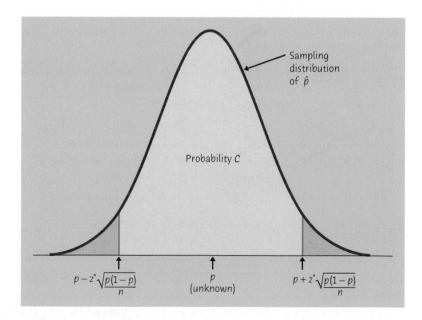

Figure 18.3 With probability C, $\hat{p}$ lies within $\pm z^* \sqrt{p(1 - p)/n}$ of the unknown population proportion p. That is, p lies within $\pm z^* \sqrt{p(1 - p)/n}$ of $\hat{p}$ in those samples.

When we estimate a mean μ, there is a separate parameter σ that describes the spread of the distribution. Estimating σ gave us the one-sample t statistic. When we estimate a proportion p, there is just one parameter, and the standard deviation of $\hat{p}$ depends on p. We don't get a t statistic—we just make the Normal approximation less accurate when we replace p by $\hat{p}$. Here is a summary of the conditions we need for inference.

CONDITIONS FOR INFERENCE ABOUT A PROPORTION

- The data are an SRS from the population of interest. This is, as usual, the most important condition.
- The population is at least 10 times as large as the sample. This condition ensures that the standard deviation of $\hat{p}$ is close to $\sqrt{p(1-p)/n}$.
- The sample size n is large enough to ensure that the distribution of z is close to standard Normal. We will see that different inference procedures require different answers to the question "how large is large enough?"

APPLY YOUR KNOWLEDGE

18.6 No inference. Tonya wants to estimate what proportion of the students in her dormitory like the dorm food. She interviews an SRS of 50 of the 175 students living in the dormitory. She finds that 14 think the dorm food is good. Tonya can't use the methods of this chapter to get a confidence interval. Why not?

18.7 No inference. A television news program conducts a call-in poll about a proposed city ban on handgun ownership. Of the 2372 calls, 1921 oppose the ban. We can't use these data as the basis for inference about the proportion of all citizens who oppose the ordinance. Why not?

Large-sample confidence intervals for a proportion

Here is the basic confidence interval for a proportion. Unfortunately, this interval can be trusted only for quite large samples. Our rule of thumb for "how large" takes into account the fact that n must be larger if the sample proportion $\hat{p}$ suggests that p may be close to 0 or 1. Because $n\hat{p}$ is just the count of successes in the sample and $n(1-\hat{p})$ is the count of failures, we can use these counts to get a simple guideline for "n is large and $\hat{p}$ is not too close to 0 or 1."

LARGE-SAMPLE CONFIDENCE INTERVAL FOR A POPULATION PROPORTION

Draw an SRS of size n from a population with unknown proportion p of successes. An approximate level C confidence interval for p is

$$\hat{p} \pm z^* \sqrt{\frac{\hat{p}(1-\hat{p})}{n}}$$

where z^* is the critical value for the standard Normal density curve with area C between $-z^*$ and z^*.

Use this interval only when the counts of successes and failures in the sample are both at least 15.[4]

EXAMPLE 18.4   Estimating risky behavior

The National AIDS Behavioral Surveys found that 170 of a sample of 2673 adult heterosexuals had multiple partners. That is, $\hat{p} = 0.0636$. We will act as if the sample were an SRS.

A 99% confidence interval for the proportion p of all adult heterosexuals with multiple partners uses the standard Normal critical value $z^* = 2.576$. (Look in the bottom row of Table C for standard Normal critical values.) The confidence interval is

$$\hat{p} \pm z^* \sqrt{\frac{\hat{p}(1-\hat{p})}{n}} = 0.0636 \pm 2.576 \sqrt{\frac{(0.0636)(0.9364)}{2673}}$$

$$= 0.0636 \pm 0.0122$$

$$= 0.0514 \text{ to } 0.0758$$

We are 99% confident that the percent of adult heterosexuals who had more than one sexual partner in the past year lies between about 5% and 7.6%.

EXAMPLE 18.5   Are the conditions met?

We used the National AIDS Behavioral Surveys data to give a confidence interval for the proportion of adult heterosexuals who have had multiple sexual partners. Does the sample meet the requirements for inference?

- The sampling design was a complex stratified sample, and the survey used inference procedures for that design. The overall effect is close to an SRS, however.

- The number of adult heterosexuals (the population) is much larger than 10 times the sample size, $n = 2673$.

- The numbers of successes (170) and failures (2503) in the sample are both much larger than 15.

The second and third requirements are easily met. The first requirement, that the sample be an SRS, is only approximately met.

As usual, the practical problems of a large sample survey pose a greater threat to the AIDS survey's conclusions. Only people in households with telephones could be reached. This is acceptable for surveys of the general population, because about 94% of American households have telephones. However, some groups at high risk for AIDS, like intravenous drug users, often don't live in settled households and are underrepresented in the sample. About 30% of the people reached refused to cooperate. A nonresponse rate of 30% is not unusual in large sample surveys, but it may cause some bias if those who refuse differ systematically from those who cooperate. The survey used statistical methods that adjust for unequal response rates in different groups. Finally, some respondents may not have told the truth when asked about their sexual behavior. The survey team tried hard to make respondents feel comfortable. For example, Hispanic women were interviewed only by Hispanic women, and Spanish speakers were interviewed by Spanish speakers with the same regional accent (Cuban, Mexican, or Puerto Rican). Nonetheless, the survey report says that some bias is probably present:

> It is more likely that the present figures are underestimates; some respondents may underreport their numbers of sexual partners and intravenous drug use because of embarrassment and fear of reprisal, or they may forget or not know details of their own or of their partner's HIV risk and their antibody testing history.[5]

Reading the report of a large study like the National AIDS Behavioral Surveys reminds us that statistics in practice involves much more than recipes for inference.

APPLY YOUR KNOWLEDGE

18.8 No confidence interval. In the National AIDS Behavioral Surveys sample of 2673 adult heterosexuals, 0.2% (that's 0.002 as a decimal fraction) had both received a blood transfusion and had a sexual partner from a group at high risk of AIDS. Explain why we can't use the large-sample confidence interval to estimate the proportion p in the population who share these two risk factors.

18.9 How common is SAT coaching? A random sample of students who took the SAT college entrance examination twice found that 427 of the respondents had paid for coaching courses and that the remaining 2733 had not.[6] Give a 99% confidence interval for the proportion of coaching among students who retake the SAT.

18.10 The millennium begins with optimism. In January of the year 2000, a Gallup Poll asked a random sample of 1633 adults, "In general, are you satisfied or dissatisfied with the way things are going in the United States at this time?" It found that 1127 said that they were satisfied. Write a short report of this finding, as if you were writing for a newspaper. Be sure to include a margin of error.

18.11 **Teens and their TV sets.** The *New York Times* and CBS News conducted a nationwide survey of 1048 randomly selected 13- to 17-year-olds. Of these teenagers, 692 had a television in their room.[7]

(a) Check that we can use the large-sample confidence interval.

(b) Give a 95% confidence interval for the proportion of all teens who have a TV set in their room.

(c) The news article says, "In theory, in 19 cases out of 20, the survey results will differ by no more than three percentage points in either direction from what would have been obtained by seeking out all American teenagers." Explain how your results agree with this statement.

Accurate confidence intervals for a proportion*

The confidence interval $\hat{p} \pm z^* \sqrt{\hat{p}(1 - \hat{p})/n}$ for a sample proportion p is easy to calculate. It is also easy to understand because it rests directly on the approximately Normal distribution of $\hat{p}$. Unfortunately, this interval is often quite inaccurate unless the sample is very large. More specifically, the actual confidence level is usually *less* than the confidence level you asked for in choosing the critical value z^*. That's bad. What is worse, accuracy does not consistently get better as the sample size n increases. There are "lucky" and "unlucky" combinations of n and the true population proportion p. Here is a quote from a recent computational study (the "standard interval" is our large-sample interval):

Who is a smoker?

When estimating a proportion p, be sure you know what counts as a "success." The news says that 20% of adolescents smoke. Shocking. It turns out that this is the percent who smoked at least once in the past month. If we say that a smoker is someone who smoked on at least 20 of the past 30 days and smoked at least half a pack on those days, fewer than 4% of adolescents qualify.

> *For instance, when n is 100, the actual coverage probability of the nominal 95% standard interval is 0.952 if p is 0.106, but only 0.911 if p is 0.107. The behavior of the coverage probability can be even more erratic as a function of n. If the true p is 0.5, the actual coverage of the nominal 95% interval is 0.953 at the rather small sample size n = 14, but falls to 0.919 at the much larger sample size n = 40.[8]*

What should we do? Fortunately, there is a simple modification that is almost magically effective in improving the accuracy of the confidence interval. We call it the "plus four" method, because all you need to do is *add four imaginary observations, two successes and two failures*. If you have X successes in your n observations, act as if you had $X + 2$ successes in $n + 4$ observations. Here is the formula.[9]

---

*Although this material is optional, it is essential unless you calculate confidence intervals for a proportion only for very large samples.

PLUS FOUR CONFIDENCE INTERVAL FOR A PROPORTION

Choose an SRS of size n from a large population that contains population proportion p of "successes." The **plus four estimate** of p is

$$\tilde{p} = \frac{\text{count of successes in the sample} + 2}{n + 4}$$

An approximate level C confidence interval for p is

$$\tilde{p} \pm z^* \sqrt{\frac{\tilde{p}(1 - \tilde{p})}{n + 4}}$$

where z^* is the critical value for the standard Normal density curve with area C between $-z^*$ and z^*.

Use this interval when C is at least 90% and the sample size n is at least 10.

EXAMPLE 18.6   Shaq's free throws

Shaquille O'Neal of the Los Angeles Lakers, the dominant center in professional basketball, has one weakness: he is a poor free-throw shooter. In his career prior to the 2000 season, Shaq made just 53.3% of his free throws. Before that season, he worked with a coach on his technique. In the first two games of the following season, he made 26 out of 39 free throws.

We can consider these as an SRS of size 39 from the population of free throws shot under game conditions with the new technique. We want a 95% confidence interval for the proportion p of free throws that Shaq will make.

First, let's calculate the large-sample interval. The sample proportion is

$$\hat{p} = \frac{26}{39} = 0.667$$

and the confidence interval based on 26 successes in 39 observations is

$$\hat{p} \pm z^* \sqrt{\frac{\hat{p}(1 - \hat{p})}{n}} = 0.667 \pm 1.960 \sqrt{\frac{(0.667)(0.333)}{39}}$$

$$= 0.667 \pm 0.148$$

We can't trust this result. The plus four estimate of p is

$$\tilde{p} = \frac{26 + 2}{39 + 4} = \frac{28}{43} = 0.651$$

The plus four confidence interval is the same as the large-sample interval based on 28 successes in 43 observations. Here it is:

$$\tilde{p} \pm z^* \sqrt{\frac{\tilde{p}(1 - \tilde{p})}{n + 4}} = 0.651 \pm 1.960 \sqrt{\frac{(0.651)(0.349)}{43}}$$

$$= 0.651 \pm 0.142$$

> We estimate with 95% confidence that Shaq will make between 50.9% and 79.3% of his free throws with the new technique.
>
> The second interval is accurate because we used the plus four method. It is too wide to be very helpful because the sample is small. We should wait for more data.

How much more accurate is the plus four interval? Computer studies have asked how large n must be to guarantee that the actual probability that a 95% confidence interval covers the true parameter value is at least 0.94 for all larger samples. If $p = 0.1$, for example, the answer is $n = 646$ for the large-sample interval and $n = 11$ for the plus four interval.[10] The consensus of computational and theoretical studies is that plus four is very much better than the large-sample interval for many combinations of n and p. **We recommend that you always use the plus four interval.**

APPLY YOUR KNOWLEDGE

18.12 **Drug-detecting rats?** Dogs are big and expensive. Rats are small and cheap. Might rats be trained to replace dogs in sniffing out illegal drugs? A first study of this idea trained rats to rear up on their hind legs when they smelled simulated cocaine. To see how well rats performed after training, they were let loose on a surface with many cups sunk in it, one of which contained simulated cocaine. Four out of six trained rats succeeded in 80 out of 80 trials.[11] If a rat succeeds in every one of 80 trials, how should we estimate that rat's long-term success rate p?

(a) Explain why the estimate $\hat{p} = 80/80 = 1$ is almost certainly too high.

(b) Find the plus four estimate $\tilde{p}$. It is more reasonable. This example shows how $\tilde{p}$ improves on $\hat{p}$ when a sample has almost all successes or almost all failures. That's why the guidelines for using the plus four interval can ignore the "near 0 or 1" issue.

18.13 **High-risk behavior.** In the National AIDS Behavioral Surveys sample of 2673 adult heterosexuals, 5 respondents had both received a blood transfusion and had a sexual partner from a group at high risk of AIDS.

(a) You should not use the large-sample confidence interval for the proportion p in the population who share these two risk factors. Why not?

(b) The plus four method adds four observations, two successes and two failures. What are the sample size and the count of successes after you do this? What is the plus four estimate $\tilde{p}$ of p?

(c) Give the plus four 95% confidence interval for p.

18.14 **Fear of crime among older black women.** The elderly fear crime more than younger people, even though they are less likely to be victims of crime. One of the few studies that looked at older blacks recruited a random sample of 56 black women over the age of 65 from

Atlantic City, New Jersey. Of these women, 27 said that they "felt vulnerable" to crime.[12]

(a) Give the two estimates $\hat{p}$ and $\tilde{p}$ of the proportion p of all elderly black women in Atlantic City who feel vulnerable to crime. There is little difference between them. This is generally true when $\hat{p}$ is not close to either 0 or 1.

(b) Give both the large-sample 95% confidence interval and the plus four 95% confidence interval for p. The plus four interval is a bit narrower. This is generally true when $\hat{p}$ is not close to either 0 or 1.

18.15 **Do college students pray?** Social scientists asked 127 undergraduate students "from courses in psychology and communications" about prayer and found that 107 prayed at least a few times a year.[13]

(a) Give the plus four 99% confidence interval for the proportion p of all students who pray.

(b) To use any inference procedure, we must be willing to regard these 127 students, as far as their religious behavior goes, as an SRS from the population of all undergraduate students. Do you think it is reasonable to do this? Why or why not?

Choosing the sample size

In planning a study, we may want to choose a sample size that will allow us to estimate the parameter within a given margin of error. We saw earlier (page 331) how to do this for a population mean. The method is similar for estimating a population proportion.

The margin of error in the large-sample confidence interval for p is

$$ m = z^* \sqrt{\frac{\hat{p}(1 - \hat{p})}{n}} $$

Here z^* is the standard Normal critical value for the level of confidence we want. Because the margin of error involves the sample proportion of successes $\hat{p}$, we need to guess this value when choosing n. Call our guess p^*. Here are two ways to get p^*:

1. Use a guess p^* based on a pilot study or on past experience with similar studies. You should do several calculations that cover the range of $\hat{p}$-values you might get.

2. Use $p^* = 0.5$ as the guess. The margin of error m is largest when $\hat{p} = 0.5$, so this guess is conservative in the sense that if we get any other $\hat{p}$ when we do our study, we will get a margin of error smaller than planned.

Once you have a guess p^*, the recipe for the margin of error can be solved to give the sample size n needed. Here is the result for the large-sample confidence interval. For simplicity, use this result even if you plan to use the plus four interval.

New York, New York

New York City, they say, is bigger, richer, faster, ruder. Maybe there's something to that. The sample survey firm Zogby International says that as a national average it takes 5 telephone calls to reach a live person. When calling to New York, it takes 12 calls. Survey firms assign their best interviewers to make calls to New York and often pay them bonuses to cope with the stress.

SAMPLE SIZE FOR DESIRED MARGIN OF ERROR

The level C confidence interval for a population proportion p will have margin of error approximately equal to a specified value m when the sample size is

$$n = \left(\frac{z^*}{m}\right)^2 p^*(1 - p^*)$$

where p^* is a guessed value for the sample proportion. The margin of error will be less than or equal to m if you take the guess p^* to be 0.5.

Which method for finding the guess p^* should you use? The n you get doesn't change much when you change p^* as long as p^* is not too far from 0.5. So use the conservative guess $p^* = 0.5$ if you expect the true $\hat{p}$ to be roughly between 0.3 and 0.7. If the true $\hat{p}$ is close to 0 or 1, using $p^* = 0.5$ as your guess will give a sample much larger than you need. So try to use a better guess from a pilot study when you suspect that $\hat{p}$ will be less than 0.3 or greater than 0.7.

EXAMPLE 18.7   **Planning a poll**

Gloria Chavez and Ronald Flynn are the candidates for mayor in a large city. You are planning a sample survey to determine what percent of the voters plan to vote for Chavez. This is a population proportion p. You will contact an SRS of registered voters in the city. You want to estimate p with 95% confidence and a margin of error no greater than 3%, or 0.03. How large a sample do you need?

The winner's share in all but the most lopsided elections is between 30% and 70% of the vote. So use the guess $p^* = 0.5$. The sample size you need is

$$n = \left(\frac{1.96}{0.03}\right)^2 (0.5)(1 - 0.5) = 1067.1$$

You should round the result up to $n = 1068$. (Rounding down would give a margin of error slightly greater than 0.03.) If you want a 2.5% margin of error, we have (after rounding up)

$$n = \left(\frac{1.96}{0.025}\right)^2 (0.5)(1 - 0.5) = 1537$$

For a 2% margin of error the sample size you need is

$$n = \left(\frac{1.96}{0.02}\right)^2 (0.5)(1 - 0.5) = 2401$$

As usual, smaller margins of error call for larger samples.

APPLY YOUR KNOWLEDGE

18.16 Canadians and doctor-assisted suicide. A Gallup Poll asked a sample of Canadian adults if they thought the law should allow doctors to end the life of a patient who is in great pain and near death if the patient makes a request in writing. The poll included 270 people in Quebec, 221 of whom agreed that doctor-assisted suicide should be allowed.[14]

    (a) What is the margin of error of the large-sample 95% confidence interval for the proportion of all Quebec adults who would allow doctor-assisted suicide?

    (b) How large a sample is needed to get the common ±3 percentage point margin of error?

18.17 Can you taste PTC? PTC is a substance that has a strong bitter taste for some people and is tasteless for others. The ability to taste PTC is inherited. About 75% of Italians can taste PTC, for example. You want to estimate the proportion of Americans with at least one Italian grandparent who can taste PTC. Starting with the 75% estimate for Italians, how large a sample must you test in order to estimate the proportion of PTC tasters within ±0.04 with 90% confidence?

Significance tests for a proportion

Tests of hypotheses about a population proportion p are based on the sampling distribution of the sample proportion $\hat{p}$, *taking the null hypothesis to be true*. Because H_0 fixes a value of p, the inaccuracy that plagues the large-sample confidence interval does not affect tests. Here is the rule for tests.

SIGNIFICANCE TESTS FOR A PROPORTION

To test the hypothesis H_0: $p = p_0$, compute the z statistic

$$z = \frac{\hat{p} - p_0}{\sqrt{\dfrac{p_0(1 - p_0)}{n}}}$$

In terms of a variable Z having the standard Normal distribution, the approximate P-value for a test of H_0 against

    H_a: $p > p_0$ is $P(Z \geq z)$

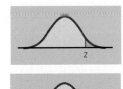

    H_a: $p < p_0$ is $P(Z \leq z)$

    H_a: $p \neq p_0$ is $2P(Z \geq |z|)$

Use this test when the sample size n is so large that both np_0 and $n(1 - p_0)$ are 10 or more.

EXAMPLE 18.8 Is this coin fair?

A coin that is balanced should come up heads half the time in the long run. The population for coin tossing contains the results of tossing the coin forever. The parameter p is the probability of a head, which is the proportion of all tosses that give a head. The tosses we actually make are an SRS from this population.

The French naturalist Count Buffon (1707–1788) tossed a coin 4040 times. He got 2048 heads. The sample proportion of heads is

$$\hat{p} = \frac{2048}{4040} = 0.5069$$

That's a bit more than one-half. Is this evidence that Buffon's coin was not balanced?

Step 1. **Hypotheses.** The null hypothesis says that the coin is balanced ($p = 0.5$). The alternative hypothesis is two-sided, because we did not suspect before seeing the data that the coin favored either heads or tails. We therefore test the hypotheses

$$H_0: p = 0.5$$
$$H_a: p \neq 0.5$$

The null hypothesis gives p the value $p_0 = 0.5$.

Step 2. **Test statistic.** The z test statistic is

$$z = \frac{\hat{p} - p_0}{\sqrt{\dfrac{p_0(1 - p_0)}{n}}}$$

$$= \frac{0.5069 - 0.5}{\sqrt{\dfrac{(0.5)(0.5)}{4040}}} = 0.88$$

Step 3. **P-value.** Because the test is two-sided, the P-value is the area under the standard Normal curve more than 0.88 away from 0 in either direction. Figure 18.4 shows this area. In Table A we read that the area below -0.88 is 0.1894. The P-value is twice this area:

$$P = 2(0.1894) = 0.3788$$

You can approximate the P-value by comparing $z = 0.88$ with the critical values in the last row of Table C. It lies between the values for tail areas 0.15 and 0.20, so the two-sided P-value lies between 0.30 and 0.40. This is enough to reach a conclusion.

Conclusion. A proportion of heads as far from one-half as Buffon's would happen more than 30% of the time when a balanced coin is tossed 4040 times. Buffon's result doesn't show that his coin is unbalanced.

In Example 18.8, we failed to find good evidence against $H_0: p = 0.5$. We *cannot* conclude that H_0 is true, that is, that the coin is perfectly balanced. No doubt p is not exactly 0.5. The test of significance shows only that the results of Buffon's 4040 tosses do not distinguish this coin from one that is perfectly

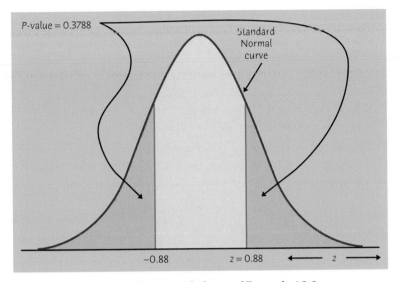

Figure 18.4 The *P*-value for the two-sided test of Example 18.8.

balanced. To see what values of *p* are consistent with the sample results, use a confidence interval.

EXAMPLE 18.9   Estimating the chance of a head

With 2048 successes in 4040 trials, the large-sample and plus four intervals will be almost identical. The 95% large-sample confidence interval for the probability *p* that Buffon's coin gives a head is

$$\hat{p} \pm z^* \sqrt{\frac{\hat{p}(1-\hat{p})}{n}} = 0.5069 \pm 1.960 \sqrt{\frac{(0.5069)(0.4931)}{4040}}$$

$$= 0.5069 \pm 0.0154$$
$$= 0.4915 \text{ to } 0.5223$$

We are 95% confident that the probability of a head is between 0.4915 and 0.5223. The confidence interval is more informative than the test in Example 18.8. We would not be surprised if the true probability of a head for Buffon's coin were something like 0.51.

APPLY YOUR KNOWLEDGE

18.18 **Spinning pennies.** Spinning a coin, unlike tossing it, may not give heads and tails equal probabilities. I spun a penny 200 times and got 83 heads. How significant is this evidence against equal probabilities? State hypotheses, give the test statistic, use Table C to approximate its *P*-value, and state your conclusion.

18.19 **Teens and their TV sets.** A random sample of 1048 13- to 17-year-olds found that 692 had a television set in their room. Is this good evidence that more than half of all teens have a TV in their room? State hypotheses, give the test statistic, use Table A to find its *P*-value, and state your conclusion.

18.20 **No test.** Explain why we can't use the z test for a proportion in these situations:

(a) You toss a coin 10 times in order to test the hypothesis $H_0: p = 0.5$ that the coin is balanced.

(b) A college president says, "99% of the alumni support my firing of Coach Boggs." You contact an SRS of 200 of the college's 15,000 living alumni to test the hypothesis $H_0 : p = 0.99$.

Chapter 18 SUMMARY

Tests and confidence intervals for a population proportion p when the data are an SRS of size n are based on the **sample proportion $\hat{p}$.**

When n is large, $\hat{p}$ has approximately the Normal distribution with mean p and standard deviation $\sqrt{p(1-p)/n}$.

The level C **large-sample confidence interval** for p is

$$\hat{p} \pm z^*\sqrt{\frac{\hat{p}(1-\hat{p})}{n}}$$

where z^* is the critical value for the standard Normal curve with area C between z^* and $-z^*$.

The true confidence level of the large-sample interval can be substantially less than the planned level C unless the sample is very large. We recommend using the plus four interval instead.

To get a more accurate confidence interval, add four imaginary observations, two successes and two failures, to your sample. Then use the same formula for the confidence interval. This is the **plus four confidence interval.** Use this interval in practice for confidence level 90% or higher and sample size n at least 10.

The **sample size** needed to obtain a confidence interval with approximate margin of error m for a population proportion is

$$n = \left(\frac{z^*}{m}\right)^2 p^*(1-p^*)$$

where p^* is a guessed value for the sample proportion $\hat{p}$, and z^* is the standard Normal critical point for the level of confidence you want. If you use $p^* = 0.5$ in this formula, the margin of error of the interval will be less than or equal to m no matter what the value of $\hat{p}$ is.

Significance tests of H_0: $p = p_0$ are based on the **z statistic**

$$z = \frac{\hat{p} - p_0}{\sqrt{\dfrac{p_0(1 - p_0)}{n}}}$$

with P-values calculated from the standard Normal distribution. Use this test when $np_0 \geq 10$ and $n(1 - p_0) \geq 10$.

Chapter 18 EXERCISES

We recommend using the plus four method for all confidence intervals for a proportion. However, the large-sample method is acceptable when the guidelines for its use are met.

18.21 **Information online.** A random sample of 1318 Internet users was asked where they will go for information the next time they need information about health or medicine; 606 said that they would use the Internet.[15] Give a 99% confidence interval for the proportion of all Internet users who feel this way. Be sure to check that the conditions for use of your method are met.

18.22 **Seat belt use.** The proportion of drivers who use seat belts depends on things like age, gender, ethnicity, and local law. As part of a broader study, investigators observed a random sample of 117 female Hispanic drivers in Boston; 68 of these drivers were wearing seat belts.[16] Give a 95% confidence interval for the proportion of all female Hispanic drivers in Boston who wear seat belts. Be sure to check that the conditions for use of your method are met.

18.23 **Attitudes toward nuclear power.** A Gallup Poll on energy use asked 512 randomly selected adults if they favored "increasing the use of nuclear power as a major source of energy." Gallup reported that 225 said "Yes." Does this poll give good evidence that fewer than half of all adults favor increased use of nuclear power? State hypotheses, give the test statistic, use Table C to approximate its P-value, and state your conclusion.

18.24 **Seat belt use, continued.** Do the data in Exercise 18.22 give good reason to conclude that more than half of Hispanic female drivers in Boston wear seat belts? State hypotheses, give the test statistic, use Table C to approximate its P-value, and state your conclusion.

18.25 **Student drinking.** The College Alcohol Study interviewed an SRS of 14,941 college students about their drinking habits. The sample was stratified using 140 colleges as strata, but the overall effect is close to an SRS of students. The response rate was between 60% and 70% at most colleges. This is quite good for a national sample, though nonresponse is as usual the biggest weakness of this survey. Of the students in the sample, 10,010 supported cracking down on underage drinking.[17] Give a 99% confidence interval for the proportion of all college students who feel this way.

18.26 Running red lights. A random-digit dialing telephone survey of 880 drivers asked, "Recalling the last ten traffic lights you drove through, how many of them were red when you entered the intersections?" Of the 880 respondents, 171 admitted that at least one light had been red.[18]

(a) Give a 95% confidence interval for the proportion of all drivers who ran one or more of the last ten red lights they met.

(b) Nonresponse is a practical problem for this survey—only 21.6% of calls that reached a live person were completed. Another practical problem is that people may not give truthful answers. What is the likely direction of the bias: do you think more or fewer than 171 of the 880 respondents really ran a red light? Why?

18.27 Detecting genetically modified soybeans. Most soybeans grown in the United States are genetically modified to, for example, resist pests and so reduce use of pesticides. Because some nations do not accept genetically modified (GM) foods, grain-handling facilities routinely test soybean shipments for the presence of GM beans. In a study of the accuracy of these tests, researchers submitted lots of soybeans containing 1% GM beans to 23 randomly selected facilities. Eighteen detected the GM beans.[19]

(a) Show that the conditions for the large-sample confidence interval are not met. Show that the conditions for the plus four interval are met.

(b) Use the plus four method to give a 90% confidence interval for the percent of all grain-handling facilities that will correctly detect 1% of GM beans in a shipment.

(Bettman/CORBIS)

18.28 Equality for women? Have efforts to promote equality for women gone far enough in the United States? A poll on this issue by the cable network MSNBC contacted 1019 adults. A newspaper article about the poll said, "Results have a margin of sampling error of plus or minus 3 percentage points."[20]

(a) Overall, 54% of the sample (550 of 1019 people) answered "Yes." Find a 95% confidence interval for the proportion in the adult population who would say "Yes" if asked. Is the report's claim about the margin of error roughly right? (Assume that the sample is an SRS.)

(b) The news article said that 65% of men, but only 43% of women, think that efforts to promote equality have gone far enough. Explain why we do not have enough information to give confidence intervals for men and women separately.

(c) Would a 95% confidence interval for women alone have a margin of error less than 0.03, about equal to 0.03, or greater than 0.03? Why? You see that the news article's statement about the margin of error for poll results is a bit misleading.

18.29 **The IRS plans an SRS.** The Internal Revenue Service plans to examine an SRS of individual federal income tax returns from each state. One variable of interest is the proportion of returns claiming itemized deductions. The total number of tax returns in a state varies from more than 13 million in California to fewer than 220,000 in Wyoming.

(a) Will the sampling variability of the sample proportion change from state to state if an SRS of 2000 tax returns is selected in each state? Explain your answer.

(b) Will the sampling variability of the sample proportion change from state to state if an SRS of 1% of all tax returns is selected in each state? Explain your answer.

18.30 **Condom usage.** The National AIDS Behavioral Surveys (Example 18.1) also interviewed a sample of adults in the cities where AIDS is most common. This sample included 803 heterosexuals who reported having more than one sexual partner in the past year. We can consider this an SRS of size 803 from the population of all heterosexuals in high-risk cities who have multiple partners. These people risk infection with the AIDS virus. Yet 304 of the respondents said they never use condoms. Is this strong evidence that more than one-third of this population never use condoms?

18.31 **Going to church.** A Gallup Poll asked a sample of 1785 adults, "Did you, yourself, happen to attend church or synagogue in the last 7 days?" Of the respondents, 750 said "Yes." Treat Gallup's sample as an SRS of all American adults. Give a 99% confidence interval for the proportion of all adults who claim that they attended church or synagogue during the week preceding the poll. (The proportion who actually attended is no doubt lower—some people say "Yes" if they usually attend, often attend, or sometimes attend.)

(Dave Bartruff/CORBIS)

18.32 **Going to church, continued.** Do the results of the poll in Exercise 18.31 provide good evidence that fewer than half of the population would claim to have attended church or synagogue?

18.33 **Going to church, continued.** How large a sample would be required to obtain a margin of error of 0.01 in a 99% confidence interval for the proportion who claim to have attended church or synagogue (see Exercise 18.31)? (Use the conservative guess $p^* = 0.5$, and explain why this method is reasonable in this situation.)

18.34 **Small-business failures.** A study of the survival of small businesses chose an SRS from the telephone directory's Yellow Pages listings of food-and-drink businesses in 12 counties in central Indiana. For various reasons, the study got no response from 45% of the businesses chosen. Interviews were completed with 148 businesses. Three years later, 22 of these businesses had failed.[21]

(a) Give a 95% confidence interval for the percent of all small businesses in this class that fail within three years.

(b) Based on the results of this study, how large a sample would you need to reduce the margin of error to 0.04?

(c) The authors hope that their findings describe the population of all small businesses. What about the study makes this unlikely? What population do you think the study findings describe?

(Daly and Newton/The Image Bank/Getty Images)

18.35 **Matched pairs.** One-sample procedures for proportions, like those for means, are used to analyze data from matched pairs designs. Here is an example.

Each of 50 subjects tastes two unmarked cups of coffee and says which he or she prefers. One cup in each pair contains instant coffee; the other, fresh-brewed coffee. Thirty-one of the subjects prefer the fresh-brewed coffee. Take p to be the proportion of the population who would prefer fresh-brewed coffee in a blind tasting.

(a) Test the claim that a majority of people prefer the taste of fresh-brewed coffee. State hypotheses and report the z statistic and its P-value. Is your result significant at the 5% level? What is your practical conclusion?

(b) Find a 90% confidence interval for p.

(c) When you do an experiment like this, in what order should you present the two cups of coffee to the subjects?

18.36 **Customer satisfaction.** An automobile manufacturer would like to know what proportion of its customers are not satisfied with the service provided by the local dealer. The customer relations department will survey a random sample of customers and compute a 99% confidence interval for the proportion who are not satisfied.

(a) Past studies suggest that this proportion will be about 0.2. Find the sample size needed if the margin of error of the confidence interval is to be about 0.015.

(b) When the sample is actually contacted, 10% of the sample say they are not satisfied. What is the margin of error of the 99% confidence interval?

18.37 **Surveying students.** You are planning a survey of students at a large university to determine what proportion favor an increase in student fees to support an expansion of the student newspaper. Using records provided by the registrar, you can select a random sample of students. You will ask each student in the sample whether he or she is in favor of the proposed increase. Your budget will allow a sample of 100 students.

(a) For a sample of size 100, construct a table of the margins of error for 95% confidence intervals when $\hat{p}$ takes the values 0.1, 0.2, 0.3, 0.4, 0.5, 0.6, 0.7, 0.8, and 0.9.

(b) A former editor of the student newspaper offers to provide funds for a sample of size 500. Repeat the margin of error calculations in (a) for the larger sample size. Then write a short thank-you note to the former editor describing how the larger sample size will improve the results of the survey.

18.38 Alternative medicine. A nationwide random survey of 1500 adults asked about attitudes toward "alternative medicine" such as acupuncture, massage therapy, and herbal therapy. Among the respondents, 660 said they would use alternative medicine if traditional medicine was not producing the results they wanted.[22]

(a) Give a 95% confidence interval for the proportion of all adults who would use alternative medicine.

(b) Write a short paragraph for a news report based on the survey results.

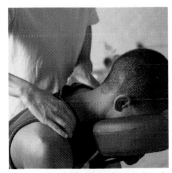

(Duncan Smith/PhotoDisc/PictureQuest)

Chapter 18 MEDIA EXERCISES

18.39 The weevils are coming. The imported longhorn weevil is a flightless insect that is a major pest of red clover. The EESEE story "Seasonal Weevil Migration" gives data from insect traps placed between a field of red clover and an adjacent woods. At one season (combining data from two years), the traps caught 55 weevils moving from the woods to the clover and 19 weevils moving from the clover to the woods. Use a confidence interval to estimate the proportion of migrating weevils in this season that move toward the clover.

(Duomo/CORBIS)

Comparing Two Proportions

In a **two-sample problem,** we want to compare two populations or the responses to two treatments based on two independent samples. When the comparison involves the *mean* of a quantitative variable, we use the two-sample t methods of Chapter 17. To compare the *standard deviations* of a variable in two groups, we use (under restrictive conditions) the F statistic also described in Chapter 17. Now we turn to methods to compare the *proportions* of successes in two groups.

Two-sample problems: proportions

We will use notation similar to that used in our study of two-sample t statistics. The groups we want to compare are Population 1 and Population 2. We have a separate SRS from each population or responses from two treatments in a randomized comparative experiment. A subscript shows which group a parameter or statistic describes. Here is our notation:

Population	Population proportion	Sample size	Sample proportion
1	p_1	n_1	$\hat{p}_1$
2	p_2	n_2	$\hat{p}_2$

We compare the populations by doing inference about the difference $p_1 - p_2$ between the population proportions. The statistic that estimates this difference is the difference between the two sample proportions, $\hat{p}_1 - \hat{p}_2$.

EXAMPLE 19.1   Does preschool help?

To study the long-term effects of preschool programs for poor children, the High/Scope Educational Research Foundation has followed two groups of Michigan children since early childhood.[1] One group of 62 attended preschool as 3- and 4-year-olds. This is a sample from Population 2, poor children who attend preschool. A control group of 61 children from the same area and similar backgrounds represents Population 1, poor children with no preschool. Thus the sample sizes are $n_1 = 61$ and $n_2 = 62$.

One response variable of interest is the need for social services as adults. In the past ten years, 38 of the preschool sample and 49 of the control sample have needed social services (mainly welfare). The sample proportions are

$$\hat{p}_1 = \frac{49}{61} = 0.803$$

$$\hat{p}_2 = \frac{38}{62} = 0.613$$

That is, about 80% of the control group uses social services, as opposed to about 61% of the preschool group.

To see if the study provides significant evidence that preschool reduces the later need for social services, we test the hypotheses

$$H_0: p_1 - p_2 = 0 \text{ (the same as } H_0: p_1 = p_2)$$
$$H_a: p_1 - p_2 > 0 \text{ (the same as } H_a: p_1 > p_2)$$

To estimate how large the reduction is, we give a confidence interval for the difference, $p_1 - p_2$. Both the test and the confidence interval start from the difference of sample proportions:

$$\hat{p}_1 - \hat{p}_2 = 0.803 - 0.613 = 0.190$$

The sampling distribution of a difference between proportions

To use $\hat{p}_1 - \hat{p}_2$ for inference, we must know its sampling distribution. Here are the facts we need:

- When the samples are large, the distribution of $\hat{p}_1 - \hat{p}_2$ is **approximately Normal.**

- The **mean** of $\hat{p}_1 - \hat{p}_2$ is $p_1 - p_2$. That is, the difference between sample proportions is an unbiased estimator of the difference between population proportions.

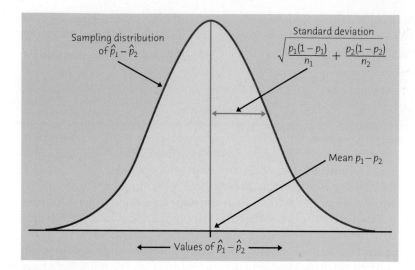

Figure 19.1 Select independent SRSs from two populations having proportions of successes p_1 and p_2. The proportions of successes in the two samples are $\hat{p}_1$ and $\hat{p}_2$. When the samples are large, the sampling distribution of the difference $\hat{p}_1 - \hat{p}_2$ is approximately Normal.

- The **standard deviation** of the difference is

$$\sqrt{\frac{p_1(1 - p_1)}{n_1} + \frac{p_2(1 - p_2)}{n_2}}$$

Figure 19.1 displays the distribution of $\hat{p}_1 - \hat{p}_2$. The standard deviation of $\hat{p}_1 - \hat{p}_2$ involves the unknown parameters p_1 and p_2. Just as in the previous chapter, we must replace these by estimates in order to do inference. And just as in the previous chapter, we do this a bit differently for confidence intervals and for tests.

Large-sample confidence intervals for comparing proportions

standard error   To obtain a confidence interval, replace the population proportions p_1 and p_2 in the standard deviation by the sample proportions. The result is the **standard error** of the statistic $\hat{p}_1 - \hat{p}_2$:

$$\text{SE} = \sqrt{\frac{\hat{p}_1(1 - \hat{p}_1)}{n_1} + \frac{\hat{p}_2(1 - \hat{p}_2)}{n_2}}$$

The confidence interval again has the form

$$\text{estimate} \pm z^* \text{SE}_{\text{estimate}}$$

LARGE-SAMPLE CONFIDENCE INTERVAL FOR COMPARING TWO PROPORTIONS

Draw an SRS of size n_1 from a population having proportion p_1 of successes and draw an independent SRS of size n_2 from another population having proportion p_2 of successes. When n_1 and n_2 are large, an approximate level C confidence interval for $p_1 - p_2$ is

$$(\hat{p}_1 - \hat{p}_2) \pm z^* SE$$

In this formula the standard error SE of $\hat{p}_1 - \hat{p}_2$ is

$$SE = \sqrt{\frac{\hat{p}_1(1 - \hat{p}_1)}{n_1} + \frac{\hat{p}_2(1 - \hat{p}_2)}{n_2}}$$

and z^* is the critical value for the standard Normal density curve with area C between $-z^*$ and z^*.

Use this interval only when the populations are at least 10 times as large as the samples and the counts of successes and failures are each 10 or more in both samples.

EXAMPLE 19.2   How much does preschool help?

Example 19.1 describes a study of the effect of preschool on later use of social services. Here are the facts:

Population	Population description	Sample size	Count of successes	Sample proportion
1	control	$n_1 = 61$	49	$\hat{p}_1 = 49/61 = 0.803$
2	preschool	$n_2 = 62$	38	$\hat{p}_2 = 38/62 = 0.613$

To check that our approximate confidence interval is safe, look at the counts of successes and failures in the two samples. The smallest of these four quantities is the count of failures in the control group, $61 - 49 = 12$. This is larger than 10, so the interval will be reasonably accurate.

The difference $p_1 - p_2$ measures the effect of preschool in reducing the proportion of people who later need social services. To compute a 95% confidence interval for $p_1 - p_2$, first find the standard error

$$SE = \sqrt{\frac{\hat{p}_1(1 - \hat{p}_1)}{n_1} + \frac{\hat{p}_2(1 - \hat{p}_2)}{n_2}}$$

$$= \sqrt{\frac{(0.803)(0.197)}{61} + \frac{(0.613)(0.387)}{62}}$$

$$= \sqrt{0.00642} = 0.0801$$

The 95% confidence interval is

$$(\hat{p}_1 - \hat{p}_2) \pm z^*\text{SE} = (0.803 - 0.613) \pm (1.960)(0.0801)$$
$$= 0.190 \pm 0.157$$
$$= 0.033 \text{ to } 0.347$$

We are 95% confident that the percent needing social services is somewhere between 3.3 and 34.7 percentage points lower among people who attended preschool. The confidence interval is wide because the samples are quite small.

The researchers in the study of Example 19.1 selected two separate samples from the two populations they wanted to compare. Many comparative studies start with just one sample, then divide it into two groups based on data gathered from the subjects. For example, an opinion poll may contact a random sample of adults, then compare the opinions of the women and the men in the sample. The two-sample z procedures for comparing proportions are valid in such situations. This is an important fact about these methods.

Using technology

Figure 19.2 displays software output for Example 19.2 from Minitab and the TI-83 calculator. As usual, you can understand the output even without knowledge of the program that produced it. The Excel spreadsheet lacks menu items for inference about proportions. You must use the spreadsheet's formula capability to program the confidence interval or test statistic and then to find the P-value of a test.

APPLY YOUR KNOWLEDGE

19.1 **Fear of crime among older blacks.** The elderly fear crime more than younger people, even though they are less likely to be victims of crime. One of the few studies that looked at older blacks recruited random samples of 56 black women and 63 black men over the age of 65 from Atlantic City, New Jersey. Of the women, 27 said they "felt vulnerable" to crime; 46 of the men said this.[2]

   (a) What proportion of women in the sample feel vulnerable? Of men? Men are victims of crime more often than women, so we expect a higher proportion of men to feel vulnerable.

   (b) Give a 95% confidence interval for the difference (men minus women).

19.2 **How to quit smoking.** Nicotine patches are often used to help smokers quit. Does giving medicine to fight depression help? A randomized double-blind experiment assigned 244 smokers who wanted to stop to

Minitab

```
┌─ Session ─────────────────────────────────────────── _|□|× ─┐
│                                                          ▲  │
│                                                             │
│  Test and CI for Two Proportions                            │
│                                                             │
│  Sample     X      N   Sample p                             │
│  1          49     61  0.803279                             │
│  2          38     62  0.612903                             │
│                                                             │
│  Estimate for p(1) - p(2):  0.190375                        │
│  95% CI for p(1) - p(2):  (0.0333680, 0.347383)             │
│  Test for p(1) - p(2) = 0 (vs not = 0):  Z = 2.38  P-Value = 0.017 │
│                                                          ▼  │
└─────────────────────────────────────────────────────────────┘
```

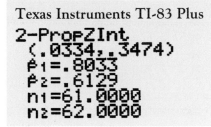

Texas Instruments TI-83 Plus

```
2-PropZInt
 (.0334,.3474)
 p̂1=.8033
 p̂2=.6129
 n1=61.0000
 n2=62.0000
```

Figure 19.2 The 95% confidence interval of Example 19.2. Output from Minitab and the TI-83 graphing calculator.

receive nicotine patches and another 245 to receive both a patch and the antidepression drug bupropion. Results: After a year, 40 subjects in the nicotine patch group had abstained from smoking, as had 87 in the patch-plus-drug group.[3] Give a 99% confidence interval for the difference (treatment minus control) in the proportion of smokers who quit.

Accurate confidence intervals for comparing proportions*

Like the large-sample confidence interval for a single proportion p, the large-sample interval for $p_1 - p_2$ generally has true confidence level less than the level you asked for. The inaccuracy is not as serious as in the one-sample case, at least if our guidelines for use are followed. Once again, a simple modification improves the accuracy, so much so that the resulting interval can be used for

---

*Although this material is optional, it is essential if you wish to calculate confidence intervals for a difference between proportions for small samples.

plus four interval

sample sizes as small as 5 in each group, with no restrictions on the counts of successes and failures.

We call this interval the **plus four interval** because you *add four imaginary observations, one success and one failure in each of the two samples.* Then just use the large-sample formula for the new data.[4] Let's apply the plus four method to the data on the effect of preschool.

Computer-assisted interviewing

The days of the interviewer with a clipboard are past. Contemporary interviewers read questions from a computer screen and use the keyboard to enter responses. The computer skips irrelevant items—once a woman says that she has no children, further questions about her children never appear. The computer can even present questions in random order to avoid bias due to always following the same order. Computer software keeps records of who has responded and prepares a file of data from the responses. The tedious process of transferring responses from paper to computer, once a source of errors, has disappeared.

| **EXAMPLE 19.3** | How much does preschool help? |

In Example 19.2, we applied the large-sample interval to a study of the effect of preschool on later use of social services. Here are the facts again:

Population	Population description	Sample size	Count of successes	Sample proportion
1	control	$n_1 = 61$	49	$\hat{p}_1 = 49/61 = 0.803$
2	preschool	$n_2 = 62$	38	$\hat{p}_2 = 38/62 = 0.613$

The plus four method adds four imaginary observations. The new facts are:

Population	Population description	Sample size	Count of successes	Plus four sample proportion
1	control	$n_1 + 2 = 63$	$49 + 1 = 50$	$\tilde{p}_1 = 50/63 = 0.794$
2	preschool	$n_2 + 2 = 64$	$38 + 1 = 39$	$\tilde{p}_2 = 39/64 = 0.609$

The standard error based on the new facts is

$$
\begin{aligned}
SE &= \sqrt{\frac{\tilde{p}_1(1 - \tilde{p}_1)}{n_1 + 2} + \frac{\tilde{p}_2(1 - \tilde{p}_2)}{n_2 + 2}} \\
&= \sqrt{\frac{(0.794)(0.206)}{63} + \frac{(0.609)(0.391)}{64}} \\
&= \sqrt{0.006317} = 0.0795
\end{aligned}
$$

The plus four 95% confidence interval is

$$
\begin{aligned}
(\tilde{p}_1 - \tilde{p}_2) \pm z^* SE &= (0.794 - 0.609) \pm (1.960)(0.0795) \\
&= 0.185 \pm 0.156 \\
&= 0.029 \text{ to } 0.341
\end{aligned}
$$

The plus four interval does not differ greatly from the large-sample interval in Example 19.2 because the sample sizes are large enough to satisfy our guidelines for use of the large-sample method.

The plus four interval may be conservative (that is, the true confidence level may be *higher* than you asked for) for very small samples and population p's close to 0 or 1. It is generally much more accurate than the large-sample interval when the samples are small.

APPLY YOUR KNOWLEDGE

19.3 Drug testing in schools. In 2002 the Supreme Court ruled that schools could require random drug tests of students participating in competitive after-school activities such as athletics. Does drug testing reduce use of illegal drugs? A study compared two similar high schools in Oregon. Wahtonka High School tested athletes at random and Warrenton High School did not. In a confidential survey, 7 of 135 athletes at Wahtonka and 27 of 141 athletes at Warrenton said they were using drugs.[5] Regard these athletes as SRSs from the populations of athletes at similar schools with and without drug testing.

  (a) You should not use the large-sample confidence interval. Why not?

  (b) The plus four method adds two observations, a success and a failure, to each sample. What are the sample sizes and the counts of drug users after you do this?

  (c) Give the plus four 95% confidence interval for the difference between the proportion of athletes using drugs at schools with and without testing.

19.4 In-line skaters. A study of injuries to in-line skaters used data from the National Electronic Injury Surveillance System, which collects data from a random sample of hospital emergency rooms. The researchers interviewed 161 people who came to emergency rooms with injuries from in-line skating. Wrist injuries (mostly fractures) were the most common.[6]

(Ron Chapple/Thinkstock/PictureQuest)

  (a) The interviews found that 53 people were wearing wrist guards and 6 of these had wrist injuries. Of the 108 who did not wear wrist guards, 45 had wrist injuries. Why should we not use the large-sample confidence interval for these data?

  (b) Give the plus four 95% confidence interval for the difference between the two population proportions of wrist injuries. State carefully what populations your inference compares. We would like to draw conclusions about all in-line skaters, but we have data only for injured skaters.

Significance tests for comparing proportions

An observed difference between two sample proportions can reflect a difference in the populations, or it may just be due to chance variation in random sampling. Significance tests help us decide if the effect we see in the samples is really there in the populations. The null hypothesis says that there is no difference between the two populations:

$$H_0: p_1 = p_2$$

The alternative hypothesis says what kind of difference we expect.

EXAMPLE 19.4 Choosing a mate

"Would you marry a person from a lower social class than your own?" Researchers asked this question of a sample of 385 black, never-married students at two historically black colleges in the South. We will consider this to be an SRS of black students at historically black colleges. Of the 149 men in the sample, 91 said "Yes." Among the 236 women, 117 said "Yes."[7] (The men and women in a single SRS can be treated as if they were separate SRSs of men and women students.)

The sample proportions who would marry someone from a lower social class are

$$\hat{p}_1 = \frac{91}{149} = 0.611 \quad \text{(men)}$$

$$\hat{p}_2 = \frac{117}{236} = 0.496 \quad \text{(women)}$$

That is, about 61% of the men but only about 50% of the women would marry beneath their class. Is this apparent difference statistically significant? We had no direction for the difference in mind before looking at the data, so we have a two-sided alternative:

$$H_0: p_1 = p_2$$
$$H_a: p_1 \neq p_2$$

To do a test, standardize $\hat{p}_1 - \hat{p}_2$ to get a z statistic. If H_0 is true, all the observations in both samples come from a single population of students of whom a single unknown proportion p would marry someone from a lower social class. So instead of estimating p_1 and p_2 separately, we pool the two samples and use the overall sample proportion to estimate the single population parameter p. Call this the **pooled sample proportion**. It is

pooled sample proportion

$$\hat{p} = \frac{\text{count of successes in both samples combined}}{\text{count of observations in both samples combined}}$$

Use $\hat{p}$ in place of both $\hat{p}_1$ and $\hat{p}_2$ in the expression for the standard error SE of $\hat{p}_1 - \hat{p}_2$ to get a z statistic that has the standard Normal distribution when H_0 is true. Here is the test.

SIGNIFICANCE TEST FOR COMPARING TWO PROPORTIONS

To test the hypothesis

$$H_0: p_1 = p_2$$

first find the pooled proportion $\hat{p}$ of successes in both samples combined. Then compute the z statistic

$$z = \frac{\hat{p}_1 - \hat{p}_2}{\sqrt{\hat{p}(1 - \hat{p})\left(\dfrac{1}{n_1} + \dfrac{1}{n_2}\right)}}$$

In terms of a variable Z having the standard Normal distribution, the P-value for a test of H_0 against

$H_a: p_1 > p_2$ is $P(Z \geq z)$

$H_a: p_1 < p_2$ is $P(Z \leq z)$

$H_a: p_1 \neq p_2$ is $2P(Z \geq |z|)$

Use this test when the populations are at least 10 times as large as the samples and the counts of successes and failures are each 5 or more in both samples.

EXAMPLE 19.5   Choosing a mate, *continued*

The pooled proportion of students who would marry beneath their own social class is

$$\hat{p} = \frac{\text{count of "Yes" responses among men and women combined}}{\text{count of subjects among men and women combined}}$$

$$= \frac{91 + 117}{149 + 236}$$

$$= \frac{208}{385} = 0.5403$$

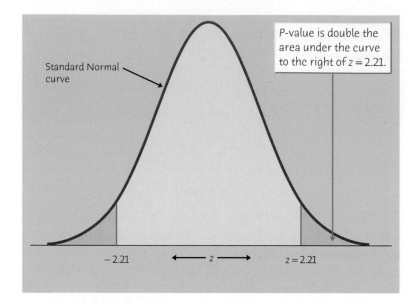

Figure 19.3 The *P*-value for the two-sided test of Example 19.5.

The *z* test statistic is

$$z = \frac{\hat{p}_1 - \hat{p}_2}{\sqrt{\hat{p}(1 - \hat{p})\left(\dfrac{1}{n_1} + \dfrac{1}{n_2}\right)}}$$

$$= \frac{0.611 - 0.496}{\sqrt{(0.5403)(0.4597)\left(\dfrac{1}{149} + \dfrac{1}{236}\right)}}$$

$$= \frac{0.115}{0.05215} = 2.21$$

The two-sided *P*-value is the area under the standard Normal curve more than 2.21 distant from 0. Figure 19.3 shows this area. Table A gives $P = 0.0272$. Alternatively, $z = 2.21$ lies between the critical values for tail areas 0.01 and 0.02 in the last row of Table C. The two-sided *P*-value is therefore between 0.02 and 0.04.

Because $P < 0.05$, the results are statistically significant at the $\alpha = 0.05$ level. There is good evidence that men are more likely than women to say they will marry someone from a lower social class.

APPLY YOUR KNOWLEDGE

19.5 Drug testing in schools, continued. Exercise 19.3 describes a study that compared the proportions of athletes who use illegal drugs in two similar high schools, one that tests for drugs and one that does not.

Drug testing is intended to reduce use of drugs. Do the data give good reason to think that drug use among athletes is lower in schools that test for drugs? State hypotheses, find the test statistic, and use Table C to approximate the P-value. Be sure to state your conclusion. (Because the study is not an experiment, the conclusion depends on the assumption that athletes in these two schools can be considered SRSs from all similar schools.)

19.6 How to quit smoking, continued. Exercise 19.2 describes a randomized comparative experiment to test whether adding medicine to fight depression increases the effectiveness of nicotine patches in helping smokers to quit. How significant is the evidence that the medicine increases the success rate? State hypotheses, calculate a test statistic, use Table A to give its P-value, and state your conclusion.

19.7 The Gold Coast. A historian examining British colonial records for the Gold Coast in Africa suspects that the death rate was higher among African miners than among European miners. In the year 1936, there were 223 deaths among 33,809 African miners and 7 deaths among 1541 European miners on the Gold Coast.[8]

Consider this year as a sample from the pre-war era in Africa. Is there good evidence that the proportion of African miners who died was higher than the proportion of European miners who died? State hypotheses, calculate a test statistic, use Table C to give a P-value, and state your conclusion.

Chapter 19 SUMMARY

We want to compare the proportions p_1 and p_2 of successes in two populations. The comparison is based on the difference $\hat{p}_1 - \hat{p}_2$ between the sample proportions of successes. When the sample sizes n_1 and n_2 are large enough, we can use z procedures because the sampling distribution of $\hat{p}_1 - \hat{p}_2$ is close to Normal.

The level C **large-sample confidence interval** for $p_1 - p_2$ is

$$(\hat{p}_1 - \hat{p}_2) \pm z^*\text{SE}$$

where the **standard error** of $\hat{p}_1 - \hat{p}_2$ is

$$\text{SE} = \sqrt{\frac{\hat{p}_1(1 - \hat{p}_1)}{n_1} + \frac{\hat{p}_2(1 - \hat{p}_2)}{n_2}}$$

and z^* is a standard Normal critical value.

The true confidence level of the large-sample interval can be substantially less than the planned level C. Use this interval only if the counts of successes and failures in both samples are 10 or greater.

To get a more accurate confidence interval, add four imaginary observations, one success and one failure in each sample. Then use the same formula for the confidence interval. This is the **plus four confidence interval.** You can use it whenever both samples have 5 or more observations.

Significance tests of H_0: $p_1 = p_2$ use the **pooled sample proportion**

$$\hat{p} = \frac{\text{count of successes in both samples combined}}{\text{count of observations in both samples combined}}$$

and the z **statistic**

$$z = \frac{\hat{p}_1 - \hat{p}_2}{\sqrt{\hat{p}(1 - \hat{p})\left(\dfrac{1}{n_1} + \dfrac{1}{n_2}\right)}}$$

P-values come from the standard Normal distribution. Use this test when there are 5 or more successes and 5 or more failures in both samples.

Chapter 19 EXERCISES

We recommend using the plus four method for all confidence intervals for proportions. However, the large-sample method is acceptable when the guidelines for its use are met.

19.8 Information online. A random-digit dialing sample of 2092 adults found that 1318 used the Internet.[9] Of the users, 1041 said that they expect businesses to have Web sites that give product information; 294 of the 774 nonusers said this.

(a) Give a 95% confidence interval for the proportion of all adults who use the Internet.

(b) Give a 95% confidence interval to compare the proportions of users and nonusers who expect businesses to have Web sites.

Call a statistician. Does involving a statistician to help with statistical methods improve the chance that a medical research paper will be published? A study of papers submitted to two medical journals found that 135 of 190 papers that lacked statistical assistance were rejected without even being reviewed in detail. In contrast, 293 of the 514 papers with statistical help were sent back without review.[10] Exercises 19.9 to 19.11 are based on this study.

19.9 Does statistical help make a difference? Is there a significant difference in the proportions of papers with and without statistical help that are rejected without review? Use Table C to get an approximate P-value. (This observational study does not establish causation: the studies that include statistical help may also be better than those that do not in other ways.)

19.10 How often are statisticians involved? Give a 95% confidence interval for the proportion of papers submitted to these journals that include help from a statistician.

19.11 How big a difference? Give a 95% confidence interval for the difference between the proportions of papers rejected without review when a statistician is and is not involved in the research.

19.12 Satisfaction with high schools. A sample survey asked 202 black parents and 201 white parents of high school children, "Are the public high schools in your state doing an excellent, good, fair or poor job, or don't you know enough to say?" The investigators suspected that black parents are generally less satisfied with their public schools than are whites. Among the black parents, 81 thought high schools were doing a "good" or "excellent" job; 103 of the white parents felt this way.[11] Is there good evidence that the proportion of all black parents who think their state's high schools are good or excellent is lower than the proportion of white parents with this opinion?

$z = -2.24$
$p = .01$

19.13 College is important. The sample survey described in the previous exercise also asked respondents if they agreed with the statement "A college education has become as important as a high school diploma used to be." In the sample, 125 of 201 white parents and 154 of 202 black parents said that they "strongly agreed." Is there good reason to think that different percents of all black and white parents would strongly agree with the statement?

19.14 Seat belt use. The proportion of drivers who use seat belts depends on things like age (young people are more likely to go unbelted) and gender (women are more likely to use belts). It also depends on local law. In New York City, police can stop a driver who is not belted. In Boston (as of late 2000), police can cite a driver for not wearing a seat belt only if the driver has been stopped for some other violation. Here are data from observing random samples of female Hispanic drivers in these two cities:[12]

City	Drivers	Belted
New York	220	183
Boston	117	68

9.49
$p = 1.16$

(Bob Llewellyn/ImageState-Pictor/ PictureQuest)

(a) Is this an experiment or an observational study? Why?

(b) Comparing local law suggests the hypothesis that a smaller proportion of drivers wear seat belts in Boston than in New York. Do the data give good evidence that this is true for female Hispanic drivers?

19.15 Ethnicity and seat belt use. Here are data from the study described in the previous exercise for Hispanic and white male drivers in Chicago:

Group	Drivers	Belted
Hispanic	539	286
White	292	164

Is there a significant difference between Hispanic and white drivers? How large is the difference? Do inference to answer both questions. Be sure to explain exactly what inference you choose to do.

19.16 Lyme disease. Lyme disease is spread in the northeastern United States by infected ticks. The ticks are infected mainly by feeding on mice, so more mice result in more infected ticks. The mouse population in turn rises and falls with the abundance of acorns, their favored food. Experimenters studied two similar forest areas in a year when the acorn crop failed. They added hundreds of thousands of acorns to one area to imitate an abundant acorn crop, while leaving the other area untouched. The next spring, 54 of the 72 mice trapped in the first area were in breeding condition, versus 10 of the 17 mice trapped in the second area.[13]

(a) The large-sample confidence interval may not be accurate for these data. Why not?

(b) Give the plus four 90% confidence interval for the difference between the proportions of mice ready to breed in good acorn years and bad acorn years.

19.17 Steroids in high school. A study by the National Athletic Trainers Association surveyed 1679 high school freshmen and 1366 high school seniors in Illinois. Results showed that 34 of the freshmen and 24 of the seniors had used anabolic steroids. Steroids, which are dangerous, are sometimes used to improve athletic performance.[14]

(a) In order to draw conclusions about all Illinois freshmen and seniors, how should the study samples be chosen?

(b) Give a 95% confidence interval for the proportion of all high school freshmen in Illinois who have used steroids.

(c) Is there a significant difference between the proportions of freshmen and seniors who have used steroids?

19.18 Detecting genetically modified soybeans. Exercise 18.27 (page 488) describes a study in which batches of soybeans containing some genetically modified (GM) beans were submitted to 23 grain-handling facilities. When batches contained 1% of GM beans, 18 of the facilities detected the presence of GM beans. Only 7 of the facilities detected GM beans when they made up one-tenth of 1% of the beans in the batches. Explain why we *cannot* use the methods of this chapter to compare the proportions of facilities that will detect the two levels of GM soybeans.

(Richard Hamilton Smith/CORBIS)

19.19 Are genetically modified foods risky? Europe and the United States differ considerably in their attitudes toward food made from crops that have been genetically modified (GM) to, for example, resist pests or contain more protein. A random sample of 12,178 European adults found that 63% thought such foods were risky. In the United States, a random sample of 863 adults who were asked the same questions found that 46% considered GM foods risky.[15]

(a) What are the counts of people in each sample who thought GM foods were risky?

(b) Give a 95% confidence interval to compare Europe and the United States.

19.20 Are urban students more successful? North Carolina State University looked at the factors that affect the success of students in a required chemical engineering course. Students must get a C or better in the course in order to continue as chemical engineering majors. There were 65 students from urban or suburban backgrounds, and 52 of these students succeeded. Another 55 students were from rural or small-town backgrounds; 30 of these students succeeded in the course.[16]

(a) Is there good evidence that the proportion of students who succeed is different for urban/suburban versus rural/small-town backgrounds? State hypotheses, give the P-value of a test, and state your conclusion.

(b) Give a 90% confidence interval for the size of the difference.

19.21 Small-business failures. The study of small-business failures described in Exercise 18.34 (page 489) looked at 148 food-and-drink businesses in central Indiana. Of these, 106 were headed by men and 42 were headed by women. During a three-year period, 15 of the men's businesses and 7 of the women's businesses failed. Is there a significant difference between the rate at which businesses headed by men and those headed by women fail?

19.22 Female and male students. The North Carolina State University study (Exercise 19.20) also looked at possible differences in the proportions of female and male students who succeeded in the course. They found that 23 of the 34 women and 60 of the 89 men succeeded. Is there evidence of a difference between the proportions of women and men who succeed?

19.23 Nobody is home in July. Nonresponse to sample surveys may differ with the season of the year. In Italy, for example, many people leave town during the summer. The Italian National Statistical Institute called random samples of telephone numbers between 7 p.m. and 10 p.m. at several seasons of the year. Here are the results for two seasons:[17]

(Dallas and John Heaton/CORBIS)

Dates	Number of calls	No answer	Total nonresponse
Jan. 1 to Apr. 13	1558	333	491
July 1 to Aug. 31	2075	861	1174

(a) How much higher is the proportion of "no answers" in July and August compared with the early part of the year? Give a 99% confidence interval.

(b) The difference between the proportions of "no answers" is so large that it is clearly statistically significant. How can you tell from your work in (a) that the difference is significant at the $\alpha = 0.01$ level?

(c) Use the information given to find the counts of calls that had nonresponse for some reason other than "no answer." Do the rates of nonresponse due to other causes also differ significantly for the two seasons?

19.24 **Who responds?** Example 17.4 (page 446) describes a study comparing college students who have volunteered for community service work with those who have not. A weakness in the study is that the response rates for the two groups appear to differ. Students were placed in the two groups on the basis of a questionnaire: 39 in the "no service" group and 71 in the "service" group. The data were gathered from a follow-up survey two years later; 17 of the 39 "no service" students responded (44%), compared with 80% response (57 of 71) in the "service" group. How significant is this difference in response rates?

19.25 **Significant does not mean important.** Never forget that even small effects can be statistically significant if the samples are large. To illustrate this fact, return to the study of 148 small businesses in Exercise 19.21.

(a) Find the proportions of failures for businesses headed by women and businesses headed by men. These sample proportions are quite close to each other. Give the P-value for the z test of the hypothesis that the same proportion of women's and men's businesses fail. (Use the two-sided alternative.) The test is very far from being significant.

(b) Now suppose that the same sample proportions came from a sample 30 times as large. That is, 210 out of 1260 businesses headed by women and 450 out of 3180 businesses headed by men fail. Verify that the proportions of failures are exactly the same as in (a). Repeat the z test for the new data, and show that it is now significant at the $\alpha = 0.05$ level.

(c) It is wise to use a confidence interval to estimate the size of an effect, rather than just giving a P-value. Give 95% confidence intervals for the difference between the proportions of women's and men's businesses that fail for the settings of both (a) and (b). What is the effect of larger samples on the confidence interval?

Chapter 19 MEDIA EXERCISES

19.26 Police radar and speeding. Do drivers reduce excessive speed when they encounter police radar? The EESEE story "Radar Detectors and Speeding" describes a study of the behavior of drivers on a rural interstate highway in Maryland where the speed limit was 55 miles per hour. The researchers measured speed with an electronic device hidden in the pavement and, to eliminate large trucks, considered only vehicles less than 20 feet long. During some time periods, police radar was set up at the measurement location. Here are some of the data:

(Joseph Sohm/ChromoSohm Inc./CORBIS)

	Number of vehicles	Number over 65 mph
No radar	12,931	5,690
Radar	3,285	1,051

(a) Give a 95% confidence interval for the proportion of vehicles going faster than 65 miles per hour when no radar is present.

(b) Give a 95% confidence interval for the effect of radar, as measured by the difference in proportions of vehicles going faster than 65 miles per hour with and without radar.

(c) The researchers chose a rural highway so that cars would be separated rather than in clusters, where some cars might slow because they see other cars slowing. Explain why such clusters might make inference invalid.

19.27 Reducing nonresponse? Telephone surveys often have high rates of nonresponse. When the call is handled by an answering machine, perhaps leaving a message on the machine will encourage people to respond when they are called again. The data below are from the EESEE story "Leave Survey after the Beep." In this study, a message was or was not left (at random) when an answering machine picked up the first call from a survey.

	Total households	Eventual contact	Completed survey
No message	100	58	33
Message	291	200	134

(a) Is there good evidence that leaving a message increases the proportion of households that are eventually contacted?

(b) Is there good evidence that leaving a message increases the proportion who complete the survey?

(c) If you find significant effects, look at their size. Do you think these effects are large enough to be important to survey takers?

Inference about Variables
Part III Review

The procedures of Chapters 16 to 19 are among the most common of all statistical inference methods. The Statistics in Summary flowchart on the next page helps you decide when to use them. It is important to do some of the review exercises because now, for the first time, you must decide which of several inference procedures to use. Learning to recognize problem settings in order to choose the right type of inference is a key step in advancing your mastery of statistics.

Statistical inference always draws conclusions about one or more parameters of a population. When you think about doing inference, ask first *what the population is* and *what type of parameter* you are interested in. You are now familiar with procedures for inference about population *means* and population *proportions*. (Chapter 17 also gave an optional discussion of inference about population *standard deviations*, but the main point was that we don't recommend those methods.) The first branch point in the Statistics in Summary flowchart asks you what type of parameter your problem concerns. Inference about means leads to t procedures based on the t distributions. Inference about proportions leads to z procedures based on the standard Normal distribution.

There are also z procedures for inference about a mean μ when we know the population standard deviation σ. We started the discussion of inference with these z procedures in Chapters 13 and 14 because they are simple, but in practice it is rare to know σ but not μ. In practice, we almost always use t methods for inference about means. One reason for their wide use is that t procedures are not strongly affected by lack of Normality.

STATISTICS IN SUMMARY
Inference about Means and Proportions

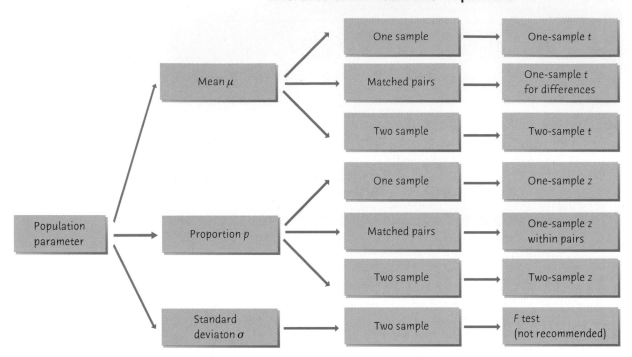

The second question that guides your choice of procedure is *what type of design* produced the data. To this point we have met procedures for a *single sample*, a *matched pairs* study, and *two independent samples*. The study may be either observational or an experiment—the inference methods are the same for both. Don't forget that experiments can give good evidence for causation and that observational studies rarely do, but this is an issue separate from choosing the right inference procedure.

The second branch point in the flowchart lays out these choices of study design and points to the corresponding methods. You must of course also decide whether your problem calls for a confidence interval, a significance test, or both. And don't forget the cautions of Chapter 15, which apply to any inference. In particular, when you do inference, you are acting as if your data came from a random sample or randomized comparative experiment. If that isn't true, be prepared to argue that it's OK to think of your subjects as a random sample from a population of interest.

You may ask, as you study the Statistics in Summary flowchart, "What if I have an experiment comparing four treatments, or samples from three populations?" The flowchart allows only one or two, not three or four or more. Be patient: methods for comparing more than two means or proportions, as well as some other settings for inference, appear in Part IV.

There is of course more to doing statistics well than choosing and carrying out the proper procedures. At the end of the book you will find a short outline of "Statistical Thinking Revisited" that reminds you of some of the big ideas that guide statistics in practice, as well as warning of some of the most common pitfalls. We suggest looking at that outline now, especially if you are near the end of your study of this book.

Part III SUMMARY

Here are the most important skills you should have acquired from reading Chapters 16 to 19.

A. RECOGNITION

1. Recognize when a problem requires inference about population means or population proportions.

2. Recognize from the design of a study whether one-sample, matched pairs, or two-sample procedures are needed.

3. Based on recognizing the problem setting, choose among the one- and two-sample t procedures for means and the one- and two-sample z procedures for proportions.

B. INFERENCE ABOUT ONE MEAN

1. Recognize when the t procedures are appropriate in practice, in particular that they are quite robust against lack of Normality as long as the distribution is not strongly skewed.

2. Also recognize when poor study design, outliers, or a small sample from a skewed distribution make the t procedures risky.

3. Use the one-sample t procedure to obtain a confidence interval at a stated level of confidence for the mean μ of a population.

4. Carry out a one-sample t test for the hypothesis that a population mean μ has a specified value against either a one-sided or a two-sided alternative. Use Table C of t distribution critical values to approximate the P-value or carry out a fixed-level α test.

5. Recognize matched pairs data and use the t procedures to obtain confidence intervals and to perform tests of significance for such data.

C. COMPARING TWO MEANS

1. Recognize when the two-sample t procedures are appropriate in practice.

2. Give a confidence interval for the difference between two means. Use the two-sample t statistic with conservative degrees of freedom if you do not have statistical software. Use software if you have it.

3. Test the hypothesis that two populations have equal means against either a one-sided or a two-sided alternative. Use the two-sample t test with conservative degrees of freedom if you do not have statistical software. Use software if you have it.

4. Know that procedures for comparing the standard deviations of two Normal populations are available, but that these procedures are risky because they are not at all robust against non-Normal distributions.

D. INFERENCE ABOUT ONE PROPORTION

1. Use the large-sample z procedure to give a confidence interval for a population proportion p. Understand that the true confidence level may be substantially less than you ask for unless the sample is very large and the true p is not close to 0 or 1.

2. (Optional) Use the plus four z procedure to give a confidence interval for p that is accurate even for small samples and for any value of p.

3. Use the z statistic to carry out a test of significance for the hypothesis H_0: $p = p_0$ about a population proportion p against either a one-sided or a two-sided alternative. Use either Table A to get the P-value or Table C to approximate P.

4. Check that you can safely use these z procedures in a particular setting.

E. COMPARING TWO PROPORTIONS

1. Use the large-sample z procedure to give a confidence interval for the difference $p_1 - p_2$ between proportions in two populations based on independent samples from the populations. Understand that the true confidence level may be less than you ask for unless the samples are quite large.

2. (Optional) Use the plus four z procedure to give a confidence interval for $p_1 - p_2$ that is accurate even for very small samples and for any values of p_1 and p_2.

3. Use a z statistic to test the hypothesis H_0: $p_1 = p_2$ that proportions in two distinct populations are equal. Use either Table A to get the P-value or Table C to approximate P.

4. Check that you can safely use these z procedures in a particular setting.

Review EXERCISES

Review exercises are short and straightforward exercises that help you solidify the basic ideas and skills in each part of this book. For tests of significance, use Table C to approximate P-values unless you use software that reports the P-value or the exercise asks you to use Table A for a z test.

III.1 Does ginkgo improve memory? Exercise 8.4 (page 205) describes a randomized comparative experiment to compare ginkgo with a placebo. Which of the procedures in the Statistics in Summary flowchart should you use to compare the ginkgo and placebo groups on

(a) whether or not the subjects thought they were taking ginkgo.

(b) score on the logical memory part of the Wechsler Memory Scale (possible scores 0 to 50).

III.2 Looking back on love. How do young adults look back on adolescent romance? Investigators interviewed 40 couples in their mid-twenties. The female and male partners were interviewed separately. Each was asked about the current relationship and also about a romantic relationship that lasted at least two months when they were aged 15 or 16.[1] One response variable was a measure on a numerical scale of how much the attractiveness of the adolescent partner mattered. Which of the tests in the Statistics in Summary flowchart should you use to compare the men and women on this measure?

(a) t for means or z for proportions?

(b) one-sample, matched pairs, or two-sample?

III.3 **Workers' absences.** The European Commission funded a study of worker absenteeism in nine countries. The study selected a sample of workers in each country and calculated the percent of total workdays (number of workers times days available for work) missed in each sample. Which of the confidence intervals in the Statistics in Summary flowchart should you use to

(a) estimate the percent of workdays missed by workers in Spain.

(b) compare the percents of workdays missed by workers in Spain and in Switzerland.

III.4 **Preventing AIDS through education.** The Multisite HIV Prevention Trial was a randomized comparative experiment to compare the effects of twice-weekly small-group AIDS discussion sessions (the treatment) with a single one-hour session (the control).[2] Which of the procedures in the Statistics in Summary flowchart should you use to compare treatment and control on the following measures?

(a) Using or not using condoms 6 months after the education sessions.

(b) Number of unprotected intercourse acts between 4.5 and 8 months after the education sessions.

(c) Infected with a sexually transmitted disease 6 months after the education sessions.

III.5 **Potato chips.** Here is the description of the design of a study that compared subjects' responses to regular and fat-free potato chips:

> *During a given 2-wk period, the participants received the same type of chip (regular or fat-free) each day. After the first 2-wk period, there was a 1-wk washout period in which no testing was performed. There was then another 2-wk period in which the participants received the alternate type of potato chip (regular or fat-free) each day under the same protocol. The order in which the chips were presented was determined by random assignment.*[3]

One response variable was the weight in grams of potato chips that a subject ate. We want to compare the amounts of regular and fat-free chips eaten.

(a) Forty-four women made up one set of subjects. Explain how to do the random assignment required by the design, and use Table B at line 101 to assign the first 5 women.

(b) Which test from the Statistics in Summary flowchart should you use to test the null hypothesis of no difference between regular and fat-free chips?

III.6 Mouse endurance. A study of the inheritance of speed and endurance in mice found a trade-off between these two characteristics, both of which help mice survive. To test endurance, mice were made to swim in a bucket with a weight attached to their tails. (The mice were rescued when exhausted.) Here are data on endurance in minutes for female and male mice:[4]

Group	n	Mean	Std. dev.
Female	162	11.4	26.09
Male	135	6.7	6.69

(a) Give a 95% confidence interval for the mean endurance of female mice swimming.

(b) Give a 95% confidence interval for the mean difference (female minus male) in endurance times.

III.7 Mouse endurance, continued. Use the information in the previous exercise to answer these questions.

(a) Do female mice have significantly higher endurance on the average than male mice?

(b) The endurance data are skewed to the right. Why are t procedures reasonably accurate for these data?

III.8 Child-care workers. The Current Population Survey (CPS) is the monthly government sample survey of 55,000 households that provides data on employment in the United States. A study of child-care workers drew a sample from the CPS data tapes. We can consider this sample to be an SRS from the population of child-care workers.[5] Out of 2455 child-care workers in private households, 7% were black. Of 1191 nonhousehold child-care workers, 14% were black. Give a 99% confidence interval for the difference in the percents of these groups of workers who are black. Is the difference statistically significant at the $\alpha = 0.01$ level?

III.9 Child-care workers, continued. The study described in the previous exercise also examined how many years of school child-care workers had. For household workers, the mean and standard deviation were $\bar{x}_1 = 11.6$ years and $s_1 = 2.2$ years. For nonhousehold workers, $\bar{x}_2 = 12.2$ years and $s_2 = 2.1$ years. Give a 99% confidence interval for the difference in mean years of education for the two groups. Is the difference significant at the $\alpha = 0.01$ level?

III.10 Men versus women. The National Assessment of Educational Progress (NAEP) Young Adult Literacy Assessment Survey interviewed a random sample of 1917 people 21 to 25 years old. The sample contained 840 men, of whom 775 were fully employed. There were 1077 women, and 680 of them were fully employed.[6] Use a 99%

confidence interval to describe the difference between the proportions of young men and young women who are fully employed. Is the difference statistically significant at the 1% significance level?

III.11 Men versus women, continued. The mean and standard deviation of scores on the NAEP's test of quantitative skills were $\bar{x}_1 = 272.40$ and $s_1 = 59.2$ for the men in the sample described in the previous exercise. For the women, the results were $\bar{x}_2 = 274.73$ and $s_2 = 57.5$. Is the difference between the mean scores for men and women significant at the 1% level?

III.12 Registering handguns. The National Gun Policy Survey has been carried out regularly over several years by the National Opinion Research Center at the University of Chicago. The sample can be considered an SRS of adult residents of the United States. It is selected by calling randomly chosen telephone numbers in all 50 states. When a household answers the phone, the questions are asked of the adult with the most recent birthday. One of the questions asked is "Do you favor or oppose the mandatory registration of handguns and pistols?"[7] Of the 1176 people in the 2001 sample, 904 favored mandatory registration. Give a 95% confidence interval for the proportion of all adults who favor registration.

III.13 Registering handguns, continued. There has been a trend toward less support for gun control. The 1998 National Gun Policy Survey asked 1201 people the question quoted in the previous exercise; 1024 favored registration. What were the percents in favor in 1998 and 2001? Did a significantly smaller proportion of the population favor mandatory registration in 2001 than in 1998?

III.14 Way down under. A remarkable discovery: large lakes exist deep under the Antarctic ice cap, kept liquid by the enormous pressure of the ice above. The largest is Lake Vostok. Do these lakes contain populations of ancient bacteria adapted to the dark, cold, high-pressure environment? Drilling over 3600 meters (over 11,000 feet) down to a depth where the ice consists of water frozen from Lake Vostok did indeed find bacteria. The researchers estimated "mean bacterial biomass" in nanograms of carbon per liter of melted water. They had 5 specimen ice cores. They split each specimen into "top" and "bottom" to get two samples of size 5, which they processed separately. Separate study of the two samples is a check on the complicated process of preparing the cores for analysis. The results "are mean estimates ±SD." Here is an extract:[8]

	Top melt	Bottom melt
Mean bacterial biomass (ng of C liter^{-1})	2.9 ± 0.4	2.8 ± 0.4

Use the results for the 5 top-melt specimens to give a 90% confidence interval for the mean bacterial biomass in the ice above Lake Vostok.

III.15   Way down under, continued. Do the results reported in the previous exercise give any reason to think that the population mean biomasses as estimated from top-melt and bottom-melt samples are different? (A difference would suggest that the preparation process influences the findings.)

III.16   Spinning euros. When the new euro coins were introduced throughout Europe in 2002, curious people tried all manner of things. Two Polish mathematicians spun a Belgian euro (one side of the coin has a different design for each country) 250 times. They got 140 heads. Newspapers reported this result widely. Is it significant evidence that the coin is not balanced when spun? State hypotheses, give the test statistic, and use Table A to calculate the P-value.

III.17   Spinning euros, continued. What do the data in Exercise III.16 allow you to say with 95% confidence about the proportion of heads in spinning Belgian euro coins?

III.18   Genetically modified foods. Europeans have been more skeptical than Americans about the use of genetic engineering to improve foods. A sample survey gathered responses from random samples of 863 Americans and 12,178 Europeans.[9] (The European sample was larger because Europe is divided into many nations.) Subjects were asked to consider the following issue:

> *Using modern biotechnology in the production of foods, for example to make them higher in protein, keep longer, or change in taste.*

They were asked if they considered this "risky for society." In all, 52% of Americans and 64% of Europeans thought the application was risky.

(a) It is clear without a formal test that the proportion of the population who consider this use of technology risky is significantly higher in Europe than in the United States. Why is this?

(b) Give a 99% confidence interval for the percent difference between Europe and the United States.

III.19   Genetically modified foods, continued. Is there convincing evidence that more than half of all adult Americans consider applying biotechnology to the production of foods risky? Use Table A to calculate the P-value.

III.20   Genetically modified foods, continued. Give a 95% confidence interval for the proportion of all European adults who consider the use of biotechnology in food production risky.

III.21   Butterflies mating. Here's how butterflies mate: a male conveys to a female a packet of sperm called a spermatophore. Females may mate several times. Will they remate sooner if the first spermatophore they receive is small? Among 20 females who received a large spermatophore (greater than 25 milligrams), the mean time to the

(Tim Zurowski/CORBIS)

next mating was 5.15 days, with standard deviation 0.18 day. For 21 females who received a small spermatophore (about 7 milligrams), the mean was 4.33 days and the standard deviation was 0.31 day.[10] How significant is the observed difference in means? State hypotheses, find the test statistic and its *P*-value, and give your conclusion.

III.22 Very-low-birth-weight babies. Starting in the 1970s, medical technology allowed babies with very low birth weight (VLBW, less than 1500 grams, about 3.3 pounds) to survive without major handicaps. It was noticed that these children nonetheless had difficulties in school and as adults. A long study has followed 242 VLBW babies to age 20 years, along with a control group of 233 babies from the same population who had normal birth weight.[11]

(a) Is this an experiment or an observational study? Why?

(b) At age 20, 179 of the VLBW group and 193 of the control group had graduated from high school. Is the graduation rate among the VLBW group significantly lower than for the normal-birth-weight controls?

(c) Give a 95% confidence interval for the proportion of VLBW babies born in the late 1970s who have graduated from high school.

III.23 Very-low-birth-weight babies, continued. IQ scores were available for 113 men in the VLBW group. The mean IQ was 87.6, and the standard deviation was 15.1. The 106 men in the control group had mean IQ 94.7, with standard deviation 14.9. Is there good evidence that mean IQ is lower among VLBW men than among controls from similar backgrounds?

III.24 Very-low-birth-weight babies, continued. Of the 126 women in the VLBW group, 37 said they had used illegal drugs; 52 of the 124 control group women had done so. The IQ scores for these VLBW women had mean 86.2 (standard deviation 13.4), and the normal-birth-weight controls had mean IQ 89.8 (standard deviation 14.0). Are either of these differences between the two groups statistically significant?

III.25 Nitrites and bacteria. Nitrites are often added to meat products as preservatives. In a study of the effect of nitrites on bacteria, researchers measured the rate of uptake of an amino acid for 60 cultures of bacteria: 30 growing in a medium to which nitrites had been added and another 30 growing in a standard medium as a control group. Table III.1 gives the data from this study. Examine each of the two samples and briefly describe their distributions. Carry out a test of the research hypothesis that nitrites decrease amino acid uptake, and report your results.

III.26 Double-decker diets. A British study compared the food and drink intake of 98 drivers and 83 conductors of London double-decker buses. The conductors' jobs require more physical activity. The article

(Corbis Images/PictureQuest)

TABLE III.1 Amino acid uptake by bacteria under two conditions

Control			Nitrite		
6,450	8,709	9,361	8,303	8,252	6,594
9,011	9,036	8,195	8,534	10,227	6,642
7,821	9,996	8,202	7,688	6,811	8,766
6,579	10,333	7,859	8,568	7,708	9,893
8,066	7,408	7,885	8,100	6,281	7,689
6,679	8,621	7,688	8,040	9,489	7,360
9,032	7,128	5,593	5,589	9,460	8,874
7,061	8,128	7,150	6,529	6,201	7,605
8,368	8,516	8,100	8,106	4,972	7,259
7,238	8,830	9,145	7,901	8,226	8,552

reporting the study gives the data as "Mean daily consumption (± s.e.)." Some of the study results appear below.[12]

	Drivers	Conductors
Total calories	2821 ± 44	2844 ± 48
Alcohol (grams)	0.24 ± 0.06	0.39 ± 0.11

(a) What does "s.e." stand for? Give $\bar{x}$ and s for each of the four sets of measurements.

(b) Is there significant evidence at the 5% level that conductors and drivers consume different numbers of calories per day?

(c) Give an 80% confidence interval for the difference in mean daily alcohol consumption between drivers and conductors.

III.27 **Double-decker diets, continued.** Use the data in the previous exercise to answer these questions.

(a) How significant is the observed difference between drivers and conductors in mean alcohol consumption?

(b) Give a 90% confidence interval for the mean daily alcohol consumption of London double-decker bus conductors.

III.28 **California counties.** Table 2.3 (page 50) gives the populations of all 58 counties in the state of California. Is it proper to apply the one-sample t method to these data to give a 95% confidence interval for the mean population of a California county? Explain your answer.

III.29 **Do fruit flies sleep?** Mammals and birds sleep. Fruit flies show a daily cycle of rest and activity, but does the rest qualify as sleep? Researchers looking at brain activity and behavior finally concluded that fruit flies do sleep. A small part of the study used an infra-red motion sensor to

see if flies moved in response to vibrations. Here are results for low
levels of vibration:[13]

	Response to vibration?	
	No	Yes
Fly was walking	10	54
Fly was resting	28	4

Analyze these results. Is there good reason to think that resting flies
respond differently than flies that are walking? (That's a sign that the
resting flies may actually be sleeping.)

III.30 Cholesterol in dogs. High levels of cholesterol in the blood are not
healthy in either humans or dogs. Because a diet rich in saturated fats
raises the cholesterol level, it is plausible that dogs owned as pets have
higher cholesterol levels than dogs owned by a veterinary research
clinic. "Normal" levels of cholesterol based on the clinic's dogs would
then be misleading. A clinic compared healthy dogs it owned with
healthy pets brought to the clinic to be neutered. The summary
statistics for blood cholesterol levels (milligrams per deciliter of
blood) appear below.[14]

(Arthur Tilley/i2i Images/PictureQuest)

Group	n	$\bar{x}$	s
Pets	26	193	68
Clinic	23	174	44

(a) Is there strong evidence that pets have a higher mean cholesterol
level than clinic dogs? State the H_0 and H_a and carry out an
appropriate test. Give the P-value and state your conclusion.

(b) Give a 95% confidence interval for the difference in mean
cholesterol levels between pets and clinic dogs.

(c) Give a 95% confidence interval for the mean cholesterol level in
pets.

(d) What assumptions must be satisfied to justify the procedures you
used in (a), (b), and (c)? Assuming that the cholesterol
measurements have no outliers and are not strongly skewed, what
is the chief threat to the validity of the results of this study?

III.31 The density of the earth. Exercise 2.20 (page 50) gives 29
measurements of the density of the earth, made in 1798 by Henry
Cavendish. Cavendish reported the mean of these measurements, but
confidence intervals had not yet been invented. Display the data
graphically to check for skewness and outliers. Then give an estimate
for the density of the earth from Cavendish's data and a margin of
error for your estimate.

Supplementary EXERCISES

Supplementary exercises apply the skills you have learned in ways that require more thought or more elaborate use of technology.

III.32 **Falling through the ice.** Table I.3 (page 165) gives the dates on which a wooden tripod fell through the ice of the Tanana River in Alaska, thus deciding the winner of the Nenana Ice Classic contest, for the years 1917 to 2002. Give a 95% confidence interval for the mean date on which the tripod falls through the ice. After calculating the interval in the scale used in the table (days from April 20, which is Day 1), translate your result into calendar dates.

III.33 **Mouse genes.** A study of genetic influences on diabetes compared normal mice with similar mice genetically altered to remove the gene called $aP2$. Mice of both types were allowed to become obese by eating a high-fat diet. The researchers then measured the levels of insulin and glucose in their blood plasma. Here are some excerpts from their findings.[15] The normal mice are called "wild-type" and the altered mice are called "$aP2^{-/-}$."

Each value is the mean $\pm$ SEM of measurements on at least 10 mice. Mean values of each plasma component are compared between $aP2^{-/-}$ mice and wild-type controls by Student's t test ($*P < 0.05$ and $**P < 0.005$).

Parameter	Wild type	$aP2^{-/-}$
Insulin (ng/ml)	5.9 ± 0.9	$0.75 \pm 0.2**$
Glucose (mg/dl)	230 ± 25	$150 \pm 17*$

Despite much greater circulating amounts of insulin, the wild-type mice had higher blood glucose than the $aP2^{-/-}$ animals. These results indicate that the absence of $aP2$ interferes with the development of dietary obesity-induced insulin resistance.

Other biologists are supposed to understand the statistics reported so tersely.

(a) What does "SEM" mean? What is the expression for SEM based on n, $\bar{x}$, and s from a sample?

(b) Which of the tests we have studied did the researchers apply?

(c) Explain to a biologist who knows no statistics what $P < 0.05$ and $P < 0.005$ mean. Which is stronger evidence of a difference between the two types of mice?

III.34 **Mouse genes, continued.** The report quoted in the previous exercise says only that the sample sizes were "at least 10." Suppose that the results are based on exactly 10 mice of each type. Use the values in the table to find $\bar{x}$ and s for the insulin concentrations in the two types of mice. Carry out a test to assess the significance of the difference in mean insulin concentration. Does your P-value confirm the claim in the report that $P < 0.005$?

III.35 Mouse endurance (Optional). The data in Exercise III.6 suggest that there is greater variation in the swimming endurance of female mice than in male mice. Suppose that we had this hypothesis in mind before looking at the data. Suppose also that the data are very close to Normally distributed.

(a) State hypotheses and carry out the test. Software can assess significance exactly, but inspection of the proper table is enough to draw a conclusion.

(b) Would the large sample sizes allow us to carry out the inference in (a) even if the distributions are not very close to Normal?

III.36 Cholesterol in dogs (Optional). Do the data in Exercise III.30 provide evidence of different standard deviations for cholesterol levels in pet dogs and clinic dogs? State the hypotheses and carry out the test. (Assume that the data follow a Normal distribution closely.) Software can assess significance exactly, but inspection of the proper table is enough to draw a conclusion.

III.37 The power of a t test (Optional). The bank in Exercise 16.16 (page 428) tested a new idea on a sample of 200 customers. The bank wants to be quite certain of detecting a mean increase of $\mu = \$100$ in the amount charged, at the $\alpha = 0.01$ significance level. Perhaps a sample of only $n = 50$ customers would accomplish this. Find the approximate power of the test with $n = 50$ against the alternative $\mu = \$100$ as follows.

(a) What is the critical value t^* for the one-sided test with $\alpha = 0.01$ and $n = 50$?

(b) Write the rule for rejecting $H_0: \mu = 0$ in terms of the t statistic. Then take $s = 108$ (an estimate based on the data in Exercise 16.16) and state the rejection rule in terms of $\bar{x}$.

(c) Assume that $\mu = 100$ (the given alternative) and that $\sigma = 108$ (an estimate from the data in Exercise 16.16). The approximate power is the probability of the event you found in (b), calculated under these assumptions. Find the power. Would you recommend that the bank do a test on 50 customers, or should more customers be included?

III.38 The power of a two-sample t test (Optional). A bank asks you to compare two ways to increase the use of their credit cards. Plan A would offer customers a cash-back rebate based on their total amount charged. Plan B would reduce the interest rate charged on card balances. The bank thinks that Plan B will be more effective. The response variable is the total amount a customer charges during the test period. You decide to offer each of Plan A and Plan B to a separate SRS of the bank's credit card customers. In the past, the mean amount charged in a six-month period has been about $1100, with a standard deviation of $400. Will a two-sample t test based on SRSs of 350 customers in each group detect a difference of $100 in the mean amounts charged under the two plans?

We will compute the approximate power of the two-sample t test of

$$H_0: \mu_B = \mu_A$$
$$H_a: \mu_B > \mu_A$$

against the specific alternative $\mu_B - \mu_A = 100$. We will use the past value \$400 as a rough estimate of both the population σ's and future sample s's.

(a) What is the approximate value of the $\alpha = 0.05$ critical value t^* for the two-sample t statistic when $n_1 = n_2 = 350$?

(b) **Step 1.** *Write the rule for rejecting H_0 in terms of $\overline{x}_B - \overline{x}_A$.* The test rejects H_0 when

$$\frac{\overline{x}_B - \overline{x}_A}{\sqrt{\dfrac{s_B^2}{n_B} + \dfrac{s_A^2}{n_A}}} \geq t^*$$

Take both s_B and s_A to be 400, and n_B and n_A to be 350. Find the number c such that the test rejects H_0 when $\overline{x}_B - \overline{x}_A \geq c$.

(c) **Step 2.** *The power is the probability of rejecting H_0 when the alternative is true.* Suppose that $\mu_B - \mu_A = 100$ and that both σ_B and σ_A are 400. The power we seek is the probability that $\overline{x}_B - \overline{x}_A \geq c$ under these assumptions. Calculate the power.

The remaining exercises concern a study of air pollution. One component of air pollution is airborne particulate matter such as dust and smoke. To measure particulate pollution, a vacuum motor draws air through a filter for 24 hours. Weigh the filter at the beginning and end of the period. The weight gained is a measure of the concentration of particles in the air. A study of air pollution made measurements every 6 days with identical instruments in the center of a small city and at a rural location 10 miles southwest of the city. Because the prevailing winds blow from the west, we suspect that the rural readings will generally be lower than the city readings, but that the city readings can be predicted from the rural readings. Table III.2 gives readings taken every 6 days over a 7-month period. The entry NA means that the reading for that date is not available, usually because of equipment failure.[16]

Missing data are common, especially in field studies like this one. We think that equipment failures are not related to pollution levels. If that is true, the missing data do not introduce bias. We can work with the data that are not missing as if they are a random sample of days. We can analyze these data in different ways to answer different questions. For each of the three exercises below, do a careful descriptive analysis with graphs and summary statistics and whatever formal inference is called for. Then present and interpret your findings.

III.39 City pollution. We want to assess the level of particulate pollution in the city center. Describe the distribution of city pollution levels, and

TABLE III.2 Particulate levels (grams) in two nearby locations

Day	Rural	City	Day	Rural	City
1	NA	39	19	43	42
2	67	68	20	39	38
3	42	42	21	NA	NA
4	33	34	22	52	57
5	46	48	23	48	50
6	NA	82	24	56	58
7	43	45	25	44	45
8	54	NA	26	51	69
9	NA	NA	27	21	23
10	NA	60	28	74	72
11	NA	57	29	48	49
12	NA	NA	30	84	86
13	38	39	31	51	51
14	88	NA	32	43	42
15	108	123	33	45	46
16	57	59	34	41	NA
17	70	71	35	47	44
18	42	41	36	35	42

estimate the mean particulate level in the city center. (All estimates should include a statistically justified margin of error.)

III.40 City versus country. We want to compare the level of particulates in the city with the rural level on the same day. We suspect that pollution is higher in the city, and we hope that a statistical test will show that there is significant evidence to confirm this suspicion. Make a graph to check for conditions that might prevent the use of the test you plan to employ. Your graph should reflect the type of procedure that you will use. Then carry out a significance test and report your conclusion. Also estimate the mean amount by which the city particulate level exceeds the rural level on the same day.

III.41 Predicting city from country. We hope to use the rural particulate level to predict the city level on the same day. Make a graph to examine the relationship. Does the graph suggest that using the least-squares regression line for prediction will give approximately correct results over the range of values appearing in the data? Calculate the least-squares line for predicting city pollution from rural pollution. What percent of the observed variation in city pollution levels does this straight-line relationship account for? On the fourteenth date in the series, the rural reading was 88 and the city reading was not available. What do you estimate the city reading to be for that date? (In Chapter 21, we will learn how to give a margin of error for the predictions we make from the regression line.)

Inference about Relationships

Statistical inference offers more methods than anyone can know well, as a glance at the offerings of any large statistical software package demonstrates. In a basic text, we must be selective. In Parts I to III we laid a foundation for understanding statistics:

- The nature and purpose of data analysis.
- The central ideas of designs for data production.
- The reasoning behind confidence intervals and significance tests.
- Experience applying these ideas in simple settings.

Each of the three chapters of Part IV offers a short introduction to a more advanced topic in statistical inference. You may choose to read any or all of them, in any order.

What makes a statistical method "more advanced"? More complex data, for one thing. In Part III, we looked only at methods for inference about a single population parameter and for comparing two parameters. All of the chapters in Part IV present methods for studying relationships between two variables. In Chapter 20, both variables are categorical, with data given as a two-way table of counts of outcomes. Chapter 21 considers inference in the setting of regressing a response variable on an explanatory variable. This is an important type of relationship between two quantitative variables. In Chapter 22 we meet methods for comparing the mean response in more than two groups. Here, the explanatory variable (group) is categorical and the response variable is quantitative. These chapters together bring our knowledge of inference to the same point that our study of data analysis reached in Chapters 1 to 6.

With greater complexity comes greater reliance on technology. In these final three chapters you will more often be interpreting the output of statistical software or using software yourself. You can do the calculations needed in

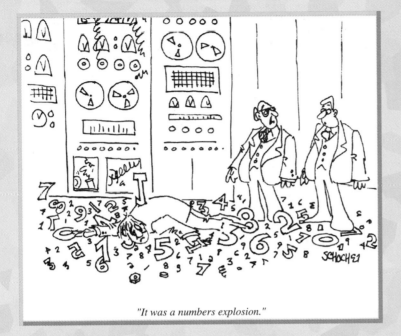

"It was a numbers explosion."

Chapter 20 without software or a specialized calculator. In Chapters 21 and 22, the pain is too great and the contribution to learning too small. Fortunately, you can grasp the ideas without step-by-step arithmetic.

Another aspect of "more advanced" methods is new concepts and ideas. This is where we draw the line in deciding what statistical topics we can master in a first course. Part IV builds elaborate methods on the foundation we have laid without introducing fundamentally new concepts. You can see that statistical practice does need additional big ideas by reading the sections on "the problem of multiple comparisons" in Chapters 20 and 22. But the ideas you already know place you among the world's statistical sophisticates.

INFERENCE ABOUT RELATIONSHIPS

(Michael Busselle/Stone/Getty Images)

Two Categorical Variables: The Chi-Square Test

The two-sample z procedures of Chapter 19 allow us to compare the proportions of successes in two groups, either two populations or two treatment groups in an experiment. What if we want to compare more than two groups? Or what if there are more than the two possible outcomes success and failure? We need a new statistical test. The new test addresses a general question: *is there a relationship between two categorical variables?* In many examples, one variable is "group" or "treatment" and the other variable is "outcome." The first example in Chapter 19 (page 493) compared two groups of children (with and without preschool) by looking at two possible outcomes (did and did not use social services by age 20). We found that preschool reduced later use of social services. At the time, we did not think of this as an example of a relationship between two categorical variables. Now we need the more general framework.

Two-way tables

The new test starts by presenting the data in a **two-way table** of counts. As we saw in Chapter 6 (page 134), a two-way table describes the relationships between two categorical variables. That's exactly what we want. Here is an example.

two-way table

EXAMPLE 20.1 Health care: Canada and the United States

Canada has universal health care. The United States does not, but often offers more elaborate treatment to patients with access. How do the two systems compare in treating heart attacks? A comparison of random samples of 2600 U.S. and 400 Canadian heart attack patients found: "The Canadian patients typically stayed in the hospital one day longer ($P = 0.009$) than the U.S. patients but had a much lower rate of cardiac catheterization (25 percent vs. 72 percent, $P < 0.001$), coronary angioplasty (11 percent vs. 29 percent, $P < 0.001$), and coronary bypass surgery (3 percent vs. 14 percent, $P < 0.001$)."[1]

The study then looked at many outcomes a year after the heart attack. There was no significant difference in the patients' survival. Another key outcome was the patients' own assessment of their quality of life relative to what it had been before the heart attack. Here are the data for the patients who survived a year:

Quality of life	Canada	United States
Much better	75	541
Somewhat better	71	498
About the same	96	779
Somewhat worse	50	282
Much worse	19	65
Total	311	2165

The two-way table in Example 20.1 shows the relationship between two categorical variables. The explanatory variable is the patient's country, Canada or the United States. The response variable is quality of life a year after a heart attack, with 5 categories. The two-way table gives the counts for all 10 combinations of values of these variables. Each of the 10 counts occupies a **cell** of the table.

cell

It is hard to compare the counts because the U.S. sample is much larger. Here are the percents of each sample with each outcome:

Quality of life	Canada	United States
Much better	24%	25%
Somewhat better	23%	23%
About the same	31%	36%
Somewhat worse	16%	13%
Much worse	6%	3%
Total	100%	100%

In the language of Chapter 6 (page 138), these are the *conditional distributions* of outcomes, given the patients' nationality. The differences are not large, but slightly higher percents of Canadians thought their quality of life was "somewhat worse" or "much worse." Figure 20.1 compares the two distributions. We

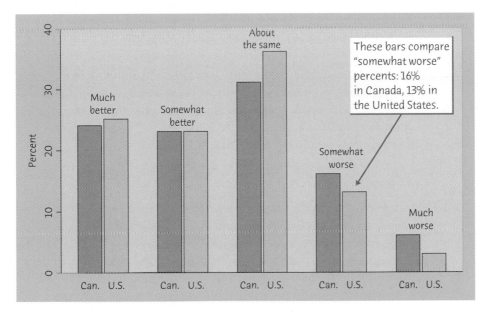

Figure 20.1 Bar graph comparing quality of life a year after a heart attack in Canada and the United States, for Example 20.1.

want to know if there is a significant difference between the two distributions of outcomes.

APPLY YOUR KNOWLEDGE

(Catherine Karnow/CORBIS)

20.1 **Smoking among French men.** Smoking remains more common in much of Europe than in the United States. In the United States, there is a strong relationship between education and smoking: well-educated people are less likely to smoke. Does a similar relationship hold in France? Here is a two-way table of the level of education and smoking status (nonsmoker, former smoker, moderate smoker, heavy smoker) of a sample of 459 French men aged 20 to 60 years.[2] The subjects are a random sample of men who visited a health center for a routine checkup. We are willing to consider them an SRS of men from their region of France.

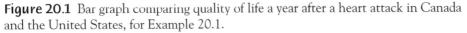

	Smoking Status			
Education	Nonsmoker	Former	Moderate	Heavy
Primary school	56	54	41	36
Secondary school	37	43	27	32
University	53	28	36	16

(a) What percent of men with a primary school education are nonsmokers? Former smokers? Moderate smokers? Heavy smokers?

These percents should add to 100% (up to roundoff error). They form the conditional distribution of smoking, given a primary education.

(b) In a similar way, find the conditional distributions of smoking among men with a secondary education and among men with a university education. Make a table that presents the three conditional distributions. Be sure to include a "Total" column showing that each row adds to 100%.

(c) Compare the three distributions. Is there any clear relationship between education and smoking?

20.2 **Extracurricular activities and grades.** North Carolina State University studied student performance in a course required by its chemical engineering major. One question of interest is the relationship between time spent in extracurricular activities and whether a student earned a C or better in the course. Here are the data for the 119 students who answered a question about extracurricular activities:[3]

Grade	Extracurricular activities (hours per week)		
	<2	2 to 12	>12
C or better	11	68	3
D or F	9	23	5

(a) Find the proportion of successful students (C or better) in each of the three extracurricular activity groups. Make a bar graph to compare the three proportions of successes.

(b) What kind of relationship between extracurricular activities and succeeding in the course do these proportions seem to show?

The problem of multiple comparisons

The null hypothesis in Example 20.1 is that there is *no difference* between the distributions of outcomes in Canada and the United States. Put another way, the null hypothesis is that there is *no relationship* between nation and outcome. The alternative hypothesis says that there *is* a relationship but does not specify any particular kind of relationship. Any difference between the Canadian and American distributions means that the null hypothesis is false and the alternative hypothesis is true. The alternative hypothesis is not one-sided or two-sided. We might call it "many-sided" because it allows any kind of difference.

With only the methods we already know, we might start by comparing the proportions of patients in the two nations with "much better" quality of life, using the two-sample z test for proportions. We could similarly compare the proportions with each of the other outcomes: five tests in all, with five P-values.

This is a bad idea: separate tests don't yield an overall conclusion with a single P-value that represents the probability that, just by chance, two entire distributions differ by as much as these two do. It may be that the "somewhat worse" proportions are significantly different if we compare just this one outcome but that it isn't surprising that the most different among five outcomes differ by this much. It is not legitimate to choose the largest of five differences and then test its significance as if it were the only comparison we had in mind. We can't safely compare many parameters by doing tests or confidence intervals for two parameters at a time.

The problem of how to do many comparisons at once with some overall measure of confidence in all our conclusions is common in statistics. This is the problem of **multiple comparisons.** Statistical methods for dealing with multiple comparisons usually have two steps:

multiple comparisons

1. An *overall test* to see if there is good evidence of *any* differences among the parameters that we want to compare.
2. A detailed *follow-up analysis* to decide which of the parameters differ and to estimate how large the differences are.

The overall test, though more complex than the tests we met earlier, is often reasonably straightforward. The follow-up analysis can be quite elaborate. In our basic introduction to statistical practice, we will concentrate on the overall test, along with data analysis that points to the nature of the differences.

Expected counts in two-way tables

Our general null hypothesis H_0 is that there is *no relationship* between the two categorical variables that label the rows and columns of a two-way table. To test H_0, we compare the observed counts in the table with the *expected counts*, the counts we would expect—except for random variation—if H_0 were true. If the observed counts are far from the expected counts, that is evidence against H_0. It is easy to find the expected counts.

EXPECTED COUNTS

The **expected count** in any cell of a two-way table when H_0 is true is

$$\text{expected count} = \frac{\text{row total} \times \text{column total}}{\text{table total}}$$

To understand why this recipe works, think first about just one proportion.

He started it!

A study of deaths in bar fights showed that in 90% of the cases, the person who died started the fight. You shouldn't believe this. If you killed someone in a fight, what would you say when the police ask you who started the fight? After all, dead men tell no tales.

EXAMPLE 20.2 *Making free throws*

Corinne is a basketball player who makes 70% of her free throws. If she shoots 10 free throws in a game, we expect her to make 70% of them, or 7 of the 10. Of course, she won't make exactly 7 every time she shoots 10 free throws in a game. There is chance variation from game to game. But in the long run, 7 of 10 is what we expect. It is, in fact, the *mean* number of shots Corinne makes when she shoots 10 times.

In more formal language, if we have n independent tries and the probability of a success on each try is p, we expect np successes. If we draw an SRS of n individuals from a population in which the proportion of successes is p, we expect np successes in the sample. That's the fact behind the formula for expected counts in a two-way table.

Let's apply this fact to the heart attack study. Here is the two-way table with row and column totals:

Quality of life	Canada	United States	Total
Much better	75	541	616
Somewhat better	71	498	569
About the same	96	779	875
Somewhat worse	50	282	332
Much worse	19	65	84
Total	311	2165	2476

What is the expected count of Canadians with much better quality of life a year after a heart attack? The proportion of all 2476 subjects with much better quality of life is

$$\frac{\text{count of successes}}{\text{table total}} = \frac{\text{row 1 total}}{\text{table total}} = \frac{616}{2476}$$

Think of this as p, the overall proportion of successes. If H_0 is true, we expect (except for random variation) this same proportion of successes in both countries. So the expected count of successes among the 311 Canadians is

$$np = (311)\left(\frac{616}{2746}\right) = 77.37$$

This expected count has the form announced in the box:

$$\frac{\text{row 1 total} \times \text{column 1 total}}{\text{table total}} = \frac{(616)(311)}{2476}$$

EXAMPLE 20.3   Observed versus expected counts

You can calculate the 10 expected counts—use your calculator to check one or two of them. Here they are:

Quality of life	Canada	United States
Much better	77.37	538.63
Somewhat better	71.47	497.53
About the same	109.91	765.09
Somewhat worse	41.70	290.30
Much worse	10.55	73.45

Compare the observed counts with these expected counts to see how the data diverge from the null hypothesis. You see, for example, that 19 Canadians reported much worse quality of life, whereas we would expect only 10.55 if the null hypothesis were true.

APPLY YOUR KNOWLEDGE

20.3 Smoking among French men, continued. The two-way table in Exercise 20.1 displays data on the education and smoking status of a sample of French men. The null hypothesis says that there is no relationship between these variables. That is, the distribution of smoking is the same for all three levels of education.

(a) Find the expected counts for each smoking status among men with a university education. This is one row of the two-way table of expected counts.

(b) We conjecture that men with a university education smoke less than the null hypothesis calls for. Does comparing the observed and expected counts in this row agree with this conjecture?

20.4 Extracurricular activities and grades, continued. Exercise 20.2 describes a study of the relationship between the extracurricular activities of students and their grades in a required course. The null hypothesis says that there is no relationship—that is, the proportion of successful students is the same in all three extracurricular activity groups in the population.

(a) Find the expected cell counts if this hypothesis is true and display them in a two-way table.

(b) Are there any large deviations between the observed counts and the expected counts? What kind of relationship between the two variables do these deviations point to?

The chi-square test

The statistical test that tells us whether the observed differences between Canada and the United States are statistically significant compares the observed and expected counts. The test statistic that makes the comparison is the *chi-square statistic*.

CHI-SQUARE STATISTIC

The **chi-square statistic** is a measure of how far the observed counts in a two-way table are from the expected counts. The formula for the statistic is

$$X^2 = \sum \frac{(\text{observed count} - \text{expected count})^2}{\text{expected count}}$$

The sum is over all cells in the table.

The chi-square statistic is a sum of terms, one for each cell in the table. In the quality-of-life example, 75 Canadian patients reported much better quality of life. The expected count for this cell is 77.37. So the term of the chi-square statistic from this cell is

$$\frac{(\text{observed count} - \text{expected count})^2}{\text{expected count}} = \frac{(75 - 77.37)^2}{77.37}$$

$$= \frac{5.617}{77.37} = 0.073$$

Think of the chi-square statistic X^2 as a measure of the distance of the observed counts from the expected counts. Like any distance, it is always zero or positive, and it is zero only when the observed counts are exactly equal to the expected counts. Large values of X^2 are evidence against H_0 because they say that the observed counts are far from what we would expect if H_0 were true. Although the alternative hypothesis H_a is many-sided, the chi-square test is one-sided because any violation of H_0 tends to produce a large value of X^2. Small values of X^2 are not evidence against H_0.

Using technology

Calculating the expected counts and then the chi-square statistic by hand is a bit time-consuming. As usual, software saves time and always gets the arithmetic right. Figure 20.2 shows output for the chi-square test for the quality-of-life data from Minitab, Excel, and the TI-83 calculator. The Minitab output is particularly helpful.

EXAMPLE 20.4 *Chi-square from software*

We entered the two-way table (the 10 counts) from Example 20.1 into Minitab statistical software and selected the chi-square test from a menu. Minitab repeats the two-way table of observed counts, puts the expected count for each cell below the

Minitab

```
Session                                                    _|□|×|

Chi-Square Test: Canada, USA

Expected counts are printed below observed counts

                        Canada        USA      Total
Much better                75         541        616
                        77.37      538.63

Somewhat better            71         498        569
                        71.47      497.53

About the same             96         779        875
                       109.91      765.09

Somewhat worse             50         282        332
                        41.70      290.30

Much worse                 19          65         84
                        10.55       73.45

Total                     311        2165       2476

Chi-Sq =   0.073 +   0.010 +
           0.003 +   0.000 +
           1.759 +   0.253 +
           1.652 +   0.237 +
           6.766 +   0.972 = 11.725
DF = 4, P-Value = 0.020
```

These are the terms of the chi-square statistic. The layout matches the two-way table.

Figure 20.2 Output from Minitab, Excel, and the TI-83 for the two-way table in the quality-of-life study.

Excel

```
Microsoft Excel - Book1                                   _|□|×|
        A          B          C          D
 1  Observed   Canada      USA
 2                  75        541
 3                  71        498
 4                  96        779
 5                  50        282
 6                  19         65
 7
 8  Expected   Canada      USA
 9               77.37     538.63
10               71.47     497.53
11              109.91     765.09
12                41.7      290.3
13               10.55      73.45
14
15
16  CHITEST(B2:C6,B9:C13)   0.019482
17
   |◄ ◄ ► ►|\Sheet1 / Sheet2 / Sheet3 / ◄|
```

TI-83

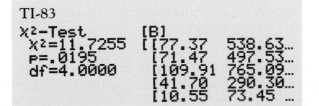

```
X²-Test        [B]
 X²=11.7255    [[77.37   538.63…
 P=.0195        [71.47   497.53…
 df=4.0000      [109.91  765.09…
                [41.70   290.30…
                [10.55    73.45 …
```

observed count, and inserts the row and column totals. Then the software calculates the chi-square statistic X^2. For these data, $X^2 = 11.725$. The P-value of the test is $P = 0.020$. There is quite good evidence that the distributions of outcomes are different in Canada and the United States.

The statistic is a sum of 10 terms, one for each cell in the table. The "Chi-Sq" display in the Minitab output shows the individual terms, as well as their sum. The first term is 0.073, just as we calculated. Look at the 10 terms. More than half the value of X^2 (6.766 out of 11.725) comes from just one cell. This points to the most important difference between the two countries: a higher proportion of Canadians report much worse quality of life. Most of the rest of X^2 comes from two other cells: more Canadians report somewhat worse quality of life, and fewer report about the same quality.

The chi-square test is the overall test for detecting relationships between two categorical variables. If the test is significant, it is important to look at the data to learn the nature of the relationship. We have three ways to look at the quality-of-life data:

- **Compare appropriate percents:** the percents of each outcome in the two countries appear in Example 20.1 and in the bar graph in Figure 20.1. This was the method we learned in Chapter 6.

- **Compare observed and expected cell counts:** which cells have more or fewer observations than we would expect if H_0 were true?

- **Look at the terms in the chi-square statistic:** which cells contribute the most to the value of X^2?

EXAMPLE 20.5 Canada and the United States: conclusions

There is a significant difference between the distributions of quality of life reported by Canadian and American patients a year after a heart attack. All three ways of comparing the distributions agree that the main difference is that a higher proportion of Canadians report that their quality of life is worse than before their heart attack. Other response variables measured in the study agree with this conclusion.

The broader conclusion, however, is controversial. Americans are likely to point to the better outcomes produced by their much more intensive treatment. Canadians reply that the differences are small, that there was no significant difference in survival, and that the American advantage comes at high cost. The resources spent on expensive treatment of heart attack victims could instead be spent on providing basic health care to the many Americans who lack it.

There is an important message here: although statistical studies shed light on issues of public policy, statistics alone rarely settles complicated questions such as "Which kind of health care system works better?"

What about the other two technologies in Figure 20.2? Neither is as helpful as Minitab. The TI-83 reports the chi-square statistic and its P-value and (on a separate screen) the table of expected cell counts. Excel lacks a menu selection for the chi-square test. You must program the spreadsheet to calculate the expected cell counts and then use the CHITEST worksheet formula. This gives the P-value but not the test statistic itself. You can of course program the spreadsheet to find the value of X^2. The Excel output in Figure 20.2 shows the observed and expected cell counts and the P-value.

APPLY YOUR KNOWLEDGE

20.5 Smoking among French men, continued. In Exercises 20.1 and 20.3, you began to analyze data on the smoking status and education of French men. Figure 20.3 displays the Minitab output for the chi-square test applied to these data.

(a) Calculate the four terms of the chi-square statistic for the bottom row (university education). Verify that your work agrees with Minitab up to roundoff error.

```
Session                                              _ □ ×

Chi-Square Test: Nonsmoker, Former, Moderate, Heavy

Expected counts are printed below observed counts

            Nonsmoke   Former   Moderate   Heavy   Total
Primary           56       54         41      36     187
               59.48    50.93      42.37   34.22

Secondary         37       43         27      32     139
               44.21    37.85      31.49   25.44

University        53       28         36      16     133
               42.31    36.22      30.14   24.34

Total            146      125        104      84     459

Chi-Sq =   0.204 +   0.186 +   0.044 +   0.092 +
           1.177 +   0.700 +   0.641 +   1.693 +
           2.704 +   1.866 +   1.141 +   2.858 = 13.305
DF = 6,  P-Value = 0.038
```

Figure 20.3 Minitab output for the two-way table of education level and smoking status among French men, for Exercise 20.5.

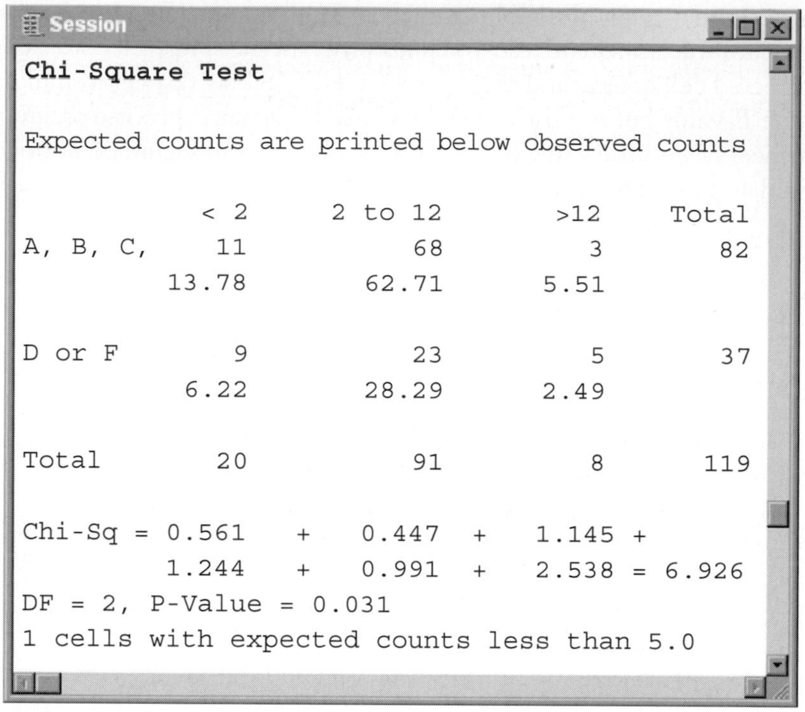

```
Session                                                    _ □ ✕

Chi-Square Test

Expected counts are printed below observed counts

                 < 2        2 to 12        >12       Total
A, B, C,         11            68            3          82
               13.78         62.71         5.51

D or F            9            23            5          37
                6.22         28.29         2.49

Total            20            91            8         119

Chi-Sq = 0.561    +    0.447    +    1.145 +
         1.244    +    0.991    +    2.538 = 6.926
DF = 2,  P-Value = 0.031
1 cells with expected counts less than 5.0
```

Figure 20.4 Minitab output for the study of extracurricular activity and success in a required course, for Exercise 20.6.

(b) According to Minitab, what is the value of the chi-square statistic X^2 and the P-value of the chi-square test?

(c) Which terms contribute the most to X^2? Write a brief summary of the nature and significance of the relationship between education and smoking.

20.6 Extracurricular activities and grades. Figure 20.4 gives Minitab output for the two-way table in Exercise 20.2.

(a) Starting from the table of expected counts, find the 6 terms of the chi-square statistic and then the statistic X^2 itself. Check your work against the computer output.

(b) What is the P-value for the test? Explain in simple language what it means to reject H_0 in this setting.

(c) Which term contributes the most to X^2? What specific relation between extracurricular activities and academic success does this term point to?

(d) Does the North Carolina State study convince you that spending more or less time on extracurricular activities *causes* changes in academic success? Explain your answer.

The chi-square distributions

Software usually finds P-values for us. The P-value for a chi-square test comes from comparing the value of the chi-square statistic with critical values for a *chi-square distribution*.

THE CHI-SQUARE DISTRIBUTIONS

The **chi-square distributions** are a family of distributions that take only positive values and are skewed to the right. A specific chi-square distribution is specified by giving its **degrees of freedom.**

The chi-square test for a two-way table with r rows and c columns uses critical values from the chi-square distribution with $(r-1)(c-1)$ degrees of freedom. The P-value is the area to the right of X^2 under the density curve of this chi-square distribution.

Figure 20.5 shows the density curves for three members of the chi-square family of distributions. As the degrees of freedom increase, the density curves become less skewed and larger values become more probable. Table E in the

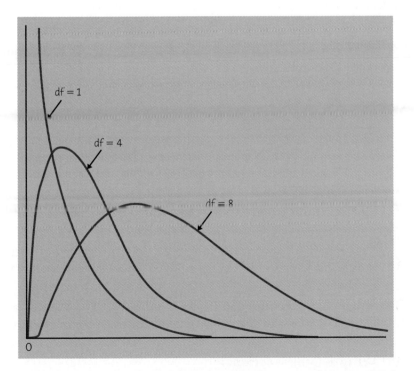

Figure 20.5 Density curves for the chi-square distributions with 1, 4, and 8 degrees of freedom. Chi-square distributions take only positive values and are right-skewed.

back of the book gives critical values for chi-square distributions. You can use Table E if software does not give you P-values for a chi-square test.

EXAMPLE 20.6   Using the chi-square table

The two-way table of 5 outcomes by 2 countries for the quality-of-life study has 5 rows and 2 columns. That is, $r = 5$ and $c = 2$. The chi-square statistic therefore has degrees of freedom

$$(r - 1)(c - 1) = (5 - 1)(2 - 1) = (4)(1) = 4$$

The Minitab and TI-83 outputs in Figure 20.2 give 4 as the degrees of freedom.

The observed value of the chi-square statistic is $X^2 = 11.725$. Look in the df $= 4$ row of Table E. The value $X^2 = 11.725$ falls between the 0.02 and 0.01 critical values of the chi-square distribution with 4 degrees of freedom. Remember that the chi-square test is always one-sided. So the P-value of $X^2 = 11.725$ is between 0.02 and 0.01. The Excel and TI-83 outputs in Figure 20.2 show that the P-value is 0.0195, close to 0.02.

df $= 4$

p	.02	.01
x^*	11.67	13.28

We know that all z and t statistics measure the size of an effect in the standard scale centered at zero. We can roughly assess the size of any z or t statistic by the 68–95–99.7 rule, though this is exact only for z. The chi-square statistic does not have any such natural interpretation. But here is a helpful fact: *the mean of any chi-square distribution is equal to its degrees of freedom*. In Example 20.6, X^2 would have mean 4 if the null hypothesis were true. The observed value $X^2 = 11.725$ is so much larger than 4 that we suspect it is significant even before we look at Table E.

APPLY YOUR KNOWLEDGE

20.7 **Smoking among French men, continued.** The Minitab output in Figure 20.3 gives the degrees of freedom for the table of education and smoking status as DF = 6.

(a) Show that this is correct for a table with 3 rows and 4 columns.

(b) Minitab gives the chi-square statistic as Chi-Sq = 13.305. Between which two entries in Table E does this value lie? Verify that Minitab's result P-Value = 0.038 lies between the tail areas for these values.

20.8 **Extracurricular activities and grades, continued.** The Minitab output in Figure 20.4 gives the degrees of freedom for the table in Exercise 20.2 as 2.

(a) Verify that this is correct.

(b) The computer gives the value of the chi-square statistic as $X^2 = 6.926$. Between what two entries in Table E does this value lie? What does the table tell you about the P-value?

(c) What would be the mean value of the statistic X^2 if the null hypothesis were true? How does the observed value of X^2 compare with the mean?

Uses of the chi-square test

Two-way tables can arise in several ways. The study of the quality of life of heart attack patients compared two independent random samples, one in Canada and the other in the United States. The design of the study fixed the sizes of the two samples. The next example illustrates a different setting, in which all the observations come from just one sample.

EXAMPLE 20.7 Marital status and job level

A study of the relationship between men's marital status and the level of their jobs used data on all 8235 male managers and professionals employed by a large manufacturing firm. Each man's job has a grade set by the company that reflects the value of that particular job to the company. The authors of the study grouped the many job grades into quarters. Grade 1 contains jobs in the lowest quarter of job grades, and grade 4 contains those in the highest quarter. Here are the data:[4]

| | Marital Status | | | |
Job Grade	Single	Married	Divorced	Widowed
1	58	874	15	8
2	222	3927	70	20
3	50	2396	34	10
4	7	533	7	4

Do these data show a statistically significant relationship between marital status and job grade?

In Example 20.7 we do not have four separate samples from the four marital statuses. We have a single group of 8235 men, each classified in two ways, by marital status and job grade. The number of men in each marital status is not fixed in advance but is known only after we have the data.

In fact, we might regard these 8235 men as an entire population rather than as a sample. The data include all the managers and professionals employed by this company. These employees are not necessarily a random sample from any larger population. Nevertheless, we would still like to decide if the relationship between marital status and job grade is statistically significant in the sense that it is too strong to happen just by chance if job grades were handed out at random to men of all marital statuses. "Not likely to happen just by chance if H_0 is true" is the usual meaning of statistical significance. In this example, that meaning makes sense even though we have data on an entire population. One of the most useful properties of chi-square is that it tests the hypothesis "the row and column variables are not related to each other" whenever this hypothesis makes sense for a two-way table.

USES OF THE CHI-SQUARE TEST

Use the chi-square test to test the null hypothesis

H_0: there is no relationship between two categorical variables

when you have a two-way table from one of these situations:

- Independent SRSs from each of several populations, with each individual classified according to one categorical variable. (The other variable says which sample the individual comes from.)
- A single SRS, with each individual classified according to both of two categorical variables.

EXAMPLE 20.8   Marital status and job level

To analyze the job grade data in Example 20.7, first do the overall chi-square test. The Minitab chi-square output appears in Figure 20.6. The observed chi-square is very large, $X^2 = 67.397$. The P-value, rounded to three decimal places, is 0. We have overwhelming evidence that job grade is related to marital status. Because the sample is so large, we are not surprised to find very significant results.

Next, do data analysis to describe the relationship. Look at the 16 terms of the chi-square statistic in Figure 20.6. The 4 terms for single men are all much larger than any other term. Together, they make up 59.887 of the total $X^2 = 67.397$. Now compare the observed and expected counts for single men. The observed counts are higher than expected in grades 1 and 2 and lower than expected in grades 3 and 4. Detailed comparison of the percent of men in each marital status whose jobs have each grade leads to the same conclusion: single men hold lower-grade jobs than married, divorced, or widowed men.

Of course, this association between marital status and job grade does not show that being single *causes* lower-grade jobs. The explanation might be as simple as the fact that single men tend to be younger and so have not yet advanced to higher grades.

Cell counts required for the chi-square test

The computer output in Figure 20.6 has one more feature. It warns us that the expected counts in 2 of the 16 cells are less than 5. The chi-square test, like the z procedures for comparing two proportions, is an approximate method that becomes more accurate as the counts in the cells of the table get larger. Fortunately, the approximation is accurate for quite modest counts. Here is a practical guideline.[5]

```
Session                                                    - □ X

Chi-Square Test

Expected counts are printed below observed counts

         Single     Married    Divorced    Widowed       Total
   1         58         874          15          8         955
          39.08      896.44       14.61       4.87

   2        222        3927          70         20        4239
         173.47     3979.05       64.86      21.62

   3         50        2396          34         10        2490
         101.90     2337.30       38.10      12.70

   4          7         533           7          4         551
          22.55      517.21        8.43       2.81

Total       337        7730         126         42        8235

Chi-Sq = 9.158   +   0.562   +   0.010   +   2.011
        13.575   +   0.681   +   0.407   +   0.121
        26.432   +   1.474   +   0.441   +   0.574
        10.722   +   0.482   +   0.243   +   0.504  =  67.397
DF = 9,  P-Value = 0.000
2 cells with expected counts less than 5.0
```

Figure 20.6 Minitab output for the two-way table of marital status and job grade in Example 20.7.

CELL COUNTS REQUIRED FOR THE CHI-SQUARE TEST

You can safely use the chi-square test with critical values from the chi-square distribution when no more than 20% of the expected counts are less than 5 and all individual expected counts are 1 or greater. In particular, all four expected counts in a 2×2 table should be 5 or greater.

Example 20.8 easily passes this test. All the expected counts are greater than 1, and only 2 out of 16 (12.5%) are less than 5.

APPLY YOUR KNOWLEDGE

20.9   Does chi-square apply? Figures 20.3 and 20.4 display Minitab output for the chi-square test applied to data on French men and on North

Carolina State University students. Using the information in the output, verify that both studies meet the cell count requirement for use of chi-square.

20.10 **Majors for men and women in business.** A study of the career plans of young women and men sent questionnaires to all 722 members of the senior class in the College of Business Administration at the University of Illinois. One question asked which major within the business program the student had chosen. Here are the data from the students who responded:[6]

	Female	Male
Accounting	68	56
Administration	91	40
Economics	5	6
Finance	61	59

This is an example of a single sample classified according to two categorical variables (gender and major).

(a) Test the null hypothesis that there is no relationship between the gender of students and their choice of major. Give a P-value and state your conclusion.

(b) Verify that the expected cell counts satisfy the requirement for use of chi-square.

(c) Describe the differences between the distributions of majors for women and men with percents, with a graph, and in words.

(d) Which two cells have the largest terms of the chi-square statistic? How do the observed and expected counts differ in these cells? (This should strengthen your conclusions in (b).)

(e) What percent of the students did not respond to the questionnaire? The nonresponse weakens conclusions drawn from these data.

The chi-square test and the z test*

One use of the chi-square test is to compare the proportions of successes in any number of groups. If the r rows of the two-way table are r groups and the columns are "success" and "failure," the counts form an $r \times 2$ table. P-values come from the chi-square distribution with $r - 1$ degrees of freedom. If $r = 2$, we are comparing just two proportions. We have two ways to do this: the z test from Chapter 19 and the chi-square test with 1 degree of freedom for a 2×2 table. *These two tests always agree.* In fact, the chi-square statistic X^2 is just the square of the z statistic, and the P-value for X^2 is exactly the same as

---

*The remainder of the material in this chapter is optional.

the two-sided P-value for z. We recommend using the z test to compare two proportions, because it gives you the choice of a one-sided test and is related to a confidence interval for the difference $p_1 - p_2$.

APPLY YOUR KNOWLEDGE

20.11 Treating ulcers. Gastric freezing was once a recommended treatment for ulcers in the upper intestine. Use of gastric freezing stopped after experiments showed it had no effect. One randomized comparative experiment found that 28 of the 82 gastric-freezing patients improved, while 30 of the 78 patients in the placebo group improved.[7] We can test the hypothesis of "no difference" between the two groups in two ways: using the two-sample z statistic or using the chi-square statistic.

(a) State the null hypothesis with a two-sided alternative and carry out the z test. What is the P-value from Table A?

(b) Present the data in a 2×2 table. Use the chi-square test to test the hypothesis from (a). Verify that the X^2 statistic is the square of the z statistic. Use software or Table E to verify that the chi-square P-value agrees with the z result (up to the accuracy of the table if you do not use software).

(c) What do you conclude about the effectiveness of gastric freezing as a treatment for ulcers?

The chi-square test for goodness of fit*

The most common and most important use of the chi-square statistic is to test the hypothesis that there is *no relationship between two categorical variables*. A variation of the statistic can be used to test a different kind of null hypothesis: that *a categorical variable has a specified distribution*. Here is an example that illustrates this use of chi-square.

More chi-square tests

There are other chi-square tests for hypotheses more specific than "no relationship." A sociologist places people in classes by social status, waits ten years, then classifies the same people again. The row and column variables are the classes at the two times. She might test the hypothesis that there has been no change in the overall distribution of social status in the group. Or she might ask if moves up in status are balanced by matching moves down. These and other null hypotheses can be tested by variations of the chi-square test.

EXAMPLE 20.9   Never on Sunday?

Births are not evenly distributed across the days of the week. Fewer babies are born on Saturday and Sunday than on other days, probably because doctors find weekend births inconvenient. Exercise 1.4 (page 9) gives national data that demonstrate this fact.

A random sample of 140 births from local records shows this distribution across the days of the week:

Day	Sun.	Mon.	Tue.	Wed.	Thu.	Fri.	Sat.
Births	13	23	24	20	27	18	15

Sure enough, the two smallest counts of births are on Saturday and Sunday. Do these data give significant evidence that local births are not equally likely on all days of the week?

The chi-square test answers the question of Example 20.9 by comparing observed counts with expected counts under the null hypothesis. The null hypothesis for births says that they *are* evenly distributed. To state the hypotheses carefully, write the discrete probability distribution for days of birth:

Day	Sun.	Mon.	Tue.	Wed.	Thu.	Fri.	Sat.
Probability	p_1	p_2	p_3	p_4	p_5	p_6	p_7

The null hypothesis says that the probabilities are the same on all days. In that case, all 7 probabilities must equal 1/7. So the null hypothesis is

$$H_0: p_1 = p_2 = p_3 = p_4 = p_5 = p_6 = p_7 = \frac{1}{7}$$

The alternative hypothesis says that days are *not* all equally probable:

$$H_a: \text{not all } p_i = \frac{1}{7}$$

As usual in chi-square tests, H_a is a "many-sided" hypothesis that simply says that H_0 is not true. The chi-square statistic is also as usual:

$$X^2 = \sum \frac{(\text{observed count} - \text{expected count})^2}{\text{expected count}}$$

The expected count for an outcome with probability p is np, exactly as in Example 20.2. Under the null hypothesis, all the probabilities p_i are the same, so all 7 expected counts are equal to

$$np_i = 140 \times \frac{1}{7} = 20$$

The chi-square statistic is therefore

$$X^2 = \sum \frac{(\text{observed count} - 20)^2}{20}$$
$$= \frac{(13 - 20)^2}{20} + \frac{(23 - 20)^2}{20} + \cdots + \frac{(15 - 20)^2}{20}$$
$$= 7.6$$

This new use of X^2 requires a different degrees of freedom. To find the P-value, compare X^2 with critical values from the chi-square distribution with degrees of freedom one fewer than the number of values the birth day can take. That's $7 - 1 = 6$ degrees of freedom. From Table E, we see that $X^2 = 7.6$ is smaller than the smallest entry in the df $= 6$ row, which is the critical value for tail area 0.25. The P-value is therefore greater than 0.25 (software gives the more exact value $P = 0.269$). These 140 births don't give convincing evidence that births are not equally likely on all days of the week.

df = 6

p	.25	.20
x^*	7.84	8.56

The chi-square test applied to the hypothesis that a categorical variable has a specified distribution is called the test for *goodness of fit*. The idea is that the test assesses whether the observed counts "fit" the distribution. The only differences between the test of fit and the test for a two-way table are that the expected counts are based on the distribution specified by the null hypothesis and that the degrees of freedom is one less than the number of possible outcomes in this distribution. Here are the details.

THE CHI-SQUARE TEST FOR GOODNESS OF FIT

A categorical variable has k possible outcomes, with probabilities p_1, p_2, p_3, ..., p_k. That is, p_i is the probability of the ith outcome. We have n independent observations from this categorical variable.

To test the null hypothesis that the probabilities have specified values

$$H_0: p_1 = p_{10}, \ p_2 = p_{20}, \ \ldots, \ p_k = p_{k0}$$

use the **chi-square statistic**

$$X^2 = \sum \frac{(\text{count of outcome } i - np_{i0})^2}{np_{i0}}$$

The P-value is the area to the right of X^2 under the density curve of the chi-square distribution with $k - 1$ degrees of freedom.

In Example 20.9, the outcomes are days of the week, with $k = 7$. The null hypothesis says that the probability of a birth on the ith day is $p_{i0} = 1/7$ for all days. We observe $n = 140$ births and count how many fall on each day. These are the counts used in the chi-square statistic.

20.12 More on birth days. Births really are not evenly distributed across the days of the week. The data in Example 20.9 failed to reject this null hypothesis because of random variation in a quite small number of births. Here are data on 700 births in the same locale:

Day	Sun.	Mon.	Tue.	Wed.	Thu.	Fri.	Sat.
Births	84	110	124	104	94	112	72

(a) The null hypothesis is that all days are equally probable. What are the probabilities specified by this null hypothesis? What are the expected counts for each day in 700 births?

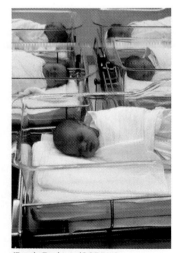
(Randy Duchaine/CORBIS)

(b) Calculate the chi-square statistic for goodness of fit.

(c) What are the degrees of freedom for this statistic? Do 700 births give significant evidence that births are not equally probable on all days of the week?

20.13 **Course grades.** Most students in a large statistics course are taught by teaching assistants (TAs). One section is taught by the course supervisor, a senior professor. The distribution of grades for the hundreds of students taught by TAs this semester was

Grade	A	B	C	D/F
Probability	0.32	0.41	0.20	0.07

The grades assigned by the professor to students in his section were

Grade	A	B	C	D/F
Count	22	38	20	11

(These data are real. We won't say when and where, but the professor was not the author of this book.)

(a) What percents of each grade did students in the professor's section earn? In what ways does this distribution of grades differ from the TA distribution?

(b) Because the TA distribution is based on hundreds of students, we are willing to regard it as a fixed probability distribution. If the professor's grading follows this distribution, what are the expected counts of each grade in his section?

(c) Does the chi-square test for goodness of fit give good evidence that the professor follows a different grade distribution? (Give the test statistic, its P-value, and your conclusion.)

20.14 **Benford's law.** The first digits of legitimate records such as invoices and expense account claims often follow the distribution known as Benford's law:[8]

First digit	1	2	3	4	5	6	7	8	9
Probability	0.301	0.176	0.125	0.097	0.079	0.067	0.058	0.051	0.046

A purchasing manager is faking invoices to steer payments to a company owned by her brother. She cleverly generates random numbers to avoid obvious patterns. Here are the counts of first digits in a sample of 45 invoices examined by an auditor:

First digit	1	2	3	4	5	6	7	8	9
Count	6	4	6	7	3	5	6	4	4

(a) The auditor tests the null hypothesis that the first digits follow Benford's law. What are the expected counts for each of the 9 possible first digits in a sample of 45 invoices?

(b) Find the chi-square statistic for goodness of fit.

(c) Do the sample invoices give good evidence that first digits do not follow Benford's law in the population of all invoices filed by this manager? By examining the terms of the chi-square statistic, describe the most important deviations of the sample from the null hypothesis.

20.15 Colors of M&M's. The M&M's candies Web site www.mms.com says that the distribution of colors for milk chocolate M&M's is

Color	Purple	Yellow	Red	Orange	Brown	Green	Blue
Probability	0.2	0.2	0.2	0.1	0.1	0.1	0.1

Open a package of M&M's: out spill 57 candies. (The count varies slightly from package to package.) The color counts are

Color	Purple	Yellow	Red	Orange	Brown	Green	Blue
Count	11	13	5	7	9	9	3

How well do the counts from this package fit the claimed distribution? Do a chi-square test of fit, report the P-value, and give a conclusion.

Chapter 20 SUMMARY

The **chi-square test** for a two-way table tests the null hypothesis H_0 that there is no relationship between the row variable and the column variable.

The test compares the observed counts of observations in the cells of the table with the counts that would be expected if H_0 were true. The **expected count** in any cell is

$$\text{expected count} = \frac{\text{row total} \times \text{column total}}{\text{table total}}$$

The **chi-square statistic** is

$$X^2 = \sum \frac{(\text{observed count} - \text{expected count})^2}{\text{expected count}}$$

The chi-square test compares the value of the statistic X^2 with critical values from the **chi-square distribution** with $(r-1)(c-1)$ **degrees of freedom.** Large values of X^2 are evidence against H_0, so the P-value is the area under the chi-square density curve to the right of X^2.

The chi-square distribution is an approximation to the distribution of the statistic X^2. You can safely use this approximation when all expected cell counts are at least 1 and no more than 20% are less than 5.

If the chi-square test finds a statistically significant relationship between the row and column variables in a two-way table, do data analysis to describe the nature of the relationship. You can do this by comparing well-chosen percents, comparing the observed counts with the expected counts, and looking for the largest **terms of the chi-square statistic.**

STATISTICS IN SUMMARY

After studying this chapter, you should be able to do the following.

A. TWO-WAY TABLES

1. Understand that the data for a chi-square test must be presented as a two-way table of counts of outcomes.

2. Use percents to describe the relationship between any two categorical variables, starting from the counts in a two-way table.

B. INTERPRETING CHI-SQUARE TESTS

1. Locate expected cell counts, the chi-square statistic, and its P-value in output from your software or calculator.

2. Explain what null hypothesis the chi-square statistic tests in a specific two-way table.

3. If the test is significant, use percents, comparison of expected and observed counts, and the terms of the chi-square statistic to see what deviations from the null hypothesis are most important.

C. DOING CHI-SQUARE TESTS (optional if you use technology)

1. Calculate the expected count for any cell from the observed counts in a two-way table.

2. Calculate the term of the chi-square statistic for any cell, as well as the overall statistic.

3. Give the degrees of freedom of a chi-square statistic. Make a quick assessment of the significance of the statistic by comparing the observed value with the degrees of freedom.

4. Use the chi-square critical values in Table E to approximate the
    P-value of a chi-square test.

Chapter 20 EXERCISES

*If you have access to software or a graphing calculator, use it to speed
your analysis of the data in these exercises.*

20.16 More on Canada versus the United States. The study described in
Example 20.1 also asked the patients to rate their physical capacity a
year after their heart attack, relative to what it was before the attack.
Here are the counts of responses:

Physical capacity	Canada	United States
Much better	37	325
Somewhat better	56	325
About the same	109	1039
Somewhat worse	78	390
Much worse	31	86
Total	311	2165

Do both data analysis and a formal test to compare the distributions
of outcomes in the two countries. Write a clear summary of your
findings.

20.17 Secondhand stores. Shopping at secondhand stores is becoming more
popular and has even attracted the attention of business schools. A
study of customers' attitudes toward secondhand stores interviewed
samples of shoppers at two secondhand stores of the same chain in two
cities. Here is the two-way table comparing the income distributions
of the shoppers in the two stores:[9]

Income	City 1	City 2
Under $10,000	70	62
$10,000 to $19,999	52	63
$20,000 to $24,999	69	50
$25,000 to $34,999	22	19
$35,000 or more	28	24

(Ryan McVay/PhotoDisc/Getty Images)

A statistical calculator gives the chi-square statistic for this table as
$X^2 = 3.955$. Is there good evidence that customers at the two stores
have different income distributions? (Give the degrees of freedom, the
P-value, and your conclusion.)

20.18 Child-care workers. A large study of child care used samples from the
data tapes of the Current Population Survey over a period of several

years. The result is close to an SRS of child-care workers. The Current Population Survey has three classes of child-care workers: private household, nonhousehold, and preschool teacher. Here are data on the number of blacks among women workers in these three classes:[10]

	Total	Black
Household	2455	172
Nonhousehold	1191	167
Teachers	659	86

(a) What percent of each class of child-care workers is black?

(b) Make a two-way table of class of worker by race (black or other).

(c) Can we safely use the chi-square test? What null and alternative hypotheses does X^2 test?

(d) The chi-square statistic for this table is $X^2 = 53.194$. What are its degrees of freedom? Use Table E to approximate the P-value.

(e) What do you conclude from these data?

20.19 **Female college professors.** Exercise 6.20 (page 147) gives these data on the rank and gender of professors at a large university:

	Female	Male	Total
Assistant professors	126	213	339
Associate professors	149	411	560
Professors	60	662	722
Total	335	1286	1621

Women are underrepresented at the full professor level. Is there a significant difference between the distribution of ranks for female and male faculty? Which cells contribute the most to the overall chi-square statistic?

20.20 **Regulating guns.** "Do you think there should be a law that would ban possession of handguns except for the police and other authorized persons?" Exercise I.14 (page 160) gives these data on the responses of a random sample of adults, broken down by level of education:

	Yes	No
Less than high school	58	58
High school graduate	84	129
Some college	169	294
College graduate	98	135
Postgraduate degree	77	99

Carry out a chi-square test, giving the statistic X^2 and its P-value. How strong is the evidence that people with different levels of education feel differently about banning private possession of handguns?

20.21 Smoking by students and their parents. How are the smoking habits of students related to their parents' smoking? Here are data from a survey of students in eight Arizona high schools:[11]

	Student smokes	Student does not smoke
Both parents smoke	400	1380
One parent smokes	416	1823
Neither parent smokes	188	1168

(a) Find the percent of students who smoke in each of the three parent groups. Make a graph to compare these percents. Describe the association between parent smoking and student smoking.

(b) Explain in words what the null hypothesis for the chi-square test says about student smoking.

(c) Find the expected counts if H_0 is true, and display them in a two-way table similar to the table of observed counts.

(d) Compare the tables of observed and expected counts. Explain how the comparison expresses the same association you saw in (a).

(e) Give the chi-square statistic and its P-value. Examine the terms of chi-square to confirm the pattern you saw in (a) and (d). What is your overall conclusion?

20.22 Python eggs. How is the hatching of water python eggs influenced by the temperature of the snake's nest? Researchers assigned newly laid eggs to one of three temperatures: hot, neutral, or cold. Hot duplicates the extra warmth provided by the mother python, and cold duplicates the absence of the mother. Here are the data on the number of eggs and the number that hatched:[12]

(David A. Northcott/CORBIS)

	Eggs	Hatched
Cold	27	16
Neutral	56	38
Hot	104	75

(a) Make a two-way table of temperature by outcome (hatched or not).

(b) Calculate the percent of eggs in each group that hatched. The researchers anticipated that eggs would not hatch at cold temperatures. Do the data support that anticipation?

(c) Are there significant differences among the proportions of eggs that hatched in the three groups?

20.23 Ebonics awareness. Ebonics, often called "black English," is a variety of English common among blacks in the United States. How aware are college students of the existence of Ebonics? Here are data from a sample of students at a racially diverse college in the South:[13]

	Aware	Not aware
Black students	121	11
White students	159	21
Other students	75	28

Do both data analysis and a formal test to compare awareness in the three groups of students. Write a clear summary of your findings.

20.24 Do you use cocaine? Sample surveys on sensitive issues can give different results depending on how the question is asked. A University of Wisconsin study divided 2400 respondents into 3 groups at random. All were asked if they had ever used cocaine. One group of 800 was interviewed by phone; 21% said they had used cocaine. Another 800 people were asked the question in a one-on-one personal interview; 25% said "Yes." The remaining 800 were allowed to make an anonymous written response; 28% said "Yes."[14] Are there statistically significant differences among these proportions? State the hypotheses, convert the information given into a two-way table, give the test statistic and its P-value, and state your conclusions.

20.25 How are schools doing? Exercise I.13 (page 159) gives these data on the responses of random samples of black, Hispanic, and white parents to the question "Are the high schools in your state doing an excellent, good, fair or poor job, or don't you know enough to say?"

	Black parents	Hispanic parents	White parents
Excellent	12	34	22
Good	69	55	81
Fair	75	61	60
Poor	24	24	24
Don't know	22	28	14
Total	202	202	201

Are the differences in the distributions of responses for the three groups of parents statistically significant? What departures from the null hypothesis "no relationship between group and response" contribute most to the value of the chi-square statistic? Write a brief conclusion based on your analysis.

20.26 Ethnicity and seat belt use. How does seat belt use vary with drivers' race or ethnic group? The answer depends on gender (males are less likely to buckle up) and also on location. Here are data on a random sample of male drivers observed in Houston:[15]

	Drivers	Belted
Black	369	273
Hispanic	540	372
White	257	193

(a) The table gives the number of drivers in each group and the number of these who were wearing seat belts. Make a two-way table of group by belted or not.

(b) Are there statistically significant differences in seat belt use among men in these three groups? If there are, describe the differences.

20.27 Did the randomization work? After randomly assigning subjects to treatments in a randomized comparative experiment, we can compare the treatment groups to see how well the randomization worked. We hope to find no significant differences among the groups. A study of how to provide premature infants with a substance essential to their development assigned infants at random to receive one of four types of supplement, called PBM, NLCP, PL-LCP, and TG-LCP.[16]

(a) The subjects were 77 premature infants. Outline the design of the experiment if 20 are assigned to the PBM group and 19 to each of the other treatments.

(b) The random assignment resulted in 9 females in the TG-LCP group and 11 females in each of the other groups. Make a two-way table of group by gender and do a chi-square test to see if there are significant differences among the groups. What do you find?

20.28 The Mediterranean diet. Cancer of the colon and rectum is less common in the Mediterranean region than in other Western countries. The Mediterranean diet contains little animal fat and lots of olive oil. Italian researchers compared 1953 patients with colon or rectal cancer with a control group of 4154 patients admitted to the same hospitals for unrelated reasons. They estimated consumption of various foods from a detailed interview, then divided the patients into three groups according to their consumption of olive oil. Here are some of the data:[17]

	Olive Oil			
	Low	Medium	High	Total
Colon cancer	398	397	430	1225
Rectal cancer	250	241	237	728
Controls	1368	1377	1409	4154

(a) Is this study an experiment? Explain your answer.

(b) Is high olive oil consumption more common among patients without cancer than in patients with colon cancer or rectal cancer?

(c) Find the chi-square statistic X^2. What would be the mean of X^2 if the null hypothesis (no relationship) were true? What does comparing the observed value of X^2 with this mean suggest? What is the P-value? What do you conclude?

(d) The investigators report that "less than 4% of cases or controls refused to participate." Why does this fact strengthen our confidence in the results?

20.29 Nobody is home in July. The success of a sample survey can depend on the season of the year. The Italian National Statistical Institute kept records of nonresponse to one of its national telephone surveys. All calls were made between 7 p.m. and 10 p.m. Here is a table of the percents of responses and of three types of nonresponse at different seasons. The percents in each row add to 100%.[18]

	Calls	Successful	Nonresponse		
Season	made	interviews	No answer	Busy signal	Refusal
Jan. 1 to Apr. 13	1558	68.5%	21.4%	5.8%	4.3%
Apr. 21 to June 20	1589	52.4%	35.8%	6.4%	5.4%
July 1 to Aug. 31	2075	43.4%	41.5%	8.6%	6.5%
Sept. 1 to Dec. 15	2638	60.0%	30.0%	5.3%	4.7%

(a) What are the degrees of freedom for the chi-square test of the hypothesis that the distribution of responses varies with the season? (Don't do the test. The sample sizes are so large that the results are sure to be highly significant.)

(b) Consider just the proportion of successful interviews. Describe how this proportion varies with the seasons, and assess the statistical significance of the changes. What do you think explains the changes? (Look at the full table for ideas.)

(c) (Optional) It is incorrect to apply the chi-square test to percents rather than to counts. If you enter the 4 × 4 table of percents above into statistical software and ask for a chi-square test, well-written software should give an error message. (Counts must be whole numbers, so the software should check that.) Try this using your software or calculator, and report the result.

20.30 **Nobody is home in July, continued.** Continue the analysis of the data in the previous exercise by considering just the proportion of people called who refused to participate. We might think that the refusal rate changes less with the season than, for example, the rate of "no answer." State the hypothesis that the refusal rate does not change with the season. Check that you can safely use the chi-square test. Carry out the test. What do you conclude?

20.31 **Titanic!** In 1912 the luxury liner *Titanic*, on its first voyage across the Atlantic, struck an iceberg and sank. Some passengers got off the ship in lifeboats, but many died. Think of the *Titanic* disaster as an experiment in how the people of that time behaved when faced with death in a situation where only some can escape. The passengers are a sample from the population of their peers. Here is information about who lived and who died, by gender and economic status. (The data leave out a few passengers whose economic status is unknown.)[19]

(Stock Montage, Inc.)

	Men			Women		
Status	Died	Survived	Status	Died	Survived
Highest	111	61	Highest	6	126
Middle	150	22	Middle	13	90
Lowest	419	85	Lowest	107	101
Total	680	168	Total	126	317

(a) Compare the percents of men and of women who died. Is there strong evidence that a higher proportion of men die in such situations? Why do you think this happened?

(b) Look only at the women. Describe how the three economic classes differ in the percent of women who died. Are these differences statistically significant?

(c) Now look only at the men and answer the same questions.

20.32 **Unhappy rats and tumors.** It seems that the attitude of cancer patients can influence the progress of their disease. We can't experiment with humans, but here is a rat experiment on this theme. Inject 60 rats with tumor cells and then divide them at random into two groups of 30. All the rats receive electric shocks, but rats in Group 1 can end the shock by pressing a lever. (Rats learn this sort of thing quickly.) The rats in Group 2 cannot control the shocks, which presumably makes them feel helpless and unhappy. We suspect that the rats in Group 1 will develop fewer tumors. The results: 11 of the Group 1 rats and 22 of the Group 2 rats developed tumors.[20]

(a) State the null and alternative hypotheses for this investigation. Explain why the z test rather than the chi-square test for a 2 × 2 table is the proper test.

(b) Carry out the test and report your conclusion.

20.33 Secondhand stores, continued. Exercise 20.17 describes a study of shoppers at secondhand stores in two cities. The breakdown of the respondents by gender is as follows:

	City 1	City 2
Men	38	68
Women	203	150
Total	241	218

Is there a significant difference between the proportions of women customers in the two cities?

(a) State the null hypothesis, find the sample proportions of women in both cities, do a two-sided z test, and give a P-value using Table A.

(b) Calculate the chi-square statistic X^2 and show that it is the square of the z statistic. Show that the P-value from Table E agrees (up to the accuracy of the table) with your result from (a).

(c) Give a 95% confidence interval for the difference between the proportions of women customers in the two cities.

Chapter 20 MEDIA EXERCISES

20.34 Alcoholism in twins. The EESEE story "Alcoholism in Twins" describes a study of possible genetic influences on alcoholism. The subjects were pairs of adult female twins, identified from the Virginia Twin Registry, which lists all twins born in Virginia. Each pair of twins was classified as identical or fraternal. Only identical twins share exactly the same genes. Based on an interview, each woman was classified as a problem drinker or not. Here are the data for the 1030 pairs of twins for which information was available:

Problem drinker	Identical	Fraternal
Neither	443	301
One	102	113
Both	45	26
Total	590	440

(a) Is there a significant relationship between type of twin and the presence of problem drinking in the twin pair? Which cells contribute heavily to the chi-square value?

(b) Your result in (a) suggests a clearer analysis. Make a 2 × 2 table of "same or different" problem-drinking behavior within a twin pair by type of twin. To do this, combine the "Neither" and "Both" categories to form the "Same behavior" category. If heredity influences behavior, we would expect a higher proportion of identical twins to show the same behavior. Is there a significant effect of this kind?

(Peter Pinnock/ImageState-Pictor/PictureQuest)

(Benelux Press/Index Stock Imagery/
PictureQuest)

scatterplot

Inference for Regression

When a scatterplot shows a linear relationship between a quantitative explanatory variable x and a quantitative response variable y, we can use the least-squares line fitted to the data to predict y for a given value of x. Now we want to do tests and confidence intervals in this setting.

EXAMPLE 21.1 Crying and IQ

Infants who cry easily may be more easily stimulated than others and this may be a sign of higher IQ. Child development researchers explored the relationship between the crying of infants four to ten days old and their later IQ test scores. A snap of a rubber band on the sole of the foot caused the infants to cry. The researchers recorded the crying and measured its intensity by the number of peaks in the most active 20 seconds. They later measured the children's IQ at age three years using the Stanford-Binet IQ test. Table 21.1 contains data on 38 infants.[1] We begin with data analysis, following the usual steps.

Plot and interpret. Figure 21.1 is a **scatterplot** of the crying data. Plot the explanatory variable (count of crying peaks) horizontally and the response variable (IQ) vertically. Look for the form, direction, and strength of the relationship as well as for outliers or other deviations. There is a moderate positive linear relationship, with no extreme outliers or potentially influential observations.

TABLE 21.1 Infants' crying and IQ scores

Crying	IQ	Crying	IQ	Crying	IQ	Crying	IQ
10	87	20	90	17	94	12	94
12	97	16	100	19	103	12	103
9	103	23	103	13	104	14	106
16	106	27	108	18	109	10	109
18	109	15	112	18	112	23	113
15	114	21	114	16	118	9	119
12	119	12	120	19	120	16	124
20	132	15	133	22	135	31	135
16	136	17	141	30	155	22	157
33	159	13	162				

Numerical summary. Because the scatterplot shows a roughly linear (straight-line) pattern, the **correlation** describes the direction and strength of the relationship. The correlation between crying and IQ is $r = 0.455$.

correlation

Mathematical model. We are interested in predicting the response from information about the explanatory variable. So we find the **least-squares regression line** for predicting IQ from crying. This line lies as close as possible to the points (in the sense of least squares) in the vertical (y) direction. The equation

least-squares line

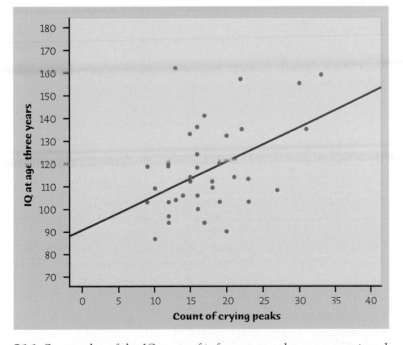

Figure 21.1 Scatterplot of the IQ score of infants at age three years against the intensity of their crying soon after birth, for Example 21.1.

May the longer name win!

Regression is far from perfect, but it beats most other ways of predicting. A writer in the early 1960s noted a simple method for predicting presidential elections: just choose the candidate with the longer name. In the 22 elections from 1876 to 1960, this method failed only once. Let's hope that the writer didn't bet the family silver on this idea. The nine elections from 1964 to 2000 presented eight tests of the "long name wins" method (the 1980 candidates and 2000 candidates had names of the same length). The longer name lost five of the eight.

of the least-squares regression line is

$$\hat{y} = a + bx$$
$$= 91.27 + 1.493x$$

We use the notation $\hat{y}$ to remind ourselves that the regression line gives *predictions* of IQ. The predictions usually won't agree exactly with the actual values of the IQ measured several years later. Drawing the least-squares line on the scatterplot helps us see the overall pattern. Because $r^2 = 0.207$, only about 21% of the variation in IQ scores is explained by crying intensity. Prediction of IQ will not be very accurate. It is nonetheless impressive that behavior soon after birth can even partly predict IQ several years later.

The regression model

The slope b and intercept a of the least-squares line are *statistics*. That is, we calculated them from the sample data. These statistics would take somewhat different values if we repeated the study with different infants. To do formal inference, we think of a and b as estimates of unknown *parameters*. The parameters appear in a mathematical model of the process that produces our data. This model lists the conditions that are needed for inference about regression.

THE REGRESSION MODEL

We have n observations on an explanatory variable x and a response variable y. Our goal is to study or predict the behavior of y for given values of x.

- For any fixed value of x, the response y varies according to a Normal distribution. Repeated responses y are independent of each other.

- The mean response μ_y has a straight-line relationship with x:

$$\mu_y = \alpha + \beta x$$

  The slope β and intercept α are unknown parameters.

- The standard deviation of y (call it σ) is the same for all values of x. The value of σ is unknown.

The regression model has three parameters, α, β, and σ.

---

true regression line

The heart of this model is that there is an "on the average" straight-line relationship between y and x. The **true regression line** $\mu_y = \alpha + \beta x$ says that the *mean* response μ_y moves along a straight line as the explanatory variable x changes. We can't observe the true regression line. The values of y that we do observe vary about their means according to a Normal distribution. If we hold x fixed and take many observations on y, the Normal pattern will eventually appear in a stemplot or histogram. In practice, we observe y for many different values of x, so that we see an overall linear pattern formed by points scattered about the true line. The standard deviation σ determines whether

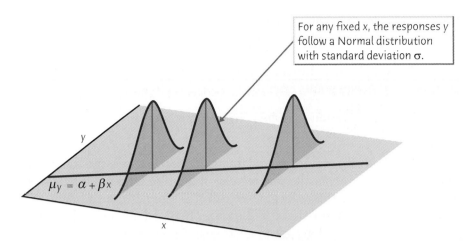

For any fixed x, the responses y follow a Normal distribution with standard deviation σ.

Figure 21.2 The regression model. The line is the true regression line, which shows how the mean response μ_y changes as the explanatory variable x changes. For any fixed value of x, the observed response y varies according to a Normal distribution having mean μ_y.

the points fall close to the true regression line (small σ) or are widely scattered (large σ).

Figure 21.2 shows the regression model in picture form. The line in the figure is the true regression line. The mean of the response y moves along this line as the explanatory variable x takes different values. The Normal curves show how y will vary when x is held fixed at different values. All of the curves have the same σ, so the variability of y is the same for all values of x. You should check the conditions for inference when you do inference about regression. We will see later how to do that.

Estimating the model parameters

The first step in inference is to estimate the unknown parameters α, β, and σ. When the regression model describes our data and we calculate the least-squares line $\hat{y} = a + bx$, **the slope b of the least-squares line is an unbiased estimator of the true slope β, and the intercept a of the least-squares line is an unbiased estimator of the true intercept α.**

EXAMPLE 21.2   Crying and IQ: slope and intercept

The data in Figure 21.1 fit the regression model of scatter about an invisible true regression line reasonably well. The least-squares line is $\hat{y} = 91.27 + 1.493x$. The slope is particularly important. *A slope is a rate of change.* The true slope β says how much higher average IQ is for children with one more peak in their crying measurement. Because $b = 1.493$ estimates the unknown β, we estimate that on the average IQ is about 1.5 points higher for each added crying peak.

We need the intercept $a = 91.27$ to draw the line, but it has no statistical meaning in this example. No child had fewer than 9 crying peaks, so we have no data near $x = 0$. We suspect that all normal children would cry when snapped with a rubber band, so that we will never observe $x = 0$.

residuals

The remaining parameter of the model is the standard deviation σ, which describes the variability of the response y about the true regression line. The least-squares line estimates the true regression line. So the **residuals** estimate how much y varies about the true line. Recall that the residuals are the vertical deviations of the data points from the least-squares line:

$$\text{residual} = \text{observed } y - \text{ predicted } y$$
$$= y - \hat{y}$$

There are n residuals, one for each data point. Because σ is the standard deviation of responses about the true regression line, we estimate it by a sample standard deviation of the residuals. We call this sample standard deviation a *standard error* to emphasize that it is estimated from data. The residuals from a least-squares line always have mean zero. That simplifies their standard error.

REGRESSION STANDARD ERROR

The **regression standard error** is

$$s = \sqrt{\frac{1}{n-2}\sum \text{residual}^2}$$
$$= \sqrt{\frac{1}{n-2}\sum (y - \hat{y})^2}$$

Use s to estimate the unknown σ in the regression model.

---

degrees of freedom

Because we use the regression standard error so often, we just call it s. Notice that s^2 is an average of the squared deviations of the data points from the line, so it qualifies as a variance. We average the squared deviations by dividing by $n - 2$, the number of data points less 2. It turns out that if we know $n - 2$ of the n residuals, the other two are determined. That is, $n - 2$ are the **degrees of freedom** of s. We first met the idea of degrees of freedom in the case of the ordinary sample standard deviation of n observations, which has $n - 1$ degrees of freedom. Now we observe two variables rather than one, and the proper degrees of freedom are $n - 2$ rather than $n - 1$.

Calculating s is unpleasant. You must find the predicted response for each x in your data set, then the residuals, and then s. In practice you will use software that does this arithmetic instantly. Nonetheless, here is an example to help you understand the standard error s.

EXAMPLE 21.3   *Crying and IQ: residuals and standard error*

Table 21.1 shows that the first infant studied had 10 crying peaks and a later IQ of 87. The predicted IQ for $x = 10$ is

$$\hat{y} = 91.27 + 1.493x$$
$$= 91.27 + 1.493(10) = 106.2$$

The residual for this observation is

$$\text{residual} = y - \hat{y}$$
$$= 87 - 106.2 = -19.2$$

That is, the observed IQ for this infant lies 19.2 points below the least-squares line on the scatterplot.

Repeat this calculation 37 more times, once for each subject. The 38 residuals are

−19.20	−31.13	−22.65	−15.18	−12.18	−15.15	−16.63	−6.18
−1.70	−22.60	−6.68	−6.17	−9.15	−23.58	−9.14	2.80
−9.14	−1.66	−6.14	−12.60	0.34	−8.62	2.85	14.30
9.82	10.82	0.37	8.85	10.87	19.34	10.89	−2.55
20.85	24.35	18.94	32.89	18.47	51.32		

Check the calculations by verifying that the sum of the residuals is zero. It is 0.04, not quite zero, because of roundoff error. Another reason to use software in regression is that roundoff errors in hand calculation can accumulate to make the results inaccurate.

The variance about the line is

$$s^2 = \frac{1}{n-2} \sum \text{residual}^2$$
$$= \frac{1}{38-2} [(-19.20)^2 + (-31.13)^2 + \cdots + (51.32)^2]$$
$$= \frac{1}{36} (11023.3) = 306.20$$

Finally, the regression standard error is

$$s = \sqrt{306.20} = 17.50$$

We will study several kinds of inference in the regression setting. The regression standard error s is the key measure of the variability of the responses in regression. It is part of the standard error of all the statistics we will use for inference.

APPLY YOUR KNOWLEDGE

21.1 An extinct beast. *Archaeopteryx* is an extinct beast having feathers like a bird but teeth and a long bony tail like a reptile. Here are the lengths in centimeters of the femur (a leg bone) and the humerus (a bone in the upper arm) for the five fossil specimens that preserve both bones:[2]

Femur	38	56	59	64	74
Humerus	41	63	70	72	84

The strong linear relationship between the lengths of the two bones helped persuade scientists that all five specimens belong to the same species.

(a) Examine the data. Make a scatterplot with femur length as the explanatory variable. Use your calculator to obtain the correlation r

(John D. Cunningham/Visuals Unlimited)

and the equation of the least-squares regression line. Do you think that femur length will allow good prediction of humerus length?

(b) Explain in words what the slope β of the true regression line says about *Archaeopteryx*. Based on the data, what are the estimates of β and the intercept α of the true regression line?

(c) Calculate the residuals for the five data points. Check that their sum is 0 (up to roundoff error). Use the residuals to estimate the standard deviation σ in the regression model. You have now estimated all three parameters in the model.

Using technology

Basic "two-variables statistics" calculators will find the slope b and intercept a of the least-squares line from keyed-in data. Inference about regression requires in addition the regression standard error s. At this point, software or a graphing calculator that includes procedures for regression inference becomes almost essential for practical work.

Figure 21.3 shows regression output for the data of Table 21.1 from the Minitab statistical software, the Excel spreadsheet, and the TI-83 graphing calculator. When we entered the data into Minitab and Excel, we called the explanatory variable "Crycount." These outputs use that label. The TI-83 just uses "x" and "y" to label the explanatory and response variables. You can locate the basic information in all three outputs. The regression slope is $b = 1.4929$ and the regression intercept is $a = 91.268$. The equation of the least-squares line is therefore (after rounding) just as given in Examples 21.2 and 21.3. The regression standard error is $s = 17.4987$ and the squared correlation is $r^2 = 0.207$. Both of these results reflect the rather wide scatter of the points in Figure 21.1 about the least-squares line.

Each output contains other information, some of which we will need shortly and some of which we don't need. In fact, we left out some output from Minitab and Excel to save space. Once you know what to look for, you can find what you want in almost any output and ignore what doesn't interest you.

(AEF-David Sharrock/The Image Bank/ Getty Images)

APPLY YOUR KNOWLEDGE

21.2 **How fast do icicles grow?** The rate at which an icicle grows depends on temperature, water flow, and wind. The data below are for an icicle grown in a cold chamber at $-11°$ C with no wind and a water flow of 11.9 milligrams per second.[3]

Time (min)	10	20	30	40	50	60	70	80	90
Length (cm)	0.6	1.8	2.9	4.0	5.0	6.1	7.9	10.1	10.9

Time (min)	100	110	120	130	140	150	160	170	180
Length (cm)	12.7	14.4	16.6	18.1	19.9	21.0	23.4	24.7	27.8

Minitab

```
Session                                                        _□×

  Regression Analysis: IQ versus Crycount

  The regression equation is
  IQ = 91.3 + 1.49 Crycount

  Predictor          Coef        SE Coef           T          P
  Constant         91.268         8.934        10.22      0.000
  Crycount         1.4929        0.4870         3.07      0.004

  S = 17.50      R-Sq = 20.7%          R-Sq(adj) = 18.5%
```

Excel

```
Microsoft Excel - Book1                                                _□×
      A                  B            C            D         E          F          G
1  SUMMARY OUTPUT
2
3     Regression Statistics
4  Multiple R           0.4550
5  R Square             0.2070
6  Adjusted R Square    0.1850
7  Standard Error      17.4987
8  Observations            38
9
10                  Coefficients  Standard Error  t Stat   P-value   Lower 95%   Upper 95%
11 Intercept           91.2683        8.9342     10.2156   3.5E-12    73.1489    109.3877
12 Crycount             1.4929        0.4870      3.0655  0.004105     0.5052      2.4806
13
 Sheet4 / Sheet1 / Sheet2 / Sheet3 /
```

TI-83

```
LinRegTTest        LinRegTTest
 y=a+bx             y=a+bx
 β≠0 and ρ≠0        β≠0 and ρ≠0
 t=3.0655           ↑b=1.4929
 p=.0041            s=17.4987
 df=36.0000         r²=.2070
 ↓a=91.2683         r=.4550
```

Figure 21.3 Output from Minitab, Excel, and the TI-83 for the regression of IQ on crying peaks, Table 21.1.

Figure 21.4 Minitab output for the icicle growth data, for Exercise 21.2.

```
Session                                                    _|□|×
━━━━━━━━━━━━━━━━━━━━━━━━━━━━━━━━━━━━━━━━━━━━━━━━━━━━━━━━━━━━━━━━━━

Regression Analysis: Length versus Time

The regression equation is
Length = -2.39 + 0.158 Time

Predictor          Coef      SE Coef           T         P
Constant        -2.3948       0.3963       -6.04     0.000
Time           0.158483     0.003661       43.29     0.000

S = 0.8059   R-Sq = 99.2%      R-Sq(adj) = 99.1%
```

We want to predict length from time. Figure 21.4 shows the Minitab regression output for these data.

(a) Make a scatterplot suitable for predicting length from time. The pattern is very linear. What is the squared correlation r^2? Time explains almost all of the change in length.

(b) The model for regression inference has three parameters, which we call α, β, and σ. From the output, what are the estimates of these parameters?

(c) What is the equation of the least-squares regression line of length on time? Add this line to your plot. We will continue the analysis of these data in later exercises.

21.3 Great Arctic rivers. One effect of global warming is to increase the flow of water into the Arctic Ocean from rivers. Such an increase might have major effects on the world's climate. Six rivers (Yenisey, Lena, Ob, Pechora, Kolyma, and Severnaya Dvina) drain two-thirds of the Arctic in Europe and Asia. Several of these are among the largest rivers on earth. Table 21.2 contains the total discharge from these rivers each year from 1936 to 1999.[4] Discharge is measured in cubic kilometers of water. Use software to analyze these data.

(a) Make a scatterplot of river discharge against time. Is there a clear increasing trend? Calculate r^2 and briefly interpret its value. There is considerable year-to-year variation, so we wonder if the trend is statistically significant.

(b) As a first step, find the least-squares line and draw it on your plot. Then find the regression standard error s, which measures scatter about this line. We will continue the analysis in later exercises.

Regression for lawyers

Jury Verdict Research (www. juryverdictresearch.com) makes money from regression. The company maintains data on more than 193,000 jury verdicts in personal-injury lawsuits, recording more than 30 variables describing each case. Multiple regression using this mass of data allows Jury Verdict Research to predict how much a jury will award in a new case. Lawyers pay for these predictions and use them to negotiate settlements with insurance companies.

Confidence intervals for the regression slope

The slope β of the true regression line is usually the most important parameter in a regression problem. The slope is the rate of change of the mean response

TABLE 21.2 Arctic river discharge (cubic kilometers), 1936 to 1999

Year	Discharge	Year	Discharge	Year	Discharge	Year	Discharge
1936	1721	1952	1829	1968	1713	1984	1823
1937	1713	1953	1652	1969	1742	1985	1822
1938	1860	1954	1589	1970	1751	1986	1860
1939	1739	1955	1656	1971	1879	1987	1732
1940	1615	1956	1721	1972	1736	1988	1906
1941	1838	1957	1762	1973	1861	1989	1932
1942	1762	1958	1936	1974	2000	1990	1861
1943	1709	1959	1906	1975	1928	1991	1801
1944	1921	1960	1736	1976	1653	1992	1793
1945	1581	1961	1970	1977	1698	1993	1845
1946	1834	1962	1849	1978	2008	1994	1902
1947	1890	1963	1774	1979	1970	1995	1842
1948	1898	1964	1606	1980	1758	1996	1849
1949	1958	1965	1735	1981	1774	1997	2007
1950	1830	1966	1883	1982	1728	1998	1903
1951	1864	1967	1642	1983	1920	1999	1970

as the explanatory variable increases. We often want to estimate β. The slope b of the least-squares line is an unbiased estimator of β. A confidence interval is more useful because it shows how accurate the estimate b is likely to be. The confidence interval for β has the familiar form

$$\text{estimate} \pm t^*\text{SE}_{\text{estimate}}$$

Because b is our estimate, the confidence interval becomes

$$b \pm t^*\text{SE}_b$$

Here are the details.

CONFIDENCE INTERVAL FOR REGRESSION SLOPE

A level C confidence interval for the slope β of the true regression line is

$$b \pm t^*\text{SE}_b$$

In this formula, the standard error of the least-squares slope b is

$$\text{SE}_b = \frac{s}{\sqrt{\sum(x - \overline{x})^2}}$$

and t^* is the critical value for the $t(n-2)$ density curve with area C between $-t^*$ and t^*.

As advertised, the standard error of b is a multiple of the regression standard error s. The degrees of freedom $n-2$ are the degrees of freedom of s. Although we give the recipe for this standard error, you should rarely have to calculate it by hand. Regression software gives the standard error SE_b along with b itself.

EXAMPLE 21.4 **Crying and IQ**: estimating the slope

The Minitab and Excel outputs in Figure 21.3 give the slope $b = 1.4929$ and also the standard error $SE_b = 0.4870$. Both outputs use a similar arrangement, a table in which each regression coefficient is followed by its standard error. The slope b and its standard error SE_b are in the row labeled with the name of the explanatory variable, Crycount. Excel also gives the lower and upper endpoints of the 95% confidence interval for the population slope β, 0.505 and 2.481.

Once we know b and SE_b, it is easy to find the confidence interval. There are 38 data points, so the degrees of freedom are $n - 2 = 36$. Table C does not have a row for df = 36, so we use the next smaller degrees of freedom, df = 30. For a 95% confidence interval for the true slope β, the critical value is $t^* = 2.042$. The interval is

$$b \pm t^*SE_b = 1.4929 \pm (2.042)(0.4870)$$
$$= 1.4929 \pm 0.9944$$
$$= 0.4985 \text{ to } 2.4873$$

Excel's result is more accurate because the software uses the critical value for 36 degrees of freedom. From either Excel or calculation, we are 95% confident that mean IQ increases by between about 0.5 and 2.5 points for each additional peak in crying.

You can find a confidence interval for the intercept α of the true regression line in the same way, using a and SE_a from the "Constant" line of the Minitab output or the "Intercept" line in Excel. We rarely need to estimate α.

APPLY YOUR KNOWLEDGE

21.4 Growth of icicles, continued. Exercise 21.2 gives data on the growth of an icicle. We want a 95% confidence interval for the slope of the true regression line. Starting from the information in the Minitab output in Figure 21.4, find this interval. Say in words what the slope of the true regression line tells us about the growth of icicles under the conditions of this experiment.

21.5 Great Arctic rivers, continued. Use the data in Table 21.2 and your work in Exercise 21.3 to give a 90% confidence interval for the slope of the true regression of Arctic river discharge on year. Does this interval convince you that discharge is actually increasing over time?

21.6 Time at the table. Does how long young children remain at the lunch table help predict how much they eat? Here are data on 20 toddlers observed over several months at a nursery school.[5] "Time" is the average number of minutes a child spent at the table when lunch was served. "Calories" is the average number of calories the child consumed during lunch, calculated from careful observation of what the child ate each day.

Time	21.4	30.8	37.7	33.5	32.8	39.5	22.8	34.1	33.9	43.8
Calories	472	498	465	456	423	437	508	431	479	454

Time	42.4	43.1	29.2	31.3	28.6	32.9	30.6	35.1	33.0	43.7
Calories	450	410	504	437	489	436	480	439	444	408

Make a scatterplot of the data and find the equation of the least-squares line for predicting calories consumed from time at the table. Describe briefly what the data show about the behavior of children. Then give a 95% confidence interval for the slope of the true regression line.

Testing the hypothesis of no linear relationship

We can also test hypotheses about the slope β. The most common hypothesis is

$$H_0: \beta = 0$$

A regression line with slope 0 is horizontal. That is, the mean of y does not change at all when x changes. So this H_0 says that there is *no true linear relationship* between x and y. Put another way, H_0 says that *straight-line dependence on x is of no value for predicting y*.

The test statistic is just the standardized version of the least-squares slope b. It is another t statistic. Here are the details.

SIGNIFICANCE TEST FOR REGRESSION SLOPE

To test the hypothesis $H_0: \beta = 0$, compute the t statistic

$$t = \frac{b}{SE_b}$$

In terms of a random variable T having the $t(n-2)$ distribution, the P-value for a test of H_0 against

$H_a: \beta > 0$ is $P(T \geq t)$

$H_a: \beta < 0$ is $P(T \leq t)$

$H_a: \beta \neq 0$ is $2P(T \geq |t|)$

EXAMPLE 21.5   Crying and IQ: testing regression slope

The hypothesis H_0: $\beta = 0$ says that crying has no straight-line relationship with IQ. Figure 21.1 shows that there is a relationship, so it is not surprising that all three outputs in Figure 21.3 give $t = 3.07$ with two-sided P-value 0.004. There is very strong evidence that there is a straight-line relationship between IQ and crying.

Some software allows you to choose between one-sided and two-sided alternatives in tests of significance. Other software always reports the P-value for the two-sided alternative. If your alternative hypothesis is one-sided, you must divide P by 2.

EXAMPLE 21.6   Beer and blood alcohol

EESEE

How well does the number of beers a student drinks predict his or her blood alcohol content? The EESEE story "Blood Alcohol Content" describes a study in which sixteen student volunteers at Ohio State University drank a randomly assigned number of cans of beer. Thirty minutes later, a police officer measured their blood alcohol content (BAC). Here are the data:

Student	1	2	3	4	5	6	7	8
Beers	5	2	9	8	3	7	3	5
BAC	0.10	0.03	0.19	0.12	0.04	0.095	0.07	0.06

Student	9	10	11	12	13	14	15	16
Beers	3	5	4	6	5	7	1	4
BAC	0.02	0.05	0.07	0.10	0.085	0.09	0.01	0.05

The students were equally divided between men and women and differed in weight and usual drinking habits. Because of this variation, many students don't believe that number of drinks predicts blood alcohol well. What do the data say?

The scatterplot in Figure 21.5 shows a clear linear relationship. Figure 21.6 gives part of the Excel regression output. The equation of the least-squares line (after rounding) is

$$\hat{y} = -0.0127 + 0.0180x$$

This is the solid line on the scatterplot. Because $r^2 = 0.7998$, number of drinks accounts for 80% of the observed variation in BAC. That is, the data say that student opinion is wrong: the number of beers you drink predicts blood alcohol level quite well. Five beers produce an average BAC of

$$\hat{y} = -0.0127 + 0.0180(5) = 0.077$$

perilously close to the legal driving limit of 0.08 in many states.

We can test the hypothesis that the number of beers has *no* effect on blood alcohol versus the one-sided alternative that more beers increases BAC. The hypotheses are

$$H_0: \beta = 0$$
$$H_a: \beta > 0$$

It is no surprise that the t statistic is $t = 7.48$ with two-sided P-value $P = 0.000$ to three decimal places. The one-sided P-value is half this value, so it is also close to 0. Check that t is the slope $b = 0.017964$ divided by its standard error, $SE_b = 0.002402$.

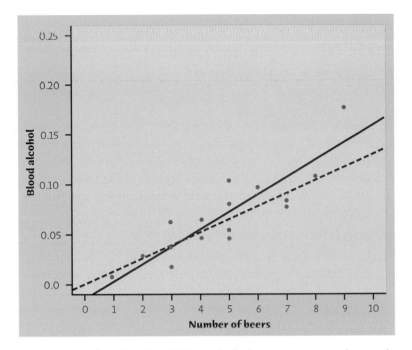

Figure 21.5 Scatterplot of students' blood alcohol content against the number of cans of beer consumed, for Example 21.6.

Figure 21.5 shows one unusual point: student number 3, who drank 9 beers. You can see that this observation lies farthest from the fitted line in the y direction. That is, this point has the largest residual. We wonder if student number 3 is influential, though the point is not extreme in the x direction. To verify that our results are not too dependent on this one observation, do the regression

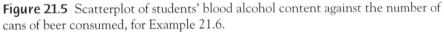

1	SUMMARY OUTPUT						
2							
3	*Regression Statistics*						
4	Multiple R	0.8943					
5	R Square	0.7998					
6	Adjusted R Square	0.7855					
7	Standard Error	0.0204					
8	Observations	16					
9							
10		*Coefficients*	*Standard Error*	*t Stat*	*P-value*	*Lower 95%*	*Upper 95%*
11	Intercept	-0.01270	0.01264	-1.005	0.3320	-0.0398	0.0144
12	Beers	0.017964	0.002402	7.480	2.97E-06	0.0128	0.0231
13							

Figure 21.6 Excel regression output for the blood alcohol content data, for Example 21.6.

again omitting student 3. The new regression line is the dashed line in Figure 21.5. Omitting student 3 decreases r^2 from 80% to 77%, and it changes the predicted BAC after 5 beers from 0.077 to 0.073. These small changes show that this observation is not very influential.

APPLY YOUR KNOWLEDGE

21.7 An extinct beast, continued. Exercise 21.1 presents data on the lengths of two bones in five fossil specimens of the extinct beast *Archaeopteryx*. In that exercise, you estimated the parameters using only a basic calculator. Software tells us that the least-squares slope is $b = 1.1969$ with standard error $SE_b = 0.0751$.

 (a) What is the t statistic for testing H_0: $\beta = 0$?

 (b) How many degrees of freedom does t have? Use Table C to approximate the P-value of t against the one-sided alternative H_a: $\beta > 0$. What do you conclude?

21.8 Great Arctic rivers, continued. The most important question we ask of the data in Table 21.2 is this: Is the increasing trend visible in your plot (Exercise 21.3) statistically significant? If so, changes in the Arctic may already be affecting earth's climate. Use software to answer this question. Give a test statistic, its P-value, and the conclusion you draw from the test.

21.9 Does fast driving waste fuel? Exercise 4.6 (page 86) gives data on the fuel consumption of a small car at various speeds from 10 to 150 kilometers per hour. Is there evidence of straight-line dependence between speed and fuel use? Make a scatterplot and use it to explain the result of your test.

Testing lack of correlation

The least-squares slope b is closely related to the correlation r between the explanatory and response variables x and y. In the same way, the slope β of the true regression line is closely related to the correlation between x and y in the population. In particular, the slope is 0 exactly when the correlation is 0.

Testing the null hypothesis H_0: $\beta = 0$ is therefore exactly the same as testing that there is *no correlation* between x and y in the population from which we drew our data. You can use the test for zero slope to test the hypothesis of zero correlation between any two quantitative variables. That's a useful trick. Do notice that testing correlation makes sense only if the subjects are a random sample from a population.

EXAMPLE 21.7 Beer and blood alcohol: testing correlation

We expect a positive correlation between the number of beers a student drinks and his or her blood alcohol content. The Excel output in Figure 21.6 reports $r = 0.8943$

for the 16 students in the study of Example 21.6. To find the P-value for the test that the population correlation is 0, look at the t statistic for the slope. Excel gives $P = 0.00000297$ for the two-sided alternative. (The E-06 in this display means move the decimal point six digits to the left.) The P-value for the one-sided alternative of positive slope is half this, about $P = 0.0000015$. **These are also the two-sided and one-sided P-values for testing correlation 0.** That is, the P-value for $r = 0.8943$ from $n = 16$ observations is $P = 0.0000015$ for the one-sided alternative. There is extremely strong evidence that the population correlation is positive.

Because correlation makes sense when there is no explanatory-response distinction, it is handy to be able to test correlation without doing regression. Table F in the back of the book gives critical values of the sample correlation r under the null hypothesis that the correlation is 0 in the population. Use this table when both variables have at least approximately Normal distributions.

EXAMPLE 21.8   An extinct beast: testing correlation

Exercise 21.1 gives data on the lengths of two bones in the 5 complete fossils of the extinct beast *Archaeopteryx*. We expect a positive correlation, but the sample size is very small. Do the data give good evidence of a positive correlation for all beasts of this species? Your calculator will give $r = 0.9941$. You can approximate the P-value from Table F without using software.

Choose the row for your sample size and compare your r with the critical values in that row. (Be sure to use sample size n, not the degrees of freedom $n - 2$.) The largest r in the $n = 5$ row is 0.9911, for tail area 0.0005. Because $r = 0.9941$ is larger than this value, the one-sided P-value is less than 0.0005. (The two-sided P-value is double the one-sided value.) The data convince us that the population correlation is positive.

APPLY YOUR KNOWLEDGE

21.10 Is wine good for your heart? There is some evidence that drinking moderate amounts of wine helps prevent heart attacks. Table 4.3 (page 101) gives data on yearly wine consumption (liters of alcohol from drinking wine, per person) and yearly deaths from heart disease (deaths per 100,000 people) in 19 developed nations. Is there statistically significant evidence that the correlation between wine consumption and heart disease deaths is negative? Use Table F.

21.11 Time at the table: testing correlation. Exercise 21.6 gives data on the time that 20 young children remained at the lunch table and the number of calories they consumed. We might think that children eat what they need either quickly or slowly, so that there is no correlation between time at the table and calories. Find the correlation r for these children and use Table F to test the hypothesis of population correlation 0 against the two-sided alternative that the correlation is not 0.

Is regression garbage?

No—but garbage can be the setting for regression. The Census Bureau once asked if weighing a neighborhood's garbage would help count its people. So 63 households had their garbage sorted and weighed. It turned out that pounds of plastic in the trash gave the best garbage prediction of the number of people in a neighborhood. The margin of error for a 95% prediction interval in a neighborhood of about 100 households, based on five weeks' worth of garbage, was about ±2.5% people. Alas, that is not accurate enough to help the Census Bureau.

prediction interval

Inference about prediction

One of the most common reasons to fit a line to data is to predict the response to a particular value of the explanatory variable. The method is simple: just substitute the value of x into the equation of the line. We saw in Example 21.6 that drinking 5 beers produces an average BAC of

$$\hat{y} = -0.0127 + 0.0180(5) = 0.077$$

We would like to give a confidence interval that describes how accurate this prediction is. To do that, you must answer these questions: Do you want to predict the *mean* BAC for *all students* who drink 5 beers? Or do you want to predict the BAC of *one individual student* who drinks 5 beers? Both of these predictions may be interesting, but they are two different problems. The actual prediction is the same, $\hat{y} = 0.077$. But the margin of error is different for the two kinds of prediction. Individual students who drink 5 beers don't all have the same BAC. So we need a larger margin of error to pin down one student's result than to estimate the mean BAC for all students who have 5 beers.

Write the given value of the explanatory variable x as x^*. In the example, $x^* = 5$. The distinction between predicting a single outcome and predicting the mean of all outcomes when $x = x^*$ determines what margin of error is correct. To emphasize the distinction, we use different terms for the two intervals.

- To estimate the *mean* response, we use a *confidence interval*. It is an ordinary confidence interval for the parameter

$$\mu_y = \alpha + \beta x^*$$

The regression model says that μ_y is the mean of responses y when x has the value x^*. It is a fixed number whose value we don't know.

- To estimate an *individual* response y, we use a **prediction interval.** A prediction interval estimates a single random response y rather than a parameter like μ_y. The response y is not a fixed number. If we took more observations with $x = x^*$, we would get different responses.

Fortunately, the meaning of a prediction interval is very much like the meaning of a confidence interval. A 95% prediction interval, like a 95% confidence interval, is right 95% of the time in repeated use. "Repeated use" now means that we take an observation on y for each of the n values of x in the original data, and then take one more observation y with $x = x^*$. Form the prediction interval from the n observations, then see if it covers the one more y. It will in 95% of all repetitions.

The interpretation of prediction intervals is a minor point. The main point is that it is harder to predict one response than to predict a mean response. Both intervals have the usual form

$$\hat{y} \pm t^* \text{SE}$$

but the prediction interval is wider than the confidence interval. Here are the details.

CONFIDENCE AND PREDICTION INTERVALS FOR REGRESSION RESPONSE

A level C **confidence interval for the mean response** μ_y when x takes the value x^* is

$$\hat{y} \pm t^* SE_{\hat{\mu}}$$

The standard error $SE_{\hat{\mu}}$ is

$$SE_{\hat{\mu}} = s\sqrt{\frac{1}{n} + \frac{(x^* - \overline{x})^2}{\sum(x - \overline{x})^2}}$$

The sum runs over all the observations on the explanatory variable x.

A level C **prediction interval for a single observation** on y when x takes the value x^* is

$$\hat{y} \pm t^* SE_{\hat{y}}$$

The standard error for prediction $SE_{\hat{y}}$ is

$$SE_{\hat{y}} = s\sqrt{1 + \frac{1}{n} + \frac{(x^* - \overline{x})^2}{\sum(x - \overline{x})^2}}$$

In both recipes, t^* is the critical value for the $t(n - 2)$ density curve with area C between $-t^*$ and t^*.

There are two standard errors: $SE_{\hat{\mu}}$ for estimating the mean response μ_y and $SE_{\hat{y}}$ for predicting an individual response y. The only difference between the two standard errors is the extra 1 under the square root sign in the standard error for prediction. The extra 1 makes the prediction interval wider. Both standard errors are multiples of the regression standard error s. The degrees of freedom are again $n - 2$, the degrees of freedom of s. Calculating these standard errors by hand is a nuisance. Statistical software such as Minitab gives both confidence and prediction intervals. If you use a spreadsheet or a graphing calculator, you will have to program the standard error and the interval.

EXAMPLE 21.9   Predicting blood alcohol

Steve thinks he can drive legally 30 minutes after he finishes drinking 5 beers. We want to predict Steve's blood alcohol content, using no information except that he drinks 5 beers. Here is the output from the prediction option in the Minitab regression command for $x^* = 5$ when we ask for 95% intervals:

```
Session                                                          _ □ ×

Predicted Values for New Observations


New Obs      Fit    SE Fit        95.0% CI            95.0% PI
1        0.07712   0.00513   ( 0.06612, 0.08812)  ( 0.03192, 0.12232)
```

The "Fit" entry gives the predicted BAC, 0.07712. This agrees with our result in Example 21.6. Minitab gives both 95% intervals. You must choose which one you want. We are predicting a single response, so the prediction interval "95.0% PI" is the right choice. We are 95% confident that Steve's BAC will fall between about 0.032 and 0.122. The upper part of that range will get him arrested if he drives. The 95% confidence interval for the mean BAC of all students after 5 beers, given as "95.0% CI," is much narrower.

APPLY YOUR KNOWLEDGE

21.12 Growth of icicles: prediction. Analysis of the data in Exercise 21.2 shows that growth of icicles is very linear. We might want to predict the mean length of icicles after 200 minutes under the same conditions of temperature, wind, and water flow. Here is the Minitab output for prediction when $x^* = 200$:

```
Session                                                          _ □ ×

Predicted Values for New Observations


New Obs      Fit    SE Fit        95.0% CI            95.0% PI
1         29.302    0.429    ( 28.393, 30.211)   ( 27.367, 31.237)
```

(a) Use the regression line from Figure 21.4 (page 570) to verify that "Fit" is the predicted value for $x^* = 200$. (Start with the results in the "Coef" column of Figure 21.4 to reduce roundoff error.)

(b) What is the 95% interval we want?

21.13 Using output. Minitab reports only one of the two standard errors used in prediction. It is $SE_{\hat{\mu}}$, the standard error for estimating the mean response. Use this fact, the output in the previous exercise, and a critical value from Table C to give a 99% confidence interval for the mean length of icicles after 200 minutes.

21.14 Time at the table: prediction. Rachel is another child at the nursery school of Exercise 21.6. Over several months, Rachel averages

40 minutes at the lunch table. Give a 95% interval to predict Rachel's average calorie consumption at lunch.

Checking the conditions for inference

You can fit a least-squares line to any set of explanatory-response data when both variables are quantitative. If the scatterplot doesn't show a roughly linear pattern, the fitted line may be almost useless. But it is still the line that fits the data best in the least-squares sense. To use regression inference, however, the data must satisfy the conditions stated by the regression model. Before we can trust the results of inference, we must check these conditions one by one.

The observations are independent. In particular, repeated observations on the same individual are not allowed. We should not use ordinary regression to make inferences about the growth of a single child over time, for example.

The true relationship is linear. We can't observe the true regression line, so we will almost never see a perfect straight-line relationship in our data. Look at the scatterplot to check that the overall pattern is roughly linear. A plot of the residuals against x magnifies any unusual pattern. Draw a horizontal line at zero on the residual plot to orient your eye. Because the sum of the residuals is always zero, zero is also the mean of the residuals.

The standard deviation of the response about the true line is the same everywhere. Look at the scatterplot again. The scatter of the data points about the line should be roughly the same over the entire range of the data. A plot of the residuals against x, with a horizontal line at zero, makes this easier to check. It is quite common to find that as the response y gets larger, so does the scatter of the points about the fitted line. Rather than remaining fixed, the standard deviation σ about the line is changing with x as the mean response changes with x. There is no fixed σ for s to estimate. We cannot trust our inference procedures when this happens.

The response varies Normally about the true regression line. We can't observe the true regression line. We can observe the least-squares line and the residuals, which show the variation of the response about the fitted line. The residuals estimate the deviations of the response from the true regression line, so they should follow a Normal distribution. Make a histogram or stemplot of the residuals and check for clear skewness or other major departures from Normality. Like other t procedures, inference for regression is (with one exception) not very sensitive to minor lack of Normality, especially when we have many observations. Do beware of influential observations, which move the regression line and can greatly affect the results of inference.

The exception is the prediction interval for a single response y. This interval relies on Normality of individual observations, not just on the approximate Normality of statistics like the slope a and intercept b of the least-squares line. The statistics a and b become more Normal as we take more observations. This contributes to the robustness of regression inference, but it isn't enough for the prediction interval. We will not study methods that carefully check

Normality of the residuals, so you should regard prediction intervals as rough approximations.

The conditions for regression inference are a bit elaborate. Fortunately, it is not hard to check for gross violations. There are ways to deal with violations of any of the regression model assumptions. If your data don't fit the regression model, get expert advice. The residuals help in checking the conditions. Most regression software will calculate and save the residuals for you.

EXAMPLE 21.10 Beer and blood alcohol: residuals

Example 21.6 shows the regression of the blood alcohol content of 16 students on the number of beers they drink. The statistical software that did the regression calculations also calculates the 16 residuals. Here they are:

| 0.0229 | 0.0068 | 0.0410 | −0.0110 | −0.0012 | −0.0180 | 0.0288 | −0.0171 |
| −0.0212 | −0.0271 | 0.0108 | 0.0049 | 0.0079 | −0.0230 | 0.0047 | −0.0092 |

A residual plot appears in Figure 21.7. The values of x are on the horizontal axis. The residuals are on the vertical axis, with a horizontal line at zero.

Examine the residual plot to check that the relationship is roughly linear and that the scatter about the line is about the same from end to end. Overall, there is no clear deviation from the even scatter about the line that should occur (except for chance variation) when the regression assumptions hold.

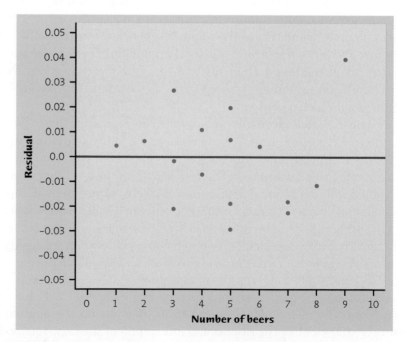

Figure 21.7 Plot of the regression residuals for the blood alcohol data against the explanatory variable, number of beers consumed, for Example 21.10. The mean of the residuals is always 0.

Now examine the distribution of the residuals for signs of strong non-Normality. Here is a stemplot of the residuals after rounding to three decimal places:

−2	731
−1	871
−0	91
0	5578
1	1
2	39
3	
4	1

Student number 3 is a mild outlier. We saw on page 576 that omitting this observation has little effect on r^2 or the fitted line. It also has little effect on inference. For example, $t = 7.58$ for the slope becomes $t = 6.57$, a change of no practical importance.

EXAMPLE 21.11 Using residual plots

The residual plots in Figure 21.8 illustrate violations of the regression assumptions that require corrective action before using regression. Both plots come from a study of the salaries of major-league baseball players.[6] Salary is the response variable. There are several explanatory variables that measure the players' past performance. Regression with more than one explanatory variable is called **multiple regression.** Although interpreting the fitted model is more complex in multiple regression, we check assumptions by examining residuals as usual.

multiple regression

Figure 21.8(a) is a plot of the residuals against the predicted salary $\hat{y}$, produced by statistical software. When points on the plot overlap, letters show how many observations each point represents. A is one observation, B stands for two observations, and so on. The plot shows a clear violation of the condition that the spread of responses about the model is everywhere the same. There is more variation among players with high salaries than among players with lower salaries.

Although we don't show a histogram, the distribution of salaries is strongly skewed to the right. Using the *logarithm* of the salary as the response variable gives a more Normal distribution and also fixes the unequal-spread problem. It is common to work with some transformation of data in order to satisfy the regression conditions. But all is not yet well. Figure 21.8(b) plots the new residuals against years in the major leagues. There is a clear curved pattern. The relationship between logarithm of salary and years in the majors is not linear but curved. The statistician must take more corrective action.

APPLY YOUR KNOWLEDGE

21.15 Crying and IQ: residuals. The residuals for the study of crying and IQ appear in Example 21.3.

(a) Make a stemplot to display the distribution of the residuals. (Round to the nearest whole number first.) Are there outliers or signs of strong departures from Normality?

Figure 21.8 Two residual plots that illustrate violations of the conditions for regression inference. **(a)** The variation of the residuals is not constant. **(b)** There is a curved relationship between the response variable and the explanatory variable.

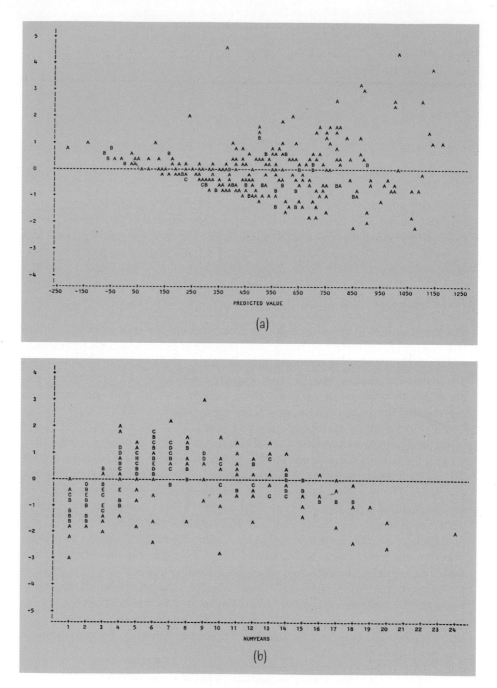

(a)

(b)

(b) Make a plot of the residuals against the explanatory variable. Draw a horizontal line at height 0 on your plot. Does the plot show a nonrandom pattern?

21.16 Growth of icicles: residuals. Figure 21.4 (page 570) gives part of the Minitab output for the data on growth of icicles in Exercise 21.2. Table 21.3 comes from another part of the output. It gives the predicted response $\hat{y}$ and the residual $y - \hat{y}$ for each of the

TABLE 21.3 Growth of icicles: predictions and residuals

Obs. number	Time x	Length y	Prediction $\hat{y}$	Residual $y - \hat{y}$
1	10	0.600	−0.810	1.410
2	20	1.800	0.775	1.025
3	30	2.900	2.360	0.540
4	40	4.000	3.945	0.055
5	50	5.000	5.529	−0.529
6	60	6.100	7.114	−1.014
7	70	7.900	8.699	−0.799
8	80	10.100	10.284	−0.184
9	90	10.900	11.869	−0.969
10	100	12.700	13.454	−0.754
11	110	14.400	15.038	−0.638
12	120	16.600	16.623	−0.023
13	130	18.100	18.208	−0.108
14	140	19.900	19.793	0.107
15	150	21.000	21.378	−0.378
16	160	23.400	22.963	0.437
17	170	24.700	24.547	0.153
18	180	27.800	26.132	1.668

18 observations. Most statistical software provides similar output. Examine the regression model conditions one by one. This example illustrates mild violations of the regression conditions that did not prevent the researchers from doing inference.

(a) **Independent observations.** The data come from the growth of a single icicle, not from a different icicle at each time. Explain why this would violate the independence condition if we had data on the growth of a child rather than of an icicle. (The researchers decided that all icicles, unlike all children, grow at the same rate if the conditions are held fixed. So one icicle can stand in for a separate icicle at each time.)

(b) **Linear relationship.** Your plot and r^2 from Exercise 21.2 show that the relationship is very linear. Residual plots magnify effects. Plot the residuals against time. What kind of deviation from a straight line is now visible? (The deviation is clear in the residual plot, but it is very small in the original scale.)

(c) **Spread about the line stays the same.** Your plot in (b) shows that it does not. (Once again, the plot greatly magnifies small deviations.)

(d) **Normal variation about the line.** Make a histogram of the residuals. With only 18 observations, no clear shape emerges. Do strong skewness or outliers suggest lack of Normality?

Chapter 21 SUMMARY

Least-squares regression fits a straight line to data in order to predict a response variable y from an explanatory variable x. Inference about regression requires more conditions.

The **regression model** says that there is a **true regression line** $\mu_y = \alpha + \beta x$ that describes how the mean response varies as x changes. The observed response y for any x has a Normal distribution with mean given by the true regression line and with the same standard deviation σ for any value of x. The **parameters of the regression model** are the intercept α, the slope β, and the standard deviation σ.

The slope a and intercept b of the least-squares line estimate the slope α and intercept β of the true regression line. To estimate σ, use the **regression standard error s.**

The standard error s has $n - 2$ **degrees of freedom.** All t procedures in regression inference have $n - 2$ degrees of freedom.

Confidence intervals for the slope of the true regression line have the form $b \pm t^*\mathrm{SE}_b$. In practice, use software to find the slope b of the least-squares line and its standard error SE_b.

To test **the hypothesis that the true slope is zero,** use the t statistic $t = b/\mathrm{SE}_b$. This null hypothesis says that straight-line dependence on x has no value for predicting y.

The t test for regression slope is also a test for **the hypothesis that the population correlation between x and y is zero.** To do this test without software, use the sample correlation r and Table F.

Confidence intervals for the mean response when x has value x^* have the form $\hat{y} \pm t^*\mathrm{SE}_\mu$. **Prediction intervals** for an individual future response y have a similar form with a larger standard error, $\hat{y} \pm t^*\mathrm{SE}_{\hat{y}}$. Software often gives these intervals.

STATISTICS IN SUMMARY

Here are the most important skills you should have developed from studying this chapter.

A. PRELIMINARIES

1. Make a scatterplot to show the relationship between an explanatory and a response variable.
2. Use a calculator or software to find the correlation and the equation of the least-squares regression line.

B. RECOGNITION

1. Recognize the regression setting: a straight-line relationship between an explanatory variable x and a response variable y.

2. Recognize which type of inference you need in a particular regression setting.

3. Inspect the data to recognize situations in which inference isn't safe: a nonlinear relationship, influential observations, strongly skewed residuals in a small sample, or nonconstant variation of the data points about the regression line.

C. DOING INFERENCE USING SOFTWARE OUTPUT

1. Explain in any specific regression setting the meaning of the slope β of the true regression line.

2. Understand software output for regression. Find in the output the slope and intercept of the least-squares line, their standard errors, and the regression standard error.

3. Use that information to carry out tests and calculate confidence intervals for β.

4. Explain the distinction between a confidence interval for the mean response and a prediction interval for an individual response.

5. If software gives output for prediction, use that output to give either confidence or prediction intervals.

Chapter 21 EXERCISES

21.17 Prey attract predators. Exercise I.24 (page 162) describes an experiment to test a theory about the relationship between predators and prey in nature. The theory says that a higher density of prey attracts more predators, who remove a higher percent of the prey and thus help keep the population stable. The explanatory variable is the number of perch (the prey) in a confined area. The response variable is the proportion of perch killed by bass (the predator) in 2 hours when the bass are allowed access to the perch. You made a scatterplot in Exercise I.24. Figure 21.9 contains Excel output for the regression.

(a) What is the equation of the least-squares line for predicting proportion killed from count of perch? What part of this equation shows that more perch does result in a higher proportion being killed by bass?

(b) What is the regression standard error s?

21.18 Prey attract predators: estimating the slope. Figure 21.9 gives Excel output for the regression of proportion of perch killed by bass on the initial number of perch. We want to estimate the slope of the true regression line.

(a) Excel gives a 95% confidence interval. What is it? Starting with Excel's values of the least-squares slope b and its standard error, verify the confidence interval.

	A	B	C	D	E	F	G
1	SUMMARY OUTPUT						
2							
3	*Regression Statistics*						
4	Multiple R	0.6821					
5	R Square	0.4652					
6	Adjusted R Square	0.4270					
7	Standard Error	0.1886					
8	Observations	16.0000					
9							
10		*Coefficients*	*Standard Error*	*t Stat*	*P-value*	*Lower 95%*	*Upper 95%*
11	Intercept	0.1205	0.0927	1.2999	0.2146	-0.0783	0.3193
12	Perch	0.0086	0.0025	3.4899	0.0036	0.0033	0.0138
13							

Sheet5 / Sheet1 / Sheet2 / Sheet3

Figure 21.9 Excel output for the regression of proportion of perch killed by bass on count of perch in a pen, for Exercises 21.17 to 21.19.

(b) Give a 90% confidence interval for the true slope. As usual, this interval is shorter than the 95% interval.

21.19 Prey attract predators: correlation. The Excel output in Figure 21.9 gives the correlation between proportion of perch killed by bass and initial count of perch as $r = 0.6821$. Use Table F to say how significant this correlation is for testing 0 correlation against positive correlation in the population. Verify that your result agrees with Excel's P-value for the test for 0 slope.

21.20 Casting aluminum. Exercise I.39 (page 167) gives data for predicting the "gate velocity" of molten metal from the thickness of the aluminum piston being cast. The data come from observing skilled workers and will be used to guide less experienced workers. Figure 21.10 displays part of the Minitab regression output. We have left out the t statistics and their P-values.

(a) Make a scatterplot suitable for predicting gate velocity from thickness. Give the value of r^2 and the equation of the least-squares line. Draw the line on your plot.

(b) Based on the information given, test the hypothesis that there is no straight-line relationship between thickness and gate velocity. Give a test statistic, its approximate P-value using a table, and your conclusion.

21.21 Casting aluminum: intervals. The output in Figure 21.10 includes prediction for $x^* = 0.5$ inch. Use the output to give 90% intervals for

(a) the slope of the true regression line of gate velocity on piston thickness.

(b) the average gate velocity for a type of piston with thickness 0.5 inch.

(IFA/eStock Photography/PictureQuest)

```
Session                                                    □ ▼

Regression Analysis: GateVel versus Thickness

The regression equation is
GateVel = 70.4 + 275 Thickness

Predictor          Coef          SE Coef
Constant          70.44            52.90
Thickness        274.78            88.18

S = 56.36        R-Sq = 49.3%     R-Sq(adj) = 44.2%

Predicted Values for New Observations

New Obs    Fit    SE Fit        95.0% CI            95.0% PI
1         207.8    17.4   ( 169.0, 246.7)   ( 76.4, 339.3)
```

Figure 21.10 Partial Minitab output for the regression of gate velocity on piston thickness in casting aluminum parts, for Exercises 21.20 and 21.21.

21.22 **Casting aluminum: residuals.** Here are the residuals (rounded to one decimal place) for the regression of gate velocity on piston thickness:

Thickness	0.248	0.359	0.366	0.400	0.524	0.552
Residual	−14.8	54.8	9.9	−75.5	14.2	1.7

Thickness	0.628	0.697	0.697	0.752	0.806	0.821
Residual	83.2	40.4	−116.8	−14.0	10.5	6.4

(a) Check the calculation of residuals by finding their sum. What should the sum be? Does the sum have that value (up to roundoff error)?

(b) Plot the residuals against thickness (the explanatory variable). Does your plot show a systematically nonlinear relationship? Does it show systematic change in the spread about the regression line?

(c) Make a histogram of the residuals. The pattern, with just 12 observations, is irregular. More advanced methods show that the distribution is reasonably Normal.

21.23 **Foot problems.** Exercises I.8 and I.9 and Table I.1 (page 158) describe the relationship between two deformities of the feet in young patients. Metatarsus adductus may help predict the severity of hallux abducto valgus. The paper that reports this study says, "Linear regression analysis, using the hallux abductus angle as the response variable, demonstrated a significant correlation between the metatarsus adductus and hallux abductus angles."[7] Do a suitable test of significance to verify this finding. (The study authors note that the scatterplot suggests that the variation in y may change as x changes, so they offer a more elaborate analysis as well.)

21.24 Age and income. Figure I.2 (page 168) is a scatterplot of the age and income of a random sample of 5712 men between the ages of 25 and 65 who have a bachelor's degree but no higher degree. We see that estimated mean income does increase with age, but not very rapidly.

(a) We know even without looking at the software output in Exercise I.40 that there is highly significant evidence that the slope of the true regression line is greater than 0. Why do we know this?

(b) The software gives the two-sided P-value for a test of $H_0: \beta = 0$. Before collecting the data, we suspected that (on the average) older men earn more. Our alternative hypothesis is therefore $H_a: \beta > 0$. How can we find the P-value for this one-sided alternative from the software output?

21.25 Age and income, continued. Give a 99% confidence interval for the slope of the true regression line of income on age for all men aged 25 to 65 with a bachelor's degree but no higher degree. Use the information in the software output in Exercise I.40 (page 168).

21.26 Sparrowhawk colonies. Exercise 4.4 (page 83) gives data on the percent of adult sparrowhawks in a colony that return from the previous year and the number of new adults that join the colony. In Example 5.2 (page 107) we used the regression line for these 13 colonies to predict the average number of new birds in colonies to which 60% of the birds return. Figure 21.11 shows part of the Minitab regression output, including prediction when $x^* = 60\%$.

(a) Write the equation of the least-squares line and use it to check that the "Fit" in the output is the predicted response for $x^* = 60\%$.

(b) Which 95% interval in the output gives us a margin of error for predicting the average number of new birds?

21.27 Sparrowhawk residuals. The regression of number of new birds that join a sparrowhawk colony on the percent of adult birds in the colony that return from the previous year is an example of data that satisfy the regression model well. Here are the residuals for the 13 colonies:

Percent return	74	66	81	52	73	62	52
Residual	−4.44	−5.87	0.69	−5.13	2.26	1.92	−0.13

Percent return	45	62	46	60	46	38
Residual	−1.25	4.92	0.05	5.31	2.05	−0.38

(a) **Independent observations.** Why are the 13 observations independent?

```
▦ Session                                                    _ □ ×

 Regression Analysis: NewBirds versus PctRet

 The regression equation is
 NewBirds = 31.9 - 0.304 PctRet

 Predictor          Coef     SE Coef         T          P
 Constant         31.934       4.838      6.60      0.000
 PctRet          -0.30402     0.08122    -3.74      0.003

 S = 3.667       R-Sq = 56.0%     R-Sq(adj) = 52.0%

 Predicted Values for New Observations

 New Obs      Fit    SE Fit        95.0% CI          95.0% PI
 1          13.69      1.03    ( 11.43, 15.95)    ( 5.31, 22.07)
```

Figure 21.11 Partial Minitab output for predicting number of new birds in a sparrowhawk colony from percent of birds returning, for Exercises 21.26 and 21.27.

(b) **Linear relationship.** Figure 5.1 (page 105) is a scatterplot of the data. Is the pattern reasonably linear? A plot of the residuals against the explanatory variable x magnifies the deviations from the least-squares line. Does the plot show any systematic pattern?

(c) **Spread about the line stays the same.** Does your plot in (b) show any systematic change in spread as x changes?

(d) **Normal variation about the line.** Make a histogram of the residuals. With only 13 observations, no clear shape emerges. Do strong skewness or outliers suggest lack of Normality?

21.28 **Beavers and beetles.** Ecologists sometimes find rather strange relationships in our environment. One study seems to show that beavers benefit beetles. The researchers laid out 23 circular plots, each 4 meters in diameter, in an area where beavers were cutting down cottonwood trees. In each plot, they measured the number of stumps from trees cut by beavers and the number of clusters of beetle larvae. Here are the data:[8]

Stumps	2	2	1	3	3	4	3	1	2	5	1	3
Beetle larvae	10	30	12	24	36	40	43	11	27	56	18	40

Stumps	2	1	2	2	1	1	4	1	2	1	4
Beetle larvae	25	8	21	14	16	6	54	9	13	14	50

(Carlyn Galati/Visuals Unlimited)

(a) Make a scatterplot that shows how the number of beaver-caused stumps influences the number of beetle larvae clusters. What does your plot show?

(b) Here is part of the Minitab regression output for these data:

```
Session                                          _ □ ×

Regression Analysis: Larvae versus Stumps

Predictor        Coef        SE Coef
Constant        -1.286         2.853
Stumps          11.894         1.136
```

Find the least-squares regression line and draw it on your plot. What percent of the observed variation in beetle larvae counts can be explained by straight-line dependence on beaver stump counts?

(c) Is there strong evidence that beaver stumps help explain beetle larvae counts? State hypotheses, give a test statistic and its P-value, and state your conclusion.

21.29 Mutual-fund performance. Exercise 4.21 (page 98) gives the percent returns for the Vanguard International Growth Fund, a mutual fund that buys foreign stocks, for each year between 1982 and 2001. The data also include the returns on the Morgan Stanley EAFE index, an average of all foreign stocks against which the mutual fund measures its performance. Use software to analyze the relationship between the two sets of returns.

(a) Make a scatterplot suitable for predicting the fund's return from the EAFE return. Describe the form, direction, and strength of the relationship and give the equation of the least-squares line. Are there any extreme outliers or observations that may be very influential?

(b) As your data analysis leads you to expect, software shows that there is a highly significant relationship. Give a 95% confidence interval for the slope of the true regression line. Explain in plain language what this interval tells us about how the fund performs.

21.30 Mutual-fund performance: prediction. Continue your analysis of the data in the previous exercise concerning the performance of a mutual fund and its benchmark index. In 2002, the EAFE index lost 15.94%. What does your regression predict to be the 2002 return on the Vanguard International Growth Fund? Give a 95% prediction interval for the 2002 return. The actual 2002 return was −17.79%. Did the prediction interval cover this return?

21.31 Mutual-fund performance: residuals. Use your software to find the residuals for the mutual-fund regression of Exercise 21.29. Plot the residuals against the predicted return if your software makes this easy; otherwise plot the residuals against the explanatory variable. Also make a histogram of the residuals. Are there signs of any major violations of the conditions for regression inference?

21.32 Weeds among the corn. Lamb's-quarter is a common weed that interferes with the growth of corn. An agriculture researcher planted corn at the same rate in 16 small plots of ground, then weeded the plots by hand to allow a fixed number of lamb's-quarter plants to grow in each meter of corn row. No other weeds were allowed to grow. Here are the yields of corn (bushels per acre) in each of the plots:[9]

Weeds per meter	Corn yield	Weeds per meter	Corn yield	Weeds per meter	Corn yield	Weeds per meter	Corn yield
0	166.7	1	166.2	3	158.6	9	162.8
0	172.2	1	157.3	3	176.4	9	142.4
0	165.0	1	166.7	3	153.1	9	162.8
0	176.9	1	161.1	3	156.0	9	162.4

(Peter Beck/CORBIS)

Use software to analyze these data.

(a) Make a scatterplot and find the least-squares line. What percent of the observed variation in corn yield can be explained by a linear relationship between yield and weeds per meter?

(b) Is there good evidence that more weeds reduce corn yield?

(c) Explain from your findings in (a) and (b) why you expect predictions based on this regression to be quite imprecise. Predict the mean corn yield under these experimental conditions when there are 6 weeds per meter of row. If your software allows, give a 95% confidence interval for this mean.

21.33 City and highway gas mileage. Table 1.2 (page 12) gives the city and highway gas mileages for 22 models of two-seater cars. The Honda Insight, a gas-electric hybrid car, is an outlier in both the x and y directions. Exercise 5.9 (page 118) asks you to investigate the influence of the Insight on the least-squares line. The influence is not large because the Insight does not lie far from the least-squares line calculated from the other 21 cars. Now you will investigate the influence of the Insight on inference. Carry out regression both with and without the Insight.

(a) How does the Insight influence the value of the t statistic for the regression slope and its P-value? Is the influence practically important?

(b) How does the Insight influence the 95% confidence interval for the regression slope? Is the influence practically important?

21.34 Fish sizes. Table 21.4 contains data on the size of perch caught in a lake in Finland.[10] Statistical software will help you analyze these data.

(a) We want to know how well we can predict the width of a perch from its length. Make a scatterplot of width against length. There

TABLE 21.4 Measurements on 56 perch

Obs. number	Weight (grams)	Length (cm)	Width (cm)	Obs. number	Weight (grams)	Length (cm)	Width (cm)
104	5.9	8.8	1.4	132	197.0	27.0	4.2
105	32.0	14.7	2.0	133	218.0	28.0	4.1
106	40.0	16.0	2.4	134	300.0	28.7	5.1
107	51.5	17.2	2.6	135	260.0	28.9	4.3
108	70.0	18.5	2.9	136	265.0	28.9	4.3
109	100.0	19.2	3.3	137	250.0	28.9	4.6
110	78.0	19.4	3.1	138	250.0	29.4	4.2
111	80.0	20.2	3.1	139	300.0	30.1	4.6
112	85.0	20.8	3.0	140	320.0	31.6	4.8
113	85.0	21.0	2.8	141	514.0	34.0	6.0
114	110.0	22.5	3.6	142	556.0	36.5	6.4
115	115.0	22.5	3.3	143	840.0	37.3	7.8
116	125.0	22.5	3.7	144	685.0	39.0	6.9
117	130.0	22.8	3.5	145	700.0	38.3	6.7
118	120.0	23.5	3.4	146	700.0	39.4	6.3
119	120.0	23.5	3.5	147	690.0	39.3	6.4
120	130.0	23.5	3.5	148	900.0	41.4	7.5
121	135.0	23.5	3.5	149	650.0	41.4	6.0
122	110.0	23.5	4.0	150	820.0	41.3	7.4
123	130.0	24.0	3.6	151	850.0	42.3	7.1
124	150.0	24.0	3.6	152	900.0	42.5	7.2
125	145.0	24.2	3.6	153	1015.0	42.4	7.5
126	150.0	24.5	3.6	154	820.0	42.5	6.6
127	170.0	25.0	3.7	155	1100.0	44.6	6.9
128	225.0	25.5	3.7	156	1000.0	45.2	7.3
129	145.0	25.5	3.8	157	1100.0	45.5	7.4
130	188.0	26.2	4.2	158	1000.0	46.0	8.1
131	180.0	26.5	3.7	159	1000.0	46.6	7.6

is a strong linear pattern, as expected. Perch number 143 had six newly eaten fish in its stomach. Find this fish on your scatterplot and circle the point. Is this fish an outlier in your plot of width against length?

(b) Find the least-squares regression line to predict width from length.

(c) The length of a typical perch is about $x^* = 27$ centimeters. Predict the mean width of such fish and give a 95% confidence interval.

(d) Examine the residuals. Is there any reason to mistrust inference? Does fish number 143 have an unusually large residual?

21.35 Transforming a variable (Optional). We can also use the data in Table 21.4 to study the prediction of the weight of a perch from its length.

(a) Make a scatterplot of weight versus length, with length as the explanatory variable. Describe the pattern of the data and any clear outliers.

(b) It is more reasonable to expect the one-third power of the weight to have a straight-line relationship with length than to expect weight itself to have a straight-line relationship with length. Explain why this is true. (Hint: What happens to weight if length, width, and height all double?)

(c) Use your software to create a new variable that is the one-third power of weight. Make a scatterplot of this new response variable against length. Describe the pattern and any clear outliers.

(d) Is the straight-line pattern in (c) stronger or weaker than that in (a)? Compare the plots and also the values of r^2.

(e) Find the least-squares regression line to predict the new weight variable from length. Predict the mean of the new variable for perch 27 centimeters long, and give a 95% confidence interval.

(f) Examine the residuals from your regressions. Does it appear that any of the conditions for regression are not met?

21.36 Standardized residuals (Optional). Software often calculates **standardized residuals** as well as the actual residuals from regression. Because the standardized residuals have the standard z-score scale, it is easier to judge whether any are extreme. Here are the standardized residuals from Exercise 21.28 (beavers and beetles), rounded to two decimal places:

standardized residuals

−1.99	1.20	0.23	−1.67	0.26	−1.06	1.38	0.06	0.72	−0.40	1.21	0.90
0.40	−0.43	−0.24	−1.36	0.88	−0.75	1.30	−0.26	−1.51	0.55	0.62	

(a) Find the mean and standard deviation of the standardized residuals. Why do you expect values close to those you obtain?

(b) Make a stemplot of the standardized residuals. Are there any striking deviations from Normality? The most extreme residual is $z = -1.99$. Would this be surprisingly large if the 23 observations had a Normal distribution? Explain your answer.

(c) Plot the standardized residuals against the explanatory variable. Are there any suspicious patterns?

21.37 Tests for the intercept (Optional). Figure 21.6 (page 575) gives Excel output for the regression of blood alcohol content (BAC) on number of beers consumed. In Example 21.6, we rejected the hypothesis that the true regression line has *slope* $\beta = 0$. The data do show a positive linear relationship between BAC and beers. We might expect the *intercept* α of the true regression line to be 0, because no beers ($x = 0$) should produce no alcohol in the blood ($y = 0$). To test

$$H_0: \alpha = 0$$
$$H_a: \alpha \neq 0$$

we use a t statistic formed by dividing the least-squares intercept a by its standard error SE_a. Locate this statistic in the output of Figure 21.6 and verify that it is in fact a divided by its standard error. What is the P-value? Do the data suggest that the intercept is not 0?

21.38 **Confidence intervals for the intercept (Optional).** The Excel output in Figure 21.6 gives 95% confidence intervals for both the slope β and the intercept α of the true regression line of BAC on beers in the population of all students.

(a) What is the 95% confidence interval for the intercept? Does it contain 0, the value we might guess for α?

(b) Confidence intervals for the intercept α have the familiar form

$$a \pm t^*SE_a$$

with degrees of freedom $n - 2$. Use the information in the output to give a 99% confidence interval for α.

(Alan Scheir Photography/CORBIS)

One-Way Analysis of Variance: Comparing Several Means

The two-sample t procedures of Chapter 17 compare the means of two populations or the mean responses to two treatments in an experiment. Of course, studies don't always compare just two groups. We need a method for comparing any number of means.

EXAMPLE 22.1 Cars, SUVs, and pickup trucks

Sport-utility vehicles (SUVs) and pickup trucks have replaced cars in many American driveways. SUVs and trucks are often larger and heavier than cars and so are under attack for reasons of safety and fuel economy. Do SUVs and trucks have lower gas mileage than cars? Table 22.1 contains data on the highway gas mileage (in miles per gallon) for 31 midsize cars, 31 SUVs, and 14 standard-size pickup trucks. The gas mileages and vehicle classifications are compiled by the Environmental Protection Agency.[1] Because the cars are two-wheel-drive, the table describes only two-wheel-drive SUVs and trucks. The popular four-wheel-drive versions get poorer mileage.

Figure 22.1 shows side-by-side stemplots of the gas mileages for the three types of vehicle. We used the same stems in all three for easier comparison. It does appear that gas mileage decreases as we move from cars to SUVs to pickups.

Here are the means, standard deviations, and five-number summaries for the three types of vehicle, from software:

	N	Mean	Median	StDev	Minimum	Maximum	Q1	Q3
Midsize	31	27.903	27.000	2.561	23.000	33.000	26.000	30.000
SUV	31	22.677	22.000	3.673	17.000	29.000	19.000	26.000
Pickup	14	21.286	20.000	2.758	17.000	26.000	19.750	23.500

We will use the mean to describe the center of the gas mileage distributions. Mean gas mileage goes down as we move from midsize cars to SUVs to pickup trucks. The differences among the means are not large. Are they statistically significant?

TABLE 22.1 Highway gas mileage for 2003 model vehicles

Midsize Cars		Sport-Utility Vehicles		Standard Pickup Trucks	
Model	MPG	Model	MPG	Model	MPG
Acura 3.5TL	29	Cadillac Escalade	18	Chevrolet Silverado	20
Audi A6	27	Chevrolet Avalanche	18	Dodge Dakota	19
Audi A8	25	Chevrolet Blazer	23	Dodge Ram	20
Buick Century	29	Chevrolet Suburban	18	Ford Explorer	20
Buick Regal	27	Chevrolet Tahoe	18	Ford F150	20
Cadillac CTS	26	Chevrolet Tracker	26	Ford Ranger	26
Cadillac Seville	27	Chevrolet Trailblazer	22	GMC Sierra	20
Chevrolet Monte Carlo	32	Chrysler PT Cruiser	25	Lincoln Blackwood	17
Chrysler Sebring	30	Dodge Durango	19	Mazda B2300	26
Honda Accord	33	Ford Escape	28	Mazda B3000	22
Hyundai Sonata	30	Ford Expedition	19	Mazda B4000	21
Hyundai XG350	26	Ford Explorer	19	Nissan Frontier	23
Infiniti I35	26	Honda CR-V	28	Toyota Tacoma	25
Jaguar S-Type	26	Hyundai Santa Fe	27	Toyota Tundra	19
Jaguar Vanden Plas	24	Infiniti QX4	21		
Kia Optima	30	Isuzu Axiom	21		
Lexus ES300	29	Isuzu Rodeo	23		
Lexus GS300	25	Jeep Grand Cherokee	21		
Lincoln LS	26	Jeep Liberty	24		
Mercedes-Benz E320	27	Kia Sorento	20		
Mercedes-Benz E500	23	Lincoln Navigator	17		
Mitsubishi Diamante	25	Mazda Tribute	28		
Mitsubishi Galant	27	Mitsubishi Montero	22		
Nissan Altima	29	Nissan Pathfinder	21		
Nissan Maxima	26	Nissan Xterra	24		
Saab 9-5	29	Saturn Vue	28		
Saturn L200	32	Suzuki Vitara	25		
Saturn L300	29	Toyota 4Runner	20		
Toyota Camry	32	Toyota Highlander	27		
Volkswagen Passat	31	Toyota Rav4	29		
Volvo S80	28	Volvo XC 90	24		

```
17 |              17 | 0          17 | 0
18 |              18 | 0000       18 |
19 |              19 | 000        19 | 00
20 |              20 | 00         20 | 00000
21 |              21 | 0000       21 | 0
22 |              22 | 00         22 | 0
23 | 0            23 | 00         23 | 0
24 | 0            24 | 000        24 |
25 | 000          25 | 00         25 | 0
26 | 000000       26 | 0          26 | 00
27 | 00000        27 | 00         27 |
28 | 0            28 | 0000       28 |
29 | 000000       29 | 0          29 |
30 | 000          30 |            30 |
31 | 0            31 |            31 |
32 | 000          32 |            32 |
33 | 0            33 |            33 |

   Midsize          SUV            Pickup
```

Figure 22.1 Side-by-side stemplots comparing the highway gas mileages of midsize cars, sport-utility vehicles, and standard-size pickup trucks, from Table 22.1.

Comparing several means

Call the mean highway gas mileages for the three populations of vehicles μ_1 for midsize cars, μ_2 for SUVs, and μ_3 for pickups. The subscript reminds us which group a parameter or statistic describes. To compare these three population means, we might use the two-sample t test several times:

- Test H_0: $\mu_1 = \mu_2$ to see if the mean miles per gallon for midsize cars differs from the mean for SUVs.
- Test H_0: $\mu_1 = \mu_3$ to see if midsize cars differ from pickup trucks.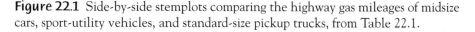
- Test H_0: $\mu_2 = \mu_3$ to see if SUVs differ from pickups.

The weakness of doing three tests is that we get three P-values, one for each test alone. That doesn't tell us how likely it is that *three* sample means are spread apart as far as these are. It may be that $\bar{x}_1 = 27.903$ and $\bar{x}_3 = 21.286$ are significantly different if we look at just two groups but not significantly different if we know that they are the smallest and the largest means in three groups. As we look at more groups, we expect the gap between the smallest and the largest sample mean to get larger. (Think of comparing the tallest and shortest person in larger and larger groups of people.) We can't safely compare many parameters by doing tests or confidence intervals for two parameters at a time.

The problem of how to do many comparisons at once with some overall measure of confidence in all our conclusions is common in statistics. This is the problem of **multiple comparisons.** Statistical methods for dealing with multiple comparisons usually have two steps:

multiple comparisons

1. An *overall test* to see if there is good evidence of *any* differences among the parameters that we want to compare.
2. A detailed *follow-up analysis* to decide which of the parameters differ and to estimate how large the differences are.

The overall test, though more complex than the tests we met earlier, is often reasonably straightforward. The follow-up analysis can be quite elaborate. In our basic introduction to statistical practice, we will concentrate on the overall test, along with data analysis that points to the nature of the differences.

The analysis of variance F test

We want to test the null hypothesis that there are *no differences* among the mean highway gas mileages for the three vehicle types:

$$H_0: \mu_1 = \mu_2 = \mu_3$$

The alternative hypothesis is that there is some difference, that not all three population means are equal:

$$H_a: \text{not all of } \mu_1, \ \mu_2, \ \text{and } \mu_3 \text{ are equal}$$

The alternative hypothesis is no longer one-sided or two-sided. It is "many-sided," because it allows any relationship other than "all three equal." For example, H_a includes the case in which $\mu_2 = \mu_3$ but μ_1 has a different value. The test of H_0 against H_a is called the **analysis of variance F test.** Analysis of variance is usually abbreviated as ANOVA. The ANOVA F test is almost always carried out by software that reports the test statistic and its P-value.

analysis of variance F test

EXAMPLE 22.2   Cars, SUVs, and pickup trucks: ANOVA

As we will soon see, software tells us that for the gas mileage data in Table 22.1, $F = 31.61$ with P-value $P < 0.001$. There is very strong evidence that the three types of vehicle do not all have the same mean gas mileage.

The F test does not say *which* of the three means are significantly different. It appears from our preliminary data analysis that SUVs and pickups have similar fuel economy and that both get distinctly poorer mileage than midsize cars. The gap of more than 5 miles per gallon between midsize cars and the other two types of vehicle is large enough to be of practical interest.

Our conclusion: There is strong evidence ($P < 0.001$) that the means are not all equal. The most important difference among the means is that midsize cars have better gas mileage than SUVs and pickups.

Example 22.2 illustrates our approach to comparing means. The ANOVA F test (done by software) assesses the evidence for *some* difference among the population means. In most cases, we expect the F test to be significant. We would not undertake a study if we did not expect to find some effect. The formal test is nonetheless important to guard against being misled by chance variation. We will not do the formal follow-up analysis that is often the most useful part of an ANOVA study. Follow-up analysis would allow us to say which means differ and by how much, with (say) 95% confidence that *all* our conclusions are correct. We rely instead on examination of the data to show what differences are present and whether they are large enough to be interesting.

APPLY YOUR KNOWLEDGE

22.1 Do fruit flies sleep? Mammals and birds sleep. Insects such as fruit flies rest, but is this rest sleep? Biologists now think that insects do sleep. One experiment gave caffeine to fruit flies to see if it affected their rest. We know that caffeine reduces sleep in mammals, so if it reduces rest in fruit flies that's another hint that the rest is really sleep. The paper reporting the study contains a graph similar to Figure 22.2 and states, "Flies given caffeine obtained less rest during the dark period in a dose-dependent fashion ($n = 36$ per group, $P < 0.0001$)."[2]

(a) The explanatory variable is amount of caffeine, in milligrams per milliliter of blood. The response variable is minutes of rest (measured by an infrared motion sensor) during a 12-hour dark period. Outline the design of this experiment.

(b) The P-value in the report comes from the ANOVA F test. What means does this test compare? State in words the null and alternative hypotheses for the test in this setting. What do the graph and the statistical test together lead you to conclude?

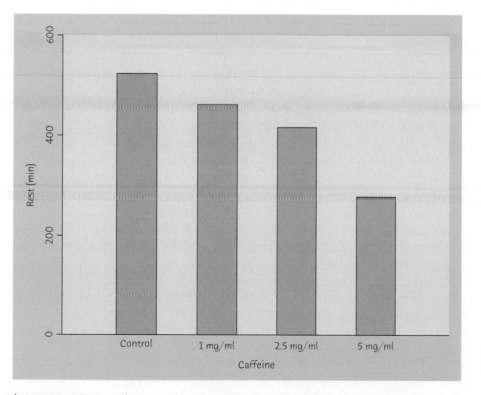

Figure 22.2 Bar graph comparing the mean rest of fruit flies given different amounts of caffeine, for Exercise 22.1.

22.2 **Road rage.** A random-digit-dialing survey asked a random sample of 1382 drivers questions about their driving. Responses to a number of questions were then combined to yield a measure of "angry/threatening driving," commonly called "road rage." What driver characteristics go with road rage? There were no significant differences among races or levels of education. What about the effect of the driver's age? Here are the mean responses for three age-groups:[3]

<30 yr	30–55 yr	>55 yr
2.22	1.33	0.66

The report says that $F = 34.96$, with $P < 0.01$.

(a) What are the null and alternative hypotheses for the ANOVA F test? Be sure to explain what means the test compares.

(b) Based on the sample means and the F test, what do you conclude?

Using technology

Any technology used for statistics should perform analysis of variance. Figure 22.3 displays ANOVA output for the data of Table 22.1 from the Minitab statistical software, the Excel spreadsheet program, and the TI-83 graphing calculator.

Minitab and Excel give the sizes of the three samples and their means. These agree with those in Example 22.1. The most important part of all three outputs reports the F test statistic, $F = 31.61$, and its P-value. Minitab sensibly reports P as 0.000, that is, 0 to three decimal places. Excel and the TI-83 show that P is about 0.0000000001. (The E-10 in these displays means move the decimal point 10 digits to the left.) There is very strong evidence that the three types of vehicle do not all have the same mean gas mileage.

All three outputs also report sums of squares (SS) and mean squares (MS). We don't need this information now. The optional section "Some Details of ANOVA" at the end of this chapter explains these terms.

Minitab also gives confidence intervals for all three means that help us see which means differ and by how much. The SUV and pickup intervals overlap, and the midsize interval lies well above them on the mileage scale. These are 95% confidence intervals for each mean separately. We are not 95% confident that *all three* intervals cover the three means. This is another example of the peril of multiple comparisons.

APPLY YOUR KNOWLEDGE

22.3 **Logging in the rainforest.** How does logging in a tropical rainforest affect the forest several years later? Researchers compared forest plots in Borneo that had never been logged (Group 1) with similar plots nearby

Minitab

```
┌─────────────────────────────────────────────────────────────────────┐
│ ▤ Session                                              _ □ ✕          │
├─────────────────────────────────────────────────────────────────────┤
│ One-way ANOVA: Midsize, SUV, Pickup                               ▲   │
│ Analysis of Variance                                                  │
│ Source      DF        SS        MS        F        P                  │
│ Factor       2     606.45    303.22    31.61    0.000                 │
│ Error       73     700.34      9.59                                   │
│ Total       75    1306.79                                             │
│                                                                       │
│                              Individual 95% CIs For Mean              │
│                              Based on Pooled StDev                    │
│ Level        N      Mean    StDev ----+---------+---------+---------+----│
│ Midsize     31    27.903    2.561                        (---*----)   │
│ SUV         31    22.677    3.673            (---*----)               │
│ Pickup      14    21.286    2.758       (---*----)                    │
│                                                                       │
│                                        ----+---------+   -----+---------+----│
│ Pooled StDev =    3.097                21.0      24.0      27.0      30.0 ▼│
│ ◄ ▶                                                              ◄ ▶  │
└─────────────────────────────────────────────────────────────────────┘
```

Excel

```
┌─────────────────────────────────────────────────────────────────────┐
│ ▨ Microsoft Excel - Book1                              _ □ ✕          │
├─────────────────────────────────────────────────────────────────────┤
│       A           B        C        D         E        F        G   ▲ │
│ 1  Anova: Single Factor                                               │
│ 2                                                                     │
│ 3  SUMMARY                                                            │
│ 4    Groups      Count     Sum    Average  Variance                   │
│ 5  Midsize        31       865    27.9032   6.5570                    │
│ 6  SUV            31       703    22.6774  13.4925                    │
│ 7  Pickup         14       298    21.2857   7.6044                    │
│ 8                                                                     │
│ 9                                                                     │
│ 10 ANOVA                                                              │
│ 11 Source of Variation  SS     df      MS       F     P-value  F crit │
│ 12 Between Groups     606.4485   2   303.2242  31.0000 1.3E-10 3.122100│
│ 13 Within Groups      700.3410  73     9.5937                         │
│ 14                                                                    │
│ 15 Total             1306.7895  75                                    │
│ 16                                                                  ▼ │
│ ◄ ▶ ▶◄ \Sheet5 / Sheet1 / Sheet2 / Sheet3 /        ◄                ▶ │
└─────────────────────────────────────────────────────────────────────┘
```

TI-83

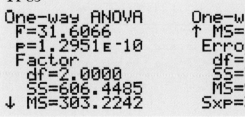

```
One-way ANOVA          One-way ANOVA
 F=31.6066             ↑ MS=303.2242
 p=1.2951E-10          Error
 Factor                 df=73.0000
  df=2.0000             SS=700.3410
  SS=606.4485           MS=9.5937
↓ MS=303.2242          Sxp=3.0974
```

Figure 22.3 Analysis of variance output for the gas mileage data from Minitab, Excel, and the TI-83.

that had been logged 1 year earlier (Group 2) and 8 years earlier (Group 3). Although the study was not an experiment, the authors explain why we can consider the plots to be randomly selected. The data appear in Table 22.2. The variable Trees is the count of trees in a plot; Species is the count of tree species in a plot. The variable Richness is the ratio of number of species to number of individual trees, calculated as Species/Trees.[4]

TABLE 22.2 Data from a study of logging in Borneo

Observation	Group	Trees	Species	Richness
1	1	27	22	0.81481
2	1	22	18	0.81818
3	1	29	22	0.75862
4	1	21	20	0.95238
5	1	19	15	0.78947
6	1	33	21	0.63636
7	1	16	13	0.81250
8	1	20	13	0.65000
9	1	24	19	0.79167
10	1	27	13	0.48148
11	1	28	19	0.67857
12	1	19	15	0.78947
13	2	12	11	0.91667
14	2	12	11	0.91667
15	2	15	14	0.93333
16	2	9	7	0.77778
17	2	20	18	0.90000
18	2	18	15	0.83333
19	2	17	15	0.88235
20	2	14	12	0.85714
21	2	14	13	0.92857
22	2	2	2	1.00000
23	2	17	15	0.88235
24	2	19	8	0.42105
25	3	18	17	0.94444
26	3	4	4	1.00000
27	3	22	18	0.81818
28	3	15	14	0.93333
29	3	18	18	1.00000
30	3	19	15	0.78947
31	3	22	15	0.68182
32	3	12	10	0.83333
33	3	12	12	1.00000

```
Microsoft Excel - Book1                                                    _ □ X
           A            B        C        D        E        F        G
 1  Anova: Single Factor
 2
 3  SUMMARY
 4        Groups        Count    Sum      Average  Variance
 5  Group 1             12       285      23.75    25.6591
 6  Group 2             12       169      14.0833  24.8106
 7  Group 3             9        142      15.7778  33.1944
 8
 9  ANOVA
10   Source of Variation  SS      df       MS       F       P-value  F crit
11  Between Groups      625.1566  2    312.57820  11.4257  0.000205  3.31583
12  Within Groups       820.7222  30    27.3574
13
14  Total               1445.879  32
15
   Sheet4 / Sheet1 / Sheet2 / Sheet3 /
```

Figure 22.4 Excel output for analysis of variance on the number of trees in forest plots, for Exercise 22.3.

(a) Make side-by-side stemplots of Trees for the three groups. Use stems 0, 1, 2, and 3 and split the stems (see page 17). What effects of logging are visible?

(b) Figure 22.4 shows Excel ANOVA output for Trees. What do the group means show about the effects of logging?

(c) What are the values of the ANOVA F statistic and its P-value? What hypotheses does F test? What conclusions about the effects of logging on number of trees do the data lead to?

22.4 Dogs, friends, and stress. If you are a dog lover, perhaps having your dog along reduces the effect of stress. To examine the effect of pets in stressful situations, researchers recruited 45 women who said they were dog lovers. The EESEE story "Stress among Pets and Friends" describes the results. Fifteen of the subjects were randomly assigned to each of three groups to do a stressful task alone (the control group), with a good friend present, or with their dog present. The subject's mean heart rate during the task is one measure of the effect of stress. Table 22.3 contains the data.

(a) Make stemplots of the heart rates for the three groups (round to the nearest whole number of beats). Do any of the groups show outliers or extreme skewness?

(b) Figure 22.5 gives the Minitab ANOVA output for these data. Do the mean heart rates for the groups appear to show that the presence of a pet or a friend reduces heart rate during a stressful task?

(c) What are the values of the ANOVA F statistic and its P-value? What hypotheses does F test? Briefly describe the conclusions you draw from these data. Did you find anything surprising?

TABLE 22.3 Mean heart rates during stress with a pet (P), with a friend (F), and for the control group (C)

Group	Rate	Group	Rate	Group	Rate
P	69.169	P	68.862	C	84.738
F	99.692	C	87.231	C	84.877
P	70.169	P	64.169	P	58.692
C	80.369	C	91.754	P	79.662
C	87.446	C	87.785	P	69.231
P	75.985	F	91.354	C	73.277
F	83.400	F	100.877	C	84.523
F	102.154	C	77.800	C	70.877
P	86.446	P	97.538	F	89.815
F	80.277	P	85.000	F	98.200
C	90.015	F	101.062	F	76.908
C	99.046	F	97.046	P	69.538
C	75.477	C	62.646	P	70.077
F	88.015	F	81.600	F	86.985
F	92.492	P	72.262	P	65.446

```
Session                                                        _ □ ×

Analysis of Variance for Beats
Source     DF        SS        MS         F         P
Group       2     2387.7    1193.8     14.08     0.000
Error      42     3561.3      84.8
Total      44     5949.0


                                  Individual 95% CIs For Mean
                                  Based on Pooled StDev

Level       N      Mean     StDev  ----+---------+---------+---------+---
Control    15    82.524     9.242              (-----*----- )
Friend     15    91.325     8.341                         (----*-----)
Pet        15    73.483     9.970   (-----*----)

                                  ---+---------+---------+---------+---
Pooled StDev =     9.208            72.0      80.0      88.0      96.0
```

Figure 22.5 Minitab output for the data in Table 22.3 on heart rates (beats per minute) during stress, for Exercise 22.4. The "Control" group worked alone, the "Friend" group had a friend present, and the "Pet" group had a pet dog present.

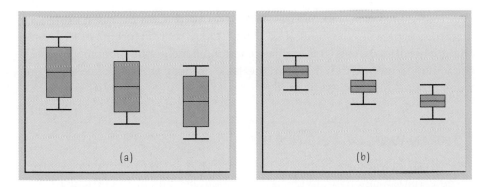

Figure 22.6 Boxplots for two sets of three samples each. The sample means are the same in (a) and (b). Analysis of variance will find a more significant difference among the means in (b) because there is less variation among the individuals within those samples.

The idea of analysis of variance

Here is the main idea for comparing means: what matters is not how far apart the sample means are but how far apart they are *relative to the variability of individual observations*. Look at the two sets of boxplots in Figure 22.6. For simplicity, these distributions are all symmetric, so that the mean and median are the same. The centerline in each boxplot is therefore the sample mean. Both sets of boxplots compare three samples with the same three means. Like the three vehicle types in Example 22.1, the means are different but not very different. Could differences this large easily arise just due to chance, or are they statistically significant?

- The boxplots in Figure 22.6(a) have tall boxes, which show lots of variation among the individuals in each group. With this much variation among individuals, we would not be surprised if another set of samples gave quite different sample means. The observed differences among the sample means could easily happen just by chance.

- The boxplots in Figure 22.6(b) have the same centers as those in Figure 22.6(a), but the boxes are much shorter. That is, there is much less variation among the individuals in each group. It is unlikely that any sample from the first group would have a mean as small as the mean of the second group. Because means as far apart as those observed would rarely arise just by chance in repeated sampling, they are good evidence of real differences among the means of the three populations we are sampling from.

You can use the *One-Way ANOVA* applet to demonstrate the analysis of variance idea for yourself. The applet allows you to change both the group means and the spread within groups. You can watch the ANOVA F statistic and its P-value change as you work.

This comparison of the two parts of Figure 22.6 is too simple in one way. It ignores the effect of the sample sizes, an effect that boxplots do not show. Small differences among sample means can be significant if the samples are very large. Large differences among sample means may fail to be significant if the samples are very small. All we can be sure of is that for the same sample size, Figure 22.6(b) will give a much smaller P-value than Figure 22.6(a). Despite this qualification, the big idea remains: if sample means are far apart relative to the variation among individuals in the same group, that's evidence that something other than chance is at work.

THE ANALYSIS OF VARIANCE IDEA

Analysis of variance compares the variation due to specific sources with the variation among individuals who should be similar. In particular, ANOVA tests whether several populations have the same mean by comparing how far apart the sample means are with how much variation there is within the samples.

It is one of the oddities of statistical language that methods for comparing means are named after the variance. The reason is that the test works by comparing two kinds of variation. Analysis of variance is a general method for studying sources of variation in responses. Comparing several means is the simplest form of ANOVA, called **one-way ANOVA.** One-way ANOVA is the only form of ANOVA that we will study.

one-way ANOVA

THE ANOVA F STATISTIC

The **analysis of variance F statistic** for testing the equality of several means has this form:

$$F = \frac{\text{variation among the sample means}}{\text{variation among individuals in the same sample}}$$

We give more detail later. Because ANOVA is in practice done by software, the idea is more important than the detail. The F statistic can only take values that are zero or positive. It is zero only when all the sample means are identical and gets larger as they move farther apart. Large values of F are evidence against the null hypothesis H_0 that all population means are the same. Although the alternative hypothesis H_a is many-sided, the ANOVA F test is one-sided because any violation of H_0 tends to produce a large value of F.

How large must F be to provide significant evidence against H_0? To answer questions of statistical significance, compare the F statistic with critical values from an **F distribution.** The F distributions are described on page 456 in Chapter 17. A specific F distribution is specified by two parameters: a numerator degrees of freedom and a denominator degrees of freedom. Table D in the back of the book contains critical values for F distributions with various degrees of freedom.

F distribution

EXAMPLE 22.3   Using the F table

Look again at the software output for the gas mileage data in Figure 22.3. All three outputs give the degrees of freedom for the F test, labeled "df" or "DF." There are 2 degrees of freedom in the numerator and 73 in the denominator.

In Table D, find the numerator degrees of freedom 2 at the top of the table. Then look for the denominator degrees of freedom 73 at the left of the table. There is no entry for 73, so we use the next smaller entry, 50 degrees of freedom. The upper critical values for 2 and 50 degrees of freedom are

p	Critical value
0.100	2.41
0.050	3.18
0.025	3.97
0.010	5.06
0.001	7.96

Figure 22.7 shows the F density curve with 2 and 50 degrees of freedom and the upper 5% critical value 3.18. The observed $F = 31.61$ lies far to the right on this curve. We see from the table that $F = 31.61$ is larger than the 0.001 critical value, so $P < 0.001$.

Software, of course, uses the F distribution with 2 and 73 degrees of freedom to calculate P-values. Excel also tells us how large F must be to be significant at the $\alpha = 0.05$ level. This is the 5% critical value for F with 2 and 73 degrees of freedom. It is 3.122. Approximating the P-value from Table D is usually adequate in practice, as it is in this example.

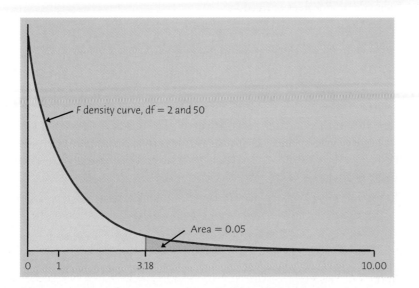

Figure 22.7 The density curve of the F distribution with 2 degrees of freedom in the numerator and 50 degrees of freedom in the denominator, for Example 22.3. The value $F^* = 3.18$ is the 0.05 critical value.

The degrees of freedom of the F statistic depend on the number of means we are comparing and the number of observations in each sample. That is, the F test takes into account the number of observations. Here are the details.

DEGREES OF FREEDOM FOR THE F TEST

We want to compare the means of I populations. We have an SRS of size n_i from the ith population, so that the total number of observations in all samples combined is

$$N = n_1 + n_2 + \cdots + n_I$$

If the null hypothesis that all population means are equal is true, the ANOVA F statistic has the F distribution with $I - 1$ degrees of freedom in the numerator and $N - I$ degrees of freedom in the denominator.

EXAMPLE 22.4   Degrees of freedom for F

In Examples 22.1 and 22.2, we compared the mean highway gas mileage for three types of vehicle, so $I = 3$. The three sample sizes are

$$n_1 = 31 \quad n_2 = 31 \quad n_3 = 14$$

The total number of observations is therefore

$$N = 31 + 31 + 14 = 76$$

The ANOVA F test has numerator degrees of freedom

$$I - 1 = 3 - 1 = 2$$

and denominator degrees of freedom

$$N - I = 76 - 3 = 73$$

These degrees of freedom are given in the outputs in Figure 22.3.

APPLY YOUR KNOWLEDGE

22.5 **Logging in the rainforest, continued.** Exercise 22.3 compares the number of tree species in rainforest plots that had never been logged (Group 1) with similar plots nearby that had been logged 1 year earlier (Group 2) and 8 years earlier (Group 3).

(a) What are I, the n_i, and N for these data? Identify these quantities in words and give their numerical values.

(b) Find the degrees of freedom for the ANOVA F statistic. Check your work against the Excel output in Figure 22.4.

(c) For these data, $F = 11.43$. What does Table D tell you about the P-value of this statistic?

22.6 **Dogs, friends, and stress, continued.** Exercise 22.4 compares the mean heart rates for women performing a stressful task under three conditions.

(a) What are I, the n_i, and N for these data? State in words what each of these numbers means.

(b) Find both degrees of freedom for the ANOVA F statistic. Check your results against the Minitab output in Figure 22.5.

(c) The output shows that $F = 14.08$. What does Table D tell you about the significance of this result?

22.7 Smoking among French men. A study of smoking categorized a sample of French men aged 20 to 60 years as nonsmokers (146 men), former smokers (125 men), moderate smokers (104 men), or heavy smokers (84 men). One set of response variables was the amount of several types of food consumed, in grams.[5]

(a) What are the degrees of freedom for an ANOVA F statistic that compares the mean weights of a food consumed in the 4 populations? Use Table D with 3 and 200 degrees of freedom to give an approximate P-value for each of the following foods:

(b) For milk, $F = 3.58$.

(c) For eggs, $F = 1.08$.

(d) For vegetables, $F = 6.02$.

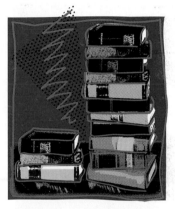

The ANOVA model

Like all inference procedures, ANOVA is valid only in some circumstances. Here is the model that describes when we can use ANOVA to compare population means.

THE ANOVA MODEL

- We have **I independent SRSs,** one from each of I populations.

- The ith population has a **Normal distribution** with unknown mean μ_i. The means may be different in the different populations. The ANOVA F statistic tests the null hypothesis that all of the populations have the same mean:

$$H_0\colon \mu_1 = \mu_2 = \cdots = \mu_I$$
$$H_a\colon \text{not all of the } \mu_i \text{ are equal}$$

- All of the populations have the **same standard deviation** σ, whose value is unknown.

The ANOVA model has $I + 1$ parameters, the I population means and the standard deviation σ.

The first two requirements are familiar from our study of the two-sample t procedures for comparing two means. As usual, the design of the data production is the most important foundation for inference. Biased sampling or confounding can make any inference meaningless. If we do not actually draw

Where are the details?

Papers reporting scientific research must be short. Brevity allows researchers to hide details about their data. Did they choose their subjects in a biased way? Did they report data on only some of their subjects? Did they try several statistical analyses and report the one that looked best? The statistician John Bailar screened more than 4000 papers for the *New England Journal of Medicine*. He says, "When it came to the statistical review, it was often clear that critical information was lacking, and the gaps nearly always had the practical effect of making the authors' conclusions look stronger than they should have."

separate SRSs from each population or carry out a randomized comparative experiment, it is often unclear to what population the conclusions of inference apply. This is the case in Example 22.1, for example. ANOVA, like other inference procedures, is often used when random samples are not available. You must judge each use on its merits, a judgment that usually requires some knowledge of the subject of the study in addition to some knowledge of statistics. We might consider the 2003 model year vehicles in Example 22.1 to be samples from vehicles of their types produced in recent years.

robustness

Because no real population has exactly a Normal distribution, the usefulness of inference procedures that assume Normality depends on how sensitive they are to departures from Normality. Fortunately, procedures for comparing means are not very sensitive to lack of Normality. The ANOVA F test, like the t procedures, is **robust.** What matters is Normality of the sample means, so ANOVA becomes safer as the sample sizes get larger, because of the central limit theorem effect. Remember to check for outliers that change the value of sample means and for extreme skewness. When there are no outliers and the distributions are roughly symmetric, you can safely use ANOVA for sample sizes as small as 4 or 5. (Don't confuse the ANOVA F, which compares several means, with the F statistic discussed in Chapter 17, which compares two standard deviations and is not robust against non-Normality.)

The third assumption is annoying: ANOVA assumes that the variability of observations, measured by the standard deviation, is the same in all populations. You may recall from Chapter 17 (page 454) that there is a special version of the two-sample t test that assumes equal standard deviations in both populations. The ANOVA F for comparing two means is exactly the square of this special t statistic. We prefer the t test that does not assume equal standard deviations, but for comparing more than two means there is no general alternative to the ANOVA F. It is not easy to check the assumption that the populations have equal standard deviations. Statistical tests for equality of standard deviations are very sensitive to lack of Normality, so much so that they are of little practical value. You must either seek expert advice or rely on the robustness of ANOVA.

How serious are unequal standard deviations? ANOVA is not too sensitive to violations of the assumption, especially when all samples have the same or similar sizes and no sample is very small. When designing a study, try to take samples of the same size from all the groups you want to compare. The sample standard deviations estimate the population standard deviations, so check before doing ANOVA that the sample standard deviations are similar to each other. We expect some variation among them due to chance. Here is a rule of thumb that is safe in almost all situations.

CHECKING STANDARD DEVIATIONS IN ANOVA

The results of the ANOVA F test are approximately correct when the largest sample standard deviation is no more than twice as large as the smallest sample standard deviation.

EXAMPLE 22.5   Do the standard deviations allow ANOVA?

In the gas mileage study, the sample standard deviations for midsize cars, SUVs, and pickups are

$$s_1 = 2.561 \quad s_2 = 3.673 \quad s_3 = 2.758$$

These standard deviations easily satisfy our rule of thumb. We can safely use ANOVA to compare the mean gas mileage for the three vehicle types.

The report from which Table 22.1 was taken also contained data on 22 subcompact cars. Can we use ANOVA to compare the means for all four types of vehicle? The standard deviation for subcompact cars is $s_4 = 5.262$ miles per gallon. This is slightly more than twice the smallest standard deviation:

$$\frac{\text{largest } s}{\text{smallest } s} = \frac{5.262}{2.561} = 2.05$$

Our rule of thumb is conservative, so many statisticians would go ahead with ANOVA in this circumstance.

A large standard deviation is often due to skewness or outliers. It turns out that the government's "subcompact" category includes three rather exotic cars, a Bentley, a Ferrari, and a Maserati. All three get poor gas mileage. The standard deviation for the remaining 19 subcompact cars is $s = 3.304$, well within the guidelines for ANOVA.

EXAMPLE 22.6   Which color attracts beetles best?

To detect the presence of harmful insects in farm fields, we can put up boards covered with a sticky material and examine the insects trapped on the boards. Which colors attract insects best? Experimenters placed six boards of each of four colors at random locations in a field of oats and measured the number of cereal leaf beetles trapped. Here are the data:[6]

Board color	Insects trapped					
Blue	16	11	20	21	14	7
Green	37	32	20	29	37	32
White	21	12	14	17	13	20
Yellow	45	59	48	46	38	47

We would like to use ANOVA to compare the mean numbers of beetles that would be trapped by all boards of each color. Because the samples are small, we plot the data in side-by-side stemplots in Figure 22.8. Minitab output for ANOVA appears

```
     Blue            Green           White           Yellow

0 | 7            0 |             0 |             0 |
1 | 146          1 |             1 | 2347        1 |
2 | 01           2 | 09          2 | 01          2 |
3 |              3 | 2277        3 |             3 | 8
4 |              4 |             4 |             4 | 5678
5 |              5 |             5 |             5 | 9
```

Figure 22.8 Side-by-side stemplots comparing the counts of insects attracted by six boards for each of four board colors, for Example 22.6.

in Figure 22.9. The yellow boards attract by far the most insects ($\bar{x}_4 = 47.167$), with green next ($\bar{x}_2 = 31.167$) and blue and white far behind.

Check that we can safely use ANOVA to test equality of the four means. The largest of the four sample standard deviations is 6.795 and the smallest is 3.764. The ratio

$$\frac{\text{largest } s}{\text{smallest } s} = \frac{6.795}{3.764} = 1.8$$

is less than 2, so these data satisfy our rule of thumb. The shapes of the four distributions are irregular, as we expect with only 6 observations in each group, but there are no outliers. The ANOVA results will be approximately correct.

There are $I = 4$ groups and $N = 24$ observations overall, so the degrees of freedom for F are

$$\text{numerator:} \quad I - 1 = 4 - 1 = 3$$
$$\text{denominator:} \quad N - I = 24 - 4 = 20$$

This agrees with Minitab's results. The F statistic is $F = 42.84$, a large F with $P < 0.001$. Despite the small samples, the experiment gives very strong evidence of differences among the colors. Yellow boards appear best at attracting leaf beetles.

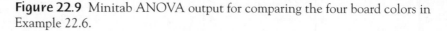

```
Session                                                          _ |□| X|

Analysis of Variance for Beetles                                      ▲
Source      DF        SS        MS        F         P
Color        3     4134.0    1378.0     42.84     0.000
Error       20      643.3      32.2
Total       23     4777.3


                               Individual 95% CIs For Mean
                               Based on Pooled StDev
Level       N       Mean      StDev  ----+---------+---------+---------+----
Blue        6     14.833     5.345   (---*----)
Green       6     31.167     6.306                  (---*----)
White       6     16.167     3.764    (---*----)
Yellow      6     47.167     6.795                             (---*---)

                                      ---+---------+---------+---------+----
Pooled StDev =     5.672              12        24        36        48
 |◄| |  |                                                           |►|  //
```

Figure 22.9 Minitab ANOVA output for comparing the four board colors in Example 22.6.

22.8 Checking standard deviations. Verify that the sample standard deviations for these sets of data do allow use of ANOVA to compare the population means.

(a) The counts of trees in Exercise 22.3 and Figure 22.4.

(b) The heart rates of Exercise 22.4 and Figure 22.5.

22.9 Species richness after logging. Table 22.2 gives data on the species richness in rainforest plots, defined as the number of tree species in a plot divided by the number of trees in the plot. ANOVA may not be trustworthy for the richness data. Do data analysis: make side-by-side stemplots to examine the distributions of the response variable in the three groups and also compare the standard deviations. What about the data makes ANOVA risky?

22.10 Marital status and salary. Married men tend to earn more than single men. An investigation of the relationship between marital status and income collected data on all 8235 men employed as managers or professionals by a large manufacturing firm in 1976. Suppose (this is risky) we regard these men as a random sample from the population of all men employed in managerial or professional positions in large companies. Here are summary statistics for the salaries of these men:[7]

	Single	Married	Divorced	Widowed
n_i	337	7,730	126	42
$\bar{x}_i$	$21,384	$26,873	$25,594	$26,936
s_i	$5,731	$7,159	$6,347	$8,119

(a) Briefly describe the relationship between marital status and salary.

(b) Do the sample standard deviations allow use of the ANOVA F test? (The distributions are skewed to the right. We expect right-skewness in income distributions. The investigators actually applied ANOVA to the logarithms of the salaries, which are more symmetric.)

(c) What are the degrees of freedom of the ANOVA F test?

(d) The F test is a formality for these data, because we are sure that the P-value will be very small. Why are we sure?

(e) Single men earn less on the average than men who are or have been married. Do the highly significant differences in mean salary show that getting married raises men's mean income? Explain your answer.

Some details of ANOVA*

Now we will give the actual formula for the ANOVA F statistic. We have SRSs from each of I populations. Subscripts from 1 to I tell us which sample a statistic refers to:

Population	Sample size	Sample mean	Sample std. dev.
1	n_1	$\overline{x}_1$	s_1
2	n_2	$\overline{x}_2$	s_2
⋮	⋮	⋮	⋮
I	n_I	$\overline{x}_I$	s_I

You can find the F statistic from just the sample sizes n_i, the sample means $\overline{x}_i$, and the sample standard deviations s_i. You don't need to go back to the individual observations.

The ANOVA F statistic has the form

$$F = \frac{\text{variation among the sample means}}{\text{variation among individuals within samples}}$$

mean squares

The measures of variation in the numerator and denominator of F are called **mean squares.** A mean square is a more general form of a sample variance. An ordinary sample variance s^2 is an average (or mean) of the squared deviations of observations from their mean, so it qualifies as a "mean square."

The numerator of F is a mean square that measures variation among the I sample means $\overline{x}_1, \overline{x}_2, \ldots, \overline{x}_I$. Call the overall mean response (the mean of all N observations together) $\overline{x}$. You can find $\overline{x}$ from the I sample means by

$$\overline{x} = \frac{n_1 \overline{x}_1 + n_2 \overline{x}_2 + \cdots + n_I \overline{x}_I}{N}$$

The sum of each mean multiplied by the number of observations it represents is the sum of all the individual observations. Dividing this sum by N, the total number of observations, gives the overall mean $\overline{x}$. The numerator mean square in F is an average of the I squared deviations of the means of the samples from $\overline{x}$. We call it the **mean square for groups,** abbreviated as MSG.

MSG

$$\text{MSG} = \frac{n_1(\overline{x}_1 - \overline{x})^2 + n_2(\overline{x}_2 - \overline{x})^2 + \cdots + n_I(\overline{x}_I - \overline{x})^2}{I - 1}$$

Each squared deviation is weighted by n_i, the number of observations it represents.

---

*This more advanced section is optional if you are using software to find the F statistic.

The mean square in the denominator of F measures variation among individual observations in the same sample. For any one sample, the sample variance s_i^2 does this job. For all I samples together, we use an average of the individual sample variances. It is again a weighted average in which each s_i^2 is weighted by one fewer than the number of observations it represents, $n_i - 1$. Another way to put this is that each s_i^2 is weighted by its degrees of freedom $n_i - 1$. The resulting mean square is called the **mean square for error,** MSE.

MSE

$$MSE = \frac{(n_1 - 1)s_1^2 + (n_2 - 1)s_2^2 + \cdots + (n_I - 1)s_I^2}{N - I}$$

"Error" doesn't mean a mistake has been made. It's a traditional term for chance variation. Here is a summary of the ANOVA test.

THE ANOVA F TEST

Draw an independent SRS from each of I populations. The ith population has the $N(\mu_i, \sigma)$ distribution, where σ is the common standard deviation in all the populations. The ith sample has size n_i, sample mean $\overline{x}_i$, and sample standard deviation s_i.

The **ANOVA F statistic** tests the null hypothesis that all I populations have the same mean:

$$H_0: \mu_1 = \mu_2 = \cdots = \mu_I$$
$$H_a: \text{not all of the } \mu_i \text{ are equal}$$

The statistic is

$$F = \frac{MSG}{MSE}$$

The numerator of F is the **mean square for groups**

$$MSG = \frac{n_1(\overline{x}_1 - \overline{x})^2 + n_2(\overline{x}_2 - \overline{x})^2 + \cdots + n_I(\overline{x}_I - \overline{x})^2}{I - 1}$$

The denominator of F is the **mean square for error**

$$MSE = \frac{(n_1 - 1)s_1^2 + (n_2 - 1)s_2^2 + \cdots + (n_I - 1)s_I^2}{N - I}$$

When H_0 is true, F has the **F distribution** with $I - 1$ and $N - I$ degrees of freedom.

The denominators in the recipes for MSG and MSE are the two degrees of freedom $I - 1$ and $N - I$ of the F test. The numerators are called **sums of squares,** from their algebraic form. It is usual to present the results of ANOVA in an **ANOVA table.** Output from software usually includes an ANOVA table.

sums of squares

ANOVA table

EXAMPLE 22.7 ANOVA calculations: software

Look again at the three outputs in Figure 22.3 (page 603). Minitab and Excel give the ANOVA table. The TI-83, with its small screen, gives the degrees of freedom, sums of squares, and mean squares separately. Each output uses slightly different language. The basic ANOVA table is:

Source of variation	df	SS	MS	F statistic
Variation among samples	2	606.45	$MSG = 303.22$	31.61
Variation within samples	73	700.34	$MSE = 9.59$	

You can check that each mean square MS is the corresponding sum of squares SS divided by its degrees of freedom df. The F statistic is MSG divided by MSE.

Because MSE is an average of the individual sample variances, it is also called the *pooled sample variance*, written as s_p^2. When all I populations have the same population variance σ^2 (ANOVA assumes that they do), s_p^2 estimates the common variance σ^2. The square root of MSE is the **pooled standard deviation** s_p. It estimates the common standard deviation σ of observations in each group. The Minitab and TI-83 outputs in Figure 22.3 give the value $s_p = 3.097$.

pooled standard deviation

The pooled standard deviation s_p is a better estimator of the common σ than any individual sample standard deviation s_i because it combines (pools) the information in all I samples. We can get a confidence interval for any of the means μ_i from the usual form

$$\text{estimate} \pm t^* SE_{\text{estimate}}$$

using s_p to estimate σ. The confidence interval for μ_i is

$$\bar{x}_i \pm t^* \frac{s_p}{\sqrt{n_i}}$$

Use the critical value t^* from the t distribution with $N - I$ degrees of freedom, because s_p has $N - I$ degrees of freedom. These are the confidence intervals that appear in Minitab ANOVA output.

EXAMPLE 22.8 ANOVA calculations: without software

We can do the ANOVA test comparing the gas mileages of midsize cars, SUVs, and pickup trucks using only the sample sizes, sample means, and sample standard deviations. These appear in Example 22.1, but it is easy to find them with a calculator. There are $I = 3$ groups with a total of $N = 76$ vehicles.

The overall mean of the 76 gas mileages in Table 22.1 (page 598) is

$$\bar{x} = \frac{n_1 \bar{x}_1 + n_2 \bar{x}_2 + n_3 \bar{x}_3}{N}$$

$$= \frac{(31)(27.903) + (31)(22.677) + (14)(21.286)}{76}$$

$$= \frac{1865.984}{76} = 24.552$$

The mean square for groups is

$$\text{MSG} = \frac{n_1(\bar{x}_1 - \bar{x})^2 + n_2(\bar{x}_2 - \bar{x})^2 + n_3(\bar{x}_3 - \bar{x})^2}{I - 1}$$

$$= \frac{1}{3 - 1}[(31)(27.903 - 24.552)^2 + (31)(22.677 - 24.552)^2$$

$$+ (14)(21.286 - 24.552)^2]$$

$$= \frac{606.424}{2} = 303.212$$

The mean square for error is

$$\text{MSE} = \frac{(n_1 - 1)s_1^2 + (n_2 - 1)s_2^2 + (n_3 - 1)s_3^2}{N - I}$$

$$= \frac{(30)(2.561^2) + (30)(3.673^2) + (13)(2.758^2)}{73}$$

$$= \frac{700.375}{73} = 9.594$$

Finally, the ANOVA test statistic is

$$F = \frac{\text{MSG}}{\text{MSE}} = \frac{303.212}{9.594} = 31.60$$

Our work differs slightly from the output in Figure 22.3 because of roundoff error. We don't recommend doing these calculations, because tedium and roundoff errors cause frequent mistakes.

APPLY YOUR KNOWLEDGE

The calculations of ANOVA use only the sample sizes n_i, the sample means $\bar{x}_i$, and the sample standard deviations s_i. You can therefore re-create the ANOVA calculations when a report gives these summaries but does not give the actual data. These optional exercises ask you to do the ANOVA calculations starting with the summary statistics.

22.11 Road rage. Exercise 22.2 describes a study of road rage. Here are the means and standard deviations for a measure of "angry/threatening driving" for random samples of drivers in three age-groups:

Age-group	n	$\bar{x}$	s
Less than 30 years	244	2.22	3.11
30 to 55 years	734	1.33	2.21
Over 55 years	364	0.66	1.60

(a) The distributions of responses are somewhat right-skewed. ANOVA is nonetheless safe for these data. Why?

(b) Check that the standard deviations satisfy the guideline for ANOVA inference.

(c) Calculate the overall mean response $\bar{x}$, the mean squares MSG and MSE, and the ANOVA F statistic.

(d) What are the degrees of freedom for F? How significant are the differences among the three mean responses?

22.12 Exercise and weight loss. What conditions help overweight people exercise regularly? Subjects were randomly assigned to three treatments: a single long exercise period 5 days per week; several 10-minute exercise periods 5 days per week; and several 10-minute periods 5 days per week on a home treadmill that was provided to the subjects. The study report contains the following information about weight loss (in kilograms) after six months of treatment:[8]

Treatment	n	$\bar{x}$	s
Long exercise periods	37	10.2	4.2
Short exercise periods	36	9.3	4.5
Short periods with equipment	42	10.2	5.2

(a) Do the standard deviations satisfy the rule of thumb for safe use of ANOVA?

(b) Calculate the overall mean response $\bar{x}$ and the mean square for groups MSG.

(c) Calculate the mean square for error MSE.

(d) Find the ANOVA F statistic and its approximate P-value. Is there evidence that the mean weight losses of people who follow the three exercise programs differ?

22.13 Weights of newly hatched pythons. A study of the effect of nest temperature on the development of water pythons separated python eggs at random into nests at three temperatures: cold, neutral, and hot. Exercise 20.22 (page 555) shows that the proportions of eggs that hatched at each temperature did not differ significantly. Now we will examine the little pythons. In all, 16 eggs hatched at the cold temperature, 38 at the neutral temperature, and 75 at the hot temperature. The report of the study summarizes the data in the common form "mean ± standard error" as follows:[9]

Temperature	n	Weight (grams) at hatching	Propensity to strike
Cold	16	28.89 ± 8.08	6.40 ± 5.67
Neutral	38	32.93 ± 5.61	5.82 ± 4.24
Hot	75	32.27 ± 4.10	4.30 ± 2.70

(a) We will compare the mean weights at hatching. Recall that the standard error of the mean is $s/\sqrt{n}$. Find the standard deviations of the weights in the three groups and verify that they satisfy our rule of thumb for using ANOVA.

(b) Starting from the sample sizes n_i, the means $\bar{x}_i$, and the standard deviations s_i, carry out an ANOVA. That is, find MSG, MSE, and the F statistic, and use Table D to approximate the P-value. Is there evidence that nest temperature affects the mean weight of newly hatched pythons?

22.14 Python strikes. The data in the previous exercise also describe the "propensity to strike" of the hatched pythons at 30 days of age. This is the number of taps on the head with a small brush until the python launches a strike. (Don't try this with adult pythons.) The data are again summarized in the form "sample mean ± standard error of the mean." Follow the outline in (a) and (b) of the previous exercise for propensity to strike. Does nest temperature appear to influence propensity to strike?

Chapter 22 SUMMARY

One-way analysis of variance (ANOVA) compares the means of several populations. The **ANOVA F test** tests the overall H_0 that all the populations have the same mean. If the F test shows significant differences, examine the data to see where the differences lie and whether they are large enough to be important.

The **ANOVA model** states that we have an **independent SRS** from each population; that each population has a **Normal distribution;** and that all populations have the **same standard deviation.**

In practice, ANOVA inference is relatively **robust** when the populations are non-Normal, especially when the samples are large. Before doing the F test, check the observations in each sample for outliers or strong skewness. Also verify that the largest sample standard deviation is no more than twice as large as the smallest standard deviation.

When the null hypothesis is true, the **ANOVA F statistic** for comparing I means from a total of N observations in all samples combined has the **F distribution** with $I - 1$ and $N - I$ degrees of freedom.

ANOVA calculations are reported in an **ANOVA table** that gives sums of squares, mean squares, and degrees of freedom for variation among groups and for variation within groups. In practice, we use software to do the calculations.

STATISTICS IN SUMMARY

After studying this chapter, you should be able to do the following.

A. RECOGNITION

1. Recognize when testing the equality of several means is helpful in understanding data.

2. Recognize that the statistical significance of differences among sample means depends on the sizes of the samples and on how much variation there is within the samples.

3. Recognize when you can safely use ANOVA to compare means. Check the data production, the presence of outliers, and the sample standard deviations for the groups you want to compare.

B. INTERPRETING ANOVA

1. Explain what null hypothesis F tests in a specific setting.

2. Locate the F statistic and its P-value on the output of analysis of variance software.

3. Find the degrees of freedom for the F statistic from the number and sizes of the samples. Use Table D of the F distributions to approximate the P-value when software does not give it.

4. If the test is significant, use graphs and descriptive statistics to see what differences among the means are most important.

Chapter 22 EXERCISES

(Shmuel Thaler/Index Stock Imager/ PictureQuest)

(Harry Sieplinga/HMS Images/The Image Bank/Getty Images)

Exercises 22.15 to 22.19 describe situations in which we want to compare the mean responses in several populations. For each setting, identify the populations and the response variable. Then give I, the n_i, and N. Finally, give the degrees of freedom of the ANOVA F test.

22.15 Smoking and sleep. A study of the effects of smoking classifies subjects as nonsmokers, moderate smokers, or heavy smokers. The investigators interview a sample of 200 people in each group. Among the questions is "How many hours do you sleep on a typical night?"

22.16 Which package design is best? A maker of detergents wants to compare the attractiveness to consumers of six package designs. Each package is shown to 120 different consumers who rate the attractiveness of the design on a 1 to 10 scale.

22.17 Growing tomatoes. Do four tomato varieties differ in mean yield? Grow 10 plants of each variety and record the yield of each plant in pounds of tomatoes.

22.18 Strong concrete. The strength of concrete depends on the mixture of sand, gravel, and cement used to prepare it. A study compares five different mixtures. Workers prepare six batches of each mixture and measure the strength of the concrete made from each batch.

22.19 Teaching sign language. Which of four methods of teaching American Sign Language is most effective? Assign 10 of the 42 students in a class at random to each of three methods. Teach the remaining 12 students by the fourth method. Record the students' scores on a standard test of sign language after a semester's study.

22.20 Plant diversity. Does the presence of more species of plants increase the productivity of a natural area, as measured by the total mass of plant material? An experiment to investigate this question started as follows:

> In a 7-year experiment, we controlled one component of diversity, the number of plant species, in 168 plots, each 9 m by 9 m. We seeded the plots, in May 1994, to have 1, 2, 4, 8, or 16 species, with 39, 35, 29, 30, and 35 replicates, respectively.[10]

Total plant mass in each plot was measured in 2000.

(a) What null and alternative hypotheses does ANOVA test here?

(b) What are the values of N, I, and the n_i?

(c) What are the degrees of freedom of the ANOVA F statistic?

22.21 Plants defend themselves. When some plants are attacked by leaf-eating insects, they release chemical compounds that attract other insects that prey on the leaf-eaters. A study carried out on plants growing naturally in the Utah desert demonstrated both the release of the compounds and that they not only repel the leaf-eaters but attract predators that act as the plants' bodyguards.[11] The investigators chose 8 plants attacked by each of three leaf-eaters and 8 more that were undamaged, 32 plants of the same species in all. They then measured emissions of several compounds during seven hours. Here are data (mean ± standard error of the mean for eight plants) for one compound. The emission rate is measured in nanograms (ng) per hour.

Group	Emission rate (ng/hr)
Control	9.22 ± 5.93
Hornworm	31.03 ± 8.75
Leaf bug	18.97 ± 6.64
Flea beetle	27.12 ± 8.62

(a) Make a graph that compares the mean emission rates for the four groups. Does it appear that emissions increase when the plant is attacked?

(b) What hypotheses does ANOVA test in this setting?

(c) We do not have the full data. What would you look for in deciding whether you can safely use ANOVA?

(d) What is the relationship between the standard error of the mean (SEM) and the standard deviation for a sample? Do the standard deviations satisfy our rule of thumb for safe use of ANOVA?

22.22 Does polyester decay? How quickly do synthetic fabrics such as polyester decay in landfills? A researcher buried polyester strips in the soil for different lengths of time, then dug up the strips and measured the force required to break them. Breaking strength is easy to measure and is a good indicator of decay; lower strength means the fabric has decayed.

Part of the study buried 20 polyester strips in well-drained soil in the summer. Five of the strips, chosen at random, were dug up after 2 weeks; another 5 were dug up after 4 weeks, 8 weeks, and 16 weeks. Here are the breaking strengths in pounds:[12]

2 weeks	118	126	126	120	129
4 weeks	130	120	114	126	128
8 weeks	122	136	128	146	140
16 weeks	124	98	110	140	110

(a) Find the mean strength for each group and plot the means against time. Does it appear that polyester loses strength consistently over time after it is buried?

(b) Find the standard deviations for each group. Do they meet our criterion for applying ANOVA?

(c) In Examples 17.2 and 17.3 (pages 441 and 445), we used the two-sample t test to compare the mean breaking strengths for strips buried for 2 weeks and for 16 weeks. The ANOVA F test extends the two-sample t to more than two groups. Explain carefully why use of the two-sample t for two of the groups was acceptable but using the F test on all four groups is not acceptable.

22.23 Can you hear these words? To test whether a hearing aid is right for a patient, audiologists play a tape on which words are pronounced at low volume. The patient tries to repeat the words. There are several different lists of words that are supposed to be equally difficult. Are the lists equally difficult when there is background noise? To find out, an experimenter had subjects with normal hearing listen to four lists with a noisy background. The response variable was the percent of the 50 words in a list that the subject repeated correctly. The data set contains 96 responses.[13] Here are two study designs that could produce these data:

(Alix Phanie, Rex Interstock/Stock Connection/PictureQuest)

Design A. The experimenter assigns 96 subjects to 4 groups at random. Each group of 24 subjects listens to one of the lists. All individuals listen and respond separately.
Design B. The experimenter has 24 subjects. Each subject listens to all four lists in random order. All individuals listen and respond separately.

Does Design A allow use of one-way ANOVA to compare the lists? Does Design B allow use of one-way ANOVA to compare the lists? Briefly explain your answers.

```
Session                                                              _ □ ×

Analysis of Variance for Percent
Source      DF         SS        MS         F         P
List         3       920.5     306.8      4.92      0.003
Error       92      5738.2      62.4
Total       95      6658.6

                                     Individual 95% CIs For Mean
                                     Based on Pooled StDev

Level        N       Mean     StDev  ----+---------+---------+---------+----
1           24     32.750     7.409                      (-------*------)
2           24     29.667     8.058                 (-------*------)
3           24     25.250     8.316       (-------*------)
4           24     25.583     7.779       (-------*------)
                                     ---+---------+---------+---------+----
Pooled StDev =     7.898             24.0      28.0      32.0      36.0
```

Figure 22.10 Minitab ANOVA output for comparing the percents heard correctly in four lists of words, for Exercise 22.24.

22.24 Can you hear these words? Figure 22.10 displays the Minitab output for one-way ANOVA applied to the hearing data described in the previous exercise. The response variable is "Percent," and "List" identifies the four lists of words. Based on this analysis, is there good reason to think that the four lists are not all equally difficult? Write a brief summary of the study findings.

22.25 How much corn should I plant? How much corn per acre should a farmer plant to obtain the highest yield? Too few plants will give a low yield. On the other hand, if there are too many plants, they will compete with each other for moisture and nutrients, and yields will fall. To find out, plant at different rates on several plots of ground and measure the harvest. (Treat all the plots the same except for the planting rate.) Use software to analyze these data from such an experiment:[14]

(Craig Aurness/CORBIS)

Plants per acre	Yield (bushels per acre)			
12,000	150.1	113.0	118.4	142.6
16,000	166.9	120.7	135.2	149.8
20,000	165.3	130.1	139.6	149.9
24,000	134.7	138.4	156.1	
28,000	119.0	150.5		

(a) Do data analysis to see what the data appear to show about the influence of plants per acre on yield and also to check the conditions for ANOVA.

(b) Carry out the ANOVA F test. State hypotheses; give F and its P-value. What do you conclude?

(c) The observed differences among the mean yields in the samples are quite large. Why are they not statistically significant?

22.26 Logging in the rainforest: species counts. Table 22.2 (page 604) gives data on the number of trees per forest plot, the number of species per plot, and species richness. Exercise 22.3 analyzed the effect of logging on number of trees. Exercise 22.9 concludes that it would be risky to use ANOVA to analyze richness. Use software to analyze the effect of logging on the number of species.

(a) Make a table of the group means and standard deviations. Do the standard deviations satisfy our rule of thumb for safe use of ANOVA? What do the means suggest about the effect of logging on the number of species?

(b) Carry out the ANOVA. Report the F statistic and its P-value and state your conclusion.

(Eye of Science/Photo Researchers)

22.27 Nematodes and tomato plants. How do nematodes (microscopic worms) affect plant growth? A botanist prepares 16 identical planting pots and then introduces different numbers of nematodes into the pots. He transplants a tomato seedling into each pot. Here are data on the increase in height of the seedlings (in centimeters) 16 days after planting:[15]

Nematodes	Seedling growth			
0	10.8	9.1	13.5	9.2
1,000	11.1	11.1	8.2	11.3
5,000	5.4	4.6	7.4	5.0
10,000	5.8	5.3	3.2	7.5

(a) Make a table of means and standard deviations for the four treatments. Make side-by-side stemplots to compare the treatments. What do the data appear to show about the effect of nematodes on growth?

(b) State H_0 and H_a for the ANOVA test for these data, and explain in words what ANOVA tests in this setting.

(c) Use software to carry out the ANOVA. Report your overall conclusions about the effect of nematodes on plant growth.

22.28 Earnings of athletic trainers (Optional). How much do newly hired athletic trainers earn? Earnings depend on the educational and other qualifications of the trainer and on the type of job. Here are summaries from a sample survey that contacted institutions advertising for trainers and asked about the people they hired:[16]

Type of institution	n	Mean salary $\overline{x}$	Std. dev. s
High school	57	$26,470	$9507
Clinic	108	$30,610	$4504
College	94	$30,019	$7158

Calculate the ANOVA table and the F statistic. Are there significant differences among the mean salaries for the three types of job in the population of all newly hired trainers? (*Comments:* The study authors did an ANOVA. As is often the case, the published work does not answer questions about the suitability of the analysis. We wonder if responses from 60% of employers who advertised positions generate a random sample of new hires. The actual data do not appear in the report. The salary distributions are no doubt right-skewed; we hope that strong skewness and outliers are absent because the subjects in each group hold similar jobs. The large samples will then justify ANOVA. Also, the sample standard deviations do not quite satisfy our rule of thumb for safe use of ANOVA.)

22.29 Plant defenses (Optional). The calculations of ANOVA use only the sample sizes n_i, the sample means $\overline{x}_i$, and the sample standard deviations s_i. You can therefore re-create the ANOVA calculations when a report gives these summaries but does not give the actual data. Use the information in Exercise 22.21 to calculate the ANOVA table (sums of squares, degrees of freedom, mean squares, and the F statistic). Note that the report gives the standard error of the mean (SEM) rather than the standard deviation. Are there significant differences among the mean emission rates for the four populations of plants?

22.30 F versus t (Optional). We have two methods to compare the means of two groups: the two-sample t test of Chapter 17 and the ANOVA F test with $I = 2$. We prefer the t test because it allows one-sided alternatives and does not assume that both populations have the same standard deviation. Let us apply both tests to the same data.

There are two types of life insurance companies. "Stock" companies have shareholders, and "mutual" companies are owned by their policyholders. Take an SRS of each type of company from those listed in a directory of the industry. Then ask the annual cost per $1000 of insurance for a $50,000 policy insuring the life of a 35-year-old man who does not smoke. Here are the data summaries:[17]

	Stock companies	Mutual companies
n_i	13	17
$\bar{x}_i$	$2.31	$2.37
s_i	$0.38	$0.58

(a) Calculate the two-sample t statistic for testing $H_0: \mu_1 = \mu_2$ against the two-sided alternative. Use the conservative method to find the P-value.

(b) Calculate MSG, MSE, and the ANOVA F statistic for the same hypotheses. What is the P-value of F?

(c) How close are the two P-values? (The square root of the F statistic is a t statistic with $N - I = n_1 + n_2 - 2$ degrees of freedom. This is the "pooled two-sample t" mentioned on page 454. So F for $I = 2$ is exactly equivalent to a t statistic, but it is a slightly different t from the one we use.)

22.31 **Exercising to lose weight (Optional).** Exercise 22.12 describes a randomized, comparative experiment that assigned subjects to three types of exercise program intended to help them lose weight. Some of the results of this study were analyzed using the chi-square test for two-way tables, and some others were analyzed using one-way ANOVA. For each of the following excerpts from the study report, say which analysis is appropriate and explain how you made your choice.

(a) "Overall, 115 subjects (78% of 148 subjects randomized) completed 18 months of treatment, with no significant difference in attrition rates between the groups ($P = .12$)."

(b) "In analyses using only the 115 subjects who completed 18 months of treatment, there were no significant differences in weight loss at 6 months among the groups."

(c) "The duration of exercise for weeks 1 through 4 was significantly greater in the SB compared with both LB and SBEQ groups ($P < .05$). ... However, exercise duration was greater in SBEQ compared with both LB and SB groups for months 13 through 18 ($P < .05$)."

Chapter 22 MEDIA EXERCISES

Analysis of variance works by comparing the differences among the sample means in several groups to the spread of observations within the groups. A set of means is significantly different if they are farther apart than random variation would allow to happen just by chance. The extent of random variation is assessed by the random spread among observations in the same group. These exercises use an applet to demonstrate this ANOVA idea.

22.32 ANOVA compares several means. The *One-Way ANOVA* applet displays the observations in three groups, with the group means highlighted by black dots. When you open or reset the applet, the scale at the bottom of the display shows that for these groups the ANOVA F statistic is $F = 31.74$, with $P < 0.001$.

(a) The middle group has larger mean than the other two. Grab its mean point with the mouse. How small can you make F? What did you do to the mean to make F-small? Roughly how significant is your small F?

(b) Starting with the three means aligned from your configuration at the end of (a), drag any one of the group means either up or down. What happens to F? What happens to P? Convince yourself that the same thing happens if you move any one of the means, or if you move one slightly and then another slightly in the opposite direction.

22.33 ANOVA uses within-group variation. Reset the *One-Way ANOVA* applet to its original state. As in Figure 22.6(b) (page 607), the differences among the three means are highly significant (large F, small P) because the observations in each group cluster tightly about the group mean.

(a) Use the mouse to slide the Pooled Standard Error at the top of the display to the right. You see that the group means do not change, but the spread of the observations in each group increases. What happens to F and P as the spread among the observations in each group increases? What are the values of F and P when the slider is all the way to the right? This is similar to Figure 22.6(a): variation within groups hides the differences among the group means.

(b) Leave the Standard Error slider at the extreme right of its scale, so that spread within groups stays fixed. Use the mouse to move the group means apart. What happens to F and P as you do this?

Statistical Thinking Revisited

The Thinking Person's Guide to Basic Statistics

We began our study of statistics with a look at "Statistical Thinking." We end with a review in outline form of the most important content of basic statistics, combining statistical thinking with your new knowledge of statistical practice. The outline contains some important warnings: look for the word "Beware."

Data Production

- Data basics:

    Individuals (subjects).

    Variables: categorical versus quantitative, units of measurement, explanatory versus response.

    Purpose of study.

- Data production basics:

    Observation versus experiment.

    Simple random samples.

    Completely randomized experiments.

- Beware: really bad data production (voluntary response, confounding) can make interpretation impossible.

- Beware: weaknesses in data production (for example, sampling students at only one campus) can make generalizing conclusions difficult.

Data Analysis

- Always plot your data. Look for overall pattern and striking deviations.
- Add numerical descriptions based on what you see.
- Beware: averages and other simple descriptions can miss the real story.
- One quantitative variable:

    Graphs: stemplot, histogram, boxplot.

    Pattern: distribution shape, center, spread. Outliers?

    Numerical descriptions: five-number summary or $\bar{x}$ and s.

- Relationships between two quantitative variables:

    Graph: scatterplot.

    Pattern: relationship form, direction, strength. Outliers? Influential observations?

    Numerical description for linear relationships: correlation, regression line.

    Beware the lurking variable: correlation does not imply causation.

- Beware the effects of outliers and influential observations.

The Reasoning of Inference

- Inference uses data to infer conclusions about a wider population.

- When you do inference, you are acting as if your data come from random samples or randomized comparative experiments. Beware: if they don't, you may have "garbage in, garbage out."

- Always examine your data before doing inference. Inference often requires a regular pattern (such as roughly normal with no strong outliers).

- Key idea: "What would happen if we did this many times?"

- Confidence intervals: estimate a population parameter.

    95% confidence: I used a method that captures the true parameter 95% of the time in repeated use.

    Beware: confidence intervals don't take into account undercoverage, nonresponse, and other sources of error in practical use.

- Significance tests: assess evidence against H_0 in favor of H_a.

    P-value: If H_0 were true, how often would I get an outcome favoring the alternative this strongly?

    Statistical significance at, say, the 5% level: $P < 0.05$, an outcome this extreme would occur less than 5% of the time if H_0 were true.

STATISTICS IN SUMMARY
Overview of basic inference methods

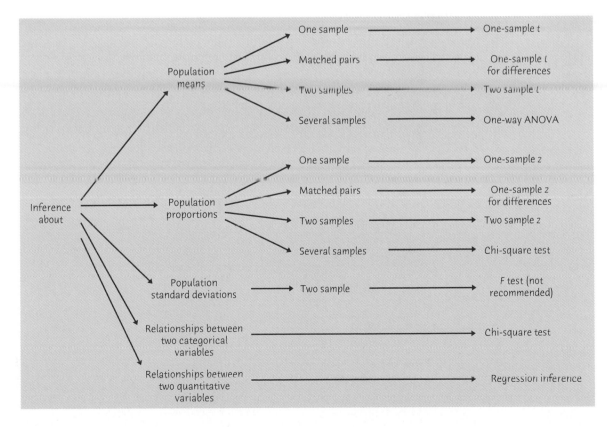

Beware: $P < 0.05$ is not sacred.

Beware: statistical significance is not the same as practical significance. Beware large samples, which can make small effects significant. Beware small samples, which can fail to declare large effects significant.

Always try to estimate the size of an effect (for example, with a confidence interval), not just its significance.

- Choose inference procedures by asking "What parameter" and "What study design." See the Statistics in Summary Figure.

Notes and Data Sources

"About This Book" Notes

1. D. S. Moore and discussants, "New pedagogy and new content: the case of statistics," *International Statistical Review*, 65 (1997), pp. 123–165. Richard Scheaffer's comment appears on page 156.

2. This summary of the committee's report was unanimously endorsed by the Board of Directors of the American Statistical Association. The full report is George Cobb, "Teaching statistics," in L. A. Steen (ed.), *Heeding the Call for Change: Suggestions for Curricular Action*, Mathematical Association of America, 1990, pp. 3–43.

3. Lawrence D. Brown, Tony Cai, and Anirban DasGupta, "Interval estimation for a binomial proportion," *Statistical Science*, 16 (2001), pp. 101–133.

4. J. D. Emerson and G. A. Colditz, "Use of statistical analysis in the *New England Journal of Medicine*," in J. C. Bailar and F. Mosteller (eds.), *Medical Uses of Statistics*, 2nd ed., NEJM Press, 1992, pp. 45–57.

"Statistical Thinking" Notes

1. E. W. Campion, "Editorial: power lines, cancer, and fear," *New England Journal of Medicine*, 337, No. 1 (1997). The study report is M. S. Linet et al., "Residential exposure to magnetic fields and acute lymphoblastic leukemia in children," in the same issue. I found these online at www.nejm.org. See also G. Taubes, "Magnetic field-cancer link: will it rest in peace?" *Science*, 277 (1997), p. 29.

2. Contributed by Marigene Arnold of Kalamazoo College.

3. See, for example, Martin Enserink, "The vanishing promises of hormone replacement," *Science*, 297 (2002), pp. 325–326; and Brian Vastag, "Hormone replacement therapy falls out of favor with expert committee," *Journal of the American Medical Association*, 287 (2002), pp. 1923–1924. A National Institutes of Heath panel's comprehensive report is *International Position Paper on Women's Health and Menopause*, NIH Publication 02-3284, 2002.

4. A. C. Nielsen, Jr., "Statistics in marketing," in *Making Statistics More Effective in Schools of Business*, Graduate School of Business, University of Chicago, 1986.

5. The data in Figure 1 are based on a component of the consumer price index, from the Bureau of Labor Statistics Web site: www.bls.gov. I converted the index number into cents per gallon using retail price information from the Energy Information Agency site: www.eia.doe.gov.

6. H. C. Sox, "Editorial: benefit and harm associated with screening for breast cancer," *New England Journal of Medicine*, 338, No. 16 (1998).

Chapter 1 Notes

1. Environmental Protection Agency, *Municipal Solid Waste in the United States: 2000 Facts and Figures*, document EPA530-R-02-001, 2002.

2. From the 2001 Dupont Automotive Color Popularity Survey, found at www.dupont.com.

3. National Center for Health Statistics, *Births: Final Data for 1999*, National Vital Statistics Reports, Vol. 49, No. 1, 2001.

4. From the Bureau of the Census Web site: www.census.gov.

5. Our eyes do respond to area, but not quite linearly. It appears that we perceive the ratio of two bars to be about the 0.7 power of the ratio of their actual areas. See W. S. Cleveland, *The Elements of Graphing Data*, Wadsworth, 1985, pp. 278–284.

6. Environmental Protection Agency, *Model Year 2002 Fuel Economy Guide*, www.fueleconomy.gov.

7. Data from Gary Community School Corporation, courtesy of Celeste Foster, Purdue University.

8. College Entrance Examination Board, *Trends in College Pricing, 2001*, www.collegeboard.com. The averages are "enrollment weighted," so that they give average tuition over *students* rather than over *colleges*.

9. See Note 1.

10. See Note 2.

11. National Center for Health Statistics, *Deaths: Preliminary Data for 2000*, National Vital Statistics Reports, Vol. 49, No. 12, 2001.

12. Bureau of the Census, *Hispanics in the U.S.A*, public use training module, 2001. Found at www.census.gov.

13. Tom Lloyd et al., "Fruit consumption, fitness, and cardiovascular health in female adolescents: the Penn State Young Women's Health Study," *American Journal of Clinical Nutrition*, 67 (1998), pp. 624–630.

14. Monthly stock returns from the Web site of Professor Kenneth French of Dartmouth: mba.tuck.dartmouth.edu/pages/faculty/ken.french.

15. See Note 3. The table adds the few babies weighing 5500 grams or more to the 5000 to 5499 class.

16. David M. Fergusson and L. John Horwood, "Cannabis use and traffic accidents in a birth cohort of young adults," *Accident Analysis and Prevention*, 33 (2001), pp. 703–711.

17. Craig Packer, Anne E. Pusey, and Lynn E. Eberly, "Egalitarianism in female African lions," *Science*, 293 (2001), pp. 690–693.

18. The Shakespeare data appear in C. B. Williams, *Style and Vocabulary: Numerical Studies*, Griffin, 1970. The student data are from a large statistics course for liberal arts students.

19. Bureau of the Census, *Poverty in the United States: 2000*, 2001.

20. *Statistical Abstract of the United States*, 2001.

21. From the Web site of the Bureau of Labor Statistics: www.bls.gov/cpi.

Chapter 2 Notes

1. From the March 2002 Annual Demographic Supplement to the Current Population Survey, on the Bureau of the Census Web site: www.census.gov. Our data are a subsample that imposed these restrictions: age 25 to 65 years and in the labor force full-time in 2001.

2. Bureau of the Census, *Money Income in the United States: 2001*, Current Population Reports P60-218, 2002.

3. See Note 6 for Chapter 1.

4. From the August 2000 supplement to the Current Population Survey, from the Bureau of the Census Web site: www.census.gov.

5. M. Ann Laskey et al., "Bone changes after 3 mo of lactation: influence of calcium intake, breast-milk output, and vitamin D–receptor genotype," *American Journal of Clinical Nutrition*, 67 (1998), pp. 685–692.

6. Cavendish's data and the background information about his work appear in S. M. Stigler, "Do robust estimators work with real data?" *Annals of Statistics*, 5 (1977), pp. 1055–1078. By the way, a poll of physicists listed Cavendish's work among the top ten experiments of all time. See George Johnson, "Here they are, science's 10 most beautiful experiments," *New York Times*, September 24, 2002.

7. From the University of Miami athletics Web site: hurricanesports.ocsn.com.

8. T. Bjerkedal, "Acquisition of resistance in guinea pigs infected with different doses of virulent tubercle bacilli," *American Journal of Hygiene*, 72 (1960), pp. 130–148.

9. Data for 1986 from David Brillinger, University of California, Berkeley. See David R. Brillinger, "Mapping aggregate birth data," in A. C. Singh and P. Whitridge (eds.), *Analysis of Data in Time*, Statistics Canada, 1990, pp. 77–83. A boxplot similar to Figure 2.5 appears in David R. Brillinger, "Some examples of random process environmental data analysis," in P. K. Sen and C. R. Rao (eds.), *Handbook of Statistics*, Vol. 18, North Holland, 2000.

Chapter 3 Notes

1. See Note 7 for Chapter 1.

2. College Entrance Examination Board, *College-Bound Seniors 2002*, at www.collegeboard.com.

3. Based on the National Health and Nutrition Examination Surveys, 1988–1994. From the Web site of the National Center for Health Statistics: www.cdc.gov/nchs.

4. Detailed data appear in P. S. Levy et al., *Total Serum Cholesterol Values for Youths 12–17 Years*, Vital and Health Statistics, Series 11, No. 155, U.S. National Center for Health Statistics, 1976.

Chapter 4 Notes

1. From a plot in Bernt-Erik Saether, Steiner Engen, and Erik Mattysen, "Demographic characteristics and population dynamical patterns of solitary birds," *Science*, 295 (2002), pp. 2070–2073.

2. Chris Carbone and John L. Gittleman, "A common rule for the scaling of carnivore density," *Science*, 295 (2002), pp. 2273–2276.

3. Based on T. N. Lam, "Estimating fuel consumption from engine size," *Journal of Transportation Engineering*, 111 (1985), pp. 339–357. The data for 10 to 50 km/h are measured; those for 60 and higher are calculated from a model given in the paper and are therefore smoothed.

4. N. Maeno et al., "Growth rates of icicles," *Journal of Glaciology*, 40 (1994), pp. 319–326.

5. A careful study of this phenomenon is W. S. Cleveland, P. Diaconis, and R. McGill, "Variables on scatterplots look more highly correlated when the scales are increased," *Science*, 216 (1982), pp. 1138–1141.

6. M. A. Houck et al. "Allometric scaling in the earliest fossil bird, *Archaeopteryx lithographica*," *Science*, 247 (1990), pp. 195–198. The authors conclude from a variety of evidence that all specimens represent the same species.

7. Data provided by Darlene Gordon, Purdue University.

8. *Consumer Reports*, June 1986, pp. 366–367. A more recent study of hot dogs appears in *Consumer Reports*, July 1993, pp. 415–419. The newer data cover few brands of poultry hot dogs and take calorie counts mainly from the package labels, resulting in suspiciously round numbers.

9. Data provided by Robert Dale, Purdue University.

10. See Note 9.

11. From a plot in Christer G. Wiklund, "Food as a mechanism of density-dependent regulation of breeding numbers in the merlin *Falco columbarius*," *Ecology*, 82 (2001), pp. 860–867.

12. From the fund description on the Vanguard Web site: www.vanguard.com.

13. Data provided by Matthew Moore.

14. Simplified from W. L. Colville and D. P. McGill, "Effect of rate and method of planting on several plant characters and yield of irrigated corn," *Agronomy Journal*, 54 (1962), pp. 235–238.

15. M. H. Criqui, University of California, San Diego, reported in the *New York Times*, December 28, 1994.

Chapter 5 Notes

1. G. L. Kooyman et al., "Diving behavior and energetics during foraging cycles in king penguins," *Ecological Monographs*, 62 (1992), pp. 143–163.

2. E. P. Hubble, "A relation between distance and radial velocity among extra-galactic nebulae," *Proceedings of the National Academy of Sciences*, 15 (1929), pp. 168–173.

3. These data were originally collected by L. M. Linde of UCLA but were first published by M. R. Mickey, O. J. Dunn, and V. Clark, "Note on the use of stepwise regression in detecting outliers," *Computers and Biomedical Research*, 1 (1967), pp. 105–111. The data have been used by several authors. I found them in N. R. Draper and J. A. John, "Influential observations and outliers in regression," *Technometrics*, 23 (1981), pp. 21–26.

4. See Note 2 for "Statistical Thinking."

5. Gannett News Service article appearing in the *Lafayette (Ind.) Journal and Courier*, April 23, 1994.

6. Laura L. Calderon et al., "Risk factors for obesity in Mexican-American girls: dietary factors, anthropometric factors, physical activity, and hours of television viewing," *Journal of the American Dietetic Association*, 96 (1996), pp. 1177–1179.

7. *The Health Consequences of Smoking: 1983*, Public Health Service, Washington, D.C., 1983.

8. Karl Pearson and A. Lee, "On the laws of inheritance in man," *Biometrika*, 2 (1902), p. 357. These data also appear in D. J. Hand et al., *A Handbook of Small Data Sets*, Chapman & Hall, 1994. This book offers more than 500 data sets that can be used in statistical exercises.

9. T. Constable and E. McBean, "BOD/TOC correlations and their application to water quality evaluation," *Water, Air, and Soil Pollution*, 11 (1979), pp. 363–375.

10. From a presentation by Charles Knauf, Monroe County (New York) Environmental Health Laboratory.

11. Frank J. Anscombe, "Graphs in statistical analysis," *The American Statistician*, 27 (1973), pp. 17–21.

12. Gary Smith, "Do statistics test scores regress toward the mean?" *Chance*, 10, No. 4 (1997), pp. 42–45.

13. Based on a plot in G. D. Martinsen, E. M. Driebe, and T. G. Whitham, "Indirect interactions mediated by changing plant chemistry: beaver browsing benefits beetles," *Ecology*, 79 (1998), pp. 192–200.

14. P. Velleman, *ActivStats 2.0*, Addison Wesley Interactive, 1997.

Chapter 6 Notes

1. *Statistical Abstract of the United States*, 2001.

2. H. Lindberg, H. Roos, and P. Gardsell, "Prevalence of coxarthritis in former soccer players," *Acta Orthopedica Scandinavica*, 64 (1993), pp. 165–167.

3. Francine D. Blau and Marianne A. Ferber, "Career plans and expectations of young women and men," *Journal of Human Resources*, 26 (1991), pp. 581–607.

4. Siem Oppe and Frank De Charro, "The effect of medical care by a helicopter trauma team on the probability of survival and the quality of life of hospitalized victims," *Accident Analysis & Prevention*, 33 (2001), pp. 129–138. The authors give the data in Example 6.5 as a "theoretical example" to illustrate the need for their more elaborate analysis of actual data using severity scores for each victim.

5. These data, from reports submitted by airlines to the Department of Transportation, appear in A. Barnett, "How numbers can trick you," *Technology Review*, October 1994, pp. 38–45.

6. M. Radelet, "Racial characteristics and imposition of the death penalty," *American Sociological Review*, 46 (1981), pp. 918–927.

7. Sanders Korenman and David Neumark, "Does marriage really make men more productive?" *Journal of Human Resources*, 26 (1991), pp. 282–307.

8. S. V. Zagona (ed.), *Studies and Issues in Smoking Behavior*, University of Arizona Press, 1967, pp. 157–180.

9. R. Shine, T. R. L. Madsen, M. J. Elphick, and P. S. Harlow, "The influence of nest temperatures and maternal brooding on hatchling phenotypes in water pythons," *Ecology*, 78 (1997), pp. 1713–1721.

10. S. W. Hargarten et al., "Characteristics of firearms involved in fatalities," *Journal of the American Medical Association*, 275 (1996), pp. 42–45.

11. D. M. Barnes, "Breaking the cycle of addiction," *Science*, 241 (1988), pp. 1029–1030.

12. Janice E. Williams et al., "Anger proneness predicts coronary heart disease risk," *Circulation*, 101 (2000), pp. 2034–2039.

13. See P. J. Bickel and J. W. O'Connell, "Is there a sex bias in graduate admissions?" *Science*, 187 (1975), pp. 398–404.

Part I Review Notes

1. Julio Mercader, Melissa Panger, and Christophe Boesch, "Excavation of a chimpanzee stone tool site in the African rainforest," *Science*, 296 (2002), pp. 1452–1455.

2. J. T. Dwyer et al., "Memory of food intake in the distant past," *American Journal of Epidemiology*, 130 (1989), pp. 1033–1046.

3. Alan S. Banks et al., "Juvenile hallux abducto valgus association with metatarsus adductus," *Journal of the American Podiatric Medical Association*, 84 (1994), pp. 219–224.

4. J. Marcus Jobe and Hutch Jobe, "A statistical approach for additional infill development," *Energy Exploration and Exploitation*, 18 (2000), pp. 89–103.

5. Data compiled from a table of percents in "Americans view higher education as key to the American dream," press release by the National Center for Public Policy and Higher Education, www.highereducation.org, May 3, 2000.

6. Based closely on Susan B. Sorenson, "Regulating firearms as a consumer product," *Science*, 286 (1999), pp. 1481–1482. Because the results in the paper were "weighted to the U.S. population," I have changed some counts slightly for consistency.

7. G. A. Sacher and E. F. Staffelt, "Relation of gestation time to brain weight for placental mammals: implications for the theory of vertebrate growth," *American Naturalist*, 108 (1974), pp. 593–613. I found the data in Fred L. Ramsey and Daniel W. Schafer, *The Statistical Sleuth: A Course in Methods of Data Analysis*, Duxbury, 1997, p. 228.

8. Todd W. Anderson, "Predator responses, prey refuges, and density-dependent mortality of a marine fish," *Ecology*, 81 (2001), pp. 245–257.

9. Scott DeCarlo with Michael Schubach and Vladimir Naumovski, "A decade of new issues," *Forbes*, March 5, 2001, at www.forbes.com.

10. Reported in the *New York Times*, July 20, 1989, from an article appearing that day in the *New England Journal of Medicine*.

11. "Dancing in step," *Economist*, March 22, 2001.

12. D. E. Powers and D. A. Rock, *Effects of Coaching on SAT I: Reasoning Test Scores*, Educational Testing Service Research Report 98-6, College Entrance Examination Board, 1998.

13. From the Nenana Ice Classic home page: www.ptialaska.net/~tripod. See Raphael Sagarin and Fiorenza Micheli, "Climate change in nontraditional data sets," *Science*, 294 (2001), p. 811, for a careful discussion.

14. James W. Grier, "Ban of DDT and subsequent recovery of reproduction in bald eagles," *Science*, 218 (1982), pp. 1232–1234.

15. From a plot in Jon J. Ramsey et al., "Energy expenditure, body composition, and glucose metabolism in lean and obese rhesus monkeys treated with ephedrine and caffeine," *American Journal of Clinical Nutrition*, 68 (1998), pp. 42–51.

16. Peter H. Chen, Neftali Herrera, and Darren Christiansen, "Relationships between gate velocity and casting features among aluminum round castings," no date. Provided by Darren Christiansen.

17. Incomes for 2000 from the March 2001 Annual Demographic Supplement to the Current Population Survey. From the Bureau of the Census Web site: www.census.gov.

18. Data provided by Samuel Phillips, Purdue University.

19. Data from many studies compiled in D. F. Greene and E. A. Johnson, "Estimating the mean annual seed production of trees," *Ecology*, 75 (1994), pp. 642–647.

20. From the Intel Web site: www.intel.com/research/silicon/mooreslaw.htm.

Chapter 7 Notes

1. See Note 3 for "Statistical Thinking."

2. John C. Barefoot et al., "Alcoholic beverage preference, diet, and health habits in the UNC Alumni Heart Study," *American Journal of Clinical Nutrition*, 76 (2002), pp. 466–472.

3. J. E. Muscat et al., "Handheld cellular telephone use and risk of brain cancer," *Journal of the American Medical Association*, 284 (2000), pp. 3001–3007.

4. See Jeffrey G. Johnson et al., "Television viewing and aggressive behavior during adolescence and adulthood," *Science*, 295 (2002), pp. 2468–2471. The authors use statistical adjustments to control for the effects of a number of lurking variables. The association between TV viewing and aggression remains significant. Statistical adjustment had been used in the observational studies that supported hormone replacement (Example 7.1) as well, a warning not to place too much trust in these methods.

5. Reported by D. Horvitz in his contribution to "Pseudo-opinion polls: SLOP or useful data?" *Chance*, 8, No. 2 (1995), pp. 16–25.

6. Based in part on Randall Rothenberger, "The trouble with mall interviewing," *New York Times*, August 16, 1989.

7. The regulations that govern seat belt survey design can be found at www.nhtsa. dor.gov. Details on the Hawaii survey are in Karl Kim et al., *Results of the 2002 Highway Seat Belt Use Survey*, at www.state.hi.us/dot.

8. The nonresponse rate for the CPS comes from "Technical notes to household survey data published in *Employment and Earnings*," found on the Bureau of Labor Statistics Web site: stats.bls.gov/cpshome.htm. The General Social Survey reports its response rate on its Web site: www.norc.org/projects/gensoc.asp. The Pew study is described in Gregory Flemming and Kimberly Parker, "Race and reluctant respondents: possible consequences of non-response for pre-election surveys," Pew Research Center for the People and the Press, 1997, found at www.people-press.org.

9. See *Report of the Executive Steering Committee for Accuracy and Coverage Evaluation Policy on Adjustment for Non-redistricting Uses*, Bureau of the Census, October 17, 2001; and Howard Hogan, *Accuracy and Coverage Evaluation: Data and Analysis to Inform the ESCAP Report*, Bureau of the Census, March 1, 2001. Both can be found at www.census.gov.

10. For more detail on the limits of memory in surveys, see N. M. Bradburn, L. J. Rips, and S. K. Shevell, "Answering autobiographical questions: the impact of memory and inference on surveys," *Science*, 236 (1987), pp. 157–161.

11. The responses on welfare are from a *New York Times*/CBS News Poll reported in the *New York Times*, July 5, 1992. Those for Scotland are from "All set for independence?" *Economist*, September 12, 1998. Many examples appear in T. W. Smith, "That which we call welfare by any other name would smell sweeter," *Public Opinion Quarterly*, 51 (1987), pp. 75–83.

12. Giuliana Coccia, "An overview of non-response in Italian telephone surveys," *Proceedings of the 99th Session of the International Statistical Institute*, 1993, Book 3, pp. 271–272.

13. K. J. Mukamal et al., "Prior alcohol consumption and mortality following acute myocardial infarction," *Journal of the American Medical Association*, 285 (2001), pp. 1965–1970.

14. L. E. Moses and F. Mosteller, "Safety of anesthetics," in J. M. Tanur et al. (eds.), *Statistics: A Guide to the Unknown*, 3rd ed., Wadsworth, 1989, pp. 15–24.

15. Mario A. Parada et al., "The validity of self-reported seatbelt use: Hispanic and non-Hispanic drivers in El Paso," *Accident Analysis and Prevention*, 33 (2001), pp. 139–143.

16. Bryan E. Porter and Thomas D. Berry, "A nationwide survey of self-reported red light running: measuring prevalence, predictors, and perceived consequences," *Accident Analysis and Prevention*, 33 (2001), pp. 735–741.

17. See Note 6 for Part I Review.

18. Information from Warren McIsaac and Vivek Goel, "Is access to physician services in Ontario equitable?" Institute for Clinical Evaluative Sciences in Ontario, October 18, 1993.

19. From the *New York Times*, August 21, 1989.

Chapter 8 Notes

1. Details of the Carolina Abecedarian Project, including references to published work, can be found online at www.fpg.unc.edu/overview/abc/abc-ov.htm.

2. Simplified from Arno J. Rethans, John L. Swasy, and Lawrence J. Marks, "Effects of television commercial repetition, receiver knowledge, and commercial length: a test of the two-factor model," *Journal of Marketing Research*, 23 (February 1986), pp. 50–61.

3. Paul R. Solomon et al., "Ginkgo for memory enhancement: a randomized controlled trial," *Journal of the American Medical Association*, 288 (2002), pp. 835–840.

4. Such an experiment is described in Geetha Thiagarajan et al., "Antioxidant properties of green and black tea, and their potential ability to retard the progression of eye lens cataract," *Experimental Eye Research*, 73 (2001), pp. 393–401.

5. K. Wang, Y. Li, and J. Erickson, "A new look at the Monday effect," *Journal of Finance*, 52 (1997), pp. 2171–2186.

6. Carol A. Warfield, "Controlled-release morphine tablets in patients with chronic cancer pain," *Cancer*, 82 (1998), pp. 2299–2306.

7. Taken from "Advertising: the cola war," *Newsweek*, August 30, 1976, p. 67.

8. E. M. Peters et al., "Vitamin C supplementation reduces the incidence of postrace symptoms of upper-respiratory tract infection in ultramarathon runners," *American Journal of Clinical Nutrition*, 57 (1993), pp. 170–174.

9. Sterling C. Hilton et al., "A randomized controlled experiment to assess technological innovations in the classroom on student outcomes: an overview of a clinical trial in education," manuscript, no date.

10. NIMH Multisite HIV Prevention Trial Group, "The NIMH multisite HIV prevention trial: reducing HIV sexual risk behavior," *Science*, 280 (1998), pp. 1889–1894.

11. Shailja V. Nigdikar et al., "Consumption of red wine polyphenols reduces the susceptibility of low-density lipoproteins to oxidation in vivo," *American Journal of Clinical Nutrition*, 68 (1998), pp. 258–265. (There were in fact only 30 subjects, some of whom received more than one treatment with a four-week period intervening.)

12. Based on Evan H. DeLucia et al., "Net primary production of a forest ecosystem with experimental CO_2 enhancement," *Science*, 284 (1999), pp. 1177–1179. The investigators used the block design.

13. The study is described in G. Kolata, "New study finds vitamins are not cancer preventers," *New York Times*, July 21, 1994. Look in the *Journal of the American Medical Association* of the same date for the details.

14. R. C. Shelton et al., "Effectiveness of St. John's wort in major depression," *Journal of the American Medical Association*, 285 (2001), pp. 1978–1986.

15. Steering Committee of the Physicians' Health Study Research Group, "Final report on the aspirin component of the ongoing Physicians' Health Study," *New England Journal of Medicine*, 321 (1989), pp. 129–135.

Chapter 9 Notes

1. You can find a mathematical explanation of Benford's law in Ted Hill, "The first-digit phenomenon," *American Scientist*, 86 (1996), pp. 358–363; and Ted Hill, "The difficulty of faking data," *Chance*, 12, No. 3 (1999), pp. 27–31. Applications in fraud detection are discussed in the second paper by Hill and in Mark A. Nigrini, "I've got your number," *Journal of Accountancy*, May 1999, available online at `www.aicpa.org/pubs/jofa/joaiss.htm`.

2. James R. Lootens, David R. Larsen, and Edward F. Loewenstein, "A matrix transition model for an uneven-aged, oak-hickory forest in the Missouri Ozark Highlands," paper presented at the Tenth Biennial Southern Silvicultural Conference, 1999.

3. Information from `www.ncsu.edu/class/grades`.

4. See Note 2 for Chapter 1.

5. From the M&M Web site: `www.mms.com`. Purple was recently added.

6. From the Bureau of the Census's 1998 American Housing Survey.

Chapter 10 Notes

1. See Note 2 for Chapter 2.

2. Strictly speaking, the recipe $\sigma/\sqrt{n}$ for the standard deviation of $\bar{x}$ assumes that we draw an SRS of size n from an *infinite* population. If the population has finite size N, the standard deviation in the recipe is multiplied by $\sqrt{1 - (n-1)/(N-1)}$. This "finite population correction" approaches 1 as N increases. When the population is at least 10 times as large as the sample, the correction factor is between about 0.95 and 1. It is reasonable to use the simpler form $\sigma/\sqrt{n}$ in these settings.

Chapter 11 Notes

1. Corey Kilgannon, "When New York is on the end of the line," *New York Times*, November 7, 1999.

2. From the March 2002 Annual Demographic Supplement to the Current Population Survey, detailed tables, "Selected characteristics of households, by total money income in 2001," at `www.census.gov`.

3. Projections by the National Center for Education Statistics, from the *Digest of Education Statistics, 2000*, at `nces.ed.gov`.

4. From the Web site of the Interactive Digital Software Association: `www.idsa.com`.

5. The probabilities given are realistic, according to the fundraising firm SCM Associates, `scmassoc.com`.

6. Probabilities from trials with 2897 people known to be free of HIV antibodies and 673 people known to be infected, reported in J. Richard George, "Alternative specimen sources: methods for confirming positives," 1998 Conference on the Laboratory Science of HIV, found online at the Centers for Disease Control and Prevention, `www.cdc.gov`.

Chapter 12 Notes

1. From a Gallup Poll taken in 2002, `www.gallup.com`.

2. The survey question is reported in Trish Hall, "Shop? Many say 'Only if I must,'" *New York Times*, November 28, 1990. In fact, 66% (1650 of 2500) in the sample said "Agree."

Chapter 13 Notes

1. Information from Francisco L. Rivera-Batiz, "Quantitative literacy and the likelihood of employment among young adults," *Journal of Human Resources*, 27 (1992), pp. 313–328.

2. This and similar results of Gallup Polls are from the Gallup Organization Web site: www.gallup.com.

3. See Note 7 for Chapter 4.

4. The values $\mu = 22$ and $\sigma = 50$ for the gains of uncoached students on the SAT mathematics exam come from a study of 2733 students reported on the College Board Web site: www.collegeboard.org.

5. See Note 5 for Chapter 2.

6. Data provided by Drina Iglesia, Purdue University. The data are part of a larger study reported in D. D. S. Iglesia, E. J. Cragoe, Jr., and J. W. Vanable, "Electric field strength and epithelization in the newt (*Notophthalmus viridescens*)," *Journal of Experimental Zoology*, 274 (1996), pp. 56–62.

Chapter 14 Notes

1. Based on Raul de la Fuente-Fernandez et al., "Expectation and dopamine release: mechanism of the placebo effect in Parkinson's disease," *Science*, 293 (2001), pp. 1164–1166.

2. Manisha Chandalia et al., "Beneficial effects of high dietary fiber intake in patients with type 2 diabetes mellitus," *New England Journal of Medicine*, 342 (2000), pp. 1392–1398.

3. Arthur Schatzkin et al., "Lack of effect of a low-fat, high-fiber diet on the recurrence of colorectal adenomas," *New England Journal of Medicine*, 342 (2000), pp. 1149–1155.

4. Seung-Ok Kim, "Burials, pigs, and political prestige in neolithic China," *Current Anthropology*, 35 (1994), pp. 119–141.

5. Sara L. Webb and Sara E. Scanga, "Windstorm disturbance without patch dynamics: twelve years of change in a Minnesota forest," *Ecology*, 82 (2001), pp. 893–897.

6. See Note 15 for Chapter 7.

Chapter 15 Notes

1. E. M. Barsamian, "The rise and fall of internal mammary artery ligation," in J. P. Bunker, B. A. Barnes, and F. Mosteller (eds.), *Costs, Risks, and Benefits of Surgery*, Oxford University Press, 1977, pp. 212–220.

2. See Note 16 for Chapter 7.

3. For a discussion of statistical significance in the legal setting, see D. H. Kaye, "Is proof of statistical significance relevant?" *Washington Law Review*, 61 (1986), pp. 1333–1365. Kaye argues: "Presenting the *P*-value without characterizing the evidence by a significance test is a step in the right direction. Interval estimation, in turn, is an improvement over *P*-values."

4. Warren E. Leary, "Cell phones: questions but no answers," *New York Times*, October 26, 1999.

5. Gabriel Gregoratos et al., "ACC/AHA guidelines for implantation of cardiac pacemakers and antiarrhythmia devices: executive summary," *Circulation*, 97 (1998), pp. 1325–1335.

6. P. H. Lewis, "Technology" column, *New York Times*, May 29, 1995.

7. Robert J. Schiller, "The volatility of stock market prices," *Science*, 235 (1987), pp. 33–36.

8. Data provided by Mugdha Gore and Joseph Thomas, Purdue University School of Pharmacy.

Part II Review Notes

1. See Note 16 for Chapter 1.

2. K. E. Hobbs et al., "Levels and patterns of persistent organochlorines in minke whale (*Balaenoptera acutorostrata*) stocks from the North Atlantic and European Arctic," *Environmental Pollution*, 121 (2003), pp. 239–252.

3. Maureen Hack et al., "Outcomes in young adulthood for very-low-birth-weight infants," *New England Journal of Medicine*, 346 (2002), pp. 149–157.

4. Pamela J. Goodwin et al., "The effect of group psychological support on survival in metastatic breast cancer," *New England Journal of Medicine*, 345 (2001), pp. 1719–1726.

5. Manuel R. Malinow et al., "Reduction of plasma homocyst(e)ine levels by breakfast food fortified with folic acid in patients with coronary heart disease," *New England Journal of Medicine*, 338 (1998), pp. 1009–1015. (The actual study design was a bit more complex.)

6. Based on a discussion of several polls by David W. Moore on the Gallup Organization Web site: www.gallup.org.

7. Virgilio P. Carnielli et al., "Intestinal absorption of long-chain polyunsaturated fatty acids in preterm infants fed breast milk or formula," *American Journal of Clinical Nutrition*, 67 (1998), pp. 97–103.

8. Based on a news item "Bee off with you," *Economist*, November 2, 2002, p. 78.

9. Jon E. Keeley, C. J. Fotheringham, and Marco Morais, "Reexamining fire suppression impacts on brushland fire regimes," *Science*, 284 (1999), pp. 1829–1831.

10. Fekri A. Hassan, "Radiocarbon chronology of predynastic Nagada settlements, Upper Egypt," *Current Anthropology*, 25 (1984), pp. 681–683.

11. Sara J. Solnick and David Hemenway, "The deadweight loss of Christmas: comment," *American Economic Review*, 86 (1996), pp. 1299–1305.

12. Charles S. Fuchs et al., "Alcohol consumption and mortality among women," *New England Journal of Medicine*, 332 (1995), pp. 1245–1250.

13. Research by Louis Chan et al., reported by Robert Schiller, *Irrational Exuberance*, Broadway Books, 2001, p. 253.

14. Based on a study reported by Alan B. Krueger, "Economic scene" column, *New York Times*, November 14, 2002.

Chapter 16 Notes

1. See Note 6 for Chapter 13.

2. See Note 11 for Chapter 8. The data are simulated from the Normal distribution with mean and standard deviation observed in this study.

3. R. A. Berner and G. P. Landis, "Gas bubbles in fossil amber as possible indicators of the major gas composition of ancient air," *Science*, 239 (1988), pp. 1406–1409.

4. Data provided by Timothy Sturm.

5. For a qualitative discussion explaining why skewness is the most serious violation of the Normal shape condition, see Dennis D. Boos and Jacquelin M. Hughes-Oliver, "How large does n have to be for the Z and t intervals?" *American Statistician*, 54 (2000), pp. 121–128. Our recommendations are based on extensive computer work. See, for example, Harry O. Posten, "The robustness of the one-sample t-test over the Pearson system," *Journal of Statistical Computation and Simulation*, 9 (1979), pp. 133–149; and E. S. Pearson and N. W.

Please, "Relation between the shape of population distribution and the robustness of four simple test statistics," *Biometrika*, 62 (1975), pp. 223–241.

6. Data provided by Jason Hamilton, University of Illinois. The study is reported in Evan H. DeLucia et al., "Net primary production of a forest ecosystem with experimental CO_2 enhancement," *Science*, 284 (1999), pp. 1177–1179.

7. See Note 4 for Part I Review.

8. The comparison interval is the bootstrap BCa interval, based on 1000 resamples and calculated by the S-PLUS software.

9. Tim Barmby and Suzyrman Sibly, "A Markov model of worker absenteeism," manuscript, November 1998.

10. This example is based on information in D. L. Shankland et al., "The effect of 5-thio-D-glucose on insect development and its absorption by insects," *Journal of Insect Physiology*, 14 (1968), pp. 63–72.

11. From the August 2000 supplement to the Current Population Survey, from the Census Bureau Web site: www.census.gov.

12. See Note 3 for Part I Review.

13. Data provided by Chris Olsen, who found the information in scuba-diving magazines.

14. See Note 1 for Chapter 14.

15. From the appendix of D. A. Kurtz (ed.), *Trace Residue Analysis*, American Chemical Society Symposium Series, No. 284, 1985.

16. See Note 6 for Chapter 13.

17. Lianng Yuh, "A biopharmaceutical example for undergraduate students," manuscript.

Chapter 17 Notes

1. Sapna Aneja, "Biodeterioration of textile fibers in soil," M. S. thesis, Purdue University, 1994.

2. Detailed information about the conservative t procedures can be found in Paul Leaverton and John J. Birch, "Small sample power curves for the two sample location problem," *Technometrics*, 11 (1969), pp. 299–307; in Henry Scheffé, "Practical solutions of the Behrens-Fisher problem," *Journal of the American Statistical Association*, 65 (1970), pp. 1501–1508; and in D. J. Best and J. C. W. Rayner, "Welch's approximate solution for the Behrens-Fisher problem," *Technometrics*, 29 (1987), pp. 205–210.

3. F. Tagliavini et al., "Effectiveness of anthracycline against experimental prion disease in Syrian hamsters," *Science*, 276 (1997), pp. 1119–1121.

4. Costas Papoulias and Panayiotis Theodossiou, "Analysis and modeling of recent business failures in Greece," *Managerial and Decision Economics*, 13 (1992), pp. 163–169.

5. Kathleen G. McKinney, "Engagement in community service among college students: is it affected by significant attachment relationships?" *Journal of Adolescence*, 25 (2002), pp. 139–154.

6. See Note 2 for Chapter 16.

7. Data provided by Charles Cannon, Duke University. The study report is C. H. Cannon, D. R. Peart, and M. Leighton, "Tree species diversity in commercially logged Bornean rainforest," *Science*, 281 (1998), pp. 1366–1367.

8. See the extensive simulation studies in Harry O. Posten, "The robustness of the two-sample *t*-test over the Pearson system," *Journal of Statistical Computation and Simulation*, 6 (1978), pp. 295–311; and in Harry O. Posten, H. Yeh, and Donald B. Owen, "Robustness of the two-sample *t*-test under violations of the homogeneity assumption," *Communications in Statistics*, 11 (1982), pp. 109–126.

9. D. L. Shankland, "Involvement of spinal cord and peripheral nerves in DDT-poisoning syndrome in albino rats," *Toxicology and Applied Pharmacology*, 6 (1964), pp. 197–213.

10. See Note 7 for Chapter 4.

11. The problem of comparing spreads is difficult even with advanced methods. Common distribution-free procedures do not offer a satisfactory alternative to the F test, because they are sensitive to unequal shapes when comparing two distributions. A good introduction to the available methods is W. J. Conover, M. E. Johnson, and M. M. Johnson, "A comparative study of tests for homogeneity of variances, with applications to outer continental shelf bidding data," *Technometrics*, 23 (1981), pp. 351–361. Modern resampling procedures often work well. See Dennis D. Boos and Colin Brownie, "Bootstrap methods for testing homogeneity of variances," *Technometrics*, 31 (1989), pp. 69–82.

12. Elisabeth Wells-Parker et al., "An exploratory study of the relationship between road rage and crash experience in a representative sample of US drivers," *Accident Analysis and Prevention*, 34 (2002), pp. 271–278.

13. Debra L. Miller et al., "Effect of fat-free potato chips with and without nutrition labels on fat and energy intakes," *American Journal of Clinical Nutrition*, 68 (1998), pp. 282–290.

14. Frank J. Elgar and Christine Arlett, "Perceived social inadequacy and depressed mood in adolescents," *Journal of Adolescence*, 25 (2002), pp. 301–305.

15. Data provided by Warren Page, New York City Technical College, from a study done by John Hudesman.

16. Fabrizio Grieco, Arie J. van Noordwijk, and Marcel E. Visser, "Evidence for the effect of learning on timing of reproduction in blue tits," *Science*, 296 (2002), pp. 136–138. The data are estimated from a plot in the paper.

17. John R. Cronin and Sandra Pizzarello, "Enantiometric excesses in meteoritic amino acids," *Science*, 275 (1997), pp. 951–955.

18. Orit E. Hetzroni, "The effects of active versus passive computer-assisted instruction on the acquisition, retention, and generalization of Blissymbols while using elements for teaching compounds," Ph.D. thesis, Purdue University, 1995.

19. See Note 7 for Chapter 4.

20. Wayne J. Camera and Donald Powers, "Coaching and the SAT I," *TIP* (online journal: www.siop.org/tip), July 1999.

Chapter 18 Notes

1. Joseph H. Catania et al., "Prevalence of AIDS-related risk factors and condom use in the United States," *Science*, 258 (1992), pp. 1101–1106.

2. The study is reported in William Celis III, "Study suggests Head Start helps beyond school," *New York Times*, April 20, 1993.

3. Strictly speaking, the recipe $\sqrt{p(1-p)/n}$ for the standard deviation of $\hat{p}$ assumes that an SRS of size n is drawn from an *infinite* population. If the population has finite size N, this

standard deviation is multiplied by $\sqrt{1 - (n-1)/(N-1)}$. This "finite population correction" approaches 1 as N increases. When the population is at least 10 times as large as the sample, the correction factor is between 0.95 and 1 and can be ignored in practice.

4. This rule of thumb is a bit arbitrary. It is based on study of computational results in the papers cited below and discussion with Alan Agresti. We strongly recommend using the plus four interval.

5. The quotation is from page 1104 of the article cited in Note 1.

6. See Note 20 for Chapter 17.

7. Laurie Goodstein and Marjorie Connelly, "Teen-age poll finds support for tradition," *New York Times*, April 30, 1998.

8. See Note 3 for "About This Book." The passage quoted appears on p. 102.

9. This interval is proposed by Alan Agresti and Brent A. Coull, "Approximate is better than 'exact' for interval estimation of binomial proportions," *The American Statistician*, 52 (1998), pp. 119–126. There are several yet more accurate but considerably more complex intervals for p that might be used in professional practice. See the paper cited in Note 3 for "About This Book." A detailed theoretical study is Lawrence D. Brown, Tony Cai, and Anirban DasGupta, "Confidence intervals for a binomial proportion and asymptotic expansions," *Annals of Statistics*, 30 (2002), pp. 160–201.

10. From the paper cited in Note 4 for Chapter 19. When can the plus four interval be safely used? The answer depends on just how much accuracy you insist on. Brown and coauthors (see Note 3 for "About This Book") recommend $n \geq 40$. Agresti and Coull (see Note 9) demonstrate that performance is almost always satisfactory in their eyes when $n \geq 5$. Our rule of thumb $n \geq 10$ allows for other confidence levels C and fits our philosophy of not insisting on very exact results in practice. The big point is that plus four is very much more accurate than the standard interval for most values of p and all but very large n.

11. James Otto, Michael F. Brown, and William Long III, "Training rats to search and alert on contraband odors," *Applied Animal Behaviour Science*, 77 (2002), pp. 217–232.

12. Janice Joseph, "Fear of crime among black elderly," *Journal of Black Studies*, 27 (1997), pp. 698–717.

13. John Paul McKinney and Kathleen G. McKinney, "Prayer in the lives of late adolescents," *Journal of Adolescence*, 22 (1999), pp. 279–290.

14. Gary Edwards and Josephine Mazzuca, "Three quarters of Canadians support doctor-assisted suicide," Gallup Poll press release, March 24, 1999, at www.gallup.com.

15. Pew Internet Project, "Counting on the Internet," December 29, 2002, at www.pewinternet.org.

16. JoAnn K. Wells, Allan F. Williams, and Charles M. Farmer, "Seat belt use among African Americans, Hispanics, and whites," *Accident Analysis and Prevention*, 34 (2002), pp. 523–529.

17. Henry Wechsler et al., *Binge Drinking on America's College Campuses*, Harvard School of Public Health, 2001.

18. See Note 16 for Chapter 7.

19. John Fagan et al., "Performance assessment under field conditions of a rapid immunological test for transgenic soybeans," *International Journal of Food Science and Technology*, 36 (2001), pp. 357–367.

20. "Poll: men, women at odds on sexual equality," Associated Press dispatch appearing in the *Lafayette (Ind.) Journal and Courier*, October 20, 1997.

21. Arne L. Kalleberg and Kevin T. Leicht, "Gender and organizational performance: determinants of small business survival and success," *The Academy of Management Journal*, 34 (1991), pp. 136–161.

22. Jane E. Brody, "Alternative medicine makes inroads," *New York Times*, April 28, 1998.

Chapter 19 Notes

1. See Note 2 for Chapter 18.

2. See Note 12 for Chapter 18.

3. Douglas E. Jorenby et al., "A controlled trial of sustained-release bupropion, a nicotine patch, or both for smoking cessation," *New England Journal of Medicine*, 340 (1999), pp. 685–691.

4. The plus four method is due to Alan Agresti and Brian Caffo, "Simple and effective confidence intervals for proportions and differences of proportions result from adding two successes and two failures," *The American Statistician*, 45 (2000), pp. 280–288.

5. From an Associated Press dispatch appearing on December 30, 2002. The study report will appear in the *Journal of Adolescent Health*.

6. Modified from Richard A. Schieber et al., "Risk factors for injuries from in-line skating and the effectiveness of safety gear," *New England Journal of Medicine*, 335 (1996), Internet summary at `content.nejm.org`.

7. Louie E. Ross, "Mate selection preferences among African American college students," *Journal of Black Studies*, 27 (1997), pp. 554–569.

8. Data courtesy of Raymond Dumett, Purdue University.

9. See Note 15 for Chapter 18.

10. Douglas G. Altman, Steven N. Goodman, and Sara Schroter, "How statistical expertise is used in medical research," *Journal of the American Medical Association*, 287 (2002), pp. 2817–2820.

11. See Note 5 for Part I Review.

12. See Note 16 for Chapter 18.

13. Clive G. Jones, Richard S. Ostfeld, Michele P. Richard, Eric M. Schauber, and Jerry O. Wolf, "Chain reactions linking acorns to gypsy moth outbreaks and Lyme disease risk," *Science*, 279 (1998), pp. 1023–1026.

14. National Athletic Trainers Association, press release dated September 30, 1994. The study was to be published in the *Journal of Athletic Training*.

15. From the online supplement to G. Gaskell et al., "Worlds apart? The reception of genetically modified foods in Europe and the U.S.," *Science*, 285 (1999), pp. 383–387.

16. Richard M. Felder et al., "Who gets it and who doesn't: a study of student performance in an introductory chemical engineering course," *1992 ASEE Annual Conference Proceedings*, American Society for Engineering Education, Washington, D.C., 1992, pp. 1516–1519.

17. See Note 12 for Chapter 7.

Part III Review Notes

1. Shmuel Shulman and Offer Kipnis, "Adolescent romantic relationships: a look from the future," *Journal of Adolescence*, 24 (2001), pp. 337–351.

2. See Note 10 for Chapter 8.

3. See Note 13 for Chapter 17.

4. Michael R. Dohm, Jack P. Hayes, and Theodore Garland, Jr., "Quantitative genetics of sprint running speed and swimming endurance in laboratory house mice (*Mus domesticus*)," *Evolution*, 50 (1996), pp. 1688–1701.

5. David M. Blau, "The child care labor market," *Journal of Human Resources*, 27 (1992), pp. 9–39.

6. See Note 1 for Chapter 13.

7. Tom W. Smith, *2001 National Gun Policy Survey of the National Opinion Research Center: Research Findings*, National Opinion Research Center, 2001.

8. D. M. Karl et al., "Microorganisms in the accreted ice of Lake Vostok, Antarctica," *Science*, 286 (1999), pp. 2144–2147.

9. See Note 15 for Chapter 19. The percents given are not exact because of rounding in the table from which they are compiled.

10. K. S. Oberhauser, "Fecundity, lifespan and egg mass in butterflies: effects of male-derived nutrients and female size," *Functional Ecology*, 11 (1997), pp. 166–175.

11. See Note 3 for Part II Review. The exercises are simplified, in that the measures reported in this paper have been statistically adjusted for "sociodemographic status."

12. From J. W. Marr and J. A. Heady, "Within- and between-person variation in dietary surveys: number of days needed to classify individuals," *Human Nutrition: Applied Nutrition*, 40A (1986), pp. 347–364.

13. Based on the online supplement to Paul J. Shaw et al., "Correlates of sleep and waking in *Drosophila melanogaster*," *Science*, 287 (2000), pp. 1834–1837.

14. From V. D. Bass, W. E. Hoffmann, and J. L. Dorner, "Normal canine lipid profiles and effects of experimentally induced pancreatitis and hepatic necrosis on lipids," *American Journal of Veterinary Research*, 37 (1976), pp. 1355–1357.

15. G. S. Hotamisligil, R. S. Johnson, R. J. Distel, R. Ellis, V. E. Papaioannou, and B. M. Spiegelman, "Uncoupling of obesity from insulin resistance through a targeted mutation in $aP2$, the adipocyte fatty acid binding protein," *Science*, 274 (1996), pp. 1377–1379.

16. Data provided by Matthew Moore.

Chapter 20 Notes

1. Daniel B. Mark et al., "Use of medical resources and quality of life after acute myocardial infarction in Canada and the United States," *New England Journal of Medicine*, 331 (1994), pp. 1130–1135. See also the discussion in the same journal, 332 (1995), pp. 469–472.

2. Karine Marangon et al., "Diet, antioxidant status, and smoking habits in French men," *American Journal of Clinical Nutrition*, 67 (1998), pp. 231–239.

3. See Note 16 for Chapter 19.

4. See Note 7 for Chapter 6.

5. There are many computer studies of the accuracy of chi-square critical values for X^2. For a brief discussion and some references, see Section 3.2.5 of David S. Moore, "Tests of chi-squared type," in Ralph B. D'Agostino and Michael A. Stephens (eds.), *Goodness-of-Fit Techniques*, Marcel Dekker, 1986, pp. 63–95. If the expected cell counts are roughly equal, the chi-square approximation is adequate when the average expected counts are as small as 1 or 2. The guideline given in the text protects against unequal expected counts. For a

survey of inference for smaller samples, see Alan Agresti, "A survey of exact inference for contingency tables," *Statistical Science*, 7 (1992), pp. 131–177.

6. See Note 3 for Chapter 6.

7. Lillian Lin Miao, "Gastric freezing: an example of the evaluation of medical therapy by randomized clinical trials," in John P. Bunker, Benjamin A. Barnes, and Frederick Mosteller (eds.), *Costs, Risks, and Benefits of Surgery*, Oxford University Press, 1977, pp. 198–211.

8. For Benford's law, see Note 1 for Chapter 9.

9. William D. Darley, "Store-choice behavior for pre-owned merchandise," *Journal of Business Research*, 27 (1993), pp. 17–31.

10. See Note 5 for Part III Review.

11. See Note 8 for Chapter 6.

12. See Note 9 for Chapter 6.

13. Sandra L. Barnes, "Ebonics and public awareness. Who knows? Who cares?" *Journal of Black Studies*, 29 (1998), pp. 17–33.

14. Modified from Felicity Barringer, "Measuring sexuality through polls can be shaky," *New York Times*, April 25, 1993.

15. See Note 16 for Chapter 18.

16. See Note 7 for Part II Review.

17. Claudia Braga et al., "Olive oil, other seasoning fats, and the risk of colorectal carcinoma," *Cancer*, 82 (1998), pp. 448–453.

18. See Note 12 for Chapter 7.

19. Robert J. M. Dawson, "The 'unusual episode' data revisited," *Journal of Statistics Education*, 3, No. 3 (1995). Electronic journal available at the American Statistical Association Web site: www.amstat.org.

20. Adapted from M. A. Visintainer, J. R. Volpicelli, and M. E. P. Seligman, "Tumor rejection in rats after inescapable or escapable shock," *Science*, 216 (1982), pp. 437–439.

Chapter 21 Notes

1. Samuel Karelitz et al., "Relation of crying activity in early infancy to speech and intellectual development at age three years," *Child Development*, 35 (1964), pp. 769–777.

2. See Note 6 for Chapter 4.

3. See Note 4 for Chapter 4.

4. Bruce J. Peterson et al., "Increasing river discharge to the Arctic Ocean," *Science*, 298 (2002), pp. 2171–2173. The data in Table 21.2 were obtained from a plot in this article.

5. Based on Marion E. Dunshee, "A study of factors affecting the amount and kind of food eaten by nursery school children," *Child Development*, 2 (1931), pp. 163–183. This article gives the means, standard deviations, and correlation for 37 children but does not give the actual data.

6. The data are for 1987 salaries and measures of past performance. They were collected and distributed by the Statistical Graphics Section of the American Statistical Association for an annual data analysis contest. The analysis here was done by Crystal Richard of Purdue University.

7. See Note 3 for Part I Review.

8. See Note 13 for Chapter 5.

9. See Note 18 for Part I Review.

10. The data in Table 21.4 are part of a larger data set in the *Journal of Statistics Education* archive, accessible via the Internet. The original source is Pekka Brofeldt, "Bidrag till kaennedom on fiskbestondet i vaara sjoear. Laengelmaevesi," in T. H. Jaervi, *Finlands Fiskeriet*, Band 4, *Meddelanden utgivna av fiskerifoereningen i Finland*, Helsinki, 1917. The data were contributed to the archive (with information in English) by Juha Puranen of the University of Helsinki.

Chapter 22 Notes

1. The data in Table 22.1 are from the Environmental Protection Agency's *Model Year 2003 Fuel Economy Guide*, available online at www.fueleconomy.gov. The table gives data for the basic engine/transmission combination for each model. Models that are essentially identical, such as the Buick Century and Chevrolet Malibu, appear only once.

2. See Note 13 for Part III Review.

3. See Note 12 for Chapter 17.

4. See Note 7 for Chapter 17.

5. See Note 2 for Chapter 20.

6. Modified from M. C. Wilson and R. E. Shade, "Relative attractiveness of various luminescent colors to the cereal leaf beetle and the meadow spittlebug," *Journal of Economic Entomology*, 60 (1967), pp. 578–580.

7. See Note 7 for Chapter 6.

8. John M. Jakicic et al., "Effects of intermittent exercise and use of home exercise equipment on adherence, weight loss, and fitness in overweight women," *Journal of the American Medical Association*, 282 (1999), pp. 1554–1560.

9. See Note 9 for Chapter 6.

10. David Tilman et al., "Diversity and productivity in a long-term grassland experiment," *Science*, 294 (2001), pp. 843–845.

11. Data from the online supplement to André Kessler and Ian T. Baldwin, "Defensive function of herbivore-induced plant volatile emissions in nature," *Science*, 291 (2001), pp. 2141–2144.

12. See Note 1 for Chapter 17.

13. The data and the full story can be found in the Data and Story Library at lib.stat.cmu.edu. The original study is by Faith Loven, "A study of interlist equivalency of the CID W-22 word list presented in quiet and in noise," M.S. thesis, University of Iowa, 1981.

14. See Note 14 for Chapter 4.

15. Data provided by Matthew Moore.

16. Based on Brent L. Arnold et al., "1994 athletic trainer employment and salary characteristics," *Journal of Athletic Training*, 31 (1996), pp. 215–218. I have updated the 1994 salaries to the equivalent in 2000 dollars, using the consumer price index, to avoid misleadingly low values.

17. Mark Kroll, Peter Wright, and Pochera Theerathorn, "Whose interests do hired managers pursue? An examination of select mutual and stock life insurers," *Journal of Business Research*, 26 (1993), pp. 133–148.

Appendix

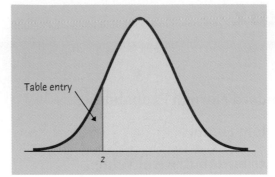

Table entry for z is the area under the standard Normal curve to the left of z.

TABLE A Standard Normal probabilities

z	.00	.01	.02	.03	.04	.05	.06	.07	.08	.09
−3.4	.0003	.0003	.0003	.0003	.0003	.0003	.0003	.0003	.0003	.0002
−3.3	.0005	.0005	.0005	.0004	.0004	.0004	.0004	.0004	.0004	.0003
−3.2	.0007	.0007	.0006	.0006	.0006	.0006	.0006	.0005	.0005	.0005
−3.1	.0010	.0009	.0009	.0009	.0008	.0008	.0008	.0008	.0007	.0007
−3.0	.0013	.0013	.0013	.0012	.0012	.0011	.0011	.0011	.0010	.0010
−2.9	.0019	.0018	.0018	.0017	.0016	.0016	.0015	.0015	.0014	.0014
−2.8	.0026	.0025	.0024	.0023	.0023	.0022	.0021	.0021	.0020	.0019
−2.7	.0035	.0034	.0033	.0032	.0031	.0030	.0029	.0028	.0027	.0026
−2.6	.0047	.0045	.0044	.0043	.0041	.0040	.0039	.0038	.0037	.0036
−2.5	.0062	.0060	.0059	.0057	.0055	.0054	.0052	.0051	.0049	.0048
−2.4	.0082	.0080	.0078	.0075	.0073	.0071	.0069	.0068	.0066	.0064
−2.3	.0107	.0104	.0102	.0099	.0096	.0094	.0091	.0089	.0087	.0084
−2.2	.0139	.0136	.0132	.0129	.0125	.0122	.0119	.0116	.0113	.0110
−2.1	.0179	.0174	.0170	.0166	.0162	.0158	.0154	.0150	.0146	.0143
−2.0	.0228	.0222	.0217	.0212	.0207	.0202	.0197	.0192	.0188	.0183
−1.9	.0287	.0281	.0274	.0268	.0262	.0256	.0250	.0244	.0239	.0233
−1.8	.0359	.0351	.0344	.0336	.0329	.0322	.0314	.0307	.0301	.0294
−1.7	.0446	.0436	.0427	.0418	.0409	.0401	.0392	.0384	.0375	.0367
−1.6	.0548	.0537	.0526	.0516	.0505	.0495	.0485	.0475	.0465	.0455
−1.5	.0668	.0655	.0643	.0630	.0618	.0606	.0594	.0582	.0571	.0559
−1.4	.0808	.0793	.0778	.0764	.0749	.0735	.0721	.0708	.0694	.0681
−1.3	.0968	.0951	.0934	.0918	.0901	.0885	.0869	.0853	.0838	.0823
−1.2	.1151	.1131	.1112	.1093	.1075	.1056	.1038	.1020	.1003	.0985
−1.1	.1357	.1335	.1314	.1292	.1271	.1251	.1230	.1210	.1190	.1170
−1.0	.1587	.1562	.1539	.1515	.1492	.1469	.1446	.1423	.1401	.1379
−0.9	.1841	.1814	.1788	.1762	.1736	.1711	.1685	.1660	.1635	.1611
−0.8	.2119	.2090	.2061	.2033	.2005	.1977	.1949	.1922	.1894	.1867
−0.7	.2420	.2389	.2358	.2327	.2296	.2266	.2236	.2206	.2177	.2148
−0.6	.2743	.2709	.2676	.2643	.2611	.2578	.2546	.2514	.2483	.2451
−0.5	.3085	.3050	.3015	.2981	.2946	.2912	.2877	.2843	.2810	.2776
−0.4	.3446	.3409	.3372	.3336	.3300	.3264	.3228	.3192	.3156	.3121
−0.3	.3821	.3783	.3745	.3707	.3669	.3632	.3594	.3557	.3520	.3483
−0.2	.4207	.4168	.4129	.4090	.4052	.4013	.3974	.3936	.3897	.3859
−0.1	.4602	.4562	.4522	.4483	.4443	.4404	.4364	.4325	.4286	.4247
−0.0	.5000	.4960	.4920	.4880	.4840	.4801	.4761	.4721	.4681	.4641

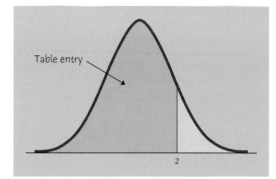

Table entry for z is the area under the standard Normal curve to the left of z.

TABLE A Standard Normal probabilities (*continued*)

z	.00	.01	.02	.03	.04	.05	.06	.07	.08	.09
0.0	.5000	.5040	.5080	.5120	.5160	.5199	.5239	.5279	.5319	.5359
0.1	.5398	.5438	.5478	.5517	.5557	.5596	.5636	.5675	.5714	.5753
0.2	.5793	.5832	.5871	.5910	.5948	.5987	.6026	.6064	.6103	.6141
0.3	.6179	.6217	.6255	.6293	.6331	.6368	.6406	.6443	.6480	.6517
0.4	.6554	.6591	.6628	.6664	.6700	.6736	.6772	.6808	.6844	.6879
0.5	.6915	.6950	.6985	.7019	.7054	.7088	.7123	.7157	.7190	.7224
0.6	.7257	.7291	.7324	.7357	.7389	.7422	.7454	.7486	.7517	.7549
0.7	.7580	.7611	.7642	.7673	.7704	.7734	.7764	.7794	.7823	.7852
0.8	.7881	.7910	.7939	.7967	.7995	.8023	.8051	.8078	.8106	.8133
0.9	.8159	.8186	.8212	.8238	.8264	.8289	.8315	.8340	.8365	.8389
1.0	.8413	.8438	.8461	.8485	.8508	.8531	.8554	.8577	.8599	.8621
1.1	.8643	.8665	.8686	.8708	.8729	.8749	.8770	.8790	.8810	.8830
1.2	.8849	.8869	.8888	.8907	.8925	.8944	.8962	.8980	.8997	.9015
1.3	.9032	.9049	.9066	.9082	.9099	.9115	.9131	.9147	.9162	.9177
1.4	.9192	.9207	.9222	.9236	.9251	.9265	.9279	.9292	.9306	.9319
1.5	.9332	.9345	.9357	.9370	.9382	.9394	.9406	.9418	.9429	.9441
1.6	.9452	.9463	.9474	.9484	.9495	.9505	.9515	.9525	.9535	.9545
1.7	.9554	.9564	.9573	.9582	.9591	.9599	.9608	.9616	.9625	.9633
1.8	.9641	.9649	.9656	.9664	.9671	.9678	.9686	.9693	.9699	.9706
1.9	.9713	.9719	.9726	.9732	.9738	.9744	.9750	.9756	.9761	.9767
2.0	.9772	.9778	.9783	.9788	.9793	.9798	.9803	.9808	.9812	.9817
2.1	.9821	.9826	.9830	.9834	.9838	.9842	.9846	.9850	.9854	.9857
2.2	.9861	.9864	.9868	.9871	.9875	.9878	.9881	.9884	.9887	.9890
2.3	.9893	.9896	.9898	.9901	.9904	.9906	.9909	.9911	.9913	.9916
2.4	.9918	.9920	.9922	.9925	.9927	.9929	.9931	.9932	.9934	.9936
2.5	.9938	.9940	.9941	.9943	.9945	.9946	.9948	.9949	.9951	.9952
2.6	.9953	.9955	.9956	.9957	.9959	.9960	.9961	.9962	.9963	.9964
2.7	.9965	.9966	.9967	.9968	.9969	.9970	.9971	.9972	.9973	.9974
2.8	.9974	.9975	.9976	.9977	.9977	.9978	.9979	.9979	.9980	.9981
2.9	.9981	.9982	.9982	.9983	.9984	.9984	.9985	.9985	.9986	.9986
3.0	.9987	.9987	.9987	.9988	.9988	.9989	.9989	.9989	.9990	.9990
3.1	.9990	.9991	.9991	.9991	.9992	.9992	.9992	.9992	.9993	.9993
3.2	.9993	.9993	.9994	.9994	.9994	.9994	.9994	.9995	.9995	.9995
3.3	.9995	.9995	.9995	.9996	.9996	.9996	.9996	.9996	.9996	.9997
3.4	.9997	.9997	.9997	.9997	.9997	.9997	.9997	.9997	.9997	.9998

TABLE B Random Digits

Line

101	19223	95034	05756	28713	96409	12531	42544	82853
102	73676	47150	99400	01927	27754	42648	82425	36290
103	45467	71709	77558	00095	32863	29485	82226	90056
104	52711	38889	93074	60227	40011	85848	48767	52573
105	95592	94007	69971	91481	60779	53791	17297	59335
106	68417	35013	15529	72765	85089	57067	50211	47487
107	82739	57890	20807	47511	81676	55300	94383	14893
108	60940	72024	17868	24943	61790	90656	87964	18883
109	36009	19365	15412	39638	85453	46816	83485	41979
110	38448	48789	18338	24697	39364	42006	76688	08708
111	81486	69487	60513	09297	00412	71238	27649	39950
112	59636	88804	04634	71197	19352	73089	84898	45785
113	62568	70206	40325	03699	71080	22553	11486	11776
114	45149	32992	75730	66280	03819	56202	02938	70915
115	61041	77684	94322	24709	73698	14526	31893	32592
116	14459	26056	31424	80371	65103	62253	50490	61181
117	38167	98532	62183	70632	23417	26185	41448	75532
118	73190	32533	04470	29669	84407	90785	65956	86382
119	95857	07118	87664	92099	58806	66979	98624	84826
120	35476	55972	39421	65850	04266	35435	43742	11937
121	71487	09984	29077	14863	61683	47052	62224	51025
122	13873	81598	95052	90908	73592	75186	87136	95761
123	54580	81507	27102	56027	55892	33063	41842	81868
124	71035	09001	43367	49497	72719	96758	27611	91596
125	96746	12149	37823	71868	18442	35119	62103	39244
126	96927	19931	36809	74192	77567	88741	48409	41903
127	43909	99477	25330	64359	40085	16925	85117	36071
128	15689	14227	06565	14374	13352	49367	81982	87209
129	36759	58984	68288	22913	18638	54303	00795	08727
130	69051	64817	87174	09517	84534	06489	87201	97245
131	05007	16632	81194	14873	04197	85576	45195	96565
132	68732	55259	84292	08796	43165	93739	31685	97150
133	45740	41807	65561	33302	07051	93623	18132	09547
134	27816	78416	18329	21337	35213	37741	04312	68508
135	66925	55658	39100	78458	11206	19876	87151	31260
136	08421	44753	77377	28744	75592	08563	79140	92454
137	53645	66812	61421	47836	12609	15373	98481	14592
138	66831	68908	40772	21558	47781	33586	79177	06928
139	55588	99404	70708	41098	43563	56934	48394	51719
140	12975	13258	13048	45144	72321	81940	00360	02428
141	96767	35964	23822	96012	94591	65194	50842	53372
142	72829	50232	97892	63408	77919	44575	24870	04178
143	88565	42628	17797	49376	61762	16953	88604	12724
144	62964	88145	83083	69453	46109	59505	69680	00900
145	19687	12633	57857	95806	09931	02150	43163	58636
146	37609	59057	66967	83401	60705	02384	90597	93600
147	54973	86278	88737	74351	47500	84552	19909	67181
148	00694	05977	19664	65441	20903	62371	22725	53340
149	71546	05233	53946	68743	72460	27601	45403	88692
150	07511	88915	41267	16853	84569	79367	32337	03316

Table entry for p and C is the critical value t^* with probability p lying to its right and probability C lying between $-t^*$ and t^*.

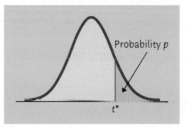

Probability p

t^*

TABLE C t distribution critical values

df	\multicolumn{12}{c}{Upper tail probability p}
	.25	.20	.15	.10	.05	.025	.02	.01	.005	.0025	.001	.0005
1	1.000	1.376	1.963	3.078	6.314	12.71	15.89	31.82	63.66	127.3	318.3	636.6
2	0.816	1.061	1.386	1.886	2.920	4.303	4.849	6.965	9.925	14.09	22.33	31.60
3	0.765	0.978	1.250	1.638	2.353	3.182	3.482	4.541	5.841	7.453	10.21	12.92
4	0.741	0.941	1.190	1.533	2.132	2.776	2.999	3.747	4.604	5.598	7.173	8.610
5	0.727	0.920	1.156	1.476	2.015	2.571	2.757	3.365	4.032	4.773	5.893	6.869
6	0.718	0.906	1.134	1.440	1.943	2.447	2.612	3.143	3.707	4.317	5.208	5.959
7	0.711	0.896	1.119	1.415	1.895	2.365	2.517	2.998	3.499	4.029	4.785	5.408
8	0.706	0.889	1.108	1.397	1.860	2.306	2.449	2.896	3.355	3.833	4.501	5.041
9	0.703	0.883	1.100	1.383	1.833	2.262	2.398	2.821	3.250	3.690	4.297	4.781
10	0.700	0.879	1.093	1.372	1.812	2.228	2.359	2.764	3.169	3.581	4.144	4.587
11	0.697	0.876	1.088	1.363	1.796	2.201	2.328	2.718	3.106	3.497	4.025	4.437
12	0.695	0.873	1.083	1.356	1.782	2.179	2.303	2.681	3.055	3.428	3.930	4.318
13	0.694	0.870	1.079	1.350	1.771	2.160	2.282	2.650	3.012	3.372	3.852	4.221
14	0.692	0.868	1.076	1.345	1.761	2.145	2.264	2.624	2.977	3.326	3.787	4.140
15	0.691	0.866	1.074	1.341	1.753	2.131	2.249	2.602	2.947	3.286	3.733	4.073
16	0.690	0.865	1.071	1.337	1.746	2.120	2.235	2.583	2.921	3.252	3.686	4.015
17	0.689	0.863	1.069	1.333	1.740	2.110	2.224	2.567	2.898	3.222	3.646	3.965
18	0.688	0.862	1.067	1.330	1.734	2.101	2.214	2.552	2.878	3.197	3.611	3.922
19	0.688	0.861	1.066	1.328	1.729	2.093	2.205	2.539	2.861	3.174	3.579	3.883
20	0.687	0.860	1.064	1.325	1.725	2.086	2.197	2.528	2.845	3.153	3.552	3.850
21	0.686	0.859	1.063	1.323	1.721	2.080	2.189	2.518	2.831	3.135	3.527	3.819
22	0.686	0.858	1.061	1.321	1.717	2.074	2.183	2.508	2.819	3.119	3.505	3.792
23	0.685	0.858	1.060	1.319	1.714	2.069	2.177	2.500	2.807	3.104	3.485	3.768
24	0.685	0.857	1.059	1.318	1.711	2.064	2.172	2.492	2.797	3.091	3.467	3.745
25	0.684	0.856	1.058	1.316	1.708	2.060	2.167	2.485	2.787	3.078	3.450	3.725
26	0.684	0.856	1.058	1.315	1.706	2.056	2.162	2.479	2.779	3.067	3.435	3.707
27	0.684	0.855	1.057	1.314	1.703	2.052	2.158	2.473	2.771	3.057	3.421	3.690
28	0.683	0.855	1.056	1.313	1.701	2.048	2.154	2.467	2.763	3.047	3.408	3.674
29	0.683	0.854	1.055	1.311	1.699	2.045	2.150	2.462	2.756	3.038	3.396	3.659
30	0.683	0.854	1.055	1.310	1.697	2.042	2.147	2.457	2.750	3.030	3.385	3.646
40	0.681	0.851	1.050	1.303	1.684	2.021	2.123	2.423	2.704	2.971	3.307	3.551
50	0.679	0.849	1.047	1.299	1.676	2.009	2.109	2.403	2.678	2.937	3.261	3.496
60	0.679	0.848	1.045	1.296	1.671	2.000	2.099	2.390	2.660	2.915	3.232	3.460
80	0.678	0.846	1.043	1.292	1.664	1.990	2.088	2.374	2.639	2.887	3.195	3.416
100	0.677	0.845	1.042	1.290	1.660	1.984	2.081	2.364	2.626	2.871	3.174	3.390
1000	0.675	0.842	1.037	1.282	1.646	1.962	2.056	2.330	2.581	2.813	3.098	3.300
z^*	0.674	0.841	1.036	1.282	1.645	1.960	2.054	2.326	2.576	2.807	3.091	3.291
	50%	60%	70%	80%	90%	95%	96%	98%	99%	99.5%	99.8%	99.9%
	\multicolumn{12}{c}{Confidence level C}

Table entry for p is the critical value F^* with probability p lying to its right.

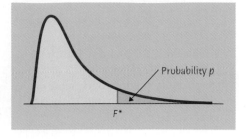

Probability p

F^*

TABLE D F distribution critical values

	p	Degrees of freedom in the numerator							
		1	2	3	4	5	6	7	8
1	0.100	39.86	49.50	53.59	55.83	57.24	58.20	58.91	59.44
	0.050	161.45	199.50	215.71	224.58	230.16	233.99	236.77	238.88
	0.025	647.79	799.50	864.16	899.58	921.85	937.11	948.22	956.66
	0.010	4052.2	4999.5	5403.4	5624.6	5763.6	5859	5928.4	5981.1
	0.001	405284	500000	540379	562500	576405	585937	592873	598144
2	0.100	8.53	9.00	9.16	9.24	9.29	9.33	.35	9.37
	0.050	18.51	19.00	19.16	19.25	19.30	19.33	19.35	19.37
	0.025	38.51	39.00	39.17	39.25	39.30	39.33	39.36	39.37
	0.010	98.50	99.00	99.17	99.25	99.30	99.33	99.36	99.37
	0.001	998.50	999.00	999.17	999.25	999.30	999.33	999.36	999.37
3	0.100	5.54	5.46	5.39	5.34	5.31	5.28	5.27	5.25
	0.050	10.13	9.55	9.28	9.12	9.01	8.94	8.89	8.85
	0.025	17.44	16.04	15.44	15.10	14.88	14.73	14.62	14.54
	0.010	34.12	30.82	29.46	28.71	28.24	27.91	27.67	27.49
	0.001	167.03	148.50	141.11	137.10	134.58	132.85	131.58	130.62
4	0.100	4.54	4.32	4.19	4.11	4.05	4.01	3.98	3.95
	0.050	7.71	6.94	6.59	6.39	6.26	6.16	6.09	6.04
	0.025	12.22	10.65	9.98	9.60	9.36	9.20	9.07	8.98
	0.010	21.20	18.00	16.69	15.98	15.52	15.21	14.98	14.80
	0.001	74.14	61.25	56.18	53.44	51.71	50.53	49.66	49.00
5	0.100	4.06	3.78	3.62	3.52	3.45	3.40	3.37	3.34
	0.050	6.61	5.79	5.41	5.19	5.05	4.95	4.88	4.82
	0.025	10.01	8.43	7.76	7.39	7.15	6.98	6.85	6.76
	0.010	16.26	13.27	12.06	11.39	10.97	10.67	10.46	10.29
	0.001	47.18	37.12	33.20	31.09	29.75	28.83	28.16	27.65
6	0.100	3.78	3.46	3.29	3.18	3.11	3.05	3.01	2.98
	0.050	5.99	5.14	4.76	4.53	4.39	4.28	4.21	4.15
	0.025	8.81	7.26	6.60	6.23	5.99	5.82	5.70	5.60
	0.010	13.75	10.92	9.78	9.15	8.75	8.47	8.26	8.10
	0.001	35.51	27.00	23.70	21.92	20.80	20.03	19.46	19.03
7	0.100	3.59	3.26	3.07	2.96	2.88	2.83	2.78	2.75
	0.050	5.59	4.74	4.35	4.12	3.97	3.87	3.79	3.73
	0.025	8.07	6.54	5.89	5.52	5.29	5.12	4.99	4.90
	0.010	12.25	9.55	8.45	7.85	7.46	7.19	6.99	6.84
	0.001	29.25	21.69	18.77	17.20	16.21	15.52	15.02	14.63
8	0.100	3.46	3.11	2.92	2.81	2.73	2.67	2.62	2.59
	0.050	5.32	4.46	4.07	3.84	3.69	3.58	3.50	3.44
	0.025	7.57	6.06	5.42	5.05	4.82	4.65	4.53	4.43
	0.010	11.26	8.65	7.59	7.01	6.63	6.37	6.18	6.03
	0.001	25.41	18.49	15.83	14.39	13.48	12.86	12.40	12.05

Degrees of freedom in the denominator

Table entry for p is the critical value F^* with probability p lying to its right.

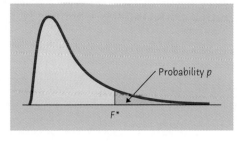

Probability p

F^*

TABLE D F distribution critical values (*continued*)

	p	\multicolumn{8}{c}{Degrees of freedom in the numerator}
		9	10	15	20	30	60	120	1000
1	0.100	59.86	60.19	61.22	61.74	62.26	62.79	63.06	63.30
	0.050	240.54	241.88	245.95	248.01	250.10	252.20	253.25	254.19
	0.025	963.28	968.63	984.87	993.10	1001.4	1009.8	1014	1017.7
	0.010	6022.5	6055.8	6157.3	6208.7	6260.6	6313	6339.4	6362.7
	0.001	602284	605621	615764	620908	626099	631337	633972	636301
2	0.100	9.38	9.39	9.42	9.44	9.46	9.47	9.48	9.49
	0.050	19.38	19.40	19.43	19.45	19.46	19.48	19.49	19.49
	0.025	39.39	39.40	39.43	39.45	39.46	39.48	39.49	39.50
	0.010	99.39	99.40	99.43	99.45	99.47	99.48	99.49	99.50
	0.001	999.39	999.40	999.43	999.45	999.47	999.48	999.49	999.50
3	0.100	5.24	5.23	5.20	5.18	5.17	5.15	5.14	5.13
	0.050	8.81	8.79	8.70	8.66	8.62	8.57	8.55	8.53
	0.025	14.47	14.42	14.25	14.17	14.08	13.99	13.95	13.91
	0.010	27.35	27.23	26.87	26.69	26.50	26.32	26.22	26.14
	0.001	129.86	129.25	127.37	126.42	125.45	124.47	123.97	123.53
4	0.100	3.94	3.92	3.87	3.84	3.82	3.79	3.78	3.76
	0.050	6.00	5.96	5.86	5.80	5.75	5.69	5.66	5.63
	0.025	8.90	8.84	8.66	8.56	8.46	8.36	8.31	8.26
	0.010	14.66	14.55	14.20	14.02	13.84	13.65	13.56	13.47
	0.001	48.47	48.05	46.76	46.10	45.43	44.75	44.40	44.09
5	0.100	3.32	3.30	3.24	3.21	3.17	3.14	3.12	3.11
	0.050	4.77	4.74	4.62	4.56	4.50	4.43	4.40	4.37
	0.025	6.68	6.62	6.43	6.33	6.23	6.12	6.07	6.02
	0.010	10.16	10.05	9.72	9.55	9.38	9.20	9.11	9.03
	0.001	27.24	26.92	25.91	25.39	24.87	24.33	24.06	23.82
6	0.100	2.96	2.94	2.87	2.84	2.80	2.76	2.74	2.72
	0.050	4.10	4.06	3.94	3.87	3.81	3.74	3.70	3.67
	0.025	5.52	5.46	5.27	5.17	5.07	4.96	4.90	4.86
	0.010	7.98	7.87	7.56	7.40	7.23	7.06	6.97	6.89
	0.001	18.69	18.41	17.56	17.12	16.67	16.21	15.98	15.77
7	0.100	2.72	2.70	2.63	2.59	2.56	2.51	2.49	2.47
	0.050	3.68	3.64	3.51	3.44	3.38	3.30	3.27	3.23
	0.025	4.82	4.76	4.57	4.47	4.36	4.25	4.20	4.15
	0.010	6.72	6.62	6.31	6.16	5.99	5.82	5.74	5.66
	0.001	14.33	14.08	13.32	12.93	12.53	12.12	11.91	11.72
8	0.100	2.56	2.54	2.46	2.42	2.38	2.34	2.32	2.30
	0.050	3.39	3.35	3.22	3.15	3.08	3.01	2.97	2.93
	0.025	4.36	4.30	4.10	4.00	3.89	3.78	3.73	3.68
	0.010	5.91	5.81	5.52	5.36	5.20	5.03	4.95	4.87
	0.001	11.77	11.54	10.84	10.48	10.11	9.73	9.53	9.36

Degree of freedom in the denominator

		Degrees of freedom in the numerator							
	p	1	2	3	4	5	6	7	8
9	0.100	3.36	3.01	2.81	2.69	2.61	2.55	2.51	2.47
	0.050	5.12	4.26	3.86	3.63	3.48	3.37	3.29	3.23
	0.025	7.21	5.71	5.08	4.72	4.48	4.32	4.20	4.10
	0.010	10.56	8.02	6.99	6.42	6.06	5.80	5.61	5.47
	0.001	22.86	16.39	13.90	12.56	11.71	11.13	10.70	10.37
10	0.100	3.29	2.92	2.73	2.61	2.52	2.46	2.41	2.38
	0.050	4.96	4.10	3.71	3.48	3.33	3.22	3.14	3.07
	0.025	6.94	5.46	4.83	4.47	4.24	4.07	3.95	3.85
	0.010	10.04	7.56	6.55	5.99	5.64	5.39	5.20	5.06
	0.001	21.04	14.91	12.55	11.28	10.48	9.93	9.52	9.20
12	0.100	3.18	2.81	2.61	2.48	2.39	2.33	2.28	2.24
	0.050	4.75	3.89	3.49	3.26	3.11	3.00	2.91	2.85
	0.025	6.55	5.10	4.47	4.12	3.89	3.73	3.61	3.51
	0.010	9.33	6.93	5.95	5.41	5.06	4.82	4.64	4.50
	0.001	18.64	12.97	10.80	9.63	8.89	8.38	8.00	7.71
15	0.100	3.07	2.70	2.49	2.36	2.27	2.21	2.16	2.12
	0.050	4.54	3.68	3.29	3.06	2.90	2.79	2.71	2.64
	0.025	6.20	4.77	4.15	3.80	3.58	3.41	3.29	3.20
	0.010	8.68	6.36	5.42	4.89	4.56	4.32	4.14	4.00
	0.001	16.59	11.34	9.34	8.25	7.57	7.09	6.74	6.47
20	0.100	2.97	2.59	2.38	2.25	2.16	2.09	2.04	2.00
	0.050	4.35	3.49	3.10	2.87	2.71	2.60	2.51	2.45
	0.025	5.87	4.46	3.86	3.51	3.29	3.13	3.01	2.91
	0.010	8.10	5.85	4.94	4.43	4.10	3.87	3.70	3.56
	0.001	14.82	9.95	8.10	7.10	6.46	6.02	5.69	5.44
25	0.100	2.92	2.53	2.32	2.18	2.09	2.02	1.97	1.93
	0.050	4.24	3.39	2.99	2.76	2.60	2.49	2.40	2.34
	0.025	5.69	4.29	3.69	3.35	3.13	2.97	2.85	2.75
	0.010	7.77	5.57	4.68	4.18	3.85	3.63	3.46	3.32
	0.001	13.88	9.22	7.45	6.49	5.89	5.46	5.15	4.91
50	0.100	2.81	2.41	2.20	2.06	1.97	1.90	1.84	1.80
	0.050	4.03	3.18	2.79	2.56	2.40	2.29	2.20	2.13
	0.025	5.34	3.97	3.39	3.05	2.83	2.67	2.55	2.46
	0.010	7.17	5.06	4.20	3.72	3.41	3.19	3.02	2.89
	0.001	12.22	7.96	6.34	5.46	4.90	4.51	4.22	4.00
100	0.100	2.76	2.36	2.14	2.00	1.91	1.83	1.78	1.73
	0.050	3.94	3.09	2.70	2.46	2.31	2.19	2.10	2.03
	0.025	5.18	3.83	3.25	2.92	2.70	2.54	2.42	2.32
	0.010	6.90	4.82	3.98	3.51	3.21	2.99	2.82	2.69
	0.001	11.50	7.41	5.86	5.02	4.48	4.11	3.83	3.61
200	0.100	2.73	2.33	2.11	1.97	1.88	1.80	1.75	1.70
	0.050	3.89	3.04	2.65	2.42	2.26	2.14	2.06	1.98
	0.025	5.10	3.76	3.18	2.85	2.63	2.47	2.35	2.26
	0.010	6.76	4.71	3.88	3.41	3.11	2.89	2.73	2.60
	0.001	11.15	7.15	5.63	4.81	4.29	3.92	3.65	3.43
1000	0.100	2.71	2.31	2.09	1.95	1.85	1.78	1.72	1.68
	0.050	3.85	3.00	2.61	2.38	2.22	2.11	2.02	1.95
	0.025	5.04	3.70	3.13	2.80	2.58	2.42	2.30	2.20
	0.010	6.66	4.63	3.80	3.34	3.04	2.82	2.66	2.53
	0.001	10.89	6.96	5.46	4.65	4.14	3.78	3.51	3.30

Degrees of freedom in the denominator

		Degrees of freedom in the numerator							
	p	9	10	15	20	30	60	120	1000
9	0.100	2.44	2.42	2.34	2.30	2.25	2.21	2.18	2.16
	0.050	3.18	3.14	3.01	2.94	2.86	2.79	2.75	2.71
	0.025	4.03	3.96	3.77	3.67	3.56	3.45	3.39	3.34
	0.010	5.35	5.26	4.96	4.81	4.65	4.48	4.40	4.32
	0.001	10.11	9.89	9.24	8.90	8.55	8.19	8.00	7.84
10	0.100	2.35	2.32	2.24	2.20	2.16	2.11	2.08	2.06
	0.050	3.02	2.98	2.85	2.77	2.70	2.62	2.58	2.54
	0.025	3.78	3.72	3.52	3.42	3.31	3.20	3.14	3.09
	0.010	4.94	4.85	4.56	4.41	4.25	4.08	4.00	3.92
	0.001	8.96	8.75	8.13	7.80	7.47	7.12	6.94	6.78
12	0.100	2.21	2.19	2.10	2.06	2.01	1.96	1.93	1.91
	0.050	2.80	2.75	2.62	2.54	2.47	2.38	2.34	2.30
	0.025	3.44	3.37	3.18	3.07	2.96	2.85	2.79	2.73
	0.010	4.39	4.30	4.01	3.86	3.70	3.54	3.45	3.37
	0.001	7.48	7.29	6.71	6.40	6.09	5.76	5.59	5.44
15	0.100	2.09	2.06	1.97	1.92	1.87	1.82	1.79	1.76
	0.050	2.59	2.54	2.40	2.33	2.25	2.16	2.11	2.07
	0.025	3.12	3.06	2.86	2.76	2.64	2.52	2.46	2.40
	0.010	3.89	3.80	3.52	3.37	3.21	3.05	2.96	2.88
	0.001	6.26	6.08	5.54	5.25	4.95	4.64	4.47	4.33
20	0.100	1.96	1.94	1.84	1.79	1.74	1.68	1.64	1.61
	0.050	2.39	2.35	2.20	2.12	2.04	1.95	1.90	1.85
	0.025	2.84	2.77	2.57	2.46	2.35	2.22	2.16	2.09
	0.010	3.46	3.37	3.09	2.94	2.78	2.61	2.52	2.43
	0.001	5.24	5.08	4.56	4.29	4.00	3.70	3.54	3.40
25	0.100	1.89	1.87	1.77	1.72	1.66	1.59	1.56	1.52
	0.050	2.28	2.24	2.09	2.01	1.92	1.82	1.77	1.72
	0.025	2.68	2.61	2.41	2.30	2.18	2.05	1.98	1.91
	0.010	3.22	3.13	2.85	2.70	2.54	2.36	2.27	2.18
	0.001	4.71	4.56	4.06	3.79	3.52	3.22	3.06	2.91
50	0.100	1.76	1.73	1.63	1.57	1.50	1.42	1.38	1.33
	0.050	2.07	2.03	1.87	1.78	1.69	1.58	1.51	1.45
	0.025	2.38	2.32	2.11	1.99	1.87	1.72	1.64	1.56
	0.010	2.78	2.70	2.42	2.27	2.10	1.91	1.80	1.70
	0.001	3.82	3.67	3.20	2.95	2.68	2.38	2.21	2.05
100	0.100	1.69	1.66	1.56	1.49	1.42	1.34	1.28	1.22
	0.050	1.97	1.93	1.77	1.60	1.57	1.45	1.38	1.30
	0.025	2.24	2.18	1.97	1.85	1.71	1.56	1.46	1.36
	0.010	2.59	2.50	2.22	2.07	1.89	1.69	1.57	1.45
	0.001	3.44	3.30	2.84	2.59	2.32	2.01	1.83	1.64
200	0.100	1.66	1.63	1.52	1.46	1.38	1.29	1.23	1.16
	0.050	1.93	1.88	1.72	1.62	1.52	1.39	1.30	1.21
	0.025	2.18	2.11	1.90	1.78	1.64	1.47	1.37	1.25
	0.010	2.50	2.41	2.13	1.97	1.79	1.58	1.45	1.30
	0.001	3.26	3.12	2.67	2.42	2.15	1.83	1.64	1.43
1000	0.100	1.64	1.61	1.49	1.43	1.35	1.25	1.18	1.08
	0.050	1.89	1.84	1.68	1.58	1.47	1.33	1.24	1.11
	0.025	2.13	2.06	1.85	1.72	1.58	1.41	1.29	1.13
	0.010	2.43	2.34	2.06	1.90	1.72	1.50	1.35	1.16
	0.001	3.13	2.99	2.54	2.30	2.02	1.69	1.49	1.22

Degrees of freedom in the denominator

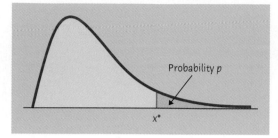

Table entry for p is the critical value x^* with probability p lying to its right.

Probability p

x^*

TABLE E Chi-square distribution critical values

df	.25	.20	.15	.10	.05	.025	.02	.01	.005	.0025	.001	.0005
1	1.32	1.64	2.07	2.71	3.84	5.02	5.41	6.63	7.88	9.14	10.83	12.12
2	2.77	3.22	3.79	4.61	5.99	7.38	7.82	9.21	10.60	11.98	13.82	15.20
3	4.11	4.64	5.32	6.25	7.81	9.35	9.84	11.34	12.84	14.32	16.27	17.73
4	5.39	5.99	6.74	7.78	9.49	11.14	11.67	13.28	14.86	16.42	18.47	20.00
5	6.63	7.29	8.12	9.24	11.07	12.83	13.39	15.09	16.75	18.39	20.51	22.11
6	7.84	8.56	9.45	10.64	12.59	14.45	15.03	16.81	18.55	20.25	22.46	24.10
7	9.04	9.80	10.75	12.02	14.07	16.01	16.62	18.48	20.28	22.04	24.32	26.02
8	10.22	11.03	12.03	13.36	15.51	17.53	18.17	20.09	21.95	23.77	26.12	27.87
9	11.39	12.24	13.29	14.68	16.92	19.02	19.68	21.67	23.59	25.46	27.88	29.67
10	12.55	13.44	14.53	15.99	18.31	20.48	21.16	23.21	25.19	27.11	29.59	31.42
11	13.70	14.63	15.77	17.28	19.68	21.92	22.62	24.72	26.76	28.73	31.26	33.14
12	14.85	15.81	16.99	18.55	21.03	23.34	24.05	26.22	28.30	30.32	32.91	34.82
13	15.98	16.98	18.20	19.81	22.36	24.74	25.47	27.69	29.82	31.88	34.53	36.48
14	17.12	18.15	19.41	21.06	23.68	26.12	26.87	29.14	31.32	33.43	36.12	38.11
15	18.25	19.31	20.60	22.31	25.00	27.49	28.26	30.58	32.80	34.95	37.70	39.72
16	19.37	20.47	21.79	23.54	26.30	28.85	29.63	32.00	34.27	36.46	39.25	41.31
17	20.49	21.61	22.98	24.77	27.59	30.19	31.00	33.41	35.72	37.95	40.79	42.88
18	21.60	22.76	24.16	25.99	28.87	31.53	32.35	34.81	37.16	39.42	42.31	44.43
19	22.72	23.90	25.33	27.20	30.14	32.85	33.69	36.19	38.58	40.88	43.82	45.97
20	23.83	25.04	26.50	28.41	31.41	34.17	35.02	37.57	40.00	42.34	45.31	47.50
21	24.93	26.17	27.66	29.62	32.67	35.48	36.34	38.93	41.40	43.78	46.80	49.01
22	26.04	27.30	28.82	30.81	33.92	36.78	37.66	40.29	42.80	45.20	48.27	50.51
23	27.14	28.43	29.98	32.01	35.17	38.08	38.97	41.64	44.18	46.62	49.73	52.00
24	28.24	29.55	31.13	33.20	36.42	39.36	40.27	42.98	45.56	48.03	51.18	53.48
25	29.34	30.68	32.28	34.38	37.65	40.65	41.57	44.31	46.93	49.44	52.62	54.95
26	30.43	31.79	33.43	35.56	38.89	41.92	42.86	45.64	48.29	50.83	54.05	56.41
27	31.53	32.91	34.57	36.74	40.11	43.19	44.14	46.96	49.64	52.22	55.48	57.86
28	32.62	34.03	35.71	37.92	41.34	44.46	45.42	48.28	50.99	53.59	56.89	59.30
29	33.71	35.14	36.85	39.09	42.56	45.72	46.69	49.59	52.34	54.97	58.30	60.73
30	34.80	36.25	37.99	40.26	43.77	46.98	47.96	50.89	53.67	56.33	59.70	62.16
40	45.62	47.27	49.24	51.81	55.76	59.34	60.44	63.69	66.77	69.70	73.40	76.09
50	56.33	58.16	60.35	63.17	67.50	71.42	72.61	76.15	79.49	82.66	86.66	89.56
60	66.98	68.97	71.34	74.40	79.08	83.30	84.58	88.38	91.95	95.34	99.61	102.7
80	88.13	90.41	93.11	96.58	101.9	106.6	108.1	112.3	116.3	120.1	124.8	128.3
100	109.1	111.7	114.7	118.5	124.3	129.6	131.1	135.8	140.2	144.3	149.4	153.2

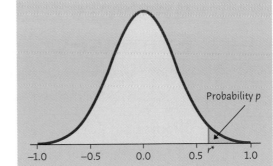

Table entry for p is the critical value r^* of the correlation coefficient r with probability p lying to its right.

TABLE F   Critical values of the correlation r

n	\.20	\.10	\.05	\.025	\.02	\.01	\.005	\.0025	\.001	\.0005
					Upper tail probability p					
3	0.8090	0.9511	0.9877	0.9969	0.9980	0.9995	0.9999	1.0000	1.0000	1.0000
4	0.6000	0.8000	0.9000	0.9500	0.9600	0.9800	0.9900	0.9950	0.9980	0.9990
5	0.4919	0.6870	0.8054	0.8783	0.8953	0.9343	0.9587	0.9740	0.9859	0.9911
6	0.4257	0.6084	0.7293	0.8114	0.8319	0.8822	0.9172	0.9417	0.9633	0.9741
7	0.3803	0.5509	0.6694	0.7545	0.7766	0.8329	0.8745	0.9056	0.9350	0.9509
8	0.3468	0.5067	0.6215	0.7067	0.7295	0.7887	0.8343	0.8697	0.9049	0.9249
9	0.3208	0.4716	0.5822	0.6664	0.6892	0.7498	0.7977	0.8359	0.8751	0.8983
10	0.2998	0.4428	0.5494	0.6319	0.6546	0.7155	0.7646	0.8046	0.8467	0.8721
11	0.2825	0.4187	0.5214	0.6021	0.6244	0.6851	0.7348	0.7759	0.8199	0.8470
12	0.2678	0.3981	0.4973	0.5760	0.5980	0.6581	0.7079	0.7496	0.7950	0.8233
13	0.2552	0.3802	0.4762	0.5529	0.5745	0.6339	0.6835	0.7255	0.7717	0.8010
14	0.2443	0.3646	0.4575	0.5324	0.5536	0.6120	0.6614	0.7034	0.7501	0.7800
15	0.2346	0.3507	0.4409	0.5140	0.5347	0.5923	0.6411	0.6831	0.7301	0.7604
16	0.2260	0.3383	0.4259	0.4973	0.5177	0.5742	0.6226	0.6643	0.7114	0.7419
17	0.2183	0.3271	0.4124	0.4821	0.5021	0.5577	0.6055	0.6470	0.6940	0.7247
18	0.2113	0.3170	0.4000	0.4683	0.4878	0.5425	0.5897	0.6308	0.6777	0.7084
19	0.2049	0.3077	0.3887	0.4555	0.4747	0.5285	0.5751	0.6158	0.6624	0.6932
20	0.1991	0.2992	0.3783	0.4438	0.4626	0.5155	0.5614	0.6018	0.6481	0.6788
21	0.1938	0.2914	0.3687	0.4329	0.4513	0.5034	0.5487	0.5886	0.6346	0.6652
22	0.1888	0.2841	0.3598	0.4227	0.4409	0.4921	0.5368	0.5763	0.6219	0.6524
23	0.1843	0.2774	0.3515	0.4132	0.4311	0.4815	0.5256	0.5647	0.6099	0.6402
24	0.1800	0.2711	0.3438	0.4044	0.4219	0.4716	0.5151	0.5537	0.5986	0.6287
25	0.1760	0.2653	0.3365	0.3961	0.4133	0.4622	0.5052	0.5434	0.5879	0.6178
26	0.1723	0.2598	0.3297	0.3882	0.4052	0.4534	0.4958	0.5336	0.5776	0.6074
27	0.1688	0.2546	0.3233	0.3809	0.3976	0.4451	0.4869	0.5243	0.5679	0.5974
28	0.1655	0.2497	0.3172	0.3739	0.3904	0.4372	0.4785	0.5154	0.5587	0.5880
29	0.1624	0.2451	0.3115	0.3673	0.3835	0.4297	0.4705	0.5070	0.5499	0.5790
30	0.1594	0.2407	0.3061	0.3610	0.3770	0.4226	0.4629	0.4990	0.5415	0.5703
40	0.1368	0.2070	0.2638	0.3120	0.3261	0.3665	0.4026	0.4353	0.4741	0.5007
50	0.1217	0.1843	0.2353	0.2787	0.2915	0.3281	0.3610	0.3909	0.4267	0.4514
60	0.1106	0.1678	0.2144	0.2542	0.2659	0.2997	0.3301	0.3578	0.3912	0.4143
80	0.0954	0.1448	0.1852	0.2199	0.2301	0.2597	0.2864	0.3109	0.3405	0.3611
100	0.0851	0.1292	0.1654	0.1966	0.2058	0.2324	0.2565	0.2786	0.3054	0.3242
1000	0.0266	0.0406	0.0520	0.0620	0.0650	0.0736	0.0814	0.0887	0.0976	0.1039

Answers to Odd-Numbered Exercises

Chapter 1

1.1 (a) Vehicles. (b) Make/model, vehicle type, transmission type (all categorical); number of cylinders, city MPG, highway MPG (all quantitative).

1.3 (a) 11.4%. (b) A pie chart could be used.

1.7 Center near $27,000, spread from $8000 to $38,000. Public: Center near $11,000, spread from $8000 to $16,000. Private: Center near $29,000, spread from $20,000 to $38,000.

1.11 About 1% are some other color. A majority of Japanese cars are gray or white.

1.13 (a) Baseball players. (b) Team and position (categorical), and age and salary (quantitative). (c) Dollars per year.

1.15 Approximately 60% Mexican and 10% Puerto Rican.

1.17 (a) Roughly symmetric. (b) Center: 0% to 2%. (c) Lowest: −16% to −14%; highest: 16% to 18%. (d) About 40%.

1.19 (a) We cannot compare counts from groups of unequal sizes. (b) Marijuana usage and accident rates rise (and fall) together.

1.21 From the top left histogram: 4, 1, 3, 2.

1.23 The distribution is slightly right-skewed, with no outliers.

1.25 Distribution is roughly symmetric; a typical year had about 46 home runs; 60 home runs is not an outlier.

1.27 (b) Women's times decreased quite rapidly from 1972 until the mid-1980s; since that time, they have been fairly consistent.

1.29 (a) States with large populations need more doctors to serve that population. (b) Right-skewed, with D.C. a high outlier.

1.37 Most eruptions last either between 1.5 and 2.5 minutes or between 4 and 5 minutes. Very few lasted around 3 minutes or more than 5 minutes.

Chapter 2

2.1 (a) 25.91 MPG. (b) 24.48 MPG.

2.3 Median $163,900, mean $210,900.

2.5 (a) 23,040, 30,315, 31,975, 32,710, and 33,650 pounds. (b) The gaps between the minimum and Q_1 and between Q_1 and M are larger than the other two gaps.

2.7   (a) We expect the mean and median to be similar. (b) $\bar{x} \doteq 54.81$ years; five-number summary: 42, 51, 55, 58, 69 years. (c) 51 to 58 years old; yes.

2.9   Both sets have $\bar{x} \doteq 7.50$ and $s \doteq 2.03$; set A is left-skewed, while set B has a high outlier.

2.11   Median $53,054, mean $72,674.

2.13   (a) $\bar{x} = 247.7$ and $M = 232$ doctors per 100,000 people. (b) Without D.C.: $\bar{x}^* = 237.5$ and $M^* = 229.5$ doctors per 100,000 people.

2.15   (a) $\bar{x} \doteq 141.06$. (b) $\bar{x}^* \doteq 137.588$. The outlier pulls the mean up.

2.17 $M = 138.5$.

2.19   Compare, for example, the means: -3.59% and 0.31%.

2.21   The distribution is strongly right-skewed with Los Angeles County a high outlier, and two other high clusters of five and two counties.

2.23   (a) The main peak occurs from 50 to 150 days; the distribution is right-skewed. (b) The five-number summary is 43, 82.5, 102.5, 151.5, and 598 days.

2.25   Five-number summary: $203,500, $266,250, $1,462,500, $3,622,500, $15,000,000. This distribution is strongly skewed, and Sammy Sosa's salary is a clear outlier.

2.27   Sammy Sosa.

2.29   (a) Any set of four identical numbers works. (b) 0, 0, 10, 10 is the only possible answer.

2.35   The mean and median always agree for two observations.

2.37   (a) Place the new point at the current median.

2.39   Nationally advertised food had a small percentage difference ($\bar{x} = 0.125\%$, $M = 2.5\%$), regional was next (25.125% and 26.5%), then local (81.75% and 70%, with high outliers).

Chapter 3

3.3   Both are 0.5.

3.5   Place 64 at the center of the curve, then mark 61.3 and 66.7 at the change-of-curvature points.

3.7   (a) 234 to 298 days. (b) Shorter than 234 days.

3.9   Women: $z = 2.96$; men: $z = 0.96$.

3.11   (a) 0.2266. (b) 0.2260. (c) Both are 0.6915.

3.13   (a) IQs below about 90. (b) About 125 or more.

3.15   (a) 50%. (b) 0.15%. (c) 16%.

3.17   About 2.5%.

3.19   (a) $z \doteq 0.84$. (b) $z \doteq 0.39$.

3.21 About 9%.

3.23 About 2.9%.

3.25 (a) About 3.13%. (b) About 1.15%.

3.27 Women's weights are left-skewed.

3.29 (a) About ± 1.28. (b) 60.5 and 67.5 inches.

3.31 $Q_1 \doteq -0.67$ and $Q_3 \doteq 0.67$.

Chapter 4

4.1 (a) Explanatory: time spent studying; response: grade. (b) Explore the relationship. (c) Explanatory: rainfall; response: crop yield. (d) Explore the relationship. (e) Explanatory: income; response: years of education.

4.3 Explanatory: weight; response: mortality rate; other variables: activity level and economic status.

4.5 Linear, negative, fairly weak; sparrowhawks are long-lived and territorial.

4.7 (b) Icicles seem to grow faster when the water runs more slowly.

4.9 (a) -0.748. (b) With Point A: -0.807; with Point B: -0.469. (c) Point A fits the association, while B weakens it.

4.11 These variables do not have a straight-line relationship; the association is neither positive nor negative.

4.13 (a) Higher IQs tend to go with higher GPAs, and lower IQs with lower GPAs. (b) Positive, roughly linear, and moderately strong (except for three outliers). (c) IQ: about 103; GPA: about 0.5.

4.15 (a) Lowest: about 107 calories (145 mg of sodium); highest: about 195 calories (510 mg of sodium). (b) Positive association: High-calorie hot dogs tend to be high in salt, and low-calorie hot dogs tend to have low sodium. (c) The lower left point in the scatterplot is an outlier. The remaining points show a moderately strong linear relationship.

4.17 (a) Positive but not near 1. (b) Closer to 1.

4.19 The increase happens for Figure 4.6.

4.21 A fairly strong, positive linear pattern; $r \doteq 0.910$; no definite outliers.

4.23 Correlation measures the strength of the association, not the slope of the line.

4.25 (a) Planting rate is explanatory. (c) The pattern is curved. The association is not linear and is neither positive nor negative. (d) 131.025, 143.15, 146.225, 143.07, and 134.75 bushels/acre. At or around 20,000 plants/acre is best.

4.27 (a) Alcohol should be on the x axis. (b) A fairly strong, linear relationship. (c) The association is negative: High wine consumption goes with fewer heart disease deaths, while low wine consumption goes with more deaths. This does not prove causation.

4.29 (a) Small-cap stocks. (b) A negative correlation.

Chapter 5

5.1   **(b)** $\bar{x} = 58.23\%$ and $s_x = 13.03\%$; $\bar{y} = 14.23$ and $s_y = 5.29$ new birds; $r = -0.748$.

5.3   **(a)** $\hat{y} = 9.14 + 0.856x$. **(b)** For every 1 MPG change in city mileage, highway mileages change by about 0.856 MPG. **(c)** 26.3 MPG.

5.5 $\bar{x} = 19.59$ MPG and $\bar{y} = 25.91$ MPG. When $x = 19.59$, $\hat{y} \doteq 25.91$.

5.7   **(b)** No; the pattern is curved, not linear. **(c)** The sum is -0.01.

5.9   **(b)** $\hat{y} = 5.01 + 1.09x$. **(c)** The predictions are 17.7, 26.3, 30.5 MPG, and 15.9, 26.9, 32.3 MPG.

5.11 For example, a student's general intelligence, problem-solving ability, or self-confidence.

5.13 Causation in the other direction makes more sense.

5.15 **(a)** $r = 0.558$. **(b)** 64.5 inches.

5.17 **(a)** On the average, BOD rises (falls) by 1.507 mg/l for every 1 mg/l increase (decrease) in TOC. **(b)** -55.43 mg/l; extrapolation.

5.19 **(b)** $\hat{y} = 71.950 + 0.38333x$. **(c)** 87.28 cm and 94.95 cm. **(d)** Sarah is growing more slowly than normal (0.38 instead of 0.5 cm/month).

5.21 $\hat{y} = 255.95$ cm, or 100.77 inches.

5.23 **(a)** $r \doteq 0.9999$, so recalibration is not necessary. **(b)** $\hat{y} = 1.6571 + 0.1133x$; when $x = 500$ mg/liter, $\hat{y} \doteq 58.31$. The relationship is strong, so the prediction should be very accurate.

5.25 **(a)** r is negative because the association is negative. The straight-line relationship explains about $r^2 \doteq 71.1\%$ of the variation in death rates. **(b)** $\hat{y} \doteq 168.8$ deaths per 100,000 people. **(c)** $b = rs_y/s_x$, and s_y and s_x are both positive.

5.27 -3184.9 deaths per 100,000 people; extrapolation.

5.29 $r = 0.40$.

5.31 Age is the lurking variable.

5.33 Lots of things change over time: students, teachers, technology, ….

5.35 Closer to 0.

5.37 **(b)** The right-hand points lie below the left-hand points. **(c)** Right hand: $\hat{y} = 99.4 + 0.0283x$ ($r = 0.305$, $r^2 = 9.3\%$). Left hand: $\hat{y} = 172 + 0.262x$ ($r = 0.318$, $r^2 = 10.1\%$). Neither is very good for prediction.

5.39 Residuals: 8.1, 12.1, -0.87, -9.9, 1.1, -3.9, -15.9, 6.1, 3.1; $r = 0$.

5.41 **(a)** Largest positive: Audi TT Roadster. Largest negative: Lamborghini Murcielago. **(b)** The Insight is an influential point. **(c)** Large positive residual means actual highway mileage is above the prediction; a negative residual means actual is less than predicted.

5.45   (a) All three plots show a positive association. (b) 0.856, 0.911, 0.393. (c) One-week lag. (d) $\hat{y} = 49.29 + 0.07222x$.

Chapter 6

6.1   The sum is 175,229 (thousand), due to rounding errors.

6.3   (a) 858 people. (b) 43 had arthritis. (c) 71 (8.3%) played elite soccer, 215 (25.1%) played non-elite soccer, and 572 (66.7%) did not play.

6.5   14.1% of elite players have arthritis, compared to 4.2% of the other two groups.

6.7   (a) For women: 30.2%, 40.4%, 2.2%, and 27.1%. For men: 34.8%, 24.8%, 3.7%, and 36.6%. More women in administration, and more men in finance. (b) 46.5%.

6.9   (a) Alaska Airlines: 13.3%. America West: 10.9%. (b) For AA: 11.1%, 5.2%, 8.6%, 16.9%, and 14.2%. For AW: 14.4%, 7.9%, 14.5%, 28.7%, 23.3%. (c) AW has many Phoenix flights and few Seattle flights, while AA has many Seattle flights and few Phoenix flights.

6.11   Marital status: 4.1%, 93.9%, 1.5%, 0.5%. Job grade: 11.6%, 51.5%, 30.2%, 6.7%.

6.13   17.2%, 65.9%, 14.8%, 2.1%.

6.15   Age is the lurking variable: married men are generally older.

6.17   (a) Hatched: 16, 38, 75; did not hatch: 11, 18, 29. (b) Cold: 59.3%; neutral: 67.9%; hot: 72.1%. Cold water did not prevent hatching, but made it less likely.

6.19   (a) No-relapse percents are 58.3% (desipramine), 25.0% (lithium), and 16.7% (placebo). (b) Because random assignment was used, causation is indicated.

6.21   The risks of CHD are 1.7%, 2.3%, 4.3%—increasing as tendency to become angry increases.

6.23   (a) Males: 490 admitted, 210 not; females: 280 admitted, 220 not. (b) Males: 70% admitted; females: 56% admitted. (c) Business school: 80% of males, 90% of females; law school: 10% of males, 33.3% of females. (d) Most male applicants apply to the business school, where admission is easier. A majority of women apply to the law school, which is more selective.

6.25   There is some evidence of a relationship between the two variables; namely, heart attack patients seem to be less likely to rate themselves as 1 or 2 on the baldness scale and more likely to respond "3."

Part I Review

I.1   (a) 479 (chimp site) and 223 (hominid site). (b) Though not identical, the two distributions are somewhat similar.

I.3   (a) Most people will "round" their answers when asked to give an estimate like this. (b) The stemplots and statistics (mean or five-number summary) suggest that women (claim to) study more than men.

I.5   (a) The distributions that are not too skewed and have no outliers. (b) The means drop by about 6 or 7 minutes, and the standard deviations drop by about 13 and 7 minutes.

I.7   Before: $M = 25°$, $\bar{x} = 25.42°$, $s = 7.47°$. After: $M = 25°$, $\bar{x} = 24.76°$, $s = 6.34°$.

I.9   (a) $\hat{y} = 19.7 + 0.339x$. (b) 28.2°. (c) $r^2 = 9.1\%$.

I.11   (a) Centimeters. (b) Centimeters. (c) Centimeters. (d) No units.

I.13   Compute and compare the percents of each group giving each rating: for example, 5.9% of black parents rated schools as excellent.

I.15   Dolphin: 180 kg, 1600 g. Hippo: 1400 kg, 600 g.

I.17   (a) Decrease. (b) Increase.

I.19   About 800 g.

I.21   Close to -1.

I.23 -56.1 grams; prediction outside the range of the available data is risky.

I.25   (a) Fidelity Technology Fund. (b) No.

I.27   This is the mean, since half the players should make more than the median.

I.29   (a) 59%. (b) 37.5% of small businesses, 59.5% of medium-sized businesses, and 80% of large businesses did not respond.

I.31   It is now more common for these stocks to rise and fall together.

I.33   Because they feel more confident because they took the course, or because they have taken the test before.

I.35   (b) $\hat{y} = 138 - 0.0624x$. (c) $r^2 = 0.0695$

I.37   (a) The time plot does suggest the ban was effective. (b) $\hat{y} = 195.56 - 0.09893x$; $\hat{y} = -84.558 + 0.04321x$. The second line could be (cautiously) used for predictions.

I.39   A scatterplot shows a moderate, positive, linear association. The regression line $\hat{y} = 70.44 + 274.78x$.

I.41   (a) Explanatory: weeds per meter; response: corn yield. (b) The median yields are 169.45, 163.65, 157.30, and 162.60 bushels per acre.

I.43   For a simple case: $(x_1 - \bar{x}) + (x_2 - \bar{x}) + (x_3 - \bar{x}) = x_1 + x_2 + x_3 - \bar{x} - \bar{x} - \bar{x} = x_1 + x_2 + x_3 - 3\bar{x}$.

I.45   (a) As given, the data show an angular negative association. (b) After taking logarithms, the scatterplot shows a reasonably linear negative relationship.

Chapter 7

7.1   This is an observational study. Explanatory variable: cell phone usage; response variable: presence/absence of brain cancer.

7.3   For example, less parental guidance or presence, less social interaction with peers.

7.5 All U.S. households.

7.7 (a) 20,135 responses. (b) We have the opinions only of those who visit this site and feel strongly enough to respond.

7.9 Number from 01 to 28 and choose 04, 10, 17, 19, 12, 13.

7.11 Label the students 01, …, 29 and choose 4 students, then label the faculty 0, …, 9 and use the table again.

7.13 The higher no-answer was probably the second period—more families are likely to be gone for vacations, etc.

7.15 Larger samples provide better information about the population.

7.17 An observational study; explanatory variable: attitude about Congress; response variable: opinion about term limits.

7.19 Population: all college students; sample: 104 students; nonresponse rate is 58.4%.

7.21 Population: small businesses; sample: those restaurants which respond.

7.23 Those who feel most strongly are more likely to respond; this result is too high.

7.25 Population: (adult) black residents of Miami. Sample: one adult from each responding black household. Some might not want to make negative comments about the police to an officer.

7.27 The results would not be random.

7.29 21 (#3002), 37 (#3018), 18 (#2011), 44 (#3025), and 23 (#3004).

7.31 1388, 0746, 0227, 4001, 1858

7.33 (a) False. (b) True. (c) False.

7.35 A stratified random sample: Select a certain number of Class I institutions, etc.

7.37 (a) Households without telephones or with unlisted numbers. Such households would likely be made up of poor individuals, those who choose not to have phones, and those who do not wish to have their phone number published. (b) Those with unlisted numbers.

7.39 (a) Dialing a randomly chosen phone number can conceivably contact any person who has a telephone. (b) "Phone-answerers" and "non-phone-answerers" might have different characteristics; a good survey should fairly represent both groups.

7.41 (a) Population: Ontario residents; sample: the 61,239 people interviewed. (b) Such a large sample should represent both men and women fairly well.

Chapter 8

8.1 Subjects: students; factor: rate structure; treatments: one flat rate and peak/off-peak rates; response variables: amount and time of use, and the effect on network congestion.

8.3 (a) Individuals: batches of the product; response variable: yield of a batch. (b) Two factors (temperature and stirring rate) and six treatments (temperature–stirring rate combinations). (c) Twelve batches.

8.5 Place 6 rats in each group. Group 1 includes rats 02, 08, 17, 10, 05, and 09; Group 2 is 06, 16, 01, 07, 18, and 15.

8.7 If this year is somehow different (for example, warmer) from last year, we cannot compare electricity consumption. Such differences would confound the effects of the treatments.

8.9 The difference in the means between the first three Mondays and the last two Mondays was so large it would rarely occur by chance. The difference from zero was small enough that it might occur by chance.

8.11 The experimenter knows which subjects were taught to meditate; if he or she has some expectations about the effect of meditation, this could influence the diagnosis.

8.13 Completely randomized: Assign 10 students to Group 1 (software) and the other 10 to Group 2 (no software). Matched pairs: Each student does the activity twice; randomly decide (for each student) whether he or she has the software the first or the second time.

8.15 For each block (pair of lecture sections), randomly assign one section to be taught using standard methods and one to be taught with multimedia. Then (at the end of the term) compare final-exam scores and students' attitudes.

8.17 (a) 52% of the control group began using condoms after a single one-hour session. (c) 1887, 2099, 3547, 0426, 3543.

8.19 Assign 6 students to each treatment. The five groups are 05, 16, 17, 20, 19, 32; 04, 25, 29, 31, 18, 07; 13, 33, 02, 36, 23, 27; 35, 21, 26, 08, 10, 11; 15, 12, 14, 09, 24, 22.

8.21 Label the men 01,...,12, and use Table B to choose 2 men (Zhang and Edwards) for treatment 1, then 2 more (Chao and Ogle) for treatment 2, etc. When all men are assigned, label the women 01,..., 24, and assign 4 (O'Brian, Trujillo, Vaughn, Denman) to treatment 1, etc.

8.23 (a) Subjects: physicians; factor: medication (aspirin or placebo); response variable: health (heart attacks or not).

8.25 (a) Explanatory: treatment method; response: survival times. (b) Women (or their doctors) chose which treatment to use. (c) Doctors may recommend a treatment based in part on how advanced the case is.

8.27 (b) Practically, it may take a long time. Ethically, can we assign subjects to different insurance plans?

8.29 Use a completely randomized design. From Table B, we assign 20, 11, 38, 31, 07, 24, 17, 09, 06; 36, 15, 23, 34, 16, 19, 18, 33, 39; 08, 30, 27, 12, 04, 35; 02, 32, 25, 14, 29, 03, 22, 26, 10; the rest drink vodka and lemonade.

8.31 (a) Randomly assign three circles to get extra carbon dioxide; the others are a control group. (b) Place pairs of circles close together, and randomly choose one of each pair for treatment.

8.33 (a) The response variable should be "number of accidents," or something similar. Designs will vary. (b) The effect of running lights may be lessened when (if) they become common enough that people no longer notice them.

8.35 (a) Individuals: chicks; response variable: weight gain. (b) Two factors (corn type and % protein); nine treatments. The diagram should be a 3 × 3 table. 90 chicks are needed. (c) Note: This diagram is quite large.

8.37 Each subject should taste both kinds of cheeseburger, in a randomly selected order, and then be asked about preference.

8.39 (a) See the definitions in this chapter.

8.45 (a) The two extra patients can be randomly assigned to two of the three groups. (b) No one involved in administering the treatments or assessing their effectiveness knew which subjects were in which group. (c) The pain scores in Group A were so much lower than the scores in Group C that such a difference would not often happen by chance. However, the difference between A and B could be due to chance.

Chapter 9

9.1 Of a large number of poker hands, about 0.4% will contain a straight.

9.7 (a) 0. (b) 1. (c) 0.01. (d) 0.6 (or 0.99).

9.9 (a) 16 possible outcomes (all possible sequences of hits and misses). (b) {0,1, 2, 3, 4}.

9.11 Soft 4: 2/36. Any 4: 3/36.

9.13 0.058.

9.15 (a) 0.155. (b) 0.609. (c) 0.660.

9.17 (a) 0.4. (b) 0.6. (c) 0.2.

9.19 (a) The density curve is a triangle. (b) 0.5. (c) 0.125.

9.21 The possible values are 2, 3, 4, ..., 12, with probabilities 1/36, 2/36, 3/36, 4/36, 5/36, 6/36, 5/36, 4/36, 3/36, 2/36, and 1/36.

9.23 (b) A personal probability might take into account specific information about your driving habits. (c) Most people believe that they are better-than-average drivers.

9.25 {0, 1, 2, ..., }.

9.27 0.253, 0.747.

9.29 (a) 0.235. (b) 0.366.

9.31 1/6.

9.33 (a) {(A,D), (A,M), (A,S), (A,R), (D,M), (D,S), (D,R), (M,S), (M,R), (S,R)}. (b) 0.1. (c) 0.4. (d) 0.3.

9.35 (a) Each arrangement has probability 1/8. (b) 0.375. (c) X can be 0, 1, 2, 3 with probabilities 0.125, 0.375, 0.375, 0.125.

9.37   Rented housing typically has fewer rooms.

9.39 Y can be $1, 2, 3, \ldots, 12$, each with probability $1/12$.

9.41   (a) 0.438 to 0.502. (b) 0.1056.

9.43   (a) $P(Y > 300) = 0.5$. (b) $P(Y > 370) = 0.025$.

9.45   (a) 1/10,000. (b) 24/10,000.

9.47   Exact match: $0.50; any order: $0.48.

9.49   Results will vary.

Chapter 10

10.1   Both are statistics.

10.3   Parameter, statistic.

10.5   If one of the 12 homes were lost, it would cost more than the collected premiums. For many policies, the average claim should be close to $250.

10.7   (a) 5.7735 mg. (b) $n = 4$. The average of several measurements is more likely to be close to the mean.

10.9   (a) 0.5, 0.16. (b) 0.5, 0.025.

10.11   (a) 0.3336. (b) 0.0013. (c) The answer to (b) is more accurate.

10.13   Center line: 75°; control limits 74.25° and 75.75°.

10.15   73.5° to 76.5°; 74° to 76°.

10.17   Statistic, parameter.

10.19   In the long run, the gambler loses an average of about 5.3 cents per bet.

10.21   0.0125.

10.23   40.125 mm and 0.001 mm.

10.25   About 133.2 mg/dl.

10.27   (a) Approximately $N(2.2, 0.1941)$. (b) 0.1515. (c) 0.0764.

10.29   0.2148, 0.1190.

10.31   (b) 141.847 days. (c) Means will vary. (d) It would be unlikely (though not impossible) for all five $\bar{x}$-values to fall on the same side of μ. (e) The mean should be μ.

10.33   4.22; 4.05 to 4.39.

10.35   3.84 to 4.60.

10.37   (a) 99.8%. (b) 95.7%.

Chapter 11

11.1   (b) 0.0125. (c) 0.0175.

11.3   Independence is not a reasonable assumption.

11.5 0.6676.

11.7 **(b)** 0.3. **(c)** 0.4.

11.9 **(a)** 922/1654. **(b)** 32/72.

11.11 **(a)** $P(B \mid A) = 0.0799$. **(b)** $P(A \mid B) = 0.48$.

11.13 0.32.

11.15 0.086.

11.17 0.015925.

11.19 0.6341.

11.21 0.5404.

11.23 **(a)** 0.4426. **(b)** 0.6899. **(c)** 0.3053.

11.25 1/4.

11.27 **(a)** 0.28. **(b)** 0.2857.

11.29 **(a)** 0.1; 0.9. **(b)** 9999 switches remain; 999 are bad; $999/9999 \doteq 0.09991$. **(c)** 9999 switches remain; 1000 are bad; $1000/9999 \doteq 0.10001$.

11.31 **(a)** 0.25. **(b)** 0.3333.

11.33 **(a)** To find $P(A \text{ or } C)$, we would need to know $P(A \text{ and } C)$. **(b)** To find $P(A \text{ and } C)$, we would need to know $P(A \text{ or } C)$.

11.35 58%.

11.37 About 62.1%.

11.39 **(a)** 0.6739. **(b)** 0.7870. **(c)** Not independent; the answers to (a) and (b) are different.

11.41 0.56.

11.43 **(a)** 0.044. **(b)** 0.088.

11.45 0.5263.

11.47 **(a)** Either B or O. **(b)** $P(B) = 0.75$, $P(O) = 0.25$.

11.49 **(a)** 0.25. **(b)** 0.015625. **(c)** 0.140625.

Chapter 12

12.1 Yes.

12.3 No, since p does not remain fixed.

12.5 **(a)** 0, 1,..., 5. **(b)** 0.2373, 0.3955, 0.2637, 0.0879, 0.0146, 0.0010.

12.7 0.4831.

12.9 **(a)** 3 calls. **(b)** 1.5492 calls. **(c)** $p = 0.08$: 1.0507 calls; $p = 0.01$: 0.3854 calls. As p gets close to 0, σ decreases.

12.11 $P(X \le 29) \doteq P(Z \le -2.94) = 0.0016$.

12.13 (a) $\mu = 180$ and $\sigma = 12.5857$ blacks. (b) $P(X \leq 170) \doteq P(Z \leq -0.79)$ $= 0.2148$.

12.15 (a) Yes: It is reasonable that each student's results are independent, and each has the same chance of passing. (b) No: Her probability of success is likely to increase. (c) No: Temperature may affect the outcome of the test.

12.17 0.4275.

12.19 (a) $n = 5$, $p = 0.65$. (b) 0, 1, ..., 5. (c) 0.00525, 0.04877, 0.18115, 0.33642, 0.31239, 0.11603. (d) $\mu = 3.25$ and $\sigma \doteq 1.0665$ years.

12.21 (a) 6 positive tests. (b) Binomial with $n = 1000$ and $p = 0.006$. (c) Because $np = 6$.

12.23 (a) There are 150 independent observations, each with response probability $p = 0.5$. (b) $\mu = 75$ responses. (c) 0.2061. (d) $n = 200$.

12.25 (a) 3.75 leaking tanks. (b) 0.0008. (c) 0.0336.

12.27 (a) 3250 dropouts. (b) $P(X \geq 3500) \doteq P(Z > 4.70) \doteq 0$.

Chapter 13

13.1 (a) 48% to 54%. (b) The method used gives correct results 95% of the time.

13.3 (a) 1.8974. (c) $m \doteq 3.8$. (d) Both intervals should be 7.6 units wide. (e) 95%.

13.5 $3.41367 \pm 0.00113 = 3.4125$ to 3.4148 grams.

13.7 (a) 0.8354 to 0.8454 g/liter. (b) 0.8275 to 0.8533 g/liter. (c) Increasing confidence makes the interval longer.

13.9 (a) 18.9 to 25.1 points. (b) 15.8 to 28.2 points. (c) 20.45 to 23.55 points. (d) 6.2, 3.1, and 1.55, respectively. Margin of error decreases with larger samples.

13.11 $n = 60$.

13.13 (b) -4.526% to -2.648%.

13.15 29,737 to 31,945 pounds.

13.17 (b) 223.973 to 224.031 mm.

13.19 (b) 22.7 to 26.8 degrees.

13.21 $n = 174$.

13.23 Larger.

13.25 Less than $\pm 3\%$.

13.27 Equal to $\pm 3\%$.

13.29 Other surveys should be close to the truth—not necessarily close to the results of this survey.

13.31 (a) Most answers: At least 40 of the 50 and between 87% and 93% of the 1000. (b) Most answers: At least 44 of the 50 and between 93% and 97% of the 1000. (c) Most answers: At least 47 of the 50 and at least 98% of the 1000.

13.33 $z^* = 1.78$; 119.2 to 128.4 bushels/acre.

Chapter 14

14.1 (a) $N(12 \text{ g/dl}, 0.2263 \text{ g/dl})$. (b) 11.3 lies far from the middle of the curve and is therefore unlikely if in fact $\mu = 12$.

14.3 $H_0: \mu = 12$; $H_a: \mu < 12$.

14.5 $H_0: \mu = 26$; $H_a: \mu > 26$.

14.7 0.95.

14.9 0.60; 1.80.

14.11 0.0010; 0.1894.

14.13 $z = 2.45$, $P = 0.0142$.

14.15 Significant at $\alpha = 0.05$ but not at $\alpha = 0.01$.

14.17 $z = -2.77$, $P = 0.0028$.

14.19 $H_0: \mu = 224 \text{ mm}$; $H_a: \mu \neq 224 \text{ mm}$; $z \doteq 0.13$, $P = 0.8966$.

14.21 $z > 2.576$.

14.23 $H_0: \mu = 0\%$; $H_a: \mu < 0\%$; $z = -9.84$, so $P \doteq 0$.

14.25 (a) No. (b) Yes.

14.27 $H_0: \mu = 100$; $H_a: \mu \neq 100$; $z = 2.17$, $P = 0.0300$.

14.29 $z = 0.35$, $P = 0.7264$.

14.31 Between 0.02 and 0.04.

14.33 The observed response was fairly substantial and would almost never occur purely by chance.

14.35 (a) $P = 0.0359$. (b) $P = 0.9641$. (c) $P = 0.0718$.

14.37 (a) Yes. (b) No.

14.39 Such a drop has only a 2% chance of happening by accident.

14.41 The differences observed would occur less than 1% of the time if pig skulls had no special meaning.

14.43 $P < 1$ always; it should be $P = 0.0918$.

14.45 No; this statement means that *if* H_0 is true, we have observed outcomes that occur about 3% of the time.

14.47 (a) Wider. (b) $16: no. $15: yes.

14.51 ± 2.54.

Chapter 15

15.1 (a) Yes. (b) No: The numbers are based on voluntary response rather than an SRS.

15.3 (a) Not included. (b) Not included. (c) Included.

15.5 (a) $z = 1.64$ is not significant. (b) $z = 1.65$ is significant.

15.7 $n = 100$: 452.24 to 503.76. $n = 1000$: 469.85 to 486.15. $n = 10,000$: 475.42 to 480.58.

15.9 (a) No: We expect to see about 5 people who have $P < 0.01$. (b) Test these four again.

15.11 (a) 0.2033. (b) 0.9927. (c) Higher, since 290 is farther from 300.

15.13 (a) H_0: Patient is ill; H_a: Patient is healthy. Type I error: clearing a patient who is ill. Type II error: sending a healthy patient to the doctor.

15.15 (a) 0.50. (b) 0.1841. (c) 0.0013.

15.17 B, A, C.

15.19 Question (b).

15.21 This is not information taken from an SRS.

15.23 When many variables are examined, "significant" results will show up by chance.

15.25 (a) About 11.3%. (b) The organizations that did not respond are (obviously) not represented in the results.

15.27 Because of the large sample size, this significant difference may not indicate a strong preference.

15.29 (a) 0.05. (b) We expect about 3 or 4 significant tests by chance.

15.31 (a) Reject H_0 if $\bar{x} \leq 0.84989$ or $\bar{x} \geq 0.87011$. (b) 0.8944. (c) 0.1056.

15.33 P(Type I error) $= 0.05$; P(Type II error) $= 0.0073$.

15.35 A low-power test will often accept H_0 when it is false simply because it is difficult to distinguish between H_0 and nearby alternatives.

Part II Review

II.1 All words in Tom Clancy's novels; the 250 words recorded; number of letters in a word.

II.3 (a) {F, M}. (b) {6,7,...,20}. (c) All numbers between 2.5 and 6 l/min. (d) Upper limits will vary.

II.5 Both are statistics.

II.7 Cut in half: $n = 56$. ±5 mg/dl: $n = 182$.

II.9 $z = -2.92$ and $P = 0.0018$; small differences can be significant with large sample sizes.

II.11   80%: 334.35 to 379.65 ng/g; 95%: 322.35 to 391.65 ng/g; margin of error grows with increasing confidence.

II.13   84.8 to 90.4.

II.15   **(a)** The treatment was not assigned. **(b)** Socioeconomic and hereditary factors may also affect IQ.

II.17 $P = 0.04$ means that such results would rarely occur by chance; $P = 0.72$ means the difference in median survival time could easily happen by chance.

II.19   Recommend either 499 or 665 μg.

II.21   **(a)** Increase. **(b)** Decrease. **(c)** Decrease. **(d)** Decrease.

II.23   **(a)** Label from 0001 to 3478. **(b)** 2940, 0769, 1481, 2975, 1315.

II.25   **(a)** 0.3707. **(b)** Mean 100, standard deviation 1.93649. **(c)** 0.0049. **(d)** The answer to (a) could be different; (b) would be the same; (c) would be fairly reliable because of the central limit theorem.

II.27   0.9389.

II.29   **(a)** 1%. **(b)** All probabilities are between 0 and 1, and they add to 1. **(c)** 0.94. **(d)** 0.86. **(e)** Either $X \geq 4$ or $X > 3$; 0.06.

II.31   The increase in fires over time would occur less than 1% of the time by chance. This straight-line relationship explains 61% of the variation in the number of fires.

II.33   **(a)** It is a convenience sample. **(b)** This use of "significant" means "quite a bit." **(c)** Results like those observed would occur less than 1% of the time if people did not truly want more money.

II.35   Placebos can provide genuine pain relief.

II.37 H_0: $p = 18/38$; H_a: $p \neq 18/48$.

II.39 $P(A) = P(B) = \cdots = P(F) = 0.12$ and $P(1) = P(2) = \cdots = P(8) = 0.035$.

II.41   **(a)** Yes. **(b)** No. **(c)** No.

II.43   Cut in half: $n = 400$. Individual scores will be as variable (or more).

II.45   \$562.50 (probability 0.25), \$975 (0.5), and \$1690 (0.25).

II.47   0.036.

II.49   **(a)** Her brother has type aa, and he got one allele from each parent. **(b)** $P(aa) = 0.25$, $P(Aa) = 0.5$, $P(AA) = 0.25$.

II.51   **(a)** $P(A) = 0.10$, $P(C) = 0.20$, and $P(A \mid C) = 0.05$. **(b)** $P(A$ and $C) = P(C)P(A \mid C) = 0.01$.

II.53   **(a)** 0.2146. **(b)** 0.2649.

II.55   0.2586.

Chapter 16

16.1   1.7898 mm.

16.3   (a) 2.015. (b) 2.518.

16.5   (a) 2.262. (b) 2.861. (c) 1.440.

16.7   (a) The distribution is slightly, but not unreasonably, left-skewed, with no outliers. (b) 54.78% to 64.40%.

16.9   (a) 24. (b) Between 1.059 ($p = 0.15$) and 1.318 ($p = 0.10$). (c) $0.20 < P < 0.30$. (d) No; no.

16.11   (a) For each subject, randomly select which knob should be used first. (b) μ is the mean of (right-hand-thread time minus left-hand-thread time); $H_0: \mu = 0$ sec; $H_a: \mu < 0$ sec. (c) $t = -2.90$; $0.0025 < P < 0.005$. Right-hand- thread times are less.

16.13 -21.2 to -5.5 sec. $\bar{x}_{RH}/\bar{x}_{LH} = 88.7\%$.

16.15   (a) 38.19 to 58.30 thousand barrels. (b) The width is similar, but the new interval is higher (by 2000 barrels).

16.17   (a) 9. (b) $0.025 < P < 0.05$.

16.19   The distribution is right-skewed, but t procedures are robust for large samples.

16.21   (a) The distribution is slightly skewed but has no apparent outliers. The t procedures are acceptable. (b) $H_0: \mu = 224$ mm; $H_a: \mu \neq 224$ mm; $t \doteq 0.13$; $P > 0.50$.

16.23   (b) $18.73 to $23.07.

16.25   About $824 million to $1.02 billion.

16.27   (a) 22.6 to 26.9 degrees. (b) This interval is narrower.

16.29   2.624.

16.31   (b) $\bar{x} = 15.59$ ft and $s = 2.550$ ft; 15.0 to 16.2 ft. (c) What population are we examining: Full-grown sharks? Male sharks?

16.33   (a) A subject's responses to the two treatments would not be independent. (b) Yes ($t = -4.41$, $P = 0.0069$).

16.35   (a) Work from the set of differences for the given data: $\bar{x} \doteq -5.71$ μm/hr, $s \doteq 10.56$ μm/hr, and $s/\sqrt{14} \doteq 2.82$ μm/hr. (b) $H_0: \mu = 0$; $H_a: \mu < 0$; $t \doteq -2.02$; $0.025 < P < 0.05$; significant at 5% but not at 1%. Altering the electric field appears to reduce the healing rate.

16.37   (a) No. (b) No ($t = 0.15$, $P = 0.88$).

Chapter 17

17.1   Matched pairs.

17.3   Single sample.

17.5 (a) IDX: $n = 10$, $\bar{x} = 116$, $s = 17.71$. Untreated: $n = 10$, $\bar{x} = 88.5$, $s = 6.01$. (b) 9.

17.7 (a) Yes ($t = 3.81$, df $= 8$, $P = 0.0026$). (b) Yes; the design seems to be sound.

17.9 With df $= 8$: 2.08% to 8.45%.

17.11 (a) Breast-feeding women and other women. (c) $H_0: \mu_1 = \mu_2$; $H_a: \mu_1 < \mu_2$; $t = 8.50$, $P < 0.001$.

17.15 $t \doteq 2.99$, $P = 0.0246$; significant at 5% level.

17.17 (a) 3.68. (b) Not significant at either level.

17.19 $H_0: \sigma_1 = \sigma_2$; $H_a: \sigma_1 \neq \sigma_2$; $F \doteq 8.68$, $0.002 < P < 0.02$. This is significant evidence that $\sigma_1 \neq \sigma_2$.

17.21 No ($F \doteq 1.63$, $P \doteq 0.45$).

17.23 (a) Very large sample sizes. (b) $-\$28,012$ to $-\$25,477$. (c) $\bar{x}$ and s are similar in size; t procedures are robust (especially for large samples).

17.25 With df $= 8$: 4.3 to 49.7 g; significant at 10%.

17.27 $t \doteq 0.255$—not even close to significant.

17.29 (a) A t procedure may be (cautiously) used, in spite of skewness and outliers, since the sum of the sample sizes is almost 40. (b) $t \doteq 2.06$, $0.025 < P < 0.05$; significant at the 5% level.

17.31 -36.57 to -3.05.

17.33 $t = -3.74$ (with df $= 5$ or df $= 10.9$).

17.35 (a) Yes ($t \doteq -4.65$, $P < 0.001$). (b) 14.13 to 40.87 days.

17.37 (a) $H_0: \mu_A = \mu_P$; $H_a: \mu_A > \mu_P$; $t \doteq 4.28$, $P < 0.0005$. Active learning results in more correct identifications. (b) We assume (but cannot check) that we have an SRS from the population of learning-impaired children. We also assume (near) Normality, although the active score show a high outlier and some skewness. Reasonably large (and equal) sample sizes make t procedures fairly reliable anyway.

17.39 (a) 3.9 to 9.2 Blissymbols. (b) 22.2 to 26.6 Blissymbols.

17.41 (a) Use a matched pairs test. (b) $t = 10.16$, $P < 0.0005$. Coached students do improve their scores. (c) About 21.5 to 36.5 points.

17.43 This was an observational study, not an experiment.

17.45 (a) $H_0: \mu = 0$; $H_a: \mu > 0$; $t \doteq 3.20$, $0.01 < P < 0.02$. This is strong evidence that low-rate exercise raises heart rate; the confidence interval for the increase is 2.60 to 13.00 bpm. (b) With the same hypotheses, $t \doteq 10.63$ and $P < 0.0005$. This is stronger evidence that medium-rate exercise raises heart rate; the confidence interval is 14.87 to 22.33 bpm. (c) $H_0: \mu_1 = \mu_2$; $H_a: \mu_1 < \mu_2$; $t = -3.60$, $0.01 < P < 0.02$. Medium-rate exercise has a greater effect than low-rate exercise.

Chapter 18

18.1   (a) p is the proportion of the population (all college students) who say they pray at least once in a while. (b) $\hat{p} = 0.8425$.

18.3   (a) Mean 0.5, standard deviation 0.0041. (b) 0.9854.

18.5   0.4714, 0.7924, and 0.9886; larger sample sizes give more accurate estimates.

18.7   This is not an SRS.

18.9   0.1194 to 0.1508.

18.11   (a) We have a large SRS from a much larger population. (b) 0.632 to 0.689. (c) Our margin of error for 95% confidence was (slightly less than) 3%.

18.13   (a) The count of successes is too small. (b) Sample size 2677, 7 successes, $\tilde{p} \doteq 0.0026$. (c) 0.00068 to 0.00455.

18.15   (a) 0.7480 to 0.9161. (b) We need to know how they were chosen. (All from the same school? Public or private? Etc.)

18.17 $n = 318$.

18.19 $z = 10.38$, P is tiny.

18.21   A large SRS from a large population, with counts 606 and 712; 0.4244 to 0.4952 (plus four: 0.4246 to 0.4952).

18.23 $z = -2.74$, $P = 0.0031$.

18.25   0.6601 to 0.6799 (plus four: 0.6600 to 0.6798).

18.27   (a) The failure count (5) is too small. (b) 0.6020 to 0.8795.

18.29   (a) The variability will be the same for all states. (b) There would be less variability for states with larger samples.

18.31   0.3901 to 0.4503 (plus four: 0.3903 to 0.4504).

18.33 $n = 16,590$. Our confidence interval shows that p is in the range 0.3 to 0.7, so that this conservative approach will not greatly inflate the sample size.

18.35   (a) $H_0: p = 0.5$; $H_a: p > 0.5$; $z \doteq 1.70$, $P = 0.0446$. (b) 0.5071 to 0.7329 (plus four: 0.5020 to 0.7202). (c) The coffee should be presented in random order.

18.37   (a) 0.0588, 0.0784, 0.0898, 0.0960, 0.0980, 0.0960, 0.0898, 0.0784, 0.0588. (b) 0.0263, 0.0351, 0.0402, 0.0429, 0.0438, 0.0429, 0.0402, 0.0351, 0.0263. The new margins of error are less than half their former size.

18.39   The 95% confidence interval would be 0.6437 to 0.8428 (plus four: 0.6323 to 0.8292).

Chapter 19

19.1   (a) $\hat{p}_w = 0.4821$, $\hat{p}_m = 0.7302$. (b) 0.0773 to 0.4187.

19.3   (a) One count is only 7. (b) 8 out of 137, and 28 out of 143. (c) 0.0614 to 0.2134.

19.5 $H_0: p_1 = p_2$; $H_a: p_1 < p_2$; $z = -3.53$, $P = 0.0002$.

19.7 $H_0: p_1 = p_2$; $H_a: p_1 > p_2$; $z \doteq 0.98$, $P = 0.1635$; we cannot conclude that death rates are different.

19.9 $H_0: p_1 = p_2$; $H_a: p_1 \neq p_2$; $z = 3.39$, $P = 0.0007$.

19.11 0.0631 to 0.2179 (plus four: 0.0614 to 0.2158).

19.13 $H_0: p_1 = p_2$; $H_a: p_1 \neq p_2$; $z = -3.06$, $P = 0.0022$.

19.15 $H_0: p_1 = p_2$; $H_a: p_1 \neq p_2$; $z = -0.86$, $P = 0.3914$. 95% confidence interval: -0.1018 to 0.0398 (plus four: -0.1044 to 0.0409).

19.17 **(a)** The samples should be randomly chosen from a variety of schools. **(b)** 0.0135 to 0.0270 (plus four: 0.0145 to 0.0283). **(c)** $z \doteq 0.54$ and $P = 0.5892$—no evidence of a difference.

19.19 **(a)** About 7672 in Europe and 397 in the United States. **(b)** 0.1356 to 0.2043 (plus four: 0.1356 to 0.2042).

19.21 $z = -0.39$ and $P = 0.6966$.

19.23 **(a)** 0.1626 to 0.2398 (plus four: 0.1623 to 0.2395). **(b)** The confidence interval does not even come close to 0, so the P-value against the two-sided alternative will be (much) smaller than 0.01. **(c)** The counts are 158 (January to April) and 313 (July and August). $H_0: p_1 = p_2$; $H_a: p_1 \neq p_2$; $z \doteq -4.39$, $P < 0.0004$; very strong evidence that other nonresponse rates also differ between the seasons.

19.25 **(a)** $\hat{p}_1 \doteq 0.1415$, $\hat{p}_2 \doteq 0.1667$; $z \doteq -0.39$; $P = 0.6966$. **(b)** $z \doteq 2.12$ and $P = 0.0340$. **(c)** For (a): -0.1559 to 0.1056 (plus four: -0.1659 to 0.0985). For (b): -0.04904 to -0.001278 (plus four: -0.0493 to -0.0016). Larger samples make the margin of error smaller.

19.27 **(a)** $z \doteq -1.95$; $P = 0.0512$. **(b)** $z \doteq -2.28$; $P = 0.0226$. **(c)** 95% confidence interval: -0.2390 to -0.0220 (plus four: -0.2352 to -0.0196).

Part III Review

III.1 **(a)** Two-sample z for proportions. **(b)** Two-sample t for means.

III.3 **(a)** One-sample z for a proportion. **(b)** Two-sample z for proportions.

III.5 **(a)** Label from 01 to 44, and choose 22 to eat regular chips first; 19, 22, 39, 34, 05. **(b)** Matched pairs (one-sample t) for means.

III.7 **(a)** $t \doteq 2.21$, $P \doteq 0.0145$. **(b)** t procedures are robust with large samples.

III.9 -0.7944 to -0.4056 years. We reject $H_0: \mu_1 = \mu_2$ at the 1% level

III.11 No: $t \doteq -0.8658$; P is close to 0.4.

III.13 1998: 85.3%. 2001: 76.9%. Yes: $z \doteq -5.23$, $P < 0.0001$.

III.15 No: $t \doteq -0.395$, $P > 0.50$.

III.17 0.4985 to 0.6215 (plus four: 0.4980 to 0.6201).

III.19 $z \doteq 1.175$, $P > 0.10$; we cannot conclude that more than half of Americans hold this opinion.

III.21   Yes: $t \doteq 10.4$, $P < 0.0001$.

III.23   Yes: $t \doteq -3.50$, $P < 0.0005$.

III.25   Both distributions are reasonably symmetrical with no extreme outliers; the nitrite group may be slightly left-skewed. $t \doteq -0.89$, $P > 0.15$.

III.27   (a) $t \doteq -1.20$, $P > 0.20$, no significant difference in alcohol consumption. (b) 0.207 to 0.573 g.

III.29   Yes: $z \doteq 6.79$, P is tiny.

III.31   The distribution is fairly symmetric, with a slightly low outlier of 4.88. $\bar{x} \doteq 5.4479$ is our best estimate; the margin of error $t^* s / \sqrt{29}$ depends on the confidence level.

III.33   (a) Standard error of the mean $= s / \sqrt{n}$. (b) Two-sample t tests. (c) The observed differences between the two groups were so large that they would rarely occur if the two groups were not different. Smaller P is stronger evidence.

III.35   (a) $H_0: \sigma_1 = \sigma_2$; $H_a: \sigma_1 > \sigma_2$; $F = 15.2$, P is tiny. (b) No: F procedures depend on Normality, and large sample sizes do not help.

III.37   (a) $t^* \doteq 2.423$. (b) Reject H_0 if $t \geq t^*$, that is, when $\bar{x} \geq 37$. (c) The power against $\mu = 100$ is $P(Z \geq -4.12) > 0.9998$. A sample of size 50 should be quite adequate.

III.39   With all observations, a 95% confidence interval is 46.82 to 61.31 g. If we discard the outlier, the interval is 46.12 to 57.26 g.

III.41 $\hat{y} = -2.580 + 1.0935x$ explains 95.1% of the variation in the data. When $x = 88$ g, $\hat{y} \doteq 93.65$ g.

Chapter 20

20.1   (a) 29.9%, 28.9%, 21.9%, 19.3%. (b) 26.6%, 30.9%, 19.4%, 23.0%; 39.8%, 21.1%, 27.1%, 12.0%. (c) There is no substantial difference between the first two groups, but university-educated men seem to be more likely to be nonsmokers or moderate smokers.

20.3   (a) 42.31, 36.22, 30.14, 24.34. (b) Yes and no: the heavy-smoker count is less than expected, but the moderate-smoker count is higher.

20.5   (b) $X^2 = 13.305$, $P = 0.038$. (c) University-educated men contribute three of the four largest terms.

20.7   (a) $(3 - 1)(4 - 1) = 6$. (b) $12.59 < X^2 < 14.45$, so $0.025 < P < 0.05$.

20.9   French smoking study: All expected cell counts are well over 5. NCSU study: One (16.7%) expected cell count is below 5.

20.11   (a) $H_0: p_1 = p_2$; $H_a: p_1 \neq p_2$; $z \doteq -0.57$, $P = 0.5686$. (b) Improved: 28, 30; no improvement: 54, 48. $X^2 = 0.322 \doteq z^2$, $P > 0.25$. (c) Gastric freezing is not significantly more (or less) effective than a placebo treatment.

20.13   (a) 24.2%, 41.8%, 22.0%, 12.1%. Fewer A's, more D/F's. (b) 29.12, 37.31, 18.20, 6.37. (c) $X^2 = 5.297$, df $= 3$, $P = 0.1513$.

20.15 $X^2 = 9.228$, df $= 6$, $P = 0.1610$.

20.17 $X^2 = 3.955$, df $= 4$, $P = 0.413$.

20.19   Yes: $X^2 = 135.592$, df $= 2$, P is tiny. The biggest contributions come from the female assistant and full professor cells.

20.21   (a) 22.5%, 18.6%, 13.9%. A student's likelihood of smoking increases with the number of smoking parents. (b) Parents' smoking habits do not affect their children. (c) For example, $(1780)(1004)/5375 \doteq 332.49$. (d) The biggest differences are in the first and last rows. (e) $X^2 = 37.566$, P is tiny.

20.23   Black students are most aware of Ebonics; $X^2 = 18.626$, df $= 2$, P is tiny.

20.25 $X^2 = 22.426$, df $= 8$, $P = 0.004$. Blacks are less likely, and Hispanics more likely, to consider schools excellent, while Hispanics and whites differ in percent considering schools good (whites are higher) and percent who "don't know" (Hispanics are higher).

20.27   (a) Label 01 to 77, and choose random digits. (b) $X^2 = 0.568$, df $= 3$, $P = 0.904$.

20.29   (a) df $= 9$. (b) The entries in the two-way table are 1067, 491; 833, 756; 901, 1174; 1583, 1055. $X^2 = 256.8$, and P is very small. The highest response rates occur from September to mid-April; the lowest occur in the summer months, when more people are likely to be on vacation.

20.31   (a) Use a 2×2 table formed by the four entries on the "Total" line. $X^2 = 332.205$; P is tiny, so the evidence is very strong. Possibly many sacrificed themselves out of a sense of chivalry. (b) $X^2 = 103.767$—a very significant result. The probability of dying decreased with increasing social status. (c) $X^2 = 34.621$—another very significant result; again, the probability of dying decreased with increasing social status.

20.33   (a) H_0: $p_1 = p_2$; $\hat{p}_1 \doteq 0.8423$, $\hat{p}_2 \doteq 0.6881$; $z \doteq 3.92$; $P < 0.0004$. (b) $X^2 = 15.334 = z^2$. Table E gives $P < 0.0005$. (c) 0.0774 to 0.2311.

Chapter 21

21.1   (a) $r = 0.994$, $\hat{y} = -3.660 + 1.1969x$. (b) β (estimated by 1.1969) represents how much we can expect the humerus length to increase with a 1-cm increase in femur length. The estimate of α is -3.660. (c) The residuals are -0.8226, -0.3668, 3.0425, -0.9420, and -0.9110. $s \doteq 1.982$.

21.3   (a) $r^2 = 11.2\%$. (b) $\hat{y} = -2057 + 1.97x$, $s = 104.0$ km^3 of water.

21.5   0.79 to 3.14 km^3/year; this interval does not contain 0.

21.7   (a) $t = 15.9374$. (b) df $= 3$, $P < 0.0005$.

21.9 $t = -0.63$, df $= 13$, $P > 0.50$. The scatterplot shows a strong curved relationship.

21.11 $r = -0.649$, $0.001 < P < 0.002$.

21.13   28.049 to 30.555 cm.

21.15   (a) One residual may be a high outlier, but the stemplot does not show any other deviations from Normality. (b) The scatterplot shows no striking features (other than the outlier).

21.17   (a) $\hat{y} = 0.1205 + 0.0086x$. The slope is positive. (b) 0.1886.

21.19 $0.6643 < r < 0.7114$, so $0.001 < P < 0.0025$. Excel's P-value is for a two-sided test.

21.21   (a) 115.0 to 434.6 (ft/sec)/inch. (b) 176.3 to 239.3 ft/sec.

21.23   With all points, $t = 1.90$ and $P = 0.065$. With the outlier omitted, $t = 2.93$ and $P = 0.006$. Perhaps the researchers omitted the outlier, or they used a one-sided alternative (in which case the first P-value is 0.033).

21.25 $733 to $1052 per year.

21.27   (a) They come from 13 unrelated colonies. (b) No obvious pattern. (c) Possibly wider in the middle but not markedly so. (d) No clear lack of Normality.

21.29   (a) $\hat{y} = 3.37 + 0.845x$. (b) 0.654 to 1.036; VIG increases between 0.65% and 1.04% for every 1% increase in the EAFE index.

21.31   No major violations.

21.33   (a) $t = 22.49$ with the Insight, and 15.56 without; no important impact, since both are quite significant. (b) 0.777 to 0.935 with the Insight, 0.946 to 1.240 without. This is a fairly substantial change.

21.35   (a) The plot shows a fairly strong curved pattern (weight increases with length). Two fish stray from the curve but would not particularly be considered outliers. (b) Weight should be roughly proportional to volume; when all dimensions change by a factor of x, the volume increases by a factor of x^3. (c) This plot shows a strong, positive, linear association, with no particular outliers. (d) The correlations reflect the increased linearity of the second plot: With weight, $r^2 = 0.9207$; with weight$^{1/3}$, $r^2 = 0.9851$. (e) $\hat{y} = -0.3283 + 0.2330x$; $\hat{y} = 5.9623$ when $x = 27$ cm; 5.886 to 6.039 g$^{1/3}$. (f) The stemplot shows no gross violations of the assumptions, except for the high outlier for fish no. 143. The scatterplot suggests that variability in weight may be greater for larger lengths. Dropping fish no. 143 changes the regression line only slightly and seems to alleviate both these problems (to some degree, at least).

21.37 $P = 0.3320$; we cannot conclude that the intercept is not 0.

Chapter 22

22.1   (a) Randomly allocate 36 flies to each of four groups, which receive varying dosages of caffeine; observe length of rest periods. (b) H_0: All groups have the same mean rest period; H_a: At least one group has a different mean rest period. We conclude that caffeine reduces the length of the rest period.

22.3   (a) The stemplots show no extreme outliers or strong skewness (given small sample sizes). (b) The means suggest that logging reduces the number of trees per plot, and that recovery is slow. (c) $F = 11.43$, $P = 0.000205$; H_0: $\mu_1 = \mu_2 = \mu_3$; H_a: not all means are the same.

22.5 (a) $I = 3$ (number of populations); $n_1 = n_2 = 12$, $n_3 = 9$ (sample sizes from each population); $N = 33$ (total sample size). (b) $I - 1 = 2$ and $N - I = 30$. (c) Since $F > 9.22$, $P < 0.001$.

22.7 (a) 3 and 455. (b) $0.010 < P < 0.025$. (c) $P > 0.100$. (d) $P < 0.001$.

22.9 The standard deviations are fine, but the distributions appear to be skewed and have outliers, especially the 1-year-ago group.

22.11 (a) The sample sizes are very large. (b) Yes (barely): The ratio is 1.94. (c) $\bar{x} \doteq 1.31$, MSG $\doteq 178.07$, MSE $\doteq 5.12$, $F \doteq 34.76$. (d) 2 and 1339 degrees of freedom; P is tiny.

22.13 (a) 32.32, 34.58, 35.51; $35.51/32.32 \doteq 1.10$. (b) MSG $\doteq 96.41$, MSE $\doteq 1216$, $F \doteq 0.08$; 2 and 126 degrees of freedom; $P > 0.100$. This is not enough evidence that nest temperature affects mean weight.

22.15 Populations: nonsmokers, moderate smokers, and heavy smokers; response variable: hours of sleep per night. $I = 3$; $n_1 = n_2 = n_3 = 200$; $N = 600$; 2 and 597 degrees of freedom.

22.17 Populations: tomato varieties; response: yield. $I = 4$; $n_1 = \cdots = n_4 = 10$; $N = 40$; 3 and 36 degrees of freedom.

22.19 Populations: students taught by different methods; response variable: test scores. $I = 4$; $n_1 = n_2 = n_3 = 10$, $n_4 = 12$; $N = 42$; 3 and 38 degrees of freedom.

22.21 (a) Yes; the mean control emission rate is half the smallest of the others. (b) H_0: All groups have the same mean emission rate. H_a: At least one group has a different mean emission rate. (c) Are the data Normally distributed? Were these random samples? (d) $s = $ SEM $\times \sqrt{8}$. The rule-of-thumb ratio is 1.4755.

22.23 Only Design A would allow use of one-way ANOVA.

22.25 (a) Yields appear to first increase with plant density, then decrease. The standard deviation ratio is 1.95. (b) H_0: $\mu_1 = \mu_2 = \mu_3 = \mu_4 = \mu_5$; H_a: not all means are the same; $F = 0.50$ and $P = 0.736$. (c) The sample sizes were small, and there is a lot of variation.

22.27 (a) Means: 10.65, 10.425, 5.60, 5.45 cm; standard deviations: 2.053, 1.486, 1.244, 1.771 cm. The means and the stemplots suggest that the presence of too many nematodes reduces growth. (b) H_0: $\mu_1 = \cdots = \mu_4$; H_a: not all means are the same. We test whether nematodes affect mean plant growth. (c) $F = 12.08$; 3 and 12 degrees of freedom; $P = 0.001$. The first two levels are similar, as are the last two. Somewhere between 1000 and 5000 nematodes, the tomato plants are hurt by the worms.

22.29 $\bar{x} \doteq 21.585$; MSG $\doteq 745.5$, MSE $\doteq 460.2$; 3 and 28 degrees of freedom; $F \doteq 1.62$—not significant.

22.31 (a) A chi-square test. (b) ANOVA. (c) ANOVA.

22.33 (a) F drops to 0.3204, while $P > 0.5$.

Data Table Index

Index